Global Climatology

Edited by Mary D'Souza

SYRAWOOD
PUBLISHING HOUSE
New York

Published by Syrawood Publishing House,
750 Third Avenue, 9th Floor,
New York, NY 10017, USA
www.syrawoodpublishinghouse.com

Global Climatology
Edited by Mary D'Souza

International Standard Book Number: 978-1-68286-603-0 (Hardback)

Cataloging-in-Publication Data

Global climatology / edited by Mary D'Souza.
p. cm.
Includes bibliographical references and index.
ISBN 978-1-68286-603-0
1. Climatology. 2. Meteorology. 3. Climatic changes. I. D'Souza, Mary.
QC981 .G56 2018
551.6--dc23

Printed in China.

TABLE OF CONTENTS

PREFACE

Every book is a source of knowledge and this one is no exception. The idea that led to the conceptualization of this book was the fact that the world is advancing rapidly; which makes it crucial to document the progress in every field. I am aware that a lot of data is already available, yet, there is a lot more to learn. Hence, I accepted the responsibility of editing this book and contributing my knowledge to the community.

Climatology studies climate, which is the aggregation of weather conditions over a defined period of time. Global climatology studies global climatic patterns by using diverse methods and models. Climate indices, which are data that has been gathered over a period of time, can be used to study changes that are occurring in global climate. This book is a valuable compilation of topics, ranging from the basic to the most complex advancements in the field of climatology. It includes contributions of experts and scientists which will provide innovative insights into this field. In this book, using case studies and examples, constant effort has been made to make the understanding of the difficult concepts of climatology as easy and informative as possible, for the readers.

While editing this book, I had multiple visions for it. Then I finally narrowed down to make every chapter a sole standing text explaining a particular topic, so that they can be used independently. However, the umbrella subject sinews them into a common theme. This makes the book a unique platform of knowledge.

I would like to give the major credit of this book to the experts from every corner of the world, who took the time to share their expertise with us. Also, I owe the completion of this book to the never-ending support of my family, who supported me throughout the project.

Editor

1

Climate Change Affects Winter Chill for Temperate Fruit and Nut Trees

Eike Luedeling[1]*, Evan H. Girvetz[2], Mikhail A. Semenov[3], Patrick H. Brown[4]

1 World Agroforestry Centre (ICRAF), Nairobi, Kenya, **2** The Nature Conservancy, Seattle, Washington, United States of America, **3** Rothamsted Research, Harpenden, United Kingdom, **4** Department of Plant Sciences, University of California Davis, Davis, California, United States of America

Abstract

Background: Temperate fruit and nut trees require adequate winter chill to produce economically viable yields. Global warming has the potential to reduce available winter chill and greatly impact crop yields.

Methodology/Principal Findings: We estimated winter chill for two past (1975 and 2000) and 18 future scenarios (mid and end 21st century; 3 Global Climate Models [GCMs]; 3 greenhouse gas emissions [GHG] scenarios). For 4,293 weather stations around the world and GCM projections, Safe Winter Chill (SWC), the amount of winter chill that is exceeded in 90% of all years, was estimated for all scenarios using the "Dynamic Model" and interpolated globally. We found that SWC ranged between 0 and about 170 Chill Portions (CP) for all climate scenarios, but that the global distribution varied across scenarios. Warm regions are likely to experience severe reductions in available winter chill, potentially threatening production there. In contrast, SWC in most temperate growing regions is likely to remain relatively unchanged, and cold regions may even see an increase in SWC. Climate change impacts on SWC differed quantitatively among GCMs and GHG scenarios, with the highest GHG leading to losses up to 40 CP in warm regions, compared to 20 CP for the lowest GHG.

Conclusions/Significance: The extent of projected changes in winter chill in many major growing regions of fruits and nuts indicates that growers of these commodities will likely experience problems in the future. Mitigation of climate change through reductions in greenhouse gas emissions can help reduce the impacts, however, adaption to changes will have to occur. To better prepare for likely impacts of climate change, efforts should be undertaken to breed tree cultivars for lower chilling requirements, to develop tools to cope with insufficient winter chill, and to better understand the temperature responses of tree crops.

Editor: Anna Traveset, Institut Mediterrani d'Estudis Avançats (CSIC/UIB), Spain

Funding: Rothamsted Research receives grant-aided support from the Biotechnology and Biological Sciences Research Council of the UK (BBSRC). The funders had no role in study design, data collection and analysis, decision to publish, or preparation of the manuscript.

Competing Interests: The authors have declared that no competing interests exist.

* E-mail: e.luedeling@cgiar.org

Introduction

Commercially successful cultivation of many fruit and nut trees requires the fulfillment of a winter chilling requirement, which is specific for every tree cultivar [1,2,3,4]. In order to avoid frost damage of sensitive tissue in the cold winters of their regions of origin, trees from temperate or cold climates evolved a period of dormancy during the cold season. After a certain duration of cold conditions (chilling), endodormancy is broken and the tree is ready to resume growth in spring. Chilling requirements vary substantially between species and cultivars from different parts of the world and commercial production of temperate tree crops requires selecting appropriate cultivars for the climatic conditions of the planned production site.

Climate change is likely to affect future winter chill and could have a major impact on the US$ 93 billion global fruit and nut industry (only species with chilling requirements, production statistics for 2005 from ref. [5], currencies converted into 2005 US$ according to ref. [6]). Temperatures are expected to increase in most parts of the world, with minimum temperatures rising most rapidly. This development may compromise the ability of many growers of temperate fruits and nuts to successfully produce the same array of crops as in the past. Climate change effects on winter chill have recently been analyzed for California [7,8], Germany [9] and high-mountain oases in Oman [10]. While conditions in Germany were relatively stable during the 20th century, winter chill was found to have declined in California and Oman, and this process was expected to continue in the future. The differences between these studies indicate that different growing regions may be differentially impacted, but to date, no estimates are available at a global scale to indicate which regions will maintain adequate winter chill for temperate fruits and nuts in the future. This study aims to fill this knowledge gap and to provide important information needed to evaluate the future viability of fruit and nut growing regions around the world.

Several models have been developed for quantifying winter chill, e.g. the Chilling Hours Model [11], the Utah Model [12] and the Dynamic Model [13,14]. These models differ greatly in their sensitivity to climate change [15], making the choice of the model a crucial determinant of the predicted extent of climate change effects on winter chill. When using different models for similar climate change scenarios, the Chilling Hours Model and the Utah

Model tend to show much stronger decreases in winter chill than the Dynamic Model, especially in warm growing regions [15]. Using the former two models probably overestimates winter chill losses, because several studies have shown the Dynamic Model to be more accurate, especially in subtropical climates [16,17,18]. One more study found it to be equal to the Utah Model in Spain [19], and another one reported failure of all models on the Tropical island of Réunion [20]. Calculating ratios between winter chill estimates with different models at warm locations shows large differences, with strong variation and strong temperature dependence [21]. For colder regions, however, such ratios tend to be fairly similar and much less variable [21]. Consequently, the Dynamic Model can be used as a proxy of winter chill in both warm and cold growing regions and, among the common winter chill models, is the one most suitable for a global analysis. The Chilling Hours and Utah Models may produce reasonably accurate results in cold regions, but are not applicable for warmer parts of the world, where their use would produce misleading overestimates of likely impacts of climate change [8]. In this study, we therefore only use the Dynamic Model (for equations, see [21]), which quantifies winter chill in Chill Portions (CP).

We quantified winter chill for the entire terrestrial globe using climate scenarios based on observed daily weather from 4293 weather stations around the world and climate projections from three Global Climate Models (GCMs). Based on this analysis we calculated the safe winter chill (SWC) metric [8], which quantifies the amount of winter chill that is exceeded in 90% of years. This metric is meaningful to fruit and nut producers, because failure to meet chilling requirements in more than 10% of years is likely to render production uneconomical. This analysis identifies important fruit and nut producing areas in the world where SWC has already decreased and is projected to decrease further, and

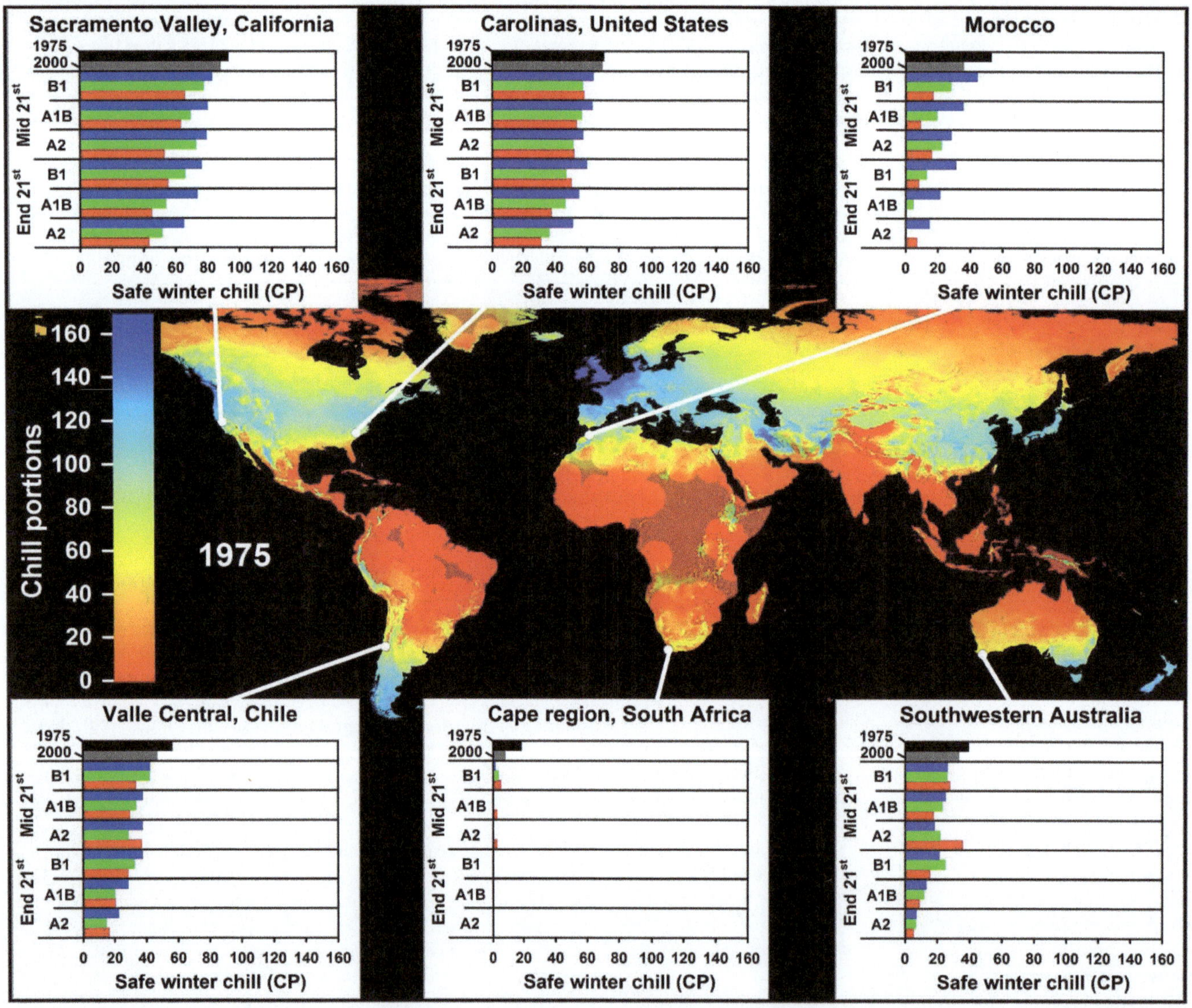

Figure 1. Modeled Safe Winter Chill around the year 1975 (large map), as well as site-specific estimates of Safe Winter Chill for six growing regions and for 20 climate scenarios, representing four points in time (1975, 2000, mid and end 21st century). Future projections include three greenhouse gas emissions scenarios (B1, A1B and A2) and three Global Climate Models (CSIRO - green bars; HADCM3 - blue bars; and MIROC - red bars). Areas that are more than 5° away from the closest weather station are shaded, because interpolated results are unreliable.

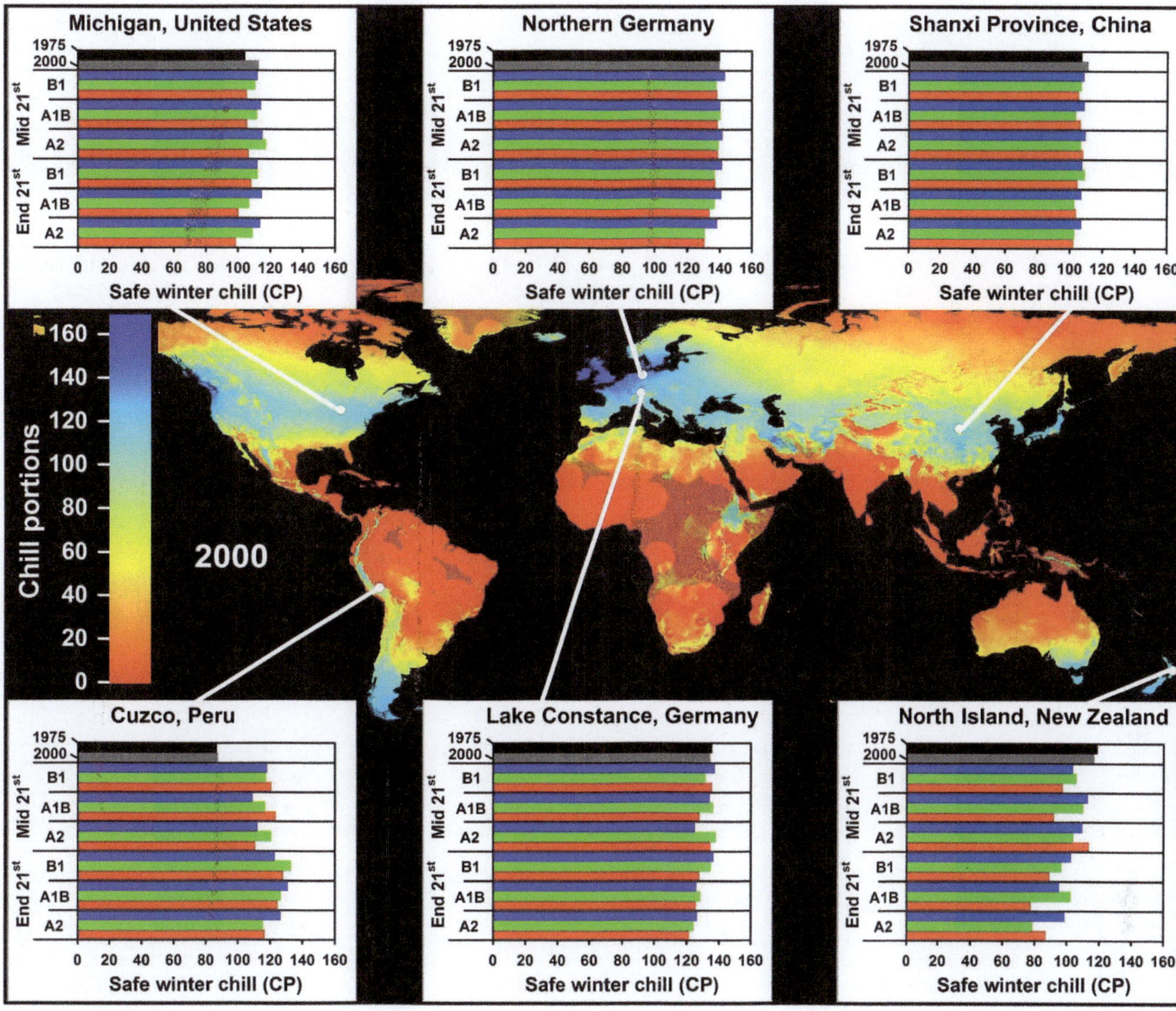

Figure 2. Modeled Safe Winter Chill around the year 2000 (large map), as well as site-specific estimates of Safe Winter Chill for six growing regions and for 20 climate scenarios, representing four points in time (1975, 2000, mid and end 21st century). Future projections include three greenhouse gas emissions scenarios (B1, A1B and A2) and three Global Climate Models (CSIRO - green bars; HADCM3 - blue bars; and MIROC - red bars). Areas that are more than 5° away from the closest weather station are shaded, because interpolated results are unreliable.

provides a comprehensive assessment of the magnitude of changes to winter chill that will likely occur.

Daily weather records for all weather stations were obtained from the National Climatic Data Center of the United States [22], subjected to a data quality filter, and used to calibrate a weather generator [23,24]. Daily weather records were then generated for 20 climate scenarios. Two scenarios represented typical climatic conditions in 1975 and 2000. Eighteen future scenarios were generated by extracting future projections from datasets assembled in the ClimateWizard tool [25]. These included statistically downscaled projections with three GCMs (MIROC3.2 (medres), UKMO-HadCM3 and CSIRO-Mk3.0; referred to as MIROC, HADCM3 and CSIRO in the following) and for three IPCC greenhouse gas emissions scenarios (B1 - global curbing of emissions over the 21st century; A1B - emissions leveling off at mid 21st century; and A2 - continually increasing rate of greenhouse gas emissions). Projections were made for the middle (2040–2059) and end (2080–2099) of the 21st century. Idealized daily temperature curves were used for converting daily to hourly weather records and allow calculation of winter chill. For each scenario, 101 years of weather records were generated and the 10% quantile of the resulting distribution of annual winter chill interpreted as Safe Winter Chill. For each scenario, SWC from all stations was spatially interpolated, and winter chill for all scenarios was extracted from the resulting layers for 24 important growing regions around the world.

Results

In all climate scenarios, estimates of Safe Winter Chill ranged from 0 CP in tropical and very cold regions to about 170 CP in maritime temperate climates of Northwestern Europe (Figs. 1–4).

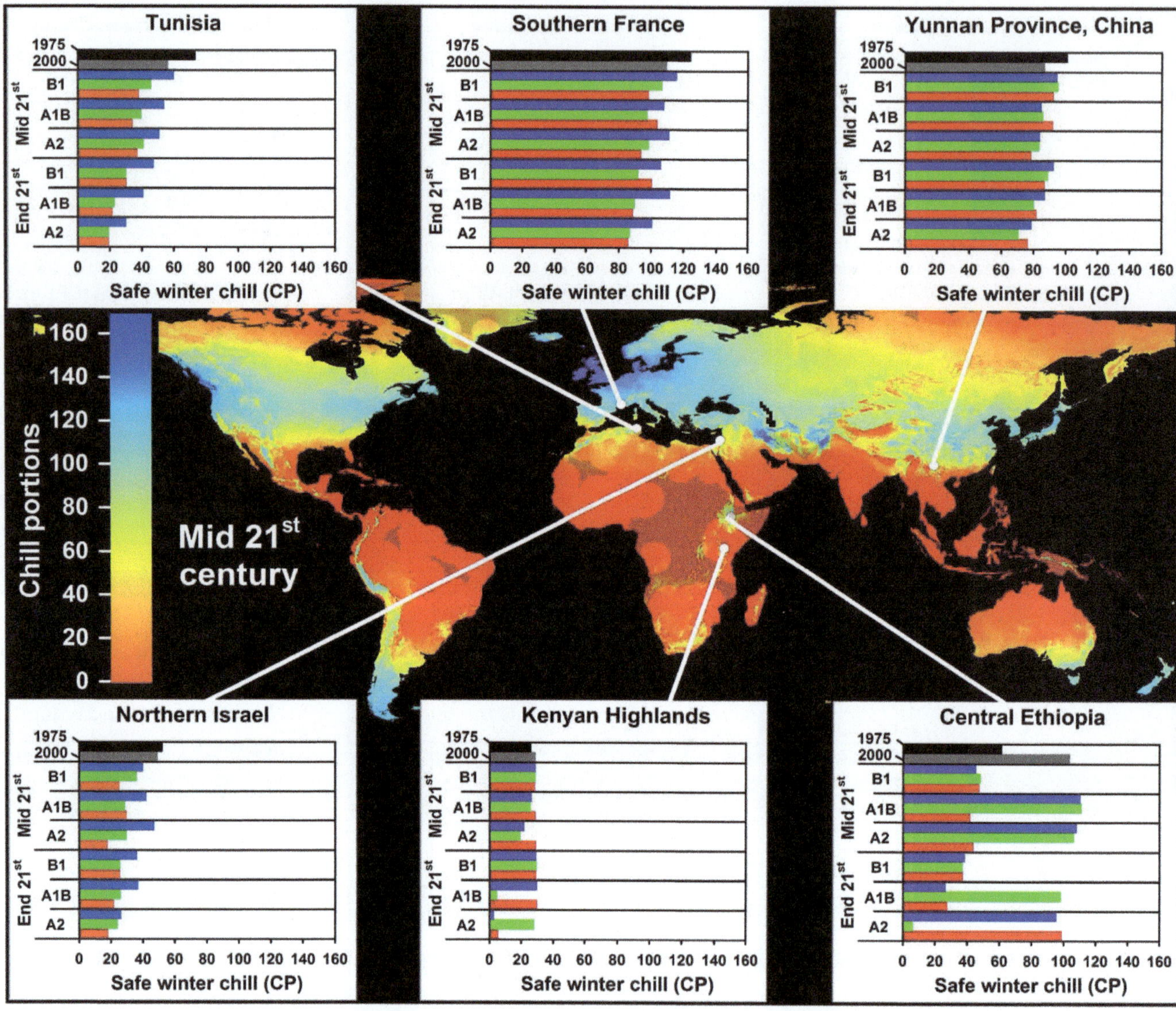

Figure 3. Modeled Safe Winter Chill around the middle of the 21st century averaged over three greenhouse gas emissions scenarios and three Global Climate Models (large map), as well as site-specific estimates of Safe Winter Chill for six growing regions and for 20 climate scenarios, representing four points in time (1975, 2000, mid and end 21st century). Future projections include three greenhouse gas emissions scenarios (B1, A1B and A2) and three Global Climate Models (CSIRO - green bars; HADCM3 - blue bars; and MIROC - red bars). Areas that are more than 5° away from the closest weather station are shaded, because interpolated results are unreliable.

While the overall range of the winter chill distribution did not change much across all scenarios, our results show changes in the global distribution of winter chill, as well as site-specific trends. Because the Dynamic Model does not consider freezing temperatures to be effective for chilling, reduced incidence of frost tends to increase the number of Chill Portions in cold regions. This process is reflected in increasing Safe Winter Chill in cold regions (Fig. 5), which may affect fruit growing regions in Canada, Southern Scandinavia and Eastern Europe (Fig. 4). Decreases are projected for warmer regions, in particular around the Mediterranean Sea (Fig. 6) and in Southwestern North America (Fig. 7), where losses up to 40 CP are expected by the end of the 21st century (Fig. 5). Many warm growing regions are projected to lose most of their winter chill, with South Africa, Southern Australia and Northern Africa particularly affected (Figs. 6 and 8).

Trends in site-specific projections for different growing regions varied substantially (site diagrams in Figs. 1–4). Most warm growing regions of temperate fruits and nuts are expected to experience decreasing winter chill, regardless of the emissions scenario or climate model used. The Sacramento Valley in California, the Southeastern United States, Chile's Valle Central, Yunnan Province in China, as well as South and Southwestern Australia are all projected to lose winter chill. This will likely require growers to transition to different species or cultivars than are grown today or to develop management practices that can help overcome shortages in winter chill. The highest losses relative to current winter chill levels occurred in Morocco, Tunisia, Israel, in the Cape region of South Africa and, for some GCMs, in the highlands of Kenya and Ethiopia. In these regions, climate change is likely to severely challenge current production systems, some of

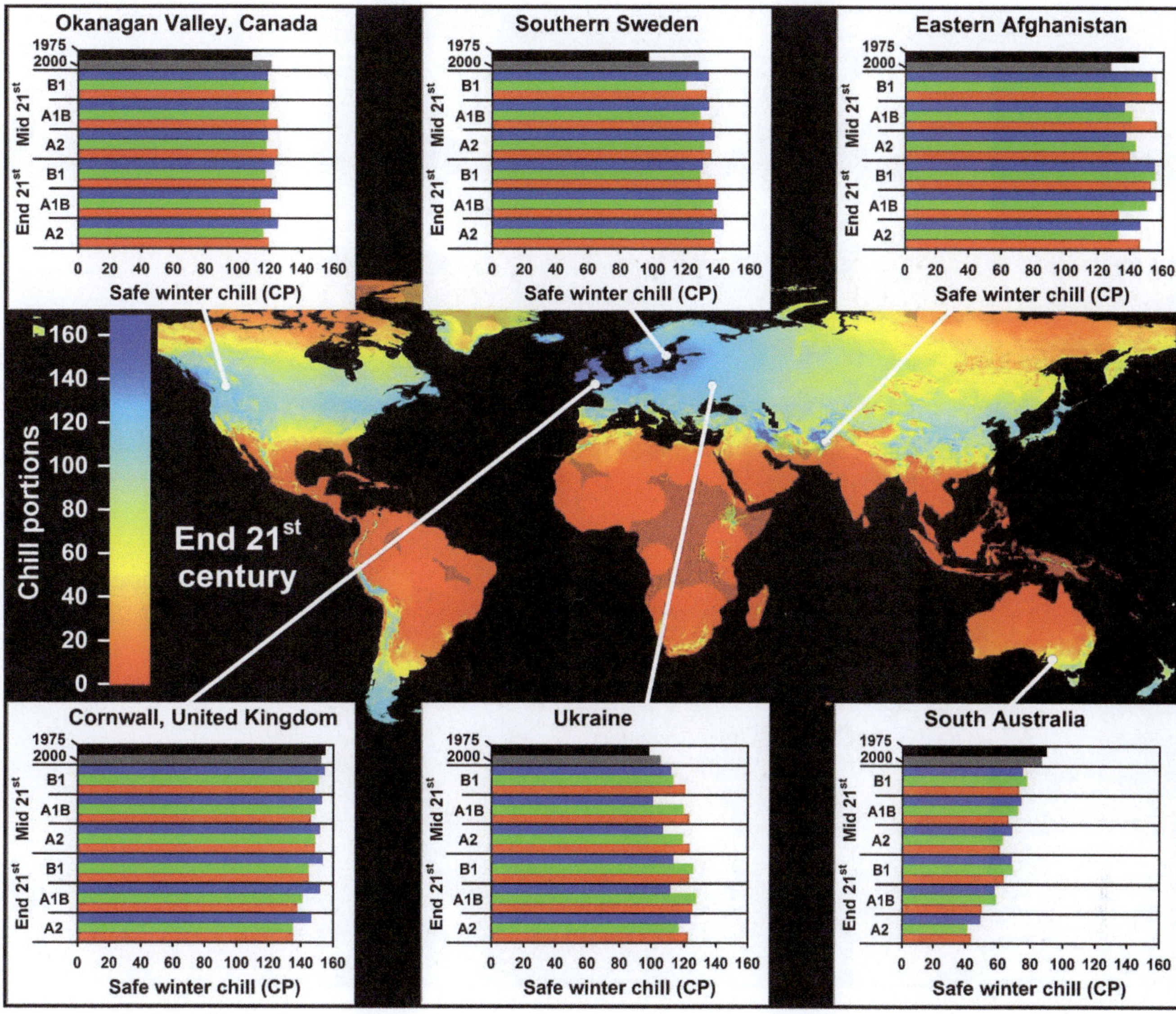

Figure 4. Modeled Safe Winter Chill around the end of the 21st century averaged over three greenhouse gas emissions scenarios and three Global Climate Models (large map), as well as site-specific estimates of Safe Winter Chill for six growing regions and for 20 climate scenarios, representing four points in time (1975, 2000, mid and end 21st century). Future projections include three greenhouse gas emissions scenarios (B1, A1B and A2) and three Global Climate Models (CSIRO - green bars; HADCM3 - blue bars; and MIROC - red bars). Areas that are more than 5° away from the closest weather station are shaded, because interpolated results are unreliable.

which already rely on cultural measures such as rest-breaking chemicals and artificial defoliation.

Cool regions are less likely to experience decreasing winter chill. Growing regions in Germany, the United Kingdom, the Midwestern United States (Fig. 7), Northern China and Central Asia are projected to see little change in SWC levels. Southern France (Fig. 6) and New Zealand (Fig. 8) may experience slight but likely insignificant losses. The coldest current growing regions (e.g. the Okanagan Valley in Canada, Southern Sweden and Eastern Europe) are expected to see more winter chill in the future. Whether these changes will require growers to adapt is currently unclear and is likely to depend more on the effects of summer warming than on winter temperatures.

In addition to the time period analyzed, the amplitude of expected changes also depended on the greenhouse gas emissions scenario. The A2 scenario consistently projected the greatest changes in winter chill, followed by the A1B scenario and the B1 scenario. If emissions are curbed to levels assumed in the B1 scenario, few growing regions are likely to see decreases by more than 20 CP by the end of the 21st century (Fig. 9). If business-as-usual emissions continue (A2 scenario), many subtropical regions will see chilling declines up to 40 CP, which can be expected to disrupt production systems.

The choice of the climate model also influenced model results. The MIROC model produced the greatest changes, followed by HADCM3 and CSIRO (Fig. 10). Within the same time period and emissions scenario, mean absolute differences between winter chill levels projected by CSIRO and HADCM3 were always smaller than for comparisons of either model with MIROC projections (Table 1). By the end of the 21st century, mean absolute differences between all modeled scenarios and historic SWC levels for 1975 were between 12.8 and 29.0 CP, on average

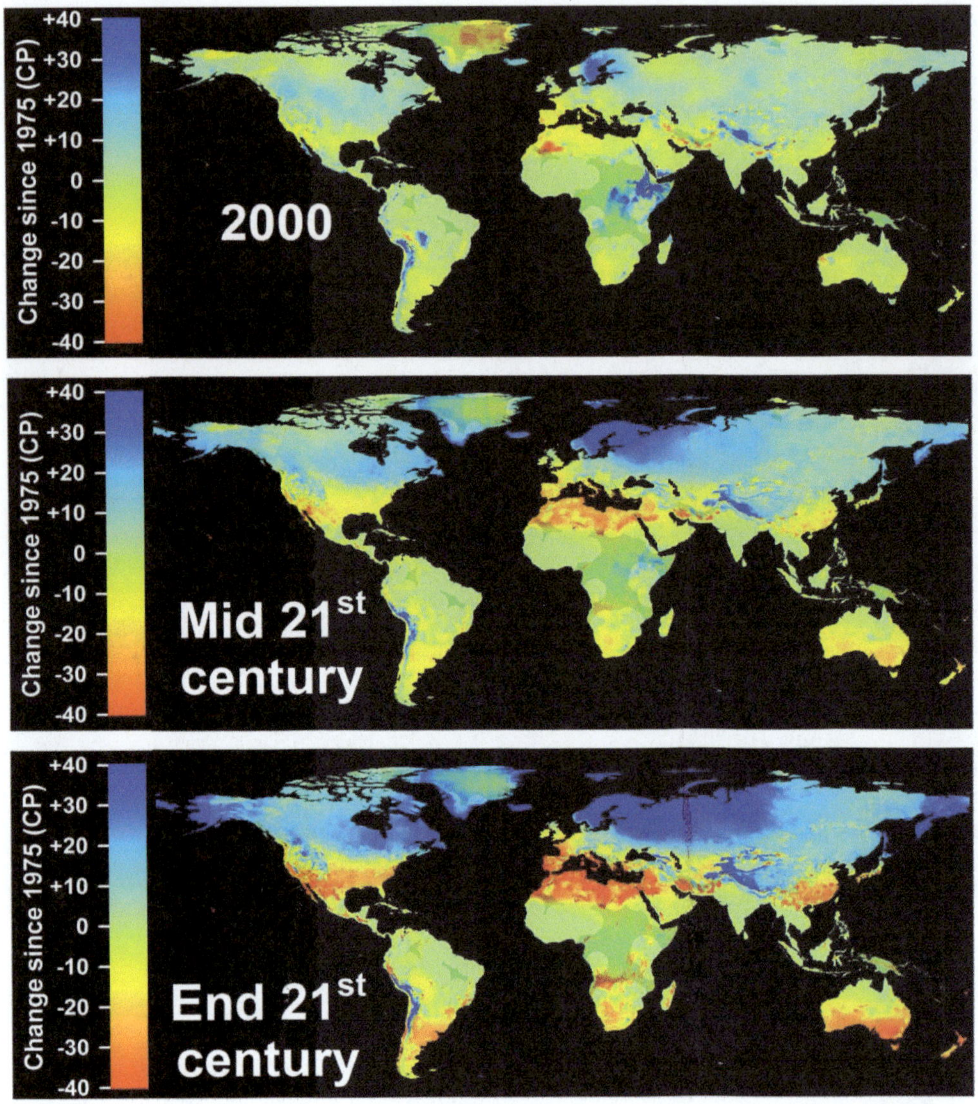

Figure 5. Modeled and projected losses in Safe Winter Chill compared to 1975 for the year 2000 (top), the middle of the 21st century (middle), and the end of the 21st century (bottom). For each point in time, results are averaged over three greenhouse gas emissions scenarios and three Global Climate Models. Areas that are more than 5° away from the closest weather station are shaded, because interpolated results are unreliable.

over all grid cells. These levels of change indicate that vegetation that relies on winter dormancy will experience very different temperature cues in the future than it does now.

Discussion

Our projections indicate that most warm growing regions will experience severe declines in Safe Winter Chill over the course of the 21st century. In contrast to this, cool regions may not see much change, because reductions in winter chill due to warming are compensated for by chilling gains caused by less frequent frost. For cold growing regions, these opposing trends play out to result in more winter chill in response to warming.

Among these changes, reduced winter chill is likely to have the most severe consequences for fruit production. Lack of winter chill can delay or prevent flowering, lead to staggered bloom, and cause various forms of anomalous growth [26,27]. Anecdotal evidence of this has been reported from various growing regions but seldom found its way into the literature.

Increases in winter chill in cold areas are less likely to lead to disruptions in fruit production, but even there, a mismatch between the chilling requirements of common species and cultivars and available winter chill could cause some problems for fruit and nut growers. In all areas, however, even in those where no changes in winter chill are projected, other manifestations of climate change are also likely to affect fruit and nut production. Plantations may be impacted by changes in rainfall, changes to summer and spring heat or increases in pest pressure due to faster reproduction of ectothermic pest organisms [28]. Due to all these additional effects of climate change, we do not attempt to predict future yields, but focus on pointing out potential problems due to lack of winter chill.

Even assuming no additional changes due to climate change, modeling the effect of changes in winter chill on crop yields is not

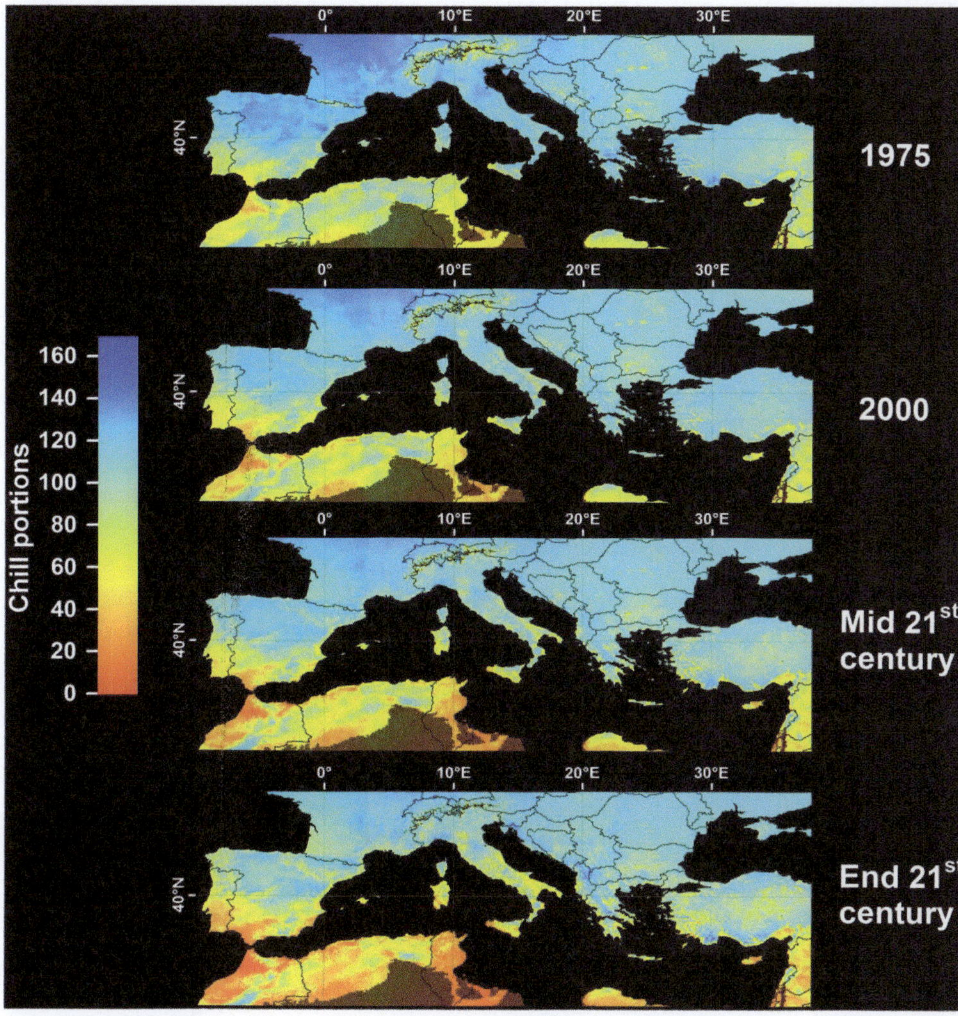

Figure 6. Modeled and projected Safe Winter Chill in the Mediterranean region, for 1975, 2000, the middle of the 21st century (middle), and the end of the 21st century (bottom). For each point in time, results are averaged over three greenhouse gas emissions scenarios and three Global Climate Models. Areas that are more than 5° away from the closest weather station, and areas with mean annual temperatures >20 or <°0C are shaded.

possible at present. In spite of two centuries of research on winter chill, it is still unclear what happens during chilling accumulation, and how exactly this process is influenced by ambient temperatures. There may also be other environmental factors, such as relative humidity, photoperiod or the Red/Far Red light ratio [29], that impact the breaking of dormancy but are not recognized in any dormancy models for tree crops. All chilling models, including the Dynamic Model, were developed to assist fruit and nut growers in selecting appropriate species and cultivars, rather than describing a biological process with scientific accuracy. They are all empirically derived rather than based on a functional understanding of the dormancy process.

Lack of knowledge about the chilling requirements of most species and cultivars, in appropriate units, also precludes detailed projections of future yields or suitable ranges. Most growers and researchers have used Chilling Hours to quantify chilling requirements, but this model has been shown to perform poorly in warm regions [16,17,18,19,20] and to be very sensitive to climate change [15]. Estimates of chilling requirements, when given in Chilling Hours, must also be adjusted before they are useful in a different location or in a warmer climate [21]. Existing lists of species-specific chilling requirements are thus of limited value for estimating future ranges of cultivars and species. Estimates in Chill Portions are less widely available, and we are not aware of a comprehensive list that compiles them. Research efforts are needed to close the pertinent knowledge gaps and allow quantitative projections of the effect that changes in winter chill will have on fruit and nut production.

As much as precise quantitative projections are impossible, changes to available winter chill and summer heat will likely change the suitable ranges of many tree crops, and it seems likely that many growing regions will become unsuitable for the cultivars that are currently produced. However, whether or not tree crops will actually be moved to cooler climates will depend on many factors, such as availability of land and critical infrastructure, land tenure and competition with other crops. For example, the ecological niche of many fruits and nuts in the Western United States is likely to move north, from California's Central Valley towards Northern California, Oregon and Washington. These new potentially suitable areas have adverse topography, poorer soils, and limited water availability compared to the Central Valley, making the economic viability of production there

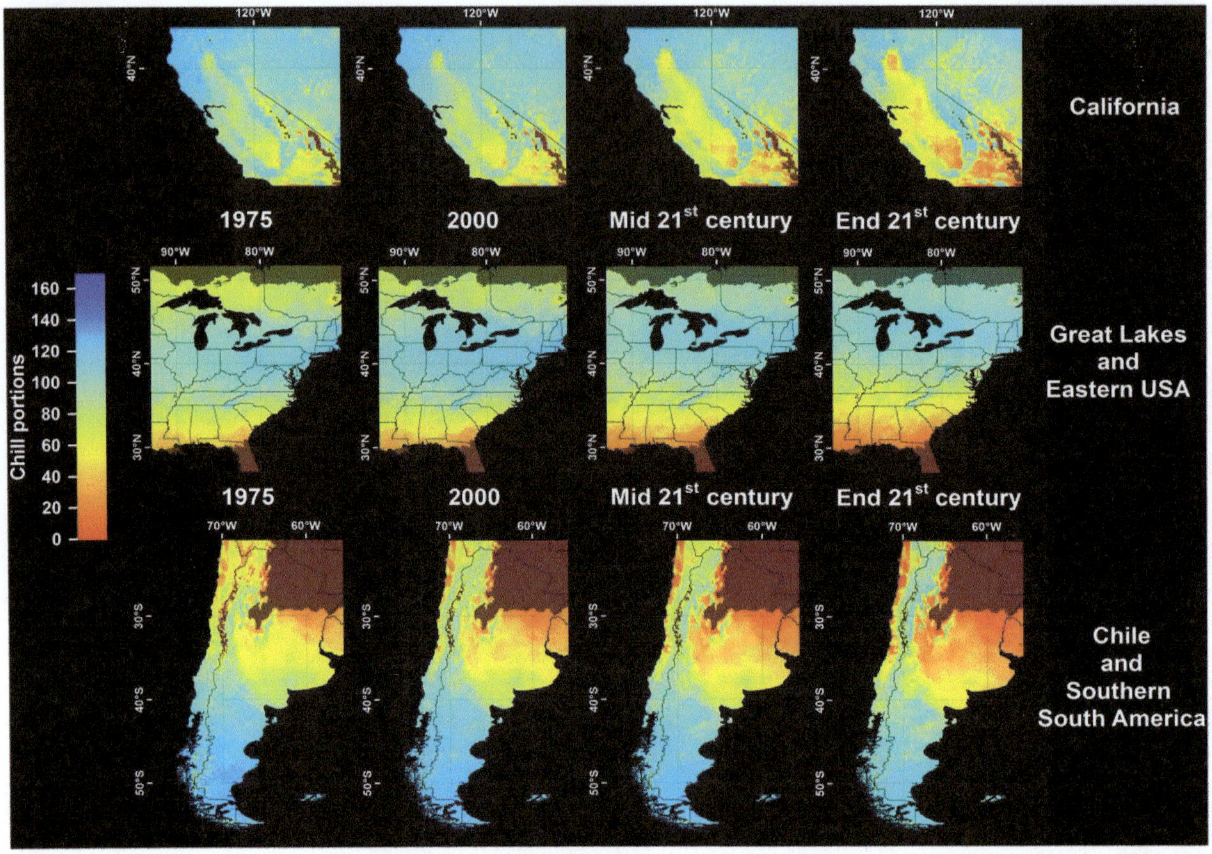

Figure 7. Modeled and projected Safe Winter Chill in California, the Eastern United States and Southern South America, for 1975, 2000, the middle of the 21st century (middle), and the end of the 21st century (bottom). For each point in time, results are averaged over three greenhouse gas emissions scenarios and three Global Climate Models. Areas that are more than 5° away from the closest weather station, and areas with mean annual temperatures >20 or <°0C are shaded.

questionable. Similarly, other regions, such as parts of Scandinavia, Canada and Siberia, that could potentially become more suitable for these tree crops, may be limited by cold winters, lack of summer heat, or adverse photoperiodic conditions. Production potential of tree crops under many of these novel conditions expected in the future has not been studied sufficiently to be discussed here.

To a certain extent, adaptation to changes in winter chill will be possible. Agrochemicals have been developed to artificially break a tree's dormancy during the later stages of chilling accumulation [30,31], irrigation and shading may influence orchard microclimates favorably [32], and other cultural practices, such as artificial defoliation [33], also have potential for reducing chilling requirements. Inclusion of low chilling requirements as an explicit target in breeding programs is likely to produce cultivars that will remain suitable in a warmer future. It will be necessary, however, to intensify efforts to develop such adaptation strategies. In particular breeding programs for low-chilling cultivars, which can take decades to produce useable results, need more attention and more resources.

The economic cost of climate change incurred by fruit and nut growers could be substantial. Many businesses may be confronted with the decision to either abandon their production or adapt as well as possible to altered climatic conditions. Applying adaptation treatments could be economically unviable. And even if crops move towards more suitable climatic zones, most small orchard operations lack the capital to move their production to a different area, potentially impacting many livelihoods.

Natural plant communities respond to similar temperature cues as fruit and nut trees and will likely be affected as well by changes in the amount of available winter chill. It seems very likely that the projected decreases in winter chill in the Subtropics, but also the increases in the colder regions, will affect local natural vegetation, potentially impacting the suitable areas for many plants. Rapid climate-driven changes in plant communities at the local scale have already been reported [34]. Most studies that have investigated climate change effects on large numbers of plant species have found that most species showed advances in spring phenology, indicating that the impact of reduced chilling is compensated, in most cases, by increases in spring heat [35,36,37]. However, the same studies include a sizeable number of species, which show an opposite trend - delayed spring phenology in response to increases in temperature. For meadow and steppe vegetation on the Tibetan Plateau, Yu et al. [38] have recently shown a clear correlation of such a delay with increases in winter temperature, implicating lack of winter chill as the cause of the delay.

Conclusion

Even under the most conservative emissions scenario examined here the projected substantial decreases in winter chill would negatively impact productivity of current cultivars and viability of fruit and nut industries in warm growing regions. Chemical, mechanical and physical methods to compensate for a loss of chilling are available but add significantly to production costs. A

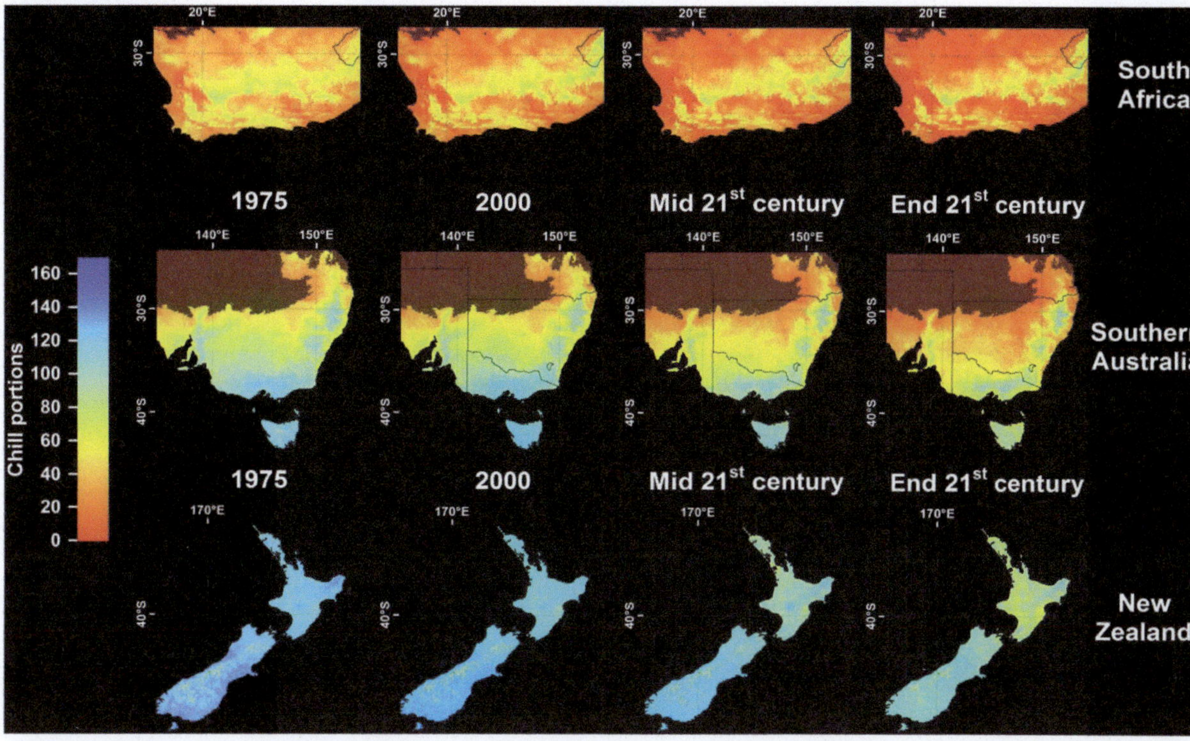

Figure 8. Modeled and projected Safe Winter Chill in South Africa, Southern Australia and New Zealand, for 1975, 2000, the middle of the 21st century (middle), and the end of the 21st century (bottom). For each point in time, results are averaged over three greenhouse gas emissions scenarios and three Global Climate Models. Areas that are more than 5° away from the closest weather station, and areas with mean annual temperatures >20 or <°0C are shaded.

shift to new production areas may be possible but would incur substantial infrastructure costs. Moreover, production viability depends on many critical factors in addition to climatic suitability, which may not be available in areas with suitable future climates. The most viable approach to adapt to climate change in deciduous fruit and nut species is through the development of new cultivars that are more productive under lower chill conditions. This will require investment in efforts to understand the biological basis for chilling, to develop better models that relate environmental cues with yield and phenology, as well as renewed emphasis on breeding programs.

Materials and Methods

Weather data

Daily temperature and rainfall records were downloaded for all 11,361 available weather stations at the National Climatic Data Center [22]. This dataset was filtered, removing all stations that had less than 5000 daily records between 1973 and 2002, and excluding all stations with more than 25% of daily minimum or maximum temperatures or 50% of daily rainfall data missing. For the remaining 5078 weather stations, all available temperature and precipitation records were used to calculate site parameters for use in the LARS-WG stochastic weather generator [23,24]. A stochastic weather generator (WG) is a model which, after calibration of site parameters with observed weather at that site, is capable of simulating synthetic time-series of daily weather that are statistically similar to observed weather [39]. By altering the site parameters of the WG using changes in climate predicted from a global climate model (GCM), it is possible to generate synthetic daily weather for the future. WGs are extensively used as a computationally inexpensive tool to produce daily site-specific climate scenarios for impact assessments of climate change [40,41,42,43]. Because the generation of daily weather is dependent on the site-specific duration of wet and dry spells, the modeling procedure required precipitation records in addition to temperatures.

Following Luedeling et al. [8], for each station we evaluated the entire weather record for all days between 1973 and 2002, calculating separate linear regression equations between time (in years) and the minimum temperature, maximum temperature and precipitation for each month of the year. We then used these regression equations to develop climate scenarios representing typical climatic conditions in 1975 and 2000. These scenarios do not represent actually observed temperatures and precipitation in these years, but rather typical conditions at these times that are representative of long-term trends over the calibration period (1973–2002). The climate scenarios contain the mean deviation of monthly minimum and maximum temperatures and precipitation from the mean of the calibration period for each station.

Using the same method, we evaluated the weather record for 18 future scenarios, based on projections by three Global Climate Models (GCMs; MIROC3.2 (medres), UKMO-HadCM3 and CSIRO-Mk3.0) that had been statistically downscaled to a 0.5 degree resolution using the CRU TS 2.0 data set to calibrate the downscaling (R. Neilson, unpublished data). Using the Climate Wizard tool (www.climatewizard.org; ref. [25]), we extracted projected minimum and maximum temperatures and precipitation for three greenhouse gas emissions scenarios of the IPCC Special Report on Emissions Scenarios [44]: the A2 scenario (continually increasing rate of greenhouse gas emissions), the A1B scenario (emissions leveling off at mid 21st century) and the B1 scenario

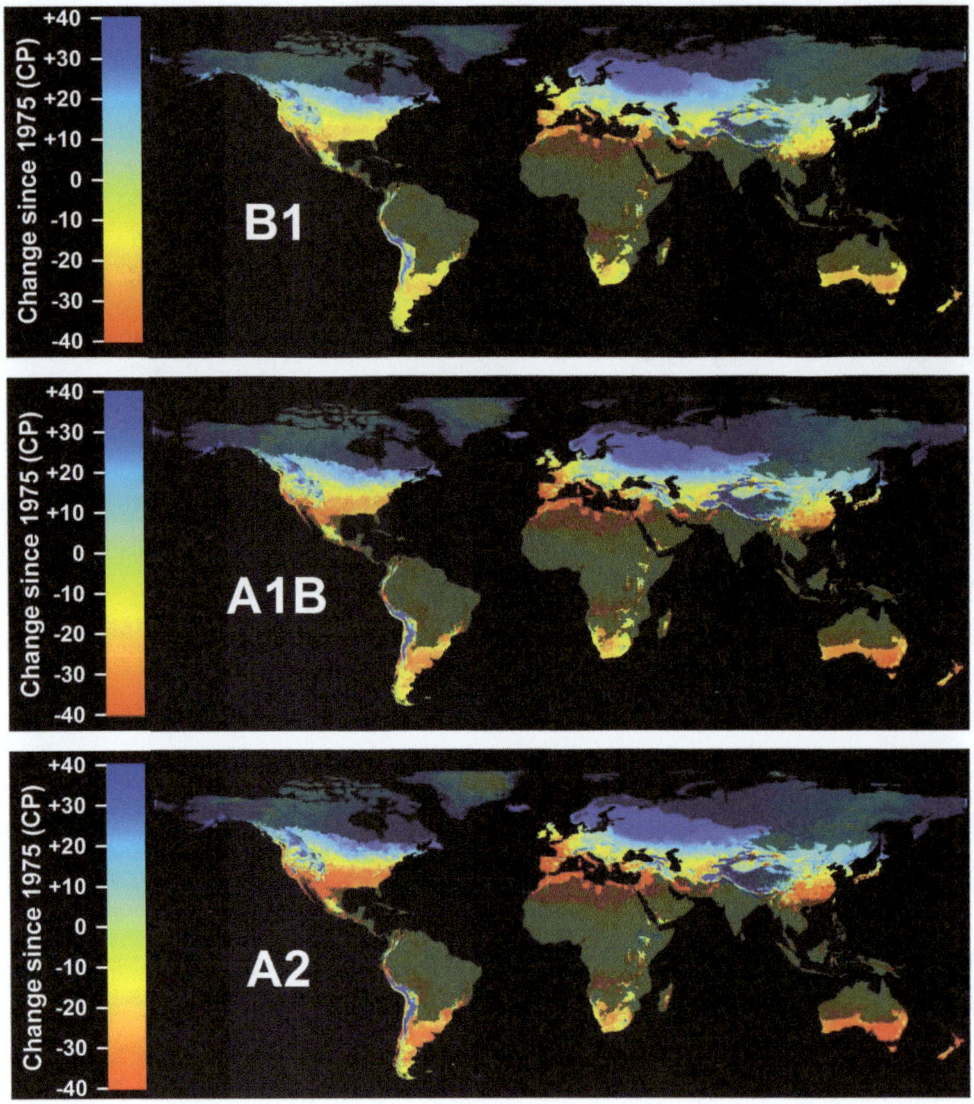

Figure 9. Projected losses in Safe Winter Chill at the end of the 21st century compared to 1975, for three greenhouse gas emissions scenarios: B1 (top), A1B (middle) and A2 (bottom). For each scenario, results are averaged over projections from three Global Climate Models. Areas that are more than 5° away from the closest weather station are shaded, because interpolated results are unreliable.

(global curbing of emissions over the 21st century). For each weather station, emissions scenario and GCM, mean monthly anomalies of minimum and maximum temperatures and precipitation relative to the calibration period were obtained for two periods of time: 2040–59 and 2080–2099, representing conditions at mid and end 21st century. For several weather stations, mostly along coast lines, no GCM projection data were available. These stations were excluded, bringing the total number of weather stations down to 4293.

For each station and for each of the 2 past and 18 future scenarios, we then generated 101 years of synthetic daily weather data using a command line version of the LARS-WG weather generator [24]. Synthetic daily maximum and minimum temperatures were converted into hourly temperatures using the idealized temperature curve proposed by Linvill [9,45]. Linvill's equations, which use a sine curve for daytime temperatures and a logarithmic decline curve for nighttime cooling, require sunset and sunrise hours, as well as daylength, as input parameters. These data were generated using equations by Spencer [46] and Almorox et al. [47]. Resulting from these processing steps were 101 years of synthetic hourly temperature for each weather station, representing typical weather conditions for each climate scenario.

Winter chill

Based on the generated hourly temperature, we calculated winter chill for 100 winters for each weather station and climate scenario, with start and end dates of the winter season set to October 1st and May 1st, respectively, for stations in the Northern Hemisphere, and to April 1st and November 1st, respectively, for stations in the Southern Hemisphere. In a global analysis, effective times of winter chill accumulation often deviate from these dates, depending on local temperature curves. Since the Dynamic Model contains a self-regulating mechanism, which only allows accumulation of Chill Portions during times with appropriate temperatures, this variability should not affect the accuracy of our results.

For the resulting distribution over 100 winters, we then calculated Safe Winter Chill (SWC), the 10% quantile of the distribution [8]. This metric is more meaningful for growers than

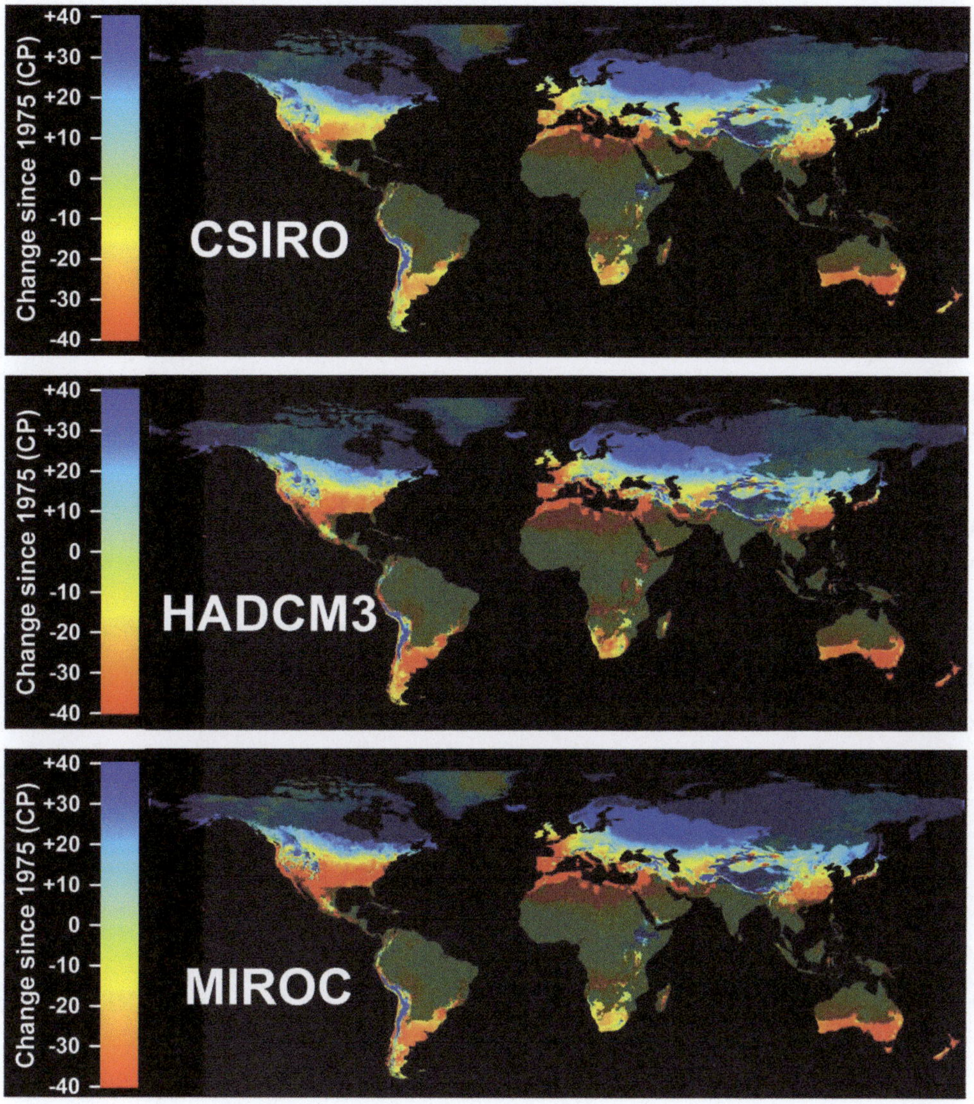

Figure 10. Projected losses in Safe Winter Chill at the end of the 21st century compared to 1975, for three Global Climate Models: CSIRO (top), HADCM3 (middle) and MIROC (bottom). For each scenario, results are averaged over projections for three greenhouse gas emissions scenarios. Areas that are more than 5° away from the closest weather station are shaded, because interpolated results are unreliable.

mean winter chill, because economic success of an orchard operation relies on fulfillment of chilling requirements in most years (e.g. 90% of all years), rather than in an average year. For all trees with lower chilling requirements than available SWC, the dormancy season should be sufficiently long and cold to allow fulfillment of tree-specific temperature needs. All data processing was implemented in JSL, the scripting language of JMP 8 (SAS Institute, Cary, NC, USA).

Spatial analysis

Based on estimates of SWC for every weather station, SWC was spatially interpolated for each climate scenario, using 12-neighbor Kriging with a spherical semivariogram at 0.1 degree spatial resolution (ArcGIS 9.3, ESRI, Redlands, CA, USA). A simple interpolation based on only the available weather stations would not be a very accurate representation of winter chill around the world, because temperatures at many locations may differ substantially from those of the closest weather station, which is often far away. We therefore used a high-resolution temperature dataset obtained from the WorldClim database [48] to correct for temperature variation that was not accounted for by the simple interpolation procedure. We calculated mean annual temperatures at a spatial resolution of 1/24 degree from monthly mean temperatures for the year 2000 given in the database. Because mean annual temperature explains much of the variation in winter chill [21], this dataset was useful for correcting the chilling estimates. To do this, we used the same Kriging procedure as for the winter chill estimates to interpolate a temperature surface from mean annual temperatures at all weather stations. The resulting grid represents the temperatures that correspond to the interpolated chill portion surface (SWC_{int}). Subtracting this grid from the original dataset of mean annual temperatures produced an estimate of the temperature variation that was not accounted for in the original interpolation of chill portion values (T_{diff}). For the correction, the effect of temperature on chill portion numbers was then estimated by a 5th order polynomial regression between mean annual temperatures at all weather stations and the amount of safe winter chill calculated for

Table 1. Mean absolute differences (in Chill Portions) between different climate scenarios [combination of time, greenhouse gas emissions scenario (GHG) and Global Climate Model (GCM)], over all relevant 0.1°×0.1° pixels of the global Safe Winter Chill projections.

Time			1975	2000	mid 21st century									end 21st century								
	GHG				B1			A1B			A2			B1			A1B			A2		
		GCM			C	H	M	C	H	M	C	H	M	C	H	M	C	H	M	C	H	M
1975				6.7	10.0	10.5	15.5	11.1	14.5	17.5	12.8	15.7	17.5	12.8	18.1	20.9	17.8	22.5	27.3	23.3	26.6	29.0
2000			6.7		7.5	7.2	11.9	6.2	9.5	13.9	7.7	10.6	13.7	9.2	14.0	17.1	13.8	18.1	23.5	18.6	22.6	24.8
mid 21st century	B1	C	10.0	7.5		4.0	6.4	4.4	6.1	8.3	4.9	6.5	9.1	3.4	8.4	11.4	8.1	13.4	18.0	13.9	17.3	20.0
		H	10.5	7.2	4.0		6.0	4.9	5.2	7.9	5.3	6.5	8.9	3.8	7.8	11.0	8.1	12.8	17.5	13.6	16.8	19.5
		M	15.5	11.9	6.4	6.0		7.7	5.2	3.1	7.0	6.0	5.5	4.5	3.6	5.9	5.5	8.4	12.4	9.1	12.1	14.6
	A1B	C	11.1	6.2	4.4	4.9	7.7		4.7	9.3	2.5	5.4	9.7	4.7	9.2	12.1	8.8	12.9	18.3	13.2	17.3	19.5
		H	14.5	9.5	6.1	5.2	5.2	4.7		6.5	4.4	2.8	7.4	4.8	5.8	8.9	7.0	9.0	14.8	9.6	14.0	15.9
		M	17.5	13.9	8.3	7.9	3.1	9.3	6.5		7.8	6.5	5.5	5.9	4.0	3.7	4.6	6.7	10.2	7.3	10.0	12.7
	A2	C	12.8	7.7	4.9	5.3	7.0	2.5	4.4	7.8		4.4	8.9	4.3	7.8	10.4	7.0	11.2	16.5	11.3	15.5	17.7
		H	15.7	10.6	6.5	6.5	6.0	5.4	2.8	6.5	4.4		7.4	5.1	5.7	8.5	6.0	8.1	14.0	8.5	12.9	15.0
		M	17.5	13.7	9.1	8.9	5.5	9.7	7.4	5.5	8.9	7.4		7.4	6.0	7.2	7.0	8.8	12.4	9.4	11.9	14.2
end 21st century	B1	C	12.8	9.2	3.4	3.8	4.5	4.7	4.8	5.9	4.3	5.1	7.4		5.8	8.8	5.4	11.0	15.4	11.3	14.6	17.5
		H	18.1	14.0	8.4	7.8	3.6	9.2	5.8	4.0	7.8	5.7	6.0	5.8		5.2	4.9	6.9	10.6	7.9	10.3	13.2
		M	20.9	17.1	11.4	11.0	5.9	12.1	8.9	3.7	10.4	8.5	7.2	8.8	5.2		5.3	5.9	6.8	5.2	8.0	9.5
	A1B	C	17.8	13.8	8.1	8.1	5.5	8.8	7.0	4.6	7.0	6.0	7.0	5.4	4.9	5.3		7.5	10.8	6.8	10.1	13.2
		H	22.5	18.1	13.4	12.8	8.4	12.9	9.0	6.7	11.2	8.1	8.8	11.0	6.9	5.9	7.5		7.7	4.9	6.8	8.0
		M	27.3	23.5	18.0	17.5	12.4	18.3	14.8	10.2	16.5	14.0	12.4	15.4	10.6	6.8	10.8	7.7		7.5	6.2	4.2
	A2	C	23.3	18.6	13.9	13.6	9.1	13.2	9.6	7.3	11.3	8.5	9.4	11.3	7.9	5.2	6.8	4.9	7.5		8.0	8.0
		H	26.6	22.6	17.3	16.8	12.1	17.3	14.0	10.0	15.5	12.9	11.9	14.6	10.3	8.0	10.1	6.8	6.2	8.0		7.1
		M	29.0	24.8	20.0	19.5	14.6	19.5	15.9	12.7	17.7	15.0	14.2	17.5	13.2	9.5	13.2	8.0	4.2	8.0	7.1	

GCMs: C – CSIRO; H – HADCM3; M – MIROC.

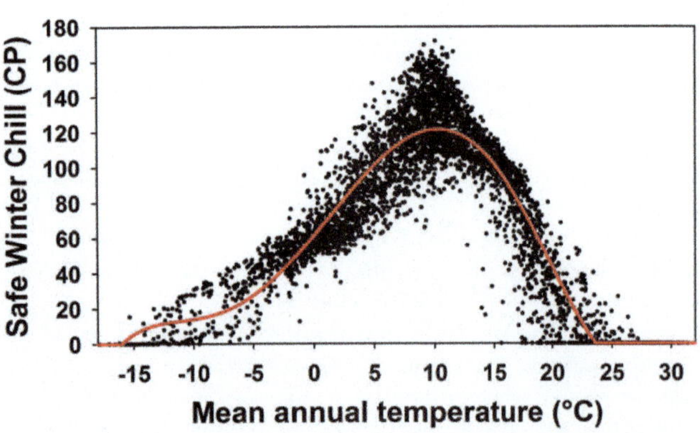

Figure 11. Correlation between mean annual temperature and modeled Safe Winter Chill for the year-2000 scenario. The red line indicates the equation used to correct for unaccounted for variation in temperature during spatial interpolation of site-specific Safe Winter Chill estimates.

them (Fig. 11). The resulting regression equation was:

$$CP(T) = 116.71 + 0.00005\cdot(T-10.005)^5 + 0.0015\cdot(T-10.005)^4 - 0.0180\cdot(T-10.005)^3 - 0.8395\cdot(T-10.005)^2 + 0.4690\cdot T$$

for $-15.996 < T < 23.594$, and $CP(T) = 0$ for temperatures outside this range. In this equation, T is the temperature and CP(T) is the corresponding number of chill portions. The final correction equation was then:

$$SWC_{corr}(x,y) = SWC_{\text{int}}(x,y) + CP(T_{mean}(x,y) + T_{diff}(x,y)) - CP(T_{mean}(x,y)),$$

with T_{mean} being a Kriging surface calculated from mean station temperatures of the respective climate scenario, and x and y the longitude and latitude of each grid cell. SWC_{corr} is the temperature corrected estimate of safe winter chill. On all maps, areas where mean annual temperatures were below 0°C or above 20°C were shaded, because such regions are not suitable for the production of fruits and nuts with chilling requirements. Likewise, all areas that were further than 5° away from the closest useable weather station were shaded, because interpolation results for such grid cells were deemed unreliable. All gridded SWC layers are available from http://treephenology.ucdavis.edu/.

Scenario evaluation

From the surfaces of safe winter chill, site-specific results were extracted for 24 point locations representing important growing regions around the world. Comparing safe winter chill estimates for different combinations of GCM, GHG emissions scenario and point in time provides an impression of the agreement between scenarios, on a case-study basis. We also evaluated differences between scenarios based on the entire distribution over all relevant grid cells (excluding all that were shaded in the maps). Because differences between models in site specific estimates can be both negative and positive, we evaluated the mean absolute difference among all grid cells. The results are indicative of the agreement between models.

Acknowledgments

We are grateful to Heiner Lieth and to the CSTARS lab at UC Davis, who contributed computing capacity, without which this study would not have been possible.

Author Contributions

Conceived and designed the experiments: EL PHB. Performed the experiments: EL. Analyzed the data: EL. Contributed reagents/materials/analysis tools: EL EHG MAS. Wrote the paper: EL EHG MAS PHB.

References

1. Knight TA (1801) Account of Some Experiments on the Ascent of the Sap in Trees. Philosophical Transactions of the Royal Society of London 91: 333–353.
2. Samish RM (1954) Dormancy in Woody Plants. Annual Review of Plant Physiology and Plant Molecular Biology 5: 183–204.
3. Saure MC (1985) Dormancy Release in Deciduous Fruit Trees. Horticultural Reviews 7: 239–300.
4. Vegis A (1961) Samenkeimung und vegetative Entwicklung der Knospen. Handbuch der Pflanzenphysiologie - Encyclopedia of Plant Physiology 16: 168–298.
5. FAOSTAT (2009) Statistical Database of the Food and Agriculture Organization of the United Nations.
6. CIA (2006) CIA World Factbook. Washington D.C., , USA: Central Intelligence Agency.
7. Baldocchi D, Wong S (2008) Accumulated winter chill is decreasing in the fruit growing regions of California. Climatic Change 87: S153–S166.
8. Luedeling E, Zhang M, Girvetz EH (2009) Climatic changes lead to declining winter chill for fruit and nut trees in California during 1950–2099. PLoS ONE 4: e6166.
9. Luedeling E, Blanke M, Gebauer J (2009) Auswirkungen des Klimawandels auf die Verfügbarkeit von Kältewirkung (Chilling) für Obstgehölze in Deutschland. Erwerbs-Obstbau 51: 81–94.
10. Luedeling E, Gebauer J, Buerkert A (2009) Climate change effects on winter chill for tree crops with chilling requirements on the Arabian Peninsula. Climatic Change 96: 219–237.
11. Chandler WH (1942) Deciduous orchards. Philadelphia, USA: Lea & Febiger. 438 p.
12. Richardson EA, Seeley SD, Walker DR (1974) A Model for Estimating the Completion of Rest for Redhaven and Elberta Peach Trees. Hortscience 9: 331–332.
13. Fishman S, Erez A, Couvillon GA (1987) The Temperature-Dependence of Dormancy Breaking in Plants - Computer-Simulation of Processes Studied under Controlled Temperatures. Journal of Theoretical Biology 126: 309–321.
14. Fishman S, Erez A, Couvillon GA (1987) The Temperature Dependence of Dormancy Breaking in Plants: Mathematical Analysis of a Two-Step Model Involving a Cooperative Transition. Journal of Theoretical Biology 124: 473–483.
15. Luedeling E, Zhang M, Luedeling V, Girvetz EH (2009) Sensitivity of winter chill models for fruit and nut trees to climate change. Agriculture, Ecosystems & Environment 133: 23–31.
16. Luedeling E, Zhang M, McGranahan G, Leslie C (2009) Validation of winter chill models using historic records of walnut phenology. Agricultural and Forest Meteorology 149: 1854–1864.
17. Allan P, Linsley-Noakes GC, Holcroft DM, Brunette SA, Burnett MJ, et al. (1997) Kiwifruit research in a subtropical area. Acta Horticulturae 444: 37–42.
18. Erez A (2000) Bud dormancy; phenomenon, problems and solutions in the tropics and subtropics. In: Erez A, ed. Temperate Fruit Crops in Warm Climates. Dordrecht, The Netherlands: Kluwer Academic. pp 17–48.
19. Ruiz D, Campoy JA, Egea J (2007) Chilling and heat requirements of apricot cultivars for flowering. Environmental and Experimental Botany 61: 254–263.
20. Balandier P, Bonhomme M, Rageau R, Capitan F, Parisot E (1993) Leaf Bud Endodormancy Release in Peach-Trees - Evaluation of Temperature Models in Temperate and Tropical Climates. Agricultural and Forest Meteorology 67: 95–113.
21. Luedeling E, Brown PH (2011) A global analysis of the comparability of winter chill models for fruit and nut trees. International Journal of Biometeorology 55: 411–421.
22. NCDC (2008) Database of the National Climatic Data Center National Oceanic and Atmospheric Administration.
23. Semenov MA, Donatelli M, Stratonovitch P, Chatzidaki E, Braruth B (in press) ELPIS: a dataset of local-scale daily climate scenarios for Europe. Climate Research.
24. Semenov MA, Stratonovitch P (2010) The use of multi-model ensembles from global climate models for impact assessments of climate change. Climate Research 41: 1–14.
25. Girvetz EH, Zganjar C, Raber GT, Maurer EP, Kareiva P, et al. (2009) Applied climate-change analysis: the Climate Wizard Tool. PLoS ONE 4: e8320.
26. Petri JL, Berenhauser Leite G (2004) Consequences of insufficient winter chilling on apple tree bud-break. Acta Horticulturae 662: 53–60.
27. Blanke M, Kunz A (2009) Einfluss rezenter Klimaveränderungen auf die Phänologie bei Kernobst am Standort Klein-Altendorf – anhand 50 jähriger Aufzeichnungen. Erwerbsobstbau 51: 101–114.
28. Luedeling E, Steinmann KP, Zhang M, Brown PH, Grant J, et al. (2011) Climate change effects on walnut pests in California. Global Change Biology 17: 228–238.
29. Erez A, Samish RM, Lavee S (1966) Role of light in leaf and flower bud break of peach (Prunus persica). Physiologia Plantarum 19: 650–659.
30. Dozier WA, Powell AA, Caylor AW, McDaniel NR, Carden EL, et al. (1990) Hydrogen Cyanamide Induces Budbreak of Peaches and Nectarines Following Inadequate Chilling. Hortscience 25: 1573–1575.
31. George AP, Lloyd J, Nissen RJ (1992) Effects of Hydrogen Cyanamide, Paclobutrazol and Pruning Date on Dormancy Release of the Low Chill Peach Cultivar Flordaprince in Subtropical Australia. Australian Journal of Experimental Agriculture 32: 89–95.
32. Erez A (1995) Means to compensate for insufficient chilling to improve bloom and leafing. Acta Horticulturae 395: 81–95.
33. Griesbach J (2007) Growing temperate fruit trees in Kenya. Nairobi, Kenya: World Agroforestry Center (ICRAF).
34. Vennetier M, Ripert C (2009) Forest flora turnover with climate change in the Mediterranean region: A case study in Southeastern France. Forest Ecology and Management 258: S56–S63.
35. Chmielewski FM, Rötzer T (2001) Response of tree phenology to climate change across Europe. Agricultural and Forest Meteorology 108: 101–112.
36. Parmesan C (2007) Influences of species, latitudes and methodologies on estimates of phenological response to global warming. Global Change Biology 13: 1860–1872.
37. Menzel A, Sparks TH, Estrella N, Koch E, Aasa A, et al. (2006) European phenological response to climate change matches the warming pattern. Global Change Biology 12: 1969–1976.
38. Yu H, Luedeling E, Xu J (2010) Winter and spring warming result in delayed spring phenology on the Tibetan Plateau. Proceedings of the National Academy of Sciences of the United States of America 107: 22151–22156.
39. Wilks DS, Wilby RL (1999) The weather generation game: a review of stochastic weather models. Progress in Physical Geography 23: 329–357.
40. Semenov MA (2009) Impacts of climate change on wheat in England and Wales. Journal of the Royal Society Interface 6: 343–350.
41. Semenov MA, Halford NG (2009) Identifying target traits and molecular mechanisms for wheat breeding under a changing climate. Journal of Experimental Botany 60: 2791–2804.
42. Evans N, Butterworth MH, Baierl A, Semenov MA, West JS, et al. (2010) The impact of climate change on disease constraints on production of oilseed rape. Food Security 2: 143–156.
43. Lazzarotto P, Calanca P, Semenov MA, Fuhrer J (2010) Transient responses to increasing CO2 and climate change in an unfertilized grass-clover sward. Climate Research 41: 221–232.
44. IPCC (2000) Special report on emissions scenarios; Nakicenovic N, Swart R, eds. Cambridge, UK: Cambridge University Press.
45. Linvill DE (1990) Calculating Chilling Hours and Chill Units from Daily Maximum and Minimum Temperature Observations. Hortscience 25: 14–16.
46. Spencer JW (1971) Fourier series representation of the position of the Sun. Search 2: 172.
47. Almorox J, Hontoria C, Benito M (2005) Statistical validation of daylength definitions for estimation of global solar radiation in Toledo, Spain. Energy Conversion and Management 46: 1465–1471.
48. Hijmans RJ, Cameron SE, Parra JL, Jones PG, Jarvis A (2005) Very high resolution interpolated climate surfaces for global land areas. International Journal of Climatology 25: 1965–1978.

2

Plasticity in Dendroclimatic Response across the Distribution Range of Aleppo Pine (*Pinus halepensis*)

Martin de Luis[1,2]*, Katarina Čufar[3], Alfredo Di Filippo[4], Klemen Novak[1,2], Andreas Papadopoulos[5], Gianluca Piovesan[4], Cyrille B. K. Rathgeber[6], José Raventós[7], Miguel Angel Saz[1,2], Kevin T. Smith[8]

1 Departamento de Geografía y Ordenación del Territorio, Universidad de Zaragoza, Zaragoza, Spain, **2** Instituto de Investigación en Ciencias Ambientales (IUCA), Universidad de Zaragoza, Zaragoza, Spain, **3** Department of Wood Science and Technology, University of Ljubljana, Ljubljana, Slovenia, **4** DendrologyLab (DAFNE), Università Degli Studi della Tuscia, Viterbo, Italy, **5** Department of Forestry and Natural Environment Management, Technological Education Institute of Lamia, Karpenissi, Greece, **6** Laboratoire d'Etude des Ressources Forèt-Bois (LERFoB), Centre INRA de Nancy, Champenoux, France, **7** Departamento de Ecología, Universidad de Alicante, San Vicente del Raspeig, Spain, **8** Northern Research Station, USDA Forest Service, Durham, North Carolina, United States of America

Abstract

We investigated the variability of the climate-growth relationship of Aleppo pine across its distribution range in the Mediterranean Basin. We constructed a network of tree-ring index chronologies from 63 sites across the region. Correlation function analysis identified the relationships of tree-ring index to climate factors for each site. We also estimated the dominant climatic gradients of the region using principal component analysis of monthly, seasonal, and annual mean temperature and total precipitation from 1,068 climatic gridpoints. Variation in ring width index was primarily related to precipitation and secondarily to temperature. However, we found that the dendroclimatic relationship depended on the position of the site along the climatic gradient. In the southern part of the distribution range, where temperature was generally higher and precipitation lower than the regional average, reduced growth was also associated with warm and dry conditions. In the northern part, where the average temperature was lower and the precipitation more abundant than the regional average, reduced growth was associated with cool conditions. Thus, our study highlights the substantial plasticity of Aleppo pine in response to different climatic conditions. These results do not resolve the source of response variability as being due to either genetic variation in provenance, to phenotypic plasticity, or a combination of factors. However, as current growth responses to inter-annual climate variability vary spatially across existing climate gradients, future climate-growth relationships will also likely be determined by differential adaptation and/or acclimation responses to spatial climatic variation. The contribution of local adaptation and/or phenotypic plasticity across populations to the persistence of species under global warming could be decisive for prediction of climate change impacts across populations. In this sense, a more complex forest dynamics modeling approach that includes the contribution of genetic variation and phenotypic plasticity can improve the reliability of the ecological inferences derived from the climate-growth relationships.

Editor: Ben Bond-Lamberty, DOE Pacific Northwest National Laboratory, United States of America

Funding: This work was partially supported by Spanish Ministry of Education and Science co-funded by FEDER program (CGL2012-31668), the European Union and the National Ministry of Education and Religion of Greece (EPEAEK- Environment – Archimedes), the Slovenian Research Agency (program P4-0015), and the USDA Forest Service. The cooperation among international partners was supported by the COST Action FP1106, STREeSS. The funders had no role in study design, data collection and analysis, decision to publish, or preparation of the manuscript.

Competing Interests: The authors have declared that no competing interests exist.

* E-mail: mdla@unizar.es

Introduction

Climate strongly influences the geographical distribution of plant species in terrestrial ecosystems [1,2]. As local temperature and water availability change with broader changes in climate, biodiversity, species distribution, and growth rates are expected to change accordingly [3,4,5]. However, the impact of climate change may vary greatly depending on the ability of species to acclimate or to adapt to future climate conditions [6].

Plant species can adjust to new environmental conditions through local adaptation or phenotypic plasticity [6]. Local adaptation may imply genetic differentiation among populations as a consequence of differential selection pressures and/or population isolation [7]. Phenotypic plasticity, defined as the range of phenotypes that a single genotype can express as a function of its environment, can also be a crucial factor for plant response to rapid climate change [6,8]. Heterogeneous responses of species to climate variability across their range are directly connected to phenomena of local adaptation and phenotypic plasticity and are the basis of potential adaptability to future climate conditions. Detailed knowledge of the relationship between climate and growth across the range of distribution of species is essential to predict and mitigate the effects of climate change.

Climate change is especially rapid and extreme in the Mediterranean basin [9,10,11]. However, due to the transitional nature of the Mediterranean climate (which ranges from near-desert to temperate regimes), Mediterranean areas contain a great variety of natural conditions which will likely affect the impact of climate change [12,13,14].

In the widely diverse set of climatic conditions contained in the Mediterranean basin, pines are by far the most widespread genus covering approximately 5% of the total land area and 25% of the forested area. The most common species is the Aleppo pine (*Pinus halepensis* Mill.) occupying large areas of the western Mediterranean as well as occurring in the eastern Mediterranean [15,16]. The ecology and biogeography of Aleppo pine is well-researched [17,18]. The dendroclimatology of Aleppo pine has been investigated at individual locations within the Mediterranean basin [19,20,21,22,23,24,25,26,27,28,29] confirming that Aleppo pine is sensitive to climate variation. However, a comprehensive dendroclimatic analysis across the distribution range of Aleppo pine is lacking.

In this paper, we investigate the variability of the climate-growth relationships of Aleppo pine using a dendroclimatic network composed of high-resolution climatic and dendrochronology databases comprising the complete climatic gradient across the Mediterranean Basin. Our objectives were to a) characterize the climatic variability within the range of Aleppo pine; b) relate high frequency variation in tree-ring width to monthly, seasonal, and annual temperature and precipitation; c) determine the dependence of the dendroclimatic relationships to underlying climatic patterns.

Methods

Ethics Statement

All sampling sites were located in public forests. The necessary permits for field sampling were issued by the forestry authorities at regional and local level. The locations were not protected areas, and the field studies did not involve endangered or protected species.

Locations and climatic data

The distribution map for Aleppo pine (Fig. 1) based on an earlier map [30] was obtained from the European Forest Genetic Resources Programme (EUFORGEN http://www.euforgen.org/distribution_maps.html). Climatic time series data for 1,068 spatial grid points within the distribution range for the period 1901–2000 were obtained from the Climatic Research Unit (CRU) of the University of East Anglia. For the western part (Spain, Algeria, France, Italy, Slovenia, Greece, and Turkey), monthly, seasonal, and annual mean temperatures and total precipitation were provided by the CRU TS 1.2 dataset (10-minute resolution) [31]. For the portion of the distribution located east of the coverage of CRU TS 1.2, we used the CRU TS 2.1 dataset (0.5° grid resolution) [32]. To characterize the spatial variability of the climate across the distribution area of Aleppo pine, while minimizing the effect of outlier values, we determined the central 95% range for each variable, defined as the values occurring between the 2.5 and 97.5 ranked percentiles.

The dendrochronological network consisted of newly collected and archived tree-ring width series from 63 Aleppo pine sites (Table 1), between 32.23° and 45.67°N latitude, 1.41°W and 36.17°E longitude, and 15 to 1650 m a.s.l. (Fig. 1). To determine the adequacy of the network to describe the range of dendroclimatic responses, climatic data for the CRU gridpoint closest to each site were compared with those of the Aleppo pine distribution as a whole.

A T-mode Principal Component Analysis (PCA) [33] was used to summarize the spatial variability of the mean climate conditions across the study area. PCA is a data reduction technique that transforms a large group of variables into a new smaller set of variables called principal components (PC), which are linear combinations of the original variables. These PC were calculated on the correlation matrix of mean monthly temperatures and mean total monthly precipitation for the period 1901–2000. The components were rotated (Varimax) to redistribute the explained variance and to obtain more stable and robust spatial patterns [34]. Selected PC, guided by Kaiser's Rule (eigenvalues >1; [35]), described climatic gradients across the Mediterranean basin were used in subsequent analysis.

Tree-ring chronology construction

The tree-ring width series were derived from a total of 1634 increment cores from 818 trees (Table 1). Tree selection, core collection, processing, and ring-width measurement were conducted using standard techniques [36,37]. Ring-width index chronologies for each site in the network were constructed using ARSTAN software version 6.05P [38] to retain the high-frequency variation that would be most sensitive to the climate variables to be tested. First, a negative exponential or linear regression function was fitted to each ring-width series. The residuals from this first detrending were then fitted with a cubic smoothing spline function (50% frequency cut-off of 30 years). The residuals from the second detrending were then fitted with an autoregressive model to reduce the autocorrelation in the model residuals which were then averaged to construct the ring-width residual index chronology (chronology RES in ARSTAN) for each site using the biweight robust mean function. An expressed population signal (EPS) of 0.85 was used as a threshold to determine the reliable part of each index chronology to use in subsequent analysis [39].

Dendroclimatic relationships across climate gradients

For our first-stage tests, the statistical relationship between monthly, seasonal and annual climate series and annual tree-ring chronologies was assessed individually for each of the 63 sites by correlation function (CF) analysis using the program DendroClim2002 [40]. The RES chronology was the dependent variable while the regressors were the monthly, seasonal and annual mean temperatures and the total precipitation for each 16-month biological year (from the previous September to the current December).

The second-stage test consisted of multiple regression analysis (forward selection) using the CF coefficients calculated in the first stage as dependent variables and the significant principal components (PC) as independent variables that characterize mean climate conditions for each of 63 study sites. In this test, dependence of CF on PC indicates that the correlation of growth to annual, seasonal, and monthly climate variability depended on the underlying climatic conditions. Then, if dendroclimatic relations were found to be climate-dependent, the variation in the climate-growth relationship may be quantified and predicted for the whole distribution area of species by resolving obtained regression models on all 1,068 points for which PC values were calculated.

Results

Climate characteristics within the distribution area of Aleppo pine

The climatic conditions varied widely from near desert to moist temperate. The central 95% of the range (bounded by the 0.025 and 0.975 percentiles) of mean annual temperatures ranged from 10.2°–17.9°C. Winter temperature (average of 7.0°C) ranged from 2.9–11.5°C and summer temperature (average of 22.3°C) ranged from 18.2–25.9°C (Fig. 2). The central 95% range of annual total precipitation (average of 570 mm) across the geographic distribution ranged from 328–957 mm. The seasonal distribution of precipitation varied greatly with winter precipitation comprising

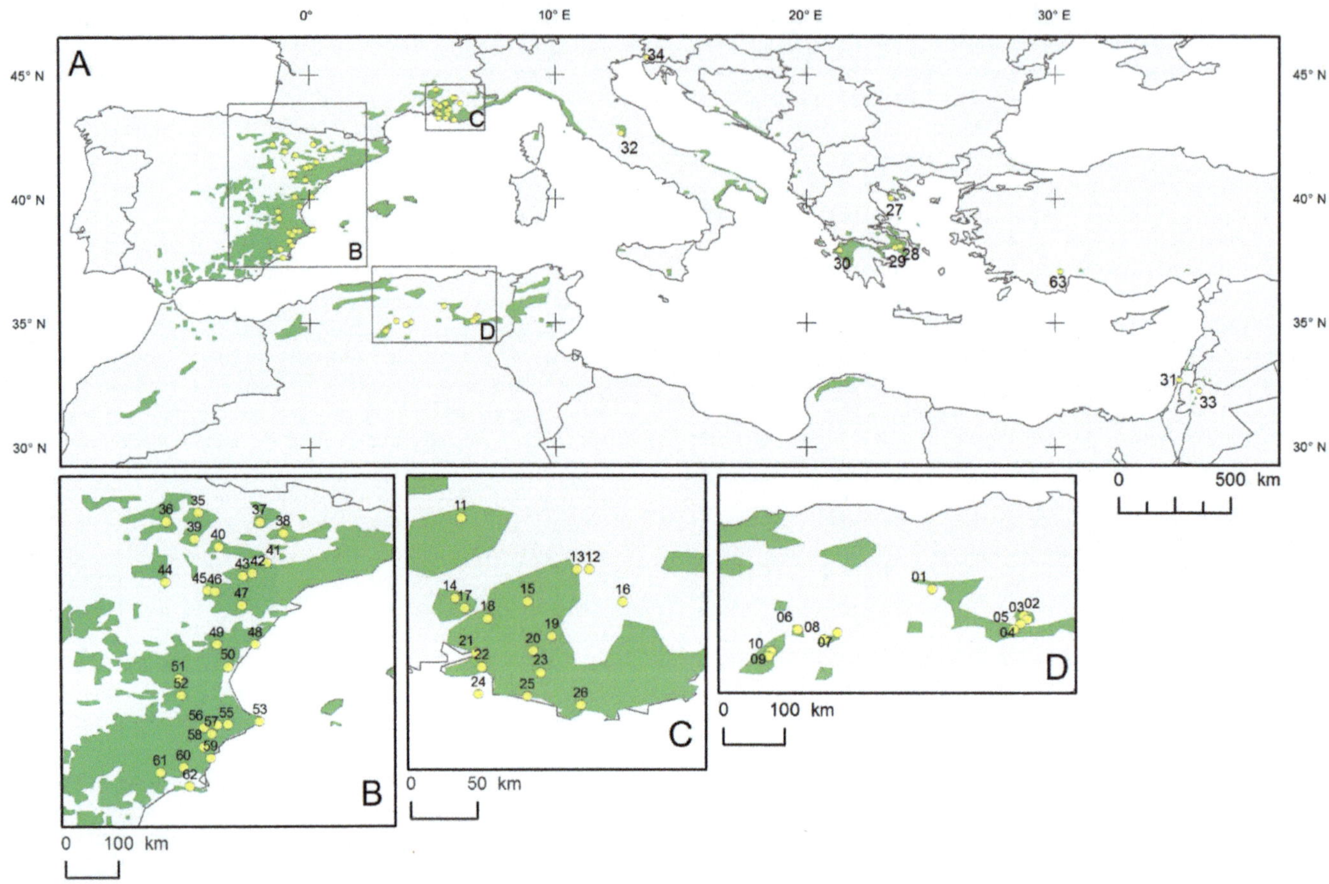

Figure 1. Map of the Mediterranean Basin and the distribution of Aleppo pine (A); details in B, C, D (bottom).

about 31% of the total annual precipitation. Among sites, winter precipitation contributed 15–66% of the annual total precipitation. Spring precipitation contributed an average of about 27% of the total annual precipitation with sites ranging from about 12–36%. Autumn precipitation contributed an average of about 30% with a range of about 13–43%. Summer was generally the dry season with an average annual contribution of about 11% of the total precipitation but ranged across sites from less than 0.1% to about 29% of the total annual precipitation (Fig. 2).

The climate variation across the distribution area of Aleppo pine can be summarized by four significant principal components (PC) which together explain about 95% of the variation. PC1 (containing 59% of the variation) represented the main climatic gradient, from cold and wet conditions to warm and dry (Fig. 3). PC2 (about 24%) related to variation in precipitation (Fig. 3). PC3 (about 8%) represented variation in the seasonal distribution of precipitation from sites where winter is the main precipitation season to sites where the contribution of summer and autumn precipitation to the total annual precipitation increases (Fig. 3). PC4 (about 3%) represented variation in temperature between winter and summer (Fig. 3).

Dendroclimatic relationships in the Aleppo pine tree-ring network

The basic statistics for 63 local chronologies are shown in Table 1. Mean correlation between trees and mean sensitivity of chronologies varied from 0.30 to 0.83 (average of 0.59) and from 0.18 to 0.65 (average of 0.34) respectively, indicating the potential for a common climate signal of varying strength among sites.

The CF analysis showed that ring-width index was primarily related to precipitation and secondarily to temperature. However, a high variability in the relationship between climate and growth was observed across the network (Fig. 4).

Tree-ring index was significantly related to annual total precipitation for all but two chronologies (FAR in France and AMA in Greece). The correlation of tree-ring index to winter temperature was usually significant and positive in colder locations, indicating reduced growth in years with cold winters. In contrast, at three network sites, where conditions are warmer, significant negative correlations indicated reduced growth in years with warm winters.

Variation in dendroclimatic relationships of Aleppo pine across its distribution

The significant regression equations of CF with the four PCs indicated that dendroclimatic relationships depended on the site position along the climatic gradient of the network (Table 2). This dependence produced a clear and coherent spatial pattern of tree growth in the Mediterranean Basin (Fig. 5). The CFs of annual temperature were inversely related to PC1 indicating that reduced growth associated with warm conditions occurred mainly in the southern part of the species distribution area where temperature was generally higher and precipitation lower than the regional average. Growth limitations due to cold conditions occurred in the northern part of the network where the average temperature was

Table 1. Study sites.

N°	Site code	Site	Country	Lat	Long	Elevation	Trees	Samples	Tree-rings	EPS>0.85	MS (res)	MC (res)
1	BOU	Bout10(***)	Algeria	35.65	5.5	1200	11	26	2463	1893–1988	0.3	0.43
2	CHE	Chelia 9(***)	Algeria	35.25	6.85	1650	11	32	4330	1819–1988	0.25	0.68
3	BEZ	Bezez 8(***)	Algeria	35.2	6.9	1300	12	34	3558	1864–1988	0.4	0.67
4	TZI	Tizi Tmellet 6(***)	Algeria	35.13	6.8	1450	10	24	3264	1831–1988	0.47	0.74
5	RAS	Ras fourar 7(***)	Algeria	35.05	6.7	1200	9	27	3342	1835–1987	0.47	0.72
6	SAH	Sahari 3(***)	Algeria	35.05	3.5	1060	11	32	3843	1849–1989	0.3	0.6
7	MES	Messaad 5(***)	Algeria	35	4.1	1100	11	31	3354	1797–1989	0.41	0.75
8	SAF	Safai 4(***)	Algeria	34.9	3.9	1100	11	28	2925	1851–1989	0.4	0.72
9	SEN	SenalbaN(***)	Algeria	34.72	3.12	1410	11	33	3322	1858–1988	0.31	0.68
10	SEW	SenalbaW(***)	Algeria	34.65	3.08	1350	11	33	3451	1859–1990	0.37	0.74
11	NYO	Nyons(*)	France	44.36	5.16	350	8	14	1171	1906–1993	0.23	0.6
12	MEE	Les Mées(*)	France	44.03	5.99	600	9	12	941	1912–1994	0.29	0.58
13	PER	Peruis(*)	France	44.03	5.91	600	8	13	1001	1924–1992	0.18	0.43
14	ROB	Robion(*)	France	43.84	5.12	300	11	20	1663	1905–1994	0.31	0.57
15	CVI	Vitrolles(*)	France	43.82	5.58	600	9	12	919	1918–1993	0.28	0.51
16	MOU	Moustier Sainte-Marie(*)	France	43.82	6.21	750	10	12	1062	1900–1994	0.32	0.57
17	FOO	Mérindol(*)	France	43.78	5.18	330	9	13	1398	1881–1994	0.35	0.69
18	RDA	Roque d'antheron(*)	France	43.71	5.33	170	13	19	1645	1896–1994	0.27	0.59
19	RIA	Rians(*)	France	43.6	5.75	450	10	14	1179	1903–1994	0.32	0.63
20	ROU	Rousset(*)	France	43.5	5.62	200	6	6	426	1920–1990	0.31	0.71
21	ROG	Rognac(*)	France	43.48	5.26	190	7	15	1346	1903–1994	0.34	0.56
22	PNM	Les Pennes Mirabeau(*)	France	43.4	5.29	150	7	14	1050	1910–1993	0.34	0.6
23	AUR	Auriol(*)	France	43.36	5.67	300	8	11	1040	1901–1993	0.28	0.51
24	MAR	Marseille(**)	France	43.22	5.27	200	7	21	2664	1833–1973	0.42	0.64
25	CIO	La Ciotat(*)	France	43.21	5.58	200	12	17	1517	1899–1994	0.31	0.5
26	FAR	Toulon(*)	France	43.15	5.94	500	9	15	1353	1904–1993	0.23	0.45
27	KAS	Kassandra	Greece	39.98	23.47	136	10	30	1369	1937–1989	0.27	0.71
28	KSY	Athens1	Greece	38.05	23.82	260	10	17	1777	1910–2003	0.31	0.48
29	DBG	Athens2	Greece	38	23.62	180	23	37	3183	1909–2001	0.35	0.37
30	AMA	Amaliada	Greece	37.9	21.4	88	12	36	1400	1948–1989	0.21	0.64
31	CAR	Carmel Mountain(*****)	Israel	32.67	35	450	8	9	613	1959–1997	0.27	0.43
32	TER	Terni	Italy	42.62	12.65	470	33	33	1985	1940–2003	0.21	0.3
33	DIB	Dibeen(*****)	Jordan	32.23	35.82	867	14	14	819	1936–1993	0.29	0.4
34	SLO	Krkavce-Dekani	Slovenia	45.67	13.7	50	25	49	2310	1949–2004	0.23	0.38
35	AYE	Ayerbe (Biel)	Spain	42.32	−0.84	924	19	33	1377	1962–2006	0.23	0.45
36	EBA	Ejea-Bardenas	Spain	42.17	−1.4	365	7	7	499	1941–2003	0.32	0.56
37	GRA	El Grado	Spain	42.16	0.2	168	15	30	1506	1950–2006	0.24	0.56
38	EST	Estopiñan del Castillo	Spain	41.97	0.61	502	14	27	1087	1965–2006	0.38	0.62
39	VLL	Villanueva de Gállego	Spain	41.88	−0.91	452	15	29	2692	1898–2006	0.4	0.58
40	CAP	Alcubierre (San Caprasio)	Spain	41.75	−0.5	738	14	27	1516	1942–2006	0.29	0.63
41	FRA	Fraga	Spain	41.47	0.32	340	14	29	3858	1857–2006	0.43	0.62
42	CAS	Caspe	Spain	41.29	0.07	166	15	28	3539	1856–2007	0.59	0.68
43	CHI	Chiprana	Spain	41.24	−0.09	160	7	9	756	1933–2003	0.32	0.45
44	DAR	Daroca	Spain	41.14	−1.41	937	14	28	1699	1936–2006	0.42	0.83
45	OLI	Oliete	Spain	40.99	−0.69	530	15	28	1099	1961–2006	0.3	0.5
46	ALL	Alloza	Spain	40.98	−0.57	595	17	31	2878	1905–2006	0.37	0.69
47	ZOR	Zorita	Spain	40.74	−0.11	857	15	30	3004	1884–2001	0.4	0.61
48	ORO	Oropesa	Spain	40.06	0.12	1	15	30	2022	1929–2003	0.32	0.54
49	MON	Montanejos	Spain	40.06	−0.54	569	16	31	1223	1956–2001	0.39	0.67

Table 1. Cont.

N°	Site code	Site	Country	Lat	Long	Elevation	Trees	Samples	Tree-rings	EPS>0.85	MS (res)	MC (res)
50	**GIL**	Gilet (Sant Espirit)	Spain	39.67	−0.35	175	14	27	2323	1910–2006	0.65	0.7
51	**REQ**	Requena	Spain	39.47	−1.2	721	15	31	3858	1816–2003	0.35	0.48
52	**JAL**	Jalance	Spain	39.19	−1.15	571	15	35	3034	1881–2003	0.41	0.65
53	**JAV**	Javea	Spain	38.73	0.19	96	15	60	3434	1927–2000	0.35	0.73
54	**ALC**	Alcoy (Penaguila)	Spain	38.68	−0.36	674	15	30	3792	1863–2001	0.32	0.51
55	**FNT**	Alcoy (Font Roja)	Spain	38.67	−0.54	1022	14	24	1960	1881–2006	0.35	0.61
56	**BIA**	Biar	Spain	38.62	−0.77	806	15	29	1798	1930–2001	0.29	0.73
57	**MAI**	Maigmo	Spain	38.52	−0.64	845	15	31	2697	1896–2000	0.4	0.65
58	**CRE**	Crevillente	Spain	38.29	−0.77	285	12	43	3715	1911–2000	0.45	0.49
59	**GUA**	Guardamar	Spain	38.1	−0.65	15	30	77	6047	1916–2006	0.36	0.59
60	**FUE**	Fuensanta	Spain	37.94	−1.12	138	14	26	2094	1908–2007	0.44	0.64
61	**SEÑ**	Sierra Espuña	Spain	37.86	−1.52	846	16	29	2670	1907–2007	0.25	0.49
62	**CAT**	Cartagena	Spain	37.61	−1.01	116	15	27	2039	1921–2007	0.42	0.61
63	**BAB**	Bayat Bademleri(****)	Turkey	37.02	30.28	700	15	15	2567	1752–2001	0.21	0.61

*DendroDB (Nicault);
**ITRDB (Serre-Bachet);
***DendroDB (Safar);
****ITRDB (Schweingruber),
*****ITRDB (Touchan),
General information on the 63 study sites in the Aleppo pine network. LAT: Latitude (°N); LON: Longitude (°E); ELE: Elevation (m a.s.l); MS: mean sensitivity; MC: mean correlation between trees.

lower and precipitation less limiting than the regional average (Fig. 5). During winter, the area where cold conditions were associated with reduced tree growth was extended to the south. In contrast, in spring and summer reduced tree growth associated with warm conditions occurred in extended areas towards the southern distribution limits of Aleppo pine (Fig. 5).

Similarly, the inverse relationship of CF to PC2 suggests that the positive effects of annual, winter, and spring precipitation on tree growth significantly increased from wetter to drier sites. The influence of winter precipitation also increased along the wet to dry gradient being especially important at sites with high annual temperatures (Fig. 5). PC3 and PC4 were also significantly related to CFs suggesting that the variation in the seasonal distribution of precipitation and temperature across the network influenced the observed dendroclimatic relationships. PC3 was positively related with summer and negatively with winter precipitation while PC4 was positively related with sites with especially high summer temperatures.

Discussion

Sensitivity of Aleppo pine to climate variations across its distribution area

The dendrochronological network of 63 sites comprises an essentially complete climatic gradient for Aleppo pine and provides information on how the growth of this species responds to climate variability across its distribution range. In the widely diverse set of climatic conditions, our study demonstrated that Aleppo pine is an excellent species for climate-growth studies. The mean correlation between trees in essentially all chronologies indicated that despite micro-environmental influences, a common pattern of year-to-year variation in ring width exists across all sites. The mean tree-ring sensitivity (the average relative difference in adjacent rings and a measure of climatic responsiveness) was generally higher than observed for other species in the Mediterranean area [41,42,43,44,45,46] highlighting the importance of Aleppo pine as a climate proxy for large regional areas. This could be especially important in the Mediterranean region where available climate observations are generally limited to the second half of the 20th century [47]. Additional sampling sites at locations predicted to contain strong climatic signals based on these results would be especially useful to improve reconstruction of past climate. Distribution maps of climate-growth relationships across the distribution range of Aleppo pine may be also especially valuable to predict the response of trees to changing climate patterns.

High sensitivity of a species to climate variations across a wide geographic distribution area or in a wide range of climate conditions as observed here and in other dendrochronological studies suggests that climate change impacts may not be restricted to the ecotone or edge but may occur across the whole distribution range [48,49]. According to this, our results for Aleppo pine suggest that in the colder sites where low winter temperatures are presently limiting, future increased temperature may have a positive effect on tree growth. However, in the drier and warmer southern Mediterranean, where current positive anomalies in temperature have negative effects on tree growth, conditions for future tree growth will become less favorable as temperatures increase.

The role of phenotypic plasticity in a species response to climate variability under global warming

The ability of a species to respond through expression of existing genetic variation or phenotypic plasticity to changing environmental conditions may play a decisive role in species persistence or expansion under future global warming [8,50]. Phenotypic

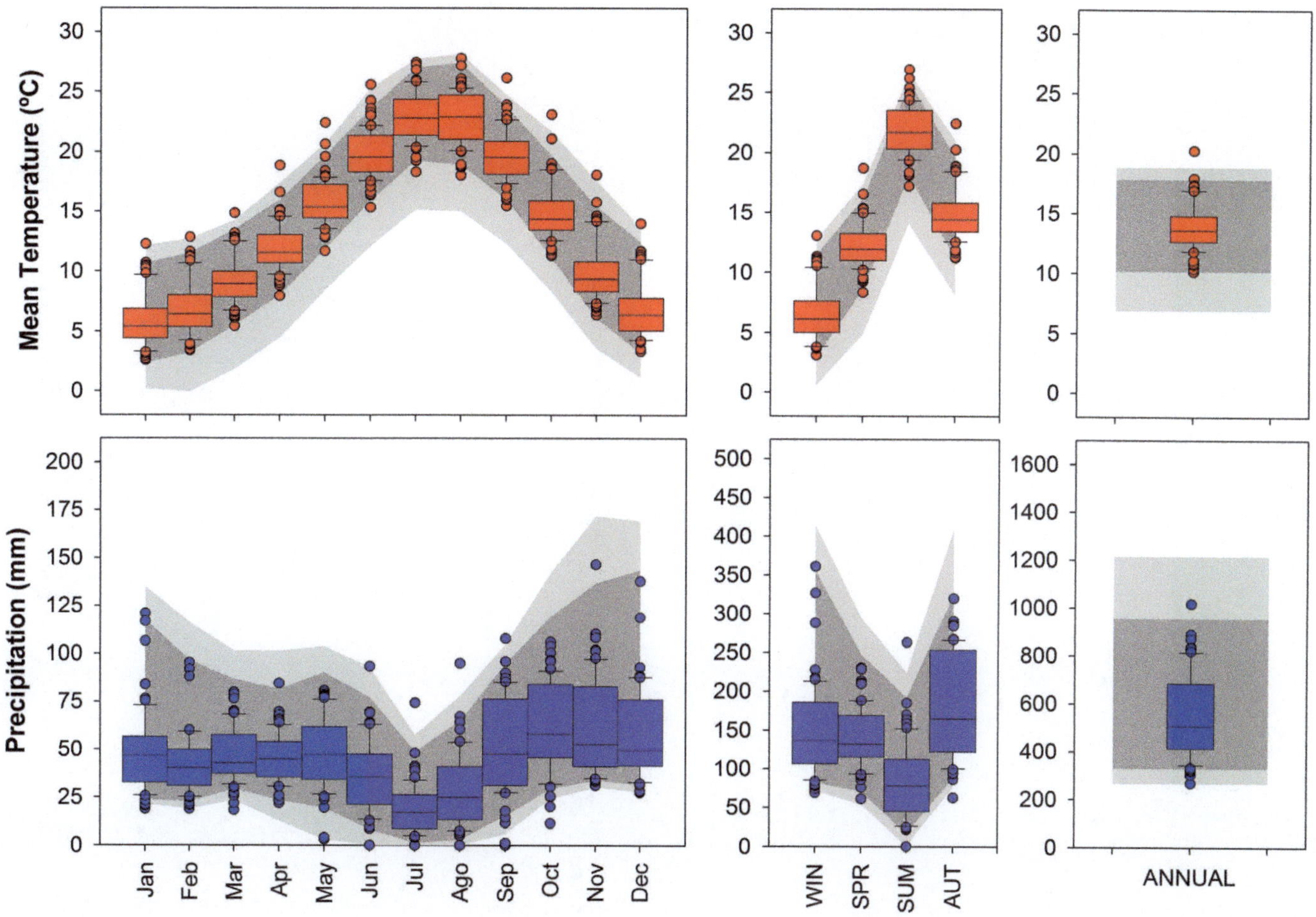

Figure 2. Climate characteristics across the distribution area of Aleppo pine. Dark grey and light grey areas represent climate envelopes defined by the 2.5th–97.5th and the 0.5th–99.5th percentiles, respectively. Box plots show the range of mean monthly air temperatures and total monthly precipitation for the *P. halepensis* dendrochronological network of 63 locations. The central vertical lines indicate median values, boxes enclose the central two quartiles, whiskers indicate the 10th and the 90th percentiles and the dots represent the full range of mean climate values.

plasticity may be especially important as an adaptive strategy in trees and woody plants since the long generation time of many perennial species implies that the same set of genotypes needs to cope with year-to-year changing environmental conditions [51].

Generally, tree species are considered to express moderate to high plasticity in their responses to environmental stress [52,53]. Highly variable species that are able to survive in a broader range of environmental conditions are expected to better adjust to future climate conditions [7].

Pinus halepensis shows a wide ecological breadth and is adapted to a large range of environmental conditions, abiotic stressors, and perturbations [54]. Several recent studies highlight substantial phenotypic plasticity of *Pinus halepensis* in relation to different anatomical, reproductive and vegetative traits [55,56,57,58,59,60,61,62,63,64,65,66,67,68,69,70].

A substantial plasticity in the annual rhythms of cambial activity of Aleppo pine in response to different climatic conditions has been also well established. Thus, some studies suggest that the cambium is able to maintain activity throughout the whole year when and where climate conditions are favourable [71]. Under other conditions, cambial activity can stop for one to three months during winter, depending on the prevalence of low temperatures [72]. In addition, cambial activity in *Pinus halepensis* can slow down, or even stop, during summer drought [59,71,72,73,74] and resume later when moisture availability increases in autumn [59,71,73,74,75,76]. In other circumstances, the trees can be subjected to "double stress" characterized by two stops in cambial activity - one during winter, caused by low temperatures, and one during the summer triggered by high temperatures and lack of precipitation [73,74,77,78].

However, despite such evidence, plasticity in species responses to climate variability has been poorly explored from a dendrochronological perspective and variability in dendroclimatic response is more often considered as due to environmental noise in the tree-ring signal rather than a consequence of the plastic character of the species.

Geographical variation in climate conditions across the distribution area of a species creates differential selection pressures, and as a consequence population responses to climate variability are likely to vary among genetic provenances as well as through phenotypic plasticity. In this sense, our results highlight significant geographical variations in growth response of Aleppo pine to interannual climate variability suggesting substantial plasticity of growth in response to various climatic conditions. These results do not resolve the source of variability as genetic variation in provenance, phenotypic plasticity or some combination of these or other factors. However, the use of isoenzymes, terpene composition, and other markers has shown that Aleppo pine contains little

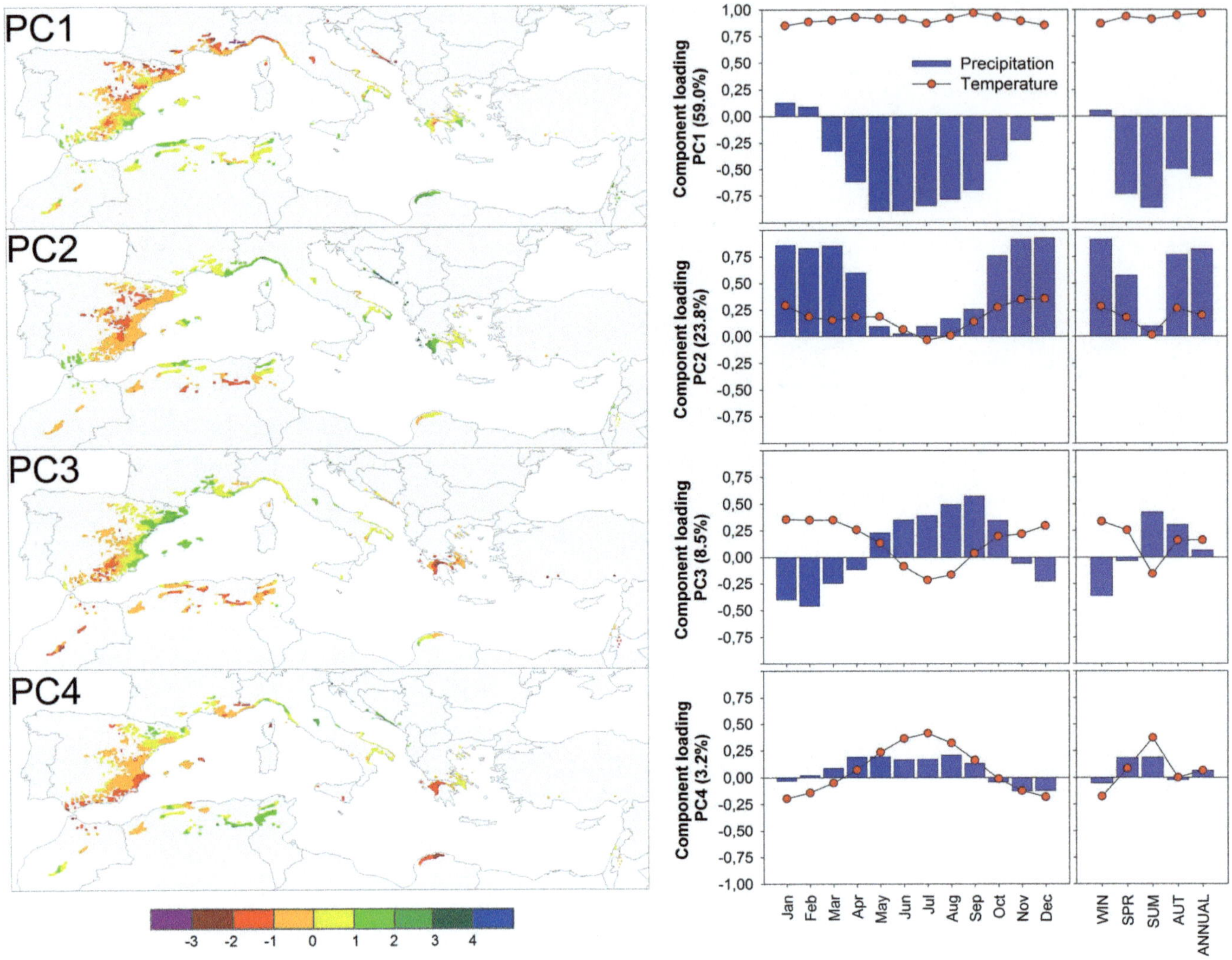

Figure 3. Spatial distribution of the first four significant principal components (PC) reflecting the spatial variability of the mean climate conditions (left). Component loading of each PC against mean monthly, seasonal and annual temperatures and precipitation for the period 1901–2000 (right).

genetic variability [79,80] and that intrapopulation genetic variability is generally much higher than interpopulation variability [81]. These findings suggest that phenotypic plasticity has a dominant role in the observed variation in dendroclimatic response across distribution range of Aleppo pine. Then, as current growth responses to inter-annual climate variability vary spatially across existing climate gradients, we expect that future climate-growth relationships will also vary with position along future climate gradients.

Implications

Historical records of biological responses to climate in situ are perhaps the most dependable data we have to reconstruct past climate conditions, to predict future climate change impacts, and to develop realistic mitigation strategies. The tree-ring record is the most important and widely used source of long-term proxy data as they are available over a wide range of temporal and spatial scales.

However, the use of tree-rings to reconstruct past climate patterns and to forecast of future forest growth within the context of global warming is based on the uniformitarian principle (UP) that assumes that the climate-limiting factors that controlled tree-ring characteristics in the past continue the present and will extend into the future [82]. Under UP, the modeled dendroclimatic relationship is viewed as stationary and consistent through time and the range of climatic variability.

To the contrary, our results show that dendroclimatic relationships significantly vary across species distribution in accordance to underlying climate conditions. Our results also suggest that such variations are likely to be related to the acclimation of trees to the new environmental conditions through a plastic response over decadal time periods. This complicates the simple projection of current relationships into the past or the future. For example, the response to winter temperature change from positive to negative moving from northern (mesic) to southern (drier) sites, with no response at intermediate sites along the thermal gradient. In this sense, the recent warming has a positive effect at the northern limit and a negative effect on the southern portion of the distribution range. At the same time, as summer drought severity increases the response to precipitation disappears, as the tree enters a quiescent and nonreactive state.

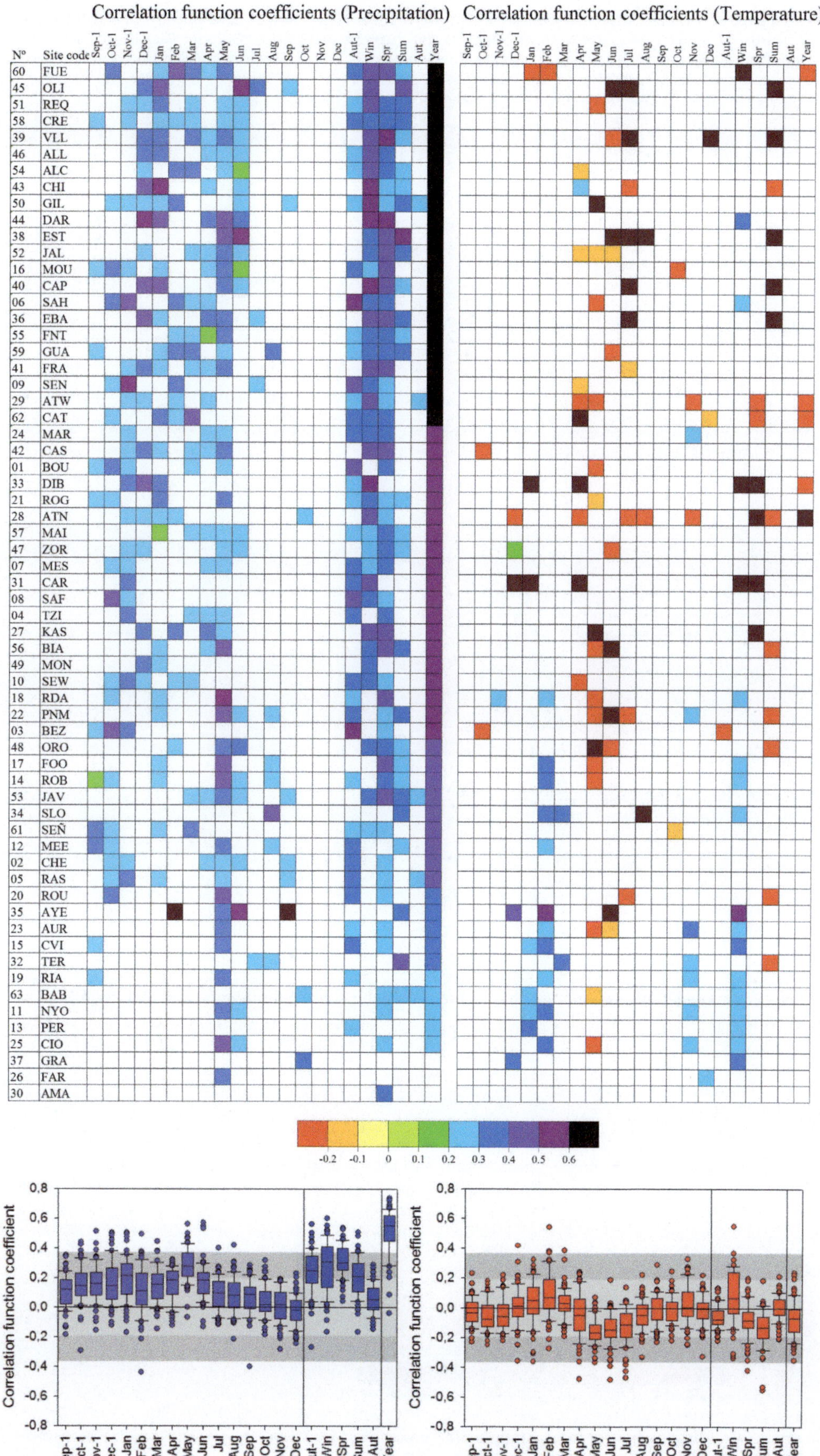
Correlation function coefficients (Precipitation)
Correlation function coefficients (Temperature)
Nº
Site code
Sep-1 Oct-1 Nov-1 Dec-1 Jan Feb Mar Apr May Jun Jul Aug Sep Oct Nov Dec Aut-1 Win Spr Sum Aut Year
60 FUE
45 OLI
51 REQ
58 CRE
39 VLL
46 ALL
54 ALC
43 CHI
50 GIL
44 DAR
38 EST
52 JAL
16 MOU
40 CAP
06 SAH
36 EBA
55 FNT
59 GUA
41 FRA
09 SEN
29 ATW
62 CAT
24 MAR
42 CAS
01 BOU
33 DIB
21 ROG
28 ATN
57 MAI
47 ZOR
07 MES
31 CAR
08 SAF
04 TZI
27 KAS
56 BIA
49 MON
10 SEW
18 RDA
22 PNM
03 BEZ
48 ORO
17 FOO
14 ROB
53 JAV
34 SLO
61 SEÑ
12 MEE
02 CHE
05 RAS
20 ROU
35 AYE
23 AUR
15 CVI
32 TER
19 RIA
63 BAB
11 NYO
13 PER
25 CIO
37 GRA
26 FAR
30 AMA
-0.2 -0.1 0 0.1 0.2 0.3 0.4 0.5 0.6
Correlation function coefficient
0,8 0,6 0,4 0,2 0,0 -0,2 -0,4 -0,6 -0,8

Figure 4. Correlation between the site chronologies of Aleppo pine and climate; monthly, seasonal, and annual precipitation and temperature from September of previous year (Sep-1) to December (Dec) of the current year (top). Only significant values are shown ($p<0.05$). Box plot showing the variability of correlation coefficients across the chronology network. The central horizontal lines indicate the median values, boxes enclose the central two quartiles, whiskers indicate the 10th and the 90th percentiles and the dots represent the full range of correlation coefficients. Dark and light grey areas indicate significance levels at 99 and 95%, respectively.

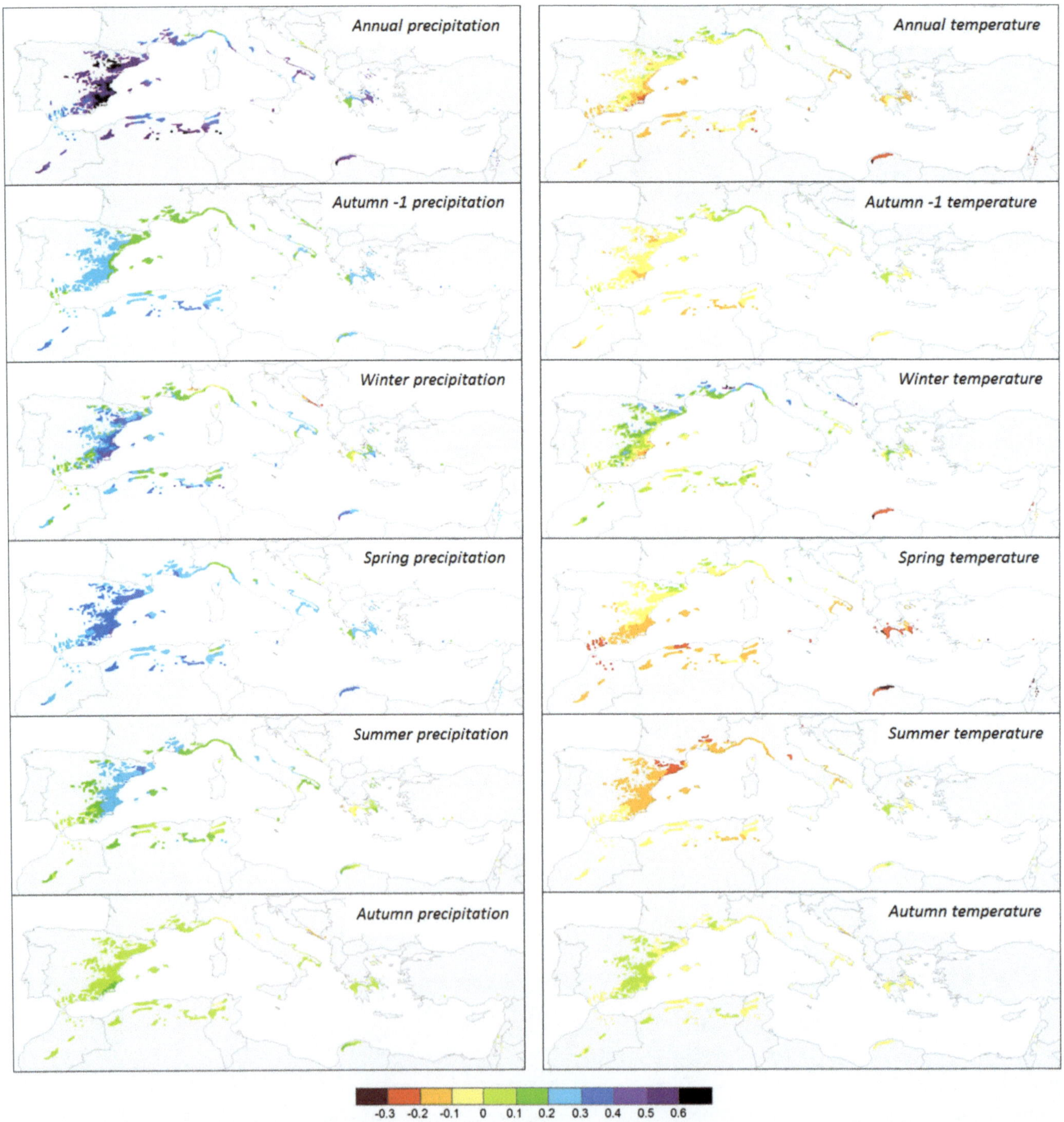

Figure 5. Correlation coefficients across the distribution area of Aleppo pine calculated by using the multiple regression models detailed in Table 2, applied to the values of the significant principal components (PC).

Table 2. Variation of dendroclimatic relationships across climate gradients.

		Intercept		PC1		PC2		PC3		PC4				
		Coef.	Sig	Coef.	Sig	Coef.	Sig	Coef.	Sig	Coef.	Sig	Adjusted r^2	F	P-value
Precipitation	**Sep-1**													ns
	Oct-1	0.1432	***	−0.0319	*	−0.0458	*					0.15	3.8	0.008
	Nov-1	0.1674	***	0.0364	*			−0.0411	**			0.27	6.6	<0.001
	Dec-1	0.1412	***			−0.0728	**			0.0389	*	0.21	5.0	0.001
	Jan	0.1633	***			−0.0644	**	0.0354	*			0.21	5.1	0.001
	Feb	0.1074	***	0.0692	***							0.17	4.1	0.005
	Mar	0.1303	***	0.047	***	−0.0666	***			−0.0245	*	0.44	13.2	<0.001
	Apr	0.1462	***	0.0307	*	−0.0531	***					0.3	7.8	<0.001
	May	0.2528	***	−0.0623	***			0.0438	***	−0.0218	*	0.45	13.9	<0.001
	Jun	0.1371	***	−0.0481	**	−0.0649	***	0.0342	*			0.26	6.5	<0.001
	Jul	0.064	***			−0.0604	***					0.19	4.7	0.002
	Aug	0.0722	***					0.0474	**			0.21	5.1	0.001
	Sep													ns
	Oct													ns
	Nov			−0.0267	*					−0.0506	***	0.24	5.8	0.001
	Dec	−0.0386	**					0.0447	***			0.16	3.9	0.007
	AUT-1	0.2373	***					−0.0408	*			0.11	2.9	0.029
	WIN	0.2348	***	0.0605	**	−0.0948	***	0.0445	*			0.33	8.7	<0.001
	SPR	0.295	***			−0.0527	***			−0.0233	*	0.17	4.2	0.005
	SUM	0.1572	***			−0.057	**	0.0662	***			0.28	6.9	<0.001
	AUT													ns
	ANNUAL	0.4666	***			−0.1166	***	0.0384	*			0.41	11.6	<0.001
Temperature	**Sep-1**			0.0292	**							0.13	3.2	0.018
	Oct-1	−0.048	***	−0.0262	*	0.0434	**			−0.0286	**	0.31	8.0	<0.001
	Nov-1	−0.0397	**	−0.0491	***	0.042	**					0.41	11.7	<0.001
	Dec-1			−0.0613	***							0.22	5.3	0.001
	Jan	0.0497	**	−0.1033	***			−0.0348	**			0.55	19.7	<0.001
	Feb	0.0921	***	−0.0975	***	0.0365	*					0.49	15.9	<0.001
	Mar									0.0275	*	0.14	3.5	0.012
	Apr	−0.0833	***	−0.1027	***			0.0476	***	0.0242	*	0.57	21.8	<0.001
	May													ns
	Jun	−0.1125	***	0.0344	**	0.051	***	−0.0374	**			0.27	6.8	<0.001
	Jul	−0.0971	***			0.0364	*	−0.0454	**			0.15	3.8	0.009
	Aug	−0.0501	***							−0.0299	**	0.2	4.8	0.002
	Sep					−0.0443	**			−0.0238	*	0.25	6.1	<0.001
	Oct													ns
	Nov			−0.0589	***							0.25	6.0	<0.001
	Dec					0.0563	***	−0.0243	*			0.19	4.7	0.002
	AUT-1	−0.0516	***	−0.0222	**	0.0432	***					0.34	9.1	<0.001
	WIN	0.0703	***	−0.1268	***							0.57	21.8	<0.001
	SPR	−0.1099	***	−0.0678	***			0.0385	***	0.0285	**	0.51	16.8	<0.001
	SUM	−0.1122	***	0.0367	*			−0.0436	**			0.16	4.1	0.006
	AUT													ns
	ANNUAL	−0.0749	***	−0.0796	***							0.48	15.1	<0.001

Multiple regression models calculated from sets of correlation coefficients from the 63 sites in the dendrochronological network and significant principal components (PC1–PC4) of spatial variability in climate factors.
*$p<0.05$;
**$p<0.01$;
***$p<0.001$.

Accordingly, the application of UP to dendrochronology may be inappropriate when the dendroclimatic responses are likely to be inherently unstable and climate-dependent as we have shown for Aleppo pine in the Mediterranean basin.

Interestingly, a number of tree-ring studies have addressed changes in tree sensitivity and/or changes in the response of tree growth to climate in coincidence with unprecedented climate warming over the recent decades [83,84,85,86,87,88,89,90]. The causes of this lack of stability in the dendroclimatic relationship are still not well understood as all potential forcing factors (e.g., changes in climate, atmospheric CO2 concentration, or nitrogen deposition) capable of driving this change covary and obscure the individual impacts [86] but a plausible explanation could be the acclimation of trees to the new environmental conditions through a plastic response over decadal time periods. Our results are related to a limited geographical area and a specific dataset and may have limited application to other species or regions. However, such patterns may occur elsewhere and if so, it may have serious implications by affecting reliability of tree-ring based climate reconstructions.

In addition, climate model validation [91], species distribution models [7,9] and forecasts of future forest growth within the context of global warming [1,22,92,93,94] whether or not based on dendrochronology also assumes a uniform response of species to climate variability and change without the competitive advantage conferred by genetic and/or phenotypic plasticity [95]. These assumptions could lead to exaggerate or underestimate species at risk under future climate change.

Accurate prediction of growth of Aleppo pine and other species in response to future climate variability requires an understanding of the plasticity of the response of growth to climate variability and change. A more complex forest dynamics modeling approach considering genetic variation and phenotypic plasticity may contribute to resolve these uncertainties and would result in a more realistic characterization of the biological processes that govern species responses to climatic changes [96,97,98]. In this sense, the reliability of old principles and assumptions requires re-examination and complementary theoretical and experimental frameworks should be developed to better understand what trees are telling us in a changing world.

Acknowledgments

The authors would like to thank the Academic editor Ben Bond-Lamberty and two anonymous reviewers that for their helpful and constructive comments on the earlier versions of the manuscript.

Author Contributions

Conceived and designed the experiments: MdL. Performed the experiments: MdL KC ADF AP GP CBKR JR MAS KTS. Analyzed the data: MdL KN JR MAS. Contributed reagents/materials/analysis tools: MdL KC ADF AP GP CBKR JR MAS. Wrote the paper: MdL KTS.

References

1. Thuiller W, Araujo MB, Lavorel S (2004) Do we need land-cover data to model species distributions in Europe?. J Biogeogr 31: 353–361.
2. Peñuelas J, Canadell JG, Ogaya R (2011) Increased water-use efficiency during the 20th century did not translate into enhanced tree growth. Global Ecol Biogeogr 20: 597–608.
3. Alcamo J, Moreno JM, Nováky B, Bindi M, Corobov R, et al. (2007) Europe. In: Parry ML, Canziani OF, Palutikof JP, van der Linden PJ, Hanson CE, editors. Climate Change 2007: Impacts, Adaptation and Vulnerability. Contribution of Working Group II to the Fourth Assessment Report of the Intergovernmental Panel on Climate Change. Cambridge, UK. Cambridge University Press, pp. 541–580.
4. Iverson LR, Prasad AM, Matthews SN, Peters M (2008) Estimating potential habitat for 134 eastern US tree species under six climate scenarios. For Ecol Manage 254: 390–406.
5. Maiorano L, Cheddadi R, Zimmermann NE, Pellissier L, Petitpierre B, et al. (2012) Building the niche through time: using 13,000 years of data to predict the effects of climate change on three tree species in Europe. Global Ecol Biogeogr 22: 302–317.
6. Nicotra AB, Atkin OK, Bonser SP, Davidson AM, Finnegan EJ, et al. (2010) Plant phenotypic plasticity in a changing climate. Trends Plant Sci 15: 684–692.
7. Benito-Garzón M, Alía R, Robson TM, Zavala MA (2011) Intra-specific variability and plasticity influence potential tree species distributions under climate change. Global Ecol Biogeogr 20: 766–778.
8. Matesanz S, Gianoli E, Valladares F (2010) Global change and the evolution of phenotypic plasticity in plants. Ann NY Acad Sci 1206: 35–55.
9. Thuiller W, Lavorel S, Araujo MB, Sykes MT, Prentice IC (2005) Climate change threats to plant diversity in Europe. Proc Natl Acad Sci USA 102: 8245.
10. Giorgi F, Lionello P (2008) Climate change projections for the Mediterranean region. Global Planet Change 63: 90–104.
11. Lindner M, Maroschek M, Netherer S, Kremer A, Barbati A, et al. (2010) Climate change impacts, adaptive capacity, and vulnerability of European forest ecosystems. For Ecol Manage 259: 698–709.
12. Peñuelas J, Boada M (2003) A global change-induced biome shift in the Montseny mountains (NE Spain). Global Change Biol 9: 131–140.
13. Schroter D, Cramer W, Leemans R, Prentice IC, Araujo MB, et al. (2005) Ecosystem service supply and vulnerability to global change in Europe. Science 310: 1333–1337.
14. Christensen JH, Hewitson B, Busuioc A, Chen A, Gao X, et al. (2007) Regional Climate Projections. In: Solomon S, Qin D, Manning M, Chen Z, Marquis M, et al. editors. Climate Change 2007: The Physical Science Basis. Contribution of Working Group I to the Fourth Assessment Report of the Intergovernmental Panel on Climate Change Cambridge University Press, Cambridge, United Kingdom and New York, NY, USA. pp. 847–940.
15. Barbéro M, Loisel R, Quézel P, Richardson DM, Romane F (1998) Pines of the Mediterranean Basin. In: Richardson DM, editor. Ecology and Biogeography of Pinus. Cambridge, UK: Cambridge University Press, pp: 153–170.
16. Quézel P (2000) Taxonomy and biogeography of Mediterranean pines (Pinus halepensis and P. brutia). In: Ne'eman G, Trabaud L, editors. Ecology, Biogeography and Management of Pinus halepensis and P. brutia Forest Ecosystems in the Mediterranean Basin Backhuys, Leiden, The Netherlands. pp. 1–12.
17. Richardson DM (1998). Ecology and Biogeography of Pinus. Cambridge University Press, Cambridge, UK. pp 3–46.
18. Ne'eman G, Trabaud L (2000) Biogeography and Management of Pinus halepensis and P. brutia Forest Ecosystems in the Mediterranean Basin. Backhuys, Leiden, The Netherlands.
19. Touchan R, Hughes MK (1999) Dendrochronology in Jordan. J Arid Environ 42: 291–303.
20. Papadopoulos A, Serre-Bachet F, Tessier L (2001) Tree ring to climate relationships of Aleppo pine (Pinus halepensis Mill.) in Greece. Ecologia Mediterranea 27: 89–98.
21. Rathgeber C, Nicault A, Kaplan JO, Guiot J (2003) Using a biogeochemistry model in simulating forests productivity responses to climatic change and [CO2] increase: example of Pinus halepensis in Provence (south-east France). Ecol Modell 166: 239–255.
22. Rathgeber C, Nicault A, Guiot J, Keller T, Guibal F, et al. (2000) Simulated responses of Pinus halepensis forest productivity to climatic change and CO2 increase using a statistical model. Global Planet Change 26: 405–421.
23. Touchan R, Xoplaki E, Funkhouser G, Luterbacher J, Hughes MK, et al. (2005) Reconstructions of spring/summer precipitation for the Eastern Mediterranean from tree-ring widths and its connection to large-scale atmospheric circulation. Clim Dyn 25: 75–98.
24. Sarris D, Christodoulakis D, Korner C (2007) Recent decline in precipitation and tree growth in the eastern Mediterranean. Global Change Biol 13: 1187–1200.
25. de Luis M, Novak K, Cufar K, Raventos J (2009) Size mediated climate-growth relationships in Pinus halepensis and Pinus pinea. Trees Struct Funct 23: 1065–1073.
26. Pasho E, Camarero JJ, de Luis M, Vicente-Serrano SM (2011) Spatial variability in large-scale and regional atmospheric drivers of Pinus halepensis growth in eastern Spain. Agric For Meteorol 151: 1106–1119.
27. Attolini MR, Calvani F, Galli M, Nanni T, Ruggiero L, et al. (1990) The relationship between climatic variables and wood structure in Pinus halepensis Mill. Theor Appl Climatol 41: 121–127.
28. Olivar J, Bogino S, Spiecker H, Bravo F (2012) Climate impact on growth dynamic and intra-annual density fluctuations in Aleppo pine (Pinus halepensis) trees of different crown classes. Dendrochronologia 30: 35–47.
29. Novak K, de Luis, M, Raventós J, Čufar K (2013) Climatic signals in tree-ring widths and wood structure of Pinus halepensis in contrasted environmental conditions. Trees Struct Funct 27: 927–936.
30. Critchfield WB, Little EL (1966) Geographic distribution of the pines of the World. USDA For Serv Misc Publ 991.

31. Mitchell TD, Carter TR, Jones PD, Hulme M, New M (2004) A comprehensive set of high-resolution grids of monthly climate for Europe and the globe: the observed record (1901–2000) and 16 scenarios (2001–2100). Tyndall Center Working Paper 55: 1–30.
32. Mitchell TD, Jones PD (2005) An improved method of constructing a database of monthly climate observations and associated high-resolution grids. Int J Climatol 25: 693–712.
33. Jolliffe IT (1986) Principal Component Analysis. Springer-Verlag, pp. 487. DOI: 10.1007/b98835. ISBN 978-0-387-95442-4.
34. Richman MB (1986) Rotation of principal components. J Clim 6: 29–35.
35. Kaiser HF (1992) On Cliff's formula, the Kaiser-Guttman rule, and the number of factors. Perceptual and Motor Skills 74: 595–598.
36. Speer JH (2010) Fundamentals of Tree-Ring Research. University of Arizona Press, Tucson, Arizona, USA, 333 pp.
37. Cook ER, Kairiukstis LA (1990) Methods of Dendrochronology. Kluwer Academic Publishers, Dordrecht/Boston/London. 394 pp.
38. Cook ER (1985) A time series analysis approach to tree-ring standardization. PhD dissertation, University of Arizona, Tucson, Arizona. 171 pp.
39. Wigley TML, Briffa KR, Jones PD (1984) On the average value of correlated time-series, with applications in dendroclimatology and hydrometeorology. J Clim Appl Meteorol 23: 201–213.
40. Biondi F, Waikul K (2004) DENDROCLIM2002: A C++ program for statistical calibration of climate signals in tree-ring chronologies. Comput Geosci 30: 303–311.
41. Andreu L, Gutierrez E, Macias M, Ribas M, Bosch O et al. (2007) Climate increases regional tree-growth variability in Iberian pine forests. Global Change Biol 13: 804–815.
42. Buntgen U, Frank D, Trouet V, Esper J (2010) Diverse climate sensitivity of Mediterranean tree-ring width and density. Trees Struct Funct 24: 261–273.
43. Carrer M, Nola P, Motta R, Urbinati C (2010) Contrasting tree-ring growth to climate responses of Abies alba toward the southern limit of its distribution area. Oikos 119: 1515–1525.
44. Lebourgeois F, Rathgeber CBK, Ulrich E (2010) Sensitivity of French temperate coniferous forests to climate variability and extreme events (Abies alba, Picea abies and Pinus sylvestris). J Veg Sci 21: 364–376.
45. Mérian P, Lebourgeois F (2011) Size-mediated climate–growth relationships in temperate forests: A multi-species analysis. For Ecol Manage 261: 1382–1391.
46. Babst F, Poulter B, Trouet V, Tan K, Neuwirth B, et al. (2013) Site- and species-specific responses of forest growth to climate across the European continent. Global Ecol Biogeogr 22: 706–717.
47. González-Hidalgo JC, Brunetti M, de Luis M (2011) A new tool for monthly precipitation analysis in Spain: MOPREDAS database (monthly precipitation trends December 1945–November 2005). Int J Climatol 31: 715–731.
48. Hamann A, Wang TL (2005) Models of climatic normals for genecology and climate change studies in British Columbia. Agric For Meteorol 128: 211–221.
49. Chen PY, Welsh C, Hamann A (2010) Geographic variation in growth response of Douglas-fir to interannual climate variability and projected climate change. Global Change Biol 16: 3374–3385.
50. Matesanz S, Valladares F (2013) Ecological and evolutionary responses of Mediterranean plants to global change. Environ Exp Bot. DOI: http://dx.doi.org/10.1016/j.envexpbot.2013.09.004.
51. Willson MF (1983) Plant Reproductive Ecology. John Wiley & Sons, New York, USA.
52. Wagner F, Below R, de Klerk P, Dilcher DL, Joosten H, et al. (1996) A natural experiment on plant acclimation: lifetime stomatal frequency response of an individual tree to annual atmospheric CO2 increase. Proc Natl Acad Sci USA 93: 11705–11708.
53. Climent JM, Aranda I, Alonso J, Pardos JA, Gil L (2006) Developmental constraints limit the response of Canary Island pine seedlings to combined shade and drought. For Ecol Manage 231: 164–168.
54. Ne'eman G, Goubitz S, Nathan R (2004) Reproductive traits of Pinus halepensis in the light of fire – a critical review. Plant Ecol 171: 69–79.
55. Baquedano FJ, Valladares F, Castillo FJ (2008) Phenotypic plasticity blurs ecotypic divergence in the response of Quercus coccifera and Pinus halepensis to water stress. Eur J Forest Res 127: 495–506.
56. Chambel MR, Climent J, Alía R (2007) Divergence among species and populations of Mediterranean pines in biomass allocation of seedlings grown under two watering regimes. Ann Forest Sci 64: 87–97.
57. Climent J, Prada MA, Calama R, Chambel R, Sánchez de Ron D, et al. (2008) To grow or to seed: ecotypic variation in reproductive allocation and cone production by young female Aleppo pine (Pinus halepensis, pinaceae). Am J Bot 95: 1–10.
58. Cuesta B, Villar-Salvador P, Puértolas J, Jacobs DF, Rey Benayas JM (2010) Why do large, nitrogen rich seedlings better resist stressful transplanting conditions? A physiological analysis in two functionally contrasting Mediterranean forest species. For Ecol Manage 260: 71–78.
59. de Luis M, Novak K, Raventós J, Gričar J, Prislan P, et al. (2011) Cambial activity, wood formation and sapling survival of Pinus halepensis exposed to different irrigation regimes. For Ecol Manage 262: 1630–1638.
60. Froux F, Ducrey M, Epron D, Dreyer E (2004) Seasonal variations and acclimation potential of the thermostability of photochemistry in four Mediterranean conifers. Ann Forest Sci 61: 235–241.
61. García-Esteban L, Martín JA, de Palacios P, García Fernández F, López R (2010) Adaptive anatomy of Pinus halepensis trees from different Mediterranean environments in Spain. Trees Struct Funct 24: 19–30.
62. Michelozzi M, Loreto F, Colom R, Rossi F, Calamassi R (2011) Drought responses in Aleppo pine seedlings from two wild provenances with different climatic features. Photosynthetica 49: 564–572.
63. Monnier Y, Vila B, Montès N, Bousquet-Mélou A, Prévosto B, et al. (2011) Fertilization and allelopathy modify Pinus halepensis saplings crown acclimation to shade. Trees Struct Funct 25: 497–507.
64. Monnier Y, Bousquet-Mélou A, Vila B, Prévosto B, Fernandez C (2013) How nutrient availability influences acclimation to shade of two (pioneer and late-successional) Mediterranean tree species?. Eur J Forest Res 132: 325–333.
65. Pardos M, Climent J, Gil L, Pardos JA (2003) Shoot growth components and flowering phenology in grafted Pinus halepensis Mill. Trees Struct Funct 17: 442–450.
66. Santos del Blanco L, Zas R, Notivol E, Chambel MR, Majada J, et al. (2010) Variation of early reproductive allocation in multi-site genetic trials of Maritime pine and Aleppo pine. Forest Systems 19: 381–392.
67. Santos del Blanco L, Bonser SP, Valladares F, Chambel MR, Climent J (2013) Plasticity in reproduction and growth among 52 range-wide populations of a Mediterranean conifer: Adaptive responses to environmental stress. J Evolution Biol 26: 1912–1924.
68. Sardans J, Rodà F, Peñuelas J (2004) Phosphorus limitation and competitive capacities of Pinus halepensis and Quercus ilex subsp. rotundifolia on different soils. Plant Ecol 174: 305–317.
69. Voltas J, Chambel M, Prada M, Ferrio J (2008) Climate-related variability in carbon and oxygen stable isotopes among populations of Aleppo pine grown in common-garden tests. Trees Struct Funct 22: 759–769.
70. Zavala MA, Espelta JM, Caspersen J, Retana J (2011) Interspecific differences in sapling performance with respect to light and aridity gradients in mediterranean pine-oak forests: Implications for species coexistence. Can J Forest Res 41: 1432–1444.
71. Liphschitz N, Lev-Yadun S, Rosen E, Waisel Y (1984) The Annual Rhythm of Activity of the Lateral Meristems (Cambium and Phellogen) in Pinus halepensis Mill and Pinus pinea L. IAWA Bulletin 5: 263–274.
72. Gindel I (1967) Cambial activity as a function of the intensity of transpiration in Pinus halepensis Mill. Proc XVI IUFRO Congr. München 1967, vol IV, sect.23: 188–206.
73. de Luis M, Gričar J, Čufar K, Raventós J (2007) Seasonal dynamics of wood formation in Pinus halepensis from dry and semi-arid ecosystems in Spain. IAWA Journal 28: 389–404.
74. de Luis M, Novak K, Raventos J, Gricar J, Prislan P, et al. (2011). Climate factors promoting intra-annual density fluctuations in Aleppo pine (Pinus halepensis) from semiarid sites. Dendrochronologia 29: 163–169.
75. Camarero JJ, Olano JM, Parras A (2010) Plastic bimodal xylogenesis in conifers from continental Mediterranean climates. New Phytol 185: 471–480.
76. Novak K, Saz Sánchez MA, Čufar K, Raventós J, de Luis M (2013) Age, climate and intra-annual density fluctuations in Pinus halepensis in Spain. IAWA Journal 34: 459–474.
77. Cherubini P, Gartner BL, Tognetti R, Braker OU, Schoch W, et al. (2003) Identification, measurement and interpretation of tree rings in woody species from Mediterranean climates. Biol Rev 78: 119–148.
78. Lev-Yadun S (2000) Wood structure and the ecology of annual growth ring formation in Pinus halepensis and P. brutia. In: Ecology, Biogeography and Management of Pinus halepensis and P. brutia Forest Ecosystems in the Mediterranean Basin (eds Ne'eman G, Trabaud L), pp. 67–78. Backhuys Publishers, Leiden, The Netherlands.
79. Schiller G, Gonkle MT, Grunwald C (1986) Local differentiation among Mediterranean populations of Aleppo pine in their isoenzymes. Silvae Genet 35: 11–19.
80. Soto A, Robledo-Arnuncio JJ, González-Martínez SC, Smouse PE, Alia R (2010) Climatic niche and neutral genetic diversity of the six Iberian pine species: a retrospective and prospective view. Mol Ecol 19: 1396–1409.
81. Gómez A, Alia R, Bueno MA (2001) Genetic diversity of Pinus halepensis Mill. populations detected by RAPD loci. Ann For Sci 58: 869–875.
82. Fritts HC (1976) Tree Rings and Climate. Academic Press, New York.
83. Briffa KR, Schweingruber FH, Jones PD, Osborn TJ, Shiyatov SG, et al. (1998) Reduced sensitivity of recent tree-growth to temperature at high northern latitudes. Nature 391: 678–682.
84. Carrer M, Urbinati C (2004) Age-dependent tree-ring growth responses to climate in Larix decidua and Pinus cembra. Ecology 85: 730–740.
85. Buntgen U, Frank D, Wilson R, Carrer M, Urbinati C, et al. (2008) Testing for tree-ring divergence in the European Alps. Glob Change Biol 14: 2443–2453.
86. D'Arrigo R, Wilson R, Liepert B, Cherubini P (2008) On the 'Divergence Problem' in Northern Forests: A review of the tree-ring evidence and possible causes. Glob Planet Change 60: 289–305.
87. Carrer M, Nola P, Eduard JL, Motta R, Urbinati C (2007) Regional variability of climate-growth relationships in Pinus cembra high elevation forests in the Alps. J Ecol 95: 1072–1083.
88. di Filippo A, Biondi F, Cufar K, de Luis M, Grabner M, et al. (2007) Bioclimatology of beech (Fagus sylvatica L.) in the Eastern Alps: spatial and altitudinal climatic signals identified through a tree-ring network. J Biogeograph 34: 1873–1892.

89. Wilmking M, Singh J (2008) Eliminating the "divergence problem" at Alaska's northern treeline. Climate Past Discuss 4: 741–759.
90. Esper J, Frank D, Buntgen U, Verstege A, Hantemirov RM, et al. (2010) Trends and uncertainties in Siberian indicators of 20th century warming. Glob Change Biol 16: 386–398.
91. Randall DA, Wood RA, Bony S, Colman R, Fichefet T, et al. (2007) Climate models and their evaluation. In: Solomon S, Qin D, Manning M, Chen Z, Marquis M, Averyt KB, et al. editors. Climate Change 2007: The Physical Science Basis. Contribution of Working Group I to the Fourth Assessment Report of the Intergovernmental Panel on Climate Change (Eds.) Cambridge University Press, Cambridge, United Kingdom and New York, NY, USA.
92. Thuiller W, Lavorel S, Sykes MT, Araujo MB (2006) Using niche-based modelling to assess the impact of climate change on tree functional diversity in Europe. Divers Distrib 12: 49–60.
93. Keenan T, Serra Diaz JM, Lloret F, Ninyerola M, Sabaté S (2011) Predicting the future of forests in the Mediterranean under climate change, with niche- and process-based models: CO2 matters!. Glob Change Biol 17: 565–579.
94. Serra Diaz JM, Keenan TF, Ninyerola M, Sabaté S, Gracia C, et al. (2013) Geographical patterns of congruence and incongruence between correlative species distribution models and a process-based ecophysiological growth model. J Biogeogr 40: 1928–1938.
95. Rodder D, Lotters S (2009) Niche shift versus niche conservatism? Climatic characteristics of the native and invasive ranges of the Mediterranean house gecko (Hemidactylus turcicus). Global Ecol Biogeogr 18: 674–687.
96. Pearson RG, Dawson TP (2003) Predicting the impacts of climate change on the distribution of species: are bioclimate envelope models useful? Global Ecol Biogeogr 12: 361–371.
97. Montoya D, Purves DW, Urbieta IR, Zavala MA (2009) Do species distribution models explain spatial structure within tree species ranges?. Global Ecol Biogeogr 18: 662–673.
98. Roberts DR, Hamann A (2012) Predicting potential climate change impacts with bioclimate envelope models: a palaeoecological perspective. Global Ecol Biogeogr 21: 121–133.

Climate Change and the Potential Distribution of an Invasive Shrub, *Lantana camara* L.

Subhashni Taylor[1]*, Lalit Kumar[1], Nick Reid[1], Darren J. Kriticos[2,3]

1 Ecosystem Management, School of Environmental and Rural Science, University of New England, Armidale, New South Wales, Australia, **2** CSIRO Ecosystem Sciences, Canberra, Australian Capital Territory, Australia, **3** E. H. Graham Centre, Charles Sturt University, Wagga Wagga, New South Wales, Australia

Abstract

The threat posed by invasive species, in particular weeds, to biodiversity may be exacerbated by climate change. *Lantana camara* L. (lantana) is a woody shrub that is highly invasive in many countries of the world. It has a profound economic and environmental impact worldwide, including Australia. Knowledge of the likely potential distribution of this invasive species under current and future climate will be useful in planning better strategies to manage the invasion. A process-oriented niche model of *L. camara* was developed using CLIMEX to estimate its potential distribution under current and future climate scenarios. The model was calibrated using data from several knowledge domains, including phenological observations and geographic distribution records. The potential distribution of lantana under historical climate exceeded the current distribution in some areas of the world, notably Africa and Asia. Under future scenarios, the climatically suitable areas for *L. camara* globally were projected to contract. However, some areas were identified in North Africa, Europe and Australia that may become climatically suitable under future climates. In South Africa and China, its potential distribution could expand further inland. These results can inform strategic planning by biosecurity agencies, identifying areas to target for eradication or containment. Distribution maps of risk of potential invasion can be useful tools in public awareness campaigns, especially in countries that have been identified as becoming climatically suitable for *L. camara* under the future climate scenarios.

Editor: Alex J. Cannon, Pacific Climate Impacts Consortium, Canada

Funding: The National Climate Change Adaptation Research Facility (NCCARF) kindly provided a grant that allowed the main author to purchase the CLIMEX software as well as attend a training workshop for CLIMEX. The funder had no role in study design, data collection and analysis, decision to publish, or preparation of the manuscript. No additional external funding was received for this study.

Competing Interests: The authors have declared that no competing interests exist.

* E-mail: btaylo36@une.edu.au

Introduction

Biotic invasions occur when organisms are transported to new areas where they may reproduce and spread. Such invasions can have far reaching ecological and economic impacts [1–4]. Along with climate, such invasions are now seen as a contributor to global change [3]. Invasive species are a major threat to the Earth's biodiversity because they often dramatically affect the structure and functioning of ecosystems [5]. Prudent management of biological invasions requires information about the expected potential distribution and relative abundance of invasive species under current and future climate scenarios. Such information is necessary for risk assessment as well as the formulation of appropriate long-term management strategies. Species distribution models (SDMs), bioclimatic models, and ecological niche models (ENMs) [6] provide many opportunities in this area. Niche modelling is underpinned by Hutchinson's [7] fundamental and realized niche concepts. However, there are conflicting views on what the models actually represent [8]. While some researchers suggest that niche models provide an estimate to the species' fundamental niche [9], others consider models as presenting a "spatial representation of the realized niche" [6,10]. The distribution of a species depends on complex interactions between a range of factors, acting with different strengths at different scales [10]. These include abiotic and biotic factors, the foremost of which is climate [11–12].

CLIMEX [13] is a useful tool for exploring the relationship between the fundamental and realized niche of species [14]. It is an eco-climatic modelling package that has been used by many researchers involved in estimating invasive species' potential distributions under current and future climate [15–18]. The realized niche of a species is the range of conditions and resources in which it can persist in the presence of competitors and predators [19] and this is represented by the native range of a species [20]. After introduction into an exotic environment, a species can commonly inhabit a broader range of climatic conditions because it is freed from many of its competitors and predators. This is potentially its fundamental niche [7]. It is important to include exotic range data when developing climate models for invasive species [21] because any predictions based on just its native range may lead to an underestimation of the species' potential range, especially if it has not had the opportunity to express its full climatic preferences in the absence of natural enemies [20].

CLIMEX allows users to model the potential distribution of organisms, drawing upon a variety of information types, including direct experimental observations of a species' growth response to temperature and soil moisture, its phenology and knowledge of its current distribution. In a review of the various climate-based packages designed to estimate potential species distributions,

Kriticos and Randall [21] found that 'CLIMEX was the most suitable climate modelling package for undertaking Weed Risk Assessments because it can support model-fitting to a global plant distribution, includes a climate change scenario mechanism, and provides an insight into the plant's ecological response to climate'. Subsequently, Webber et al. [22] found that CLIMEX was better placed than two correlative modelling methods (MaxEnt and Boosted Regression Trees) to project a species' distribution in a novel climate such as a new continent, or under a future climate scenario. Modelling the potential distribution of a species using climatic mapping has received some criticism because it assumes that climate alone limits the geographical distribution and does not include biotic interactions and dispersal [23]. However, despite its limitations, climatic mapping plays an important role in the definition of the fundamental (potential) niche of an invasive species in its exotic range [24]. Since climate is one of the major determinants of the potential range of species, climate changes could have a significant impact on species' distributions.

There is now overwhelming evidence for rapid climate change with global mean surface temperatures projected to increase by 2.4 to 6.4°C between 1990 and 2100 [25] along with various changes in rainfall patterns (increases, decreases and changes in seasonality). Together with major threats to biodiversity, agriculture and human health, climate change also has implications for invasive species. The immediate effect of climate change on such species will most likely be shifts in their distributions facilitated by changes in temperature and rainfall patterns that define their range

Table 1. The CLIMEX parameter values that were used for *L. camara*.

Parameter	Values
Limiting low temperature (DV0)	10°C
Lower optimal temperature (DV1)	25°C
Upper optimal temperature (DV2)	30°C
Limiting high temperature (DV3)	33°C
Limiting low soil moisture (SM0)	0.1
Lower optimal soil moisture (SM1)	0.5
Upper optimal soil moisture (SM2)	1.2
Limiting high soil moisture (SM3)	1.6
Cold stress temperature threshold (TTCS)	5°C
Cold stress temperature rate (THCS)	-0.004 week^{-1}
Minimum degree-day cold stress threshold (DTCS)	15°C days
Degree-day cold stress rate (DHCS)	-0.0022 week^{-1}
Heat stress temperature threshold(TTHS)	33°C
Heat stress temperature rate (THHS)	0.001 week^{-1}
Dry stress threshold (SMDS)	0.1
Dry stress rate (HDS)	-0.01 week^{-1}
Wet stress threshold (SMWS)	1.6
Wet stress rate (HWS)	0.01 week^{-1}

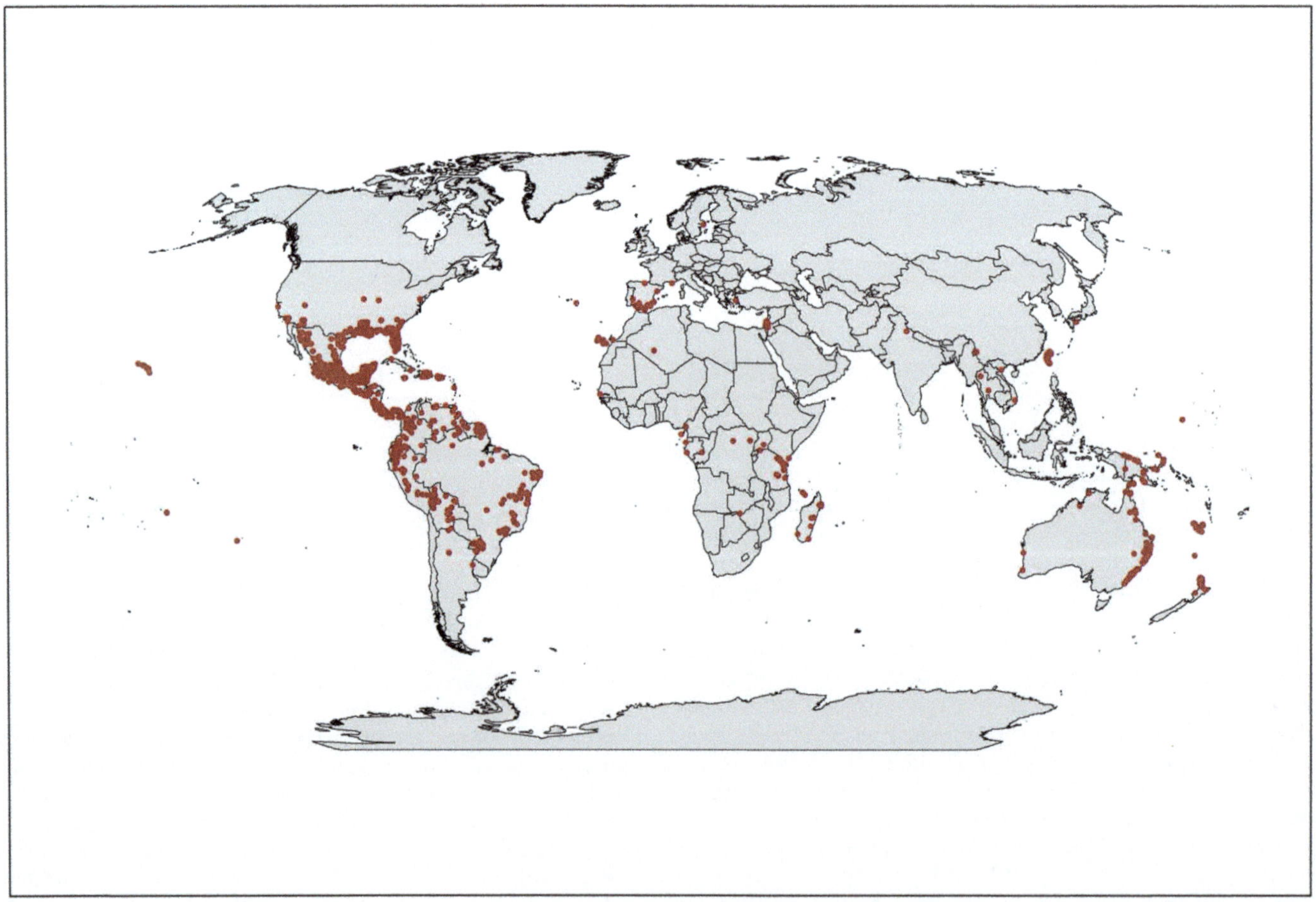

Figure 1. The current global distribution of *L. camara* taken from the Global Biodiversity Information Facility 2007. Red dots indicate occurrence records of *L. camara*.

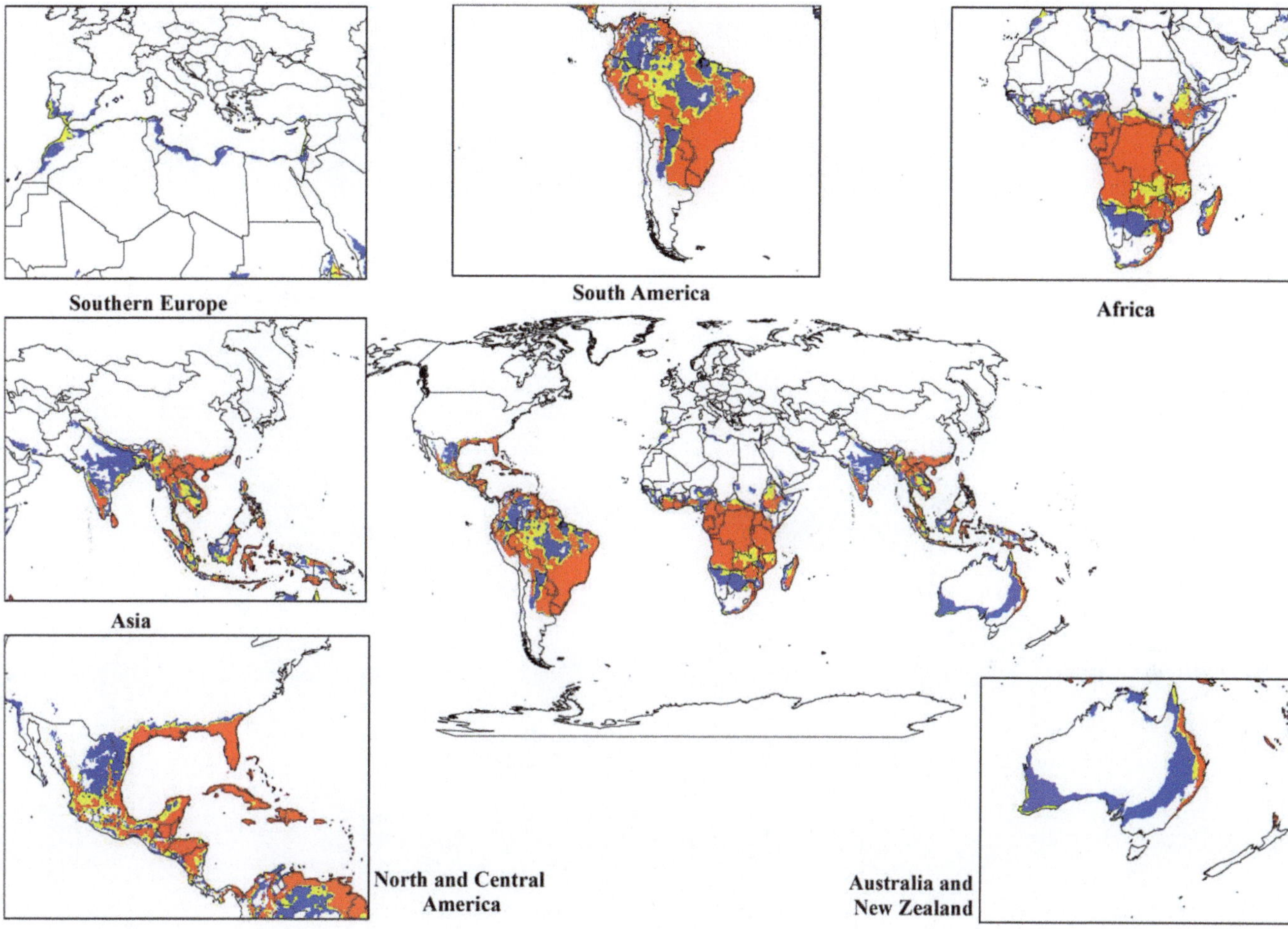

Figure 2. The climate (EI) for *L. camara* based on CLIMEX for reference climate (averaging period 1950–2000). White areas indicate unsuitable climate areas (EI = 0), blue areas indicate marginal climate areas (EI = 1–10), yellow areas indicate suitable climate areas (EI = 10–20) and red areas indicate highly suitable climate areas (EI>20).

boundaries. Species that can tolerate a wide range of climatic conditions may be favoured, and as a result they may have greater competitive success than most native species [26–28].

One such species, *Lantana camara* L. (lantana), is a major weed in many tropical and subtropical countries outside its native range of Central and northern South America and the Caribbean. Its global distribution includes approximately 60 countries or island groups between 35°N and 35°S [1]. It has a variety of impacts including a reduction in native species diversity, extinctions, decline in soil fertility, and allelopathic alteration of soil properties as well as an alteration of ecosystem processes [1]. Allelochemicals found in lantana have been shown to inhibit the growth of other species growing close to it [29–30]. Persistent lantana infestations can lead to a reduction in biodiversity because it has the potential to block succession and cause the displacement of native species [31–32]. It can cause striking changes in the structural and floristic composition of natural communities by interrupting the regeneration processes of other native species thus reducing species richness [33].

This study utilized the CLIMEX modelling package to develop a model of the climate responses of lantana based on its native distribution and invasive distribution outside Australia. This model was then used to project its potential distribution under current climate, using the extensive Australian distribution data for model validation and assess the impacts of climate change on its potential distribution using two global climate models (GCM), CSIRO-Mk3.0 and MIROC-H. These were run with the A1B and A2 SRES (Special Report on Emissions Scenarios) emission scenarios for 2030 and 2070.

Materials and Methods

CLIMEX Software

CLIMEX for Windows Version 3 [13,34–35] was used to develop a model of the potential distribution of *L. camara* under current and future climate scenarios. CLIMEX is based on the observation that the distribution of plants and poikilothermal animals is primarily determined by climate [36]. The software works on the basis of an eco-physiological model that assumes that at each location, a species may experience a favourable season with positive population growth and an unfavourable season that causes population decline [35]. The user can use the model to infer parameters that describe the species' response to climate based on its geographic range or phenological observations [35]. CLIMEX can also be used deductively to apply climate response parameters extracted from experimental observations to climatic datasets. In practice, both approaches can be applied to inform the selection of parameter values. These parameters can then be applied to novel climates to project the species' potential range in new regions or

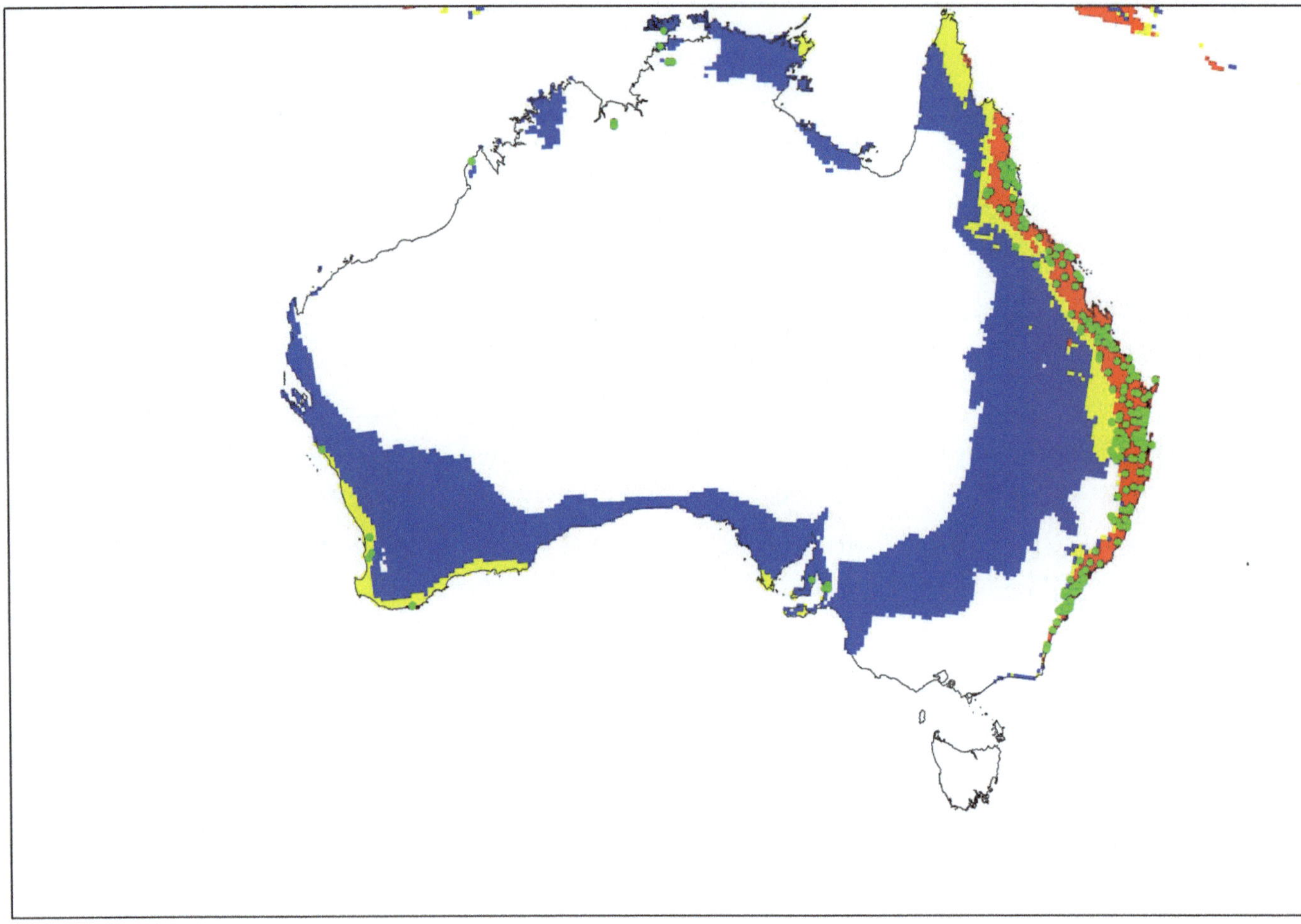

Figure 3. Current and modelled climate (EI) for *L. camara* based on CLIMEX for reference climate (averaging period 1950–2000). Data for current Australian distribution is taken from Australia's Virtual Herbarium. White areas indicate unsuitable climate areas (EI = 0), blue areas indicate marginal climate areas (EI = 1–10), yellow areas indicate suitable climate areas (EI = 10–20) and red areas indicate highly suitable climate areas (EI>20). Green dots indicate occurrence records of *L. camara*.

climate scenarios [22,37]. The potential for population growth during favourable climate conditions is described by an annual growth index (GI_A) that conforms to the law of tolerance [38] and the law of minimum [39]. Four stress indices (cold, wet, hot and dry) and up to four interaction stresses (hot–dry, hot–wet, cold–dry and cold–wet) are used to describe the probability that the population can survive unfavourable conditions. The growth and stress indices are calculated weekly and combined into an overall annual index of climatic suitability, the Ecoclimatic index (EI) which is theoretically scaled from 0 to 100. Establishment is only possible if EI>0. In practice, EI values close to the maximum are rare, and confined to species with an equatorial range, as this would imply ideal growing conditions year-round [40]. EI values close to zero indicate a low probability of conditions conducive to persistence in time and space. In such marginally suitable climates, species are likely to be restricted to favourable microhabitats, and to exhibit significant metapopulation dynamics.

Taxonomy and Native and Naturalized Distribution of *L. camara*

The genus *Lantana* L. (Verbenaceae) includes up to 150 species [41–42]. Many of these species are native to South America, Central America or southern North America, while a few species occur naturally in Africa and Asia [43]. There is considerable uncertainty associated with the taxonomy of the genus *Lantana*. Four distinct groups can be recognized within the genus [44]. These are referred to as *Lantana* sections *Calliorheas*, *Sarcolippia*, *Rhytocamara* and *Camara*. *Lantana* section *Camara* is divided into three complexes based on *L. urticifolia*, *L. hirsuta* and *L. camara*. The *L. camara* complex contains the weedy lantana generally referred to as *L. camara* L. *sensu lato*, which has a pan-tropical distribution [1]. *Lantana camara sensu stricto* is known from Jamaica, Trinidad, Mexico, Brazil and Florida [44]. It may have a wider native distribution in South America [45] extending to Argentina and Uruguay [1,46]. The present study only addresses the 'weedy taxa' of *Lantana* section *Camara* which are the most prevalent taxa in the genus. They are important due to economic and environmental impacts as they can invade natural and agricultural ecosystems [47–48]. Its environmental impacts are especially damaging in native forests that have undergone disturbance. In such cases, lantana forms a dense understorey, disrupts succession and decreases biodiversity [1,32]. In areas that have a high density of lantana, species richness is reduced [33,49] and local flora is threatened [50–51]. In natural systems, dense lantana infestations can alter fire regimes [52]. Lantana is a weed of important crops such as coffee, oil palms, coconuts, cotton, bananas, pineapples, sugarcane, tea, rubber and rice in various countries [53]. It forms

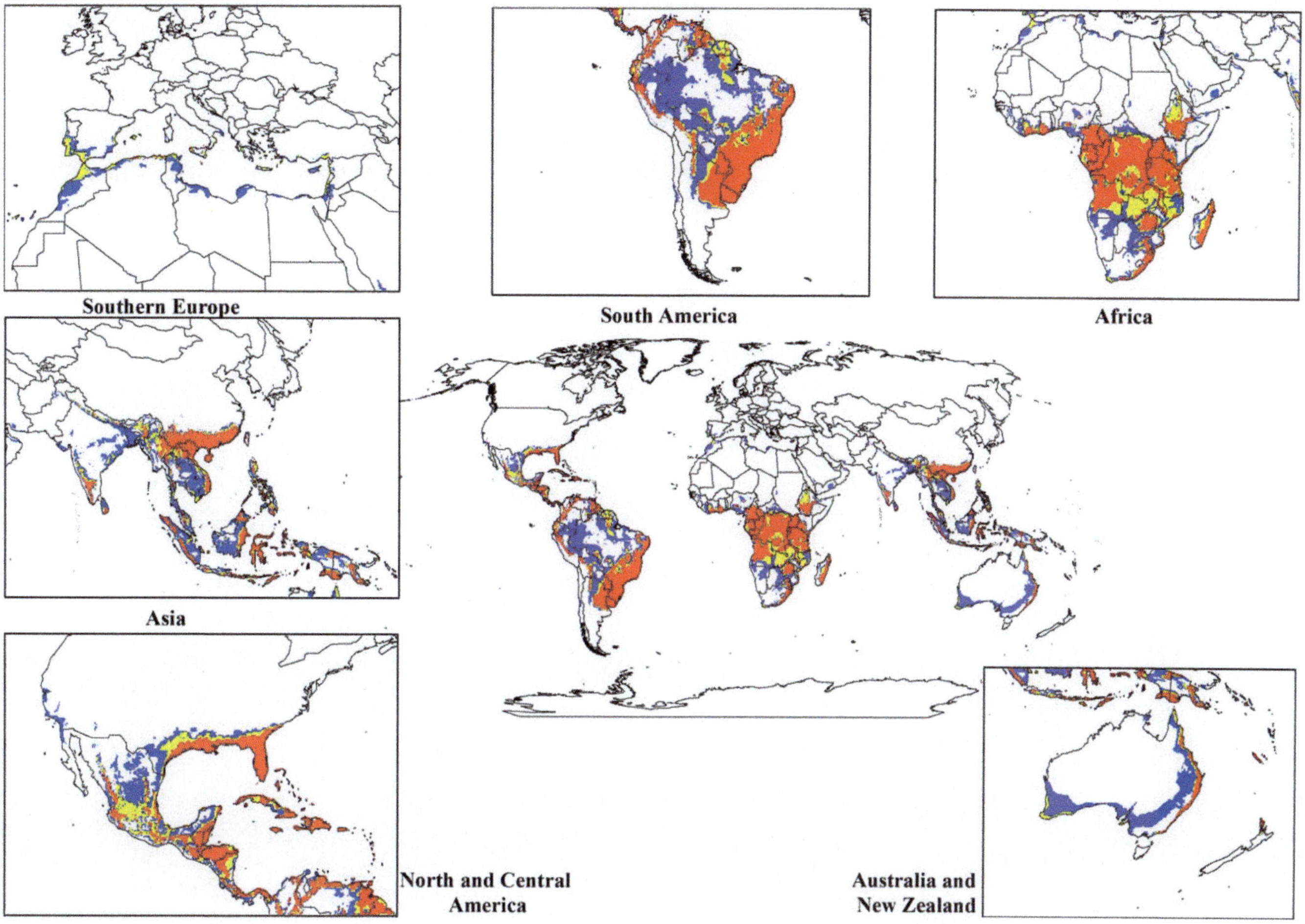

Figure 4. The climate (EI) for *L. camara* in the 2030s projected using CLIMEX under the CSIRO-Mk3.0 GCM running the SRES A1B scenario. White areas indicate unsuitable climate areas (EI = 0), blue areas indicate marginal climate areas (EI = 1–10), yellow areas indicate suitable climate areas (EI = 10–20) and red areas indicate highly suitable climate areas (EI>20).

dense thickets in pastures, outcompeting desirable pasture species and rendering infested areas useless for pasture [1,53]. Within the 'weedy taxa', there are many variants of *L. camara*, referred to here as varieties. Twenty-nine varieties are recognized in Australia [54]. The common name lantana is used in the remainder of the paper to refer to the weedy taxa of the section *Camara*.

The Global Biodiversity Information Facility (GBIF) is a database of natural history collections around the world for various species and is available for download. Information on *L. camara* distribution was downloaded [55] (Figure 1) and used in parameter fitting. Some 4126 records were downloaded but many did not have geographic coordinates and were removed, leaving 2753 records. However, many of these records were duplicates and were also removed. Thus 1740 records from the GBIF database were used in parameter fitting. Distribution data from South Africa [56] and Asia [57–61] were also obtained to assist in fitting parameters. Seasonal phenology data for the southern states of Brazil were used to fit growth parameters [62,63]. Although Winder's seasonal phenology observations were restricted to *Lantana tiliaefolia* and *L. glutinosa*, the ecology of these two species are similar to the weedy taxa of lantana, and thus these data were used in parameter fitting.

Climate Data and Climate Change Scenarios

The CliMond 10′ gridded climate data [64] were used for modelling. Average minimum monthly temperature (T_{min}), average maximum monthly temperature (T_{max}), average monthly precipitation (P_{total}) and relative humidity at 09:00 h ($RH_{09:00}$) and 15:00 h ($RH_{15:00}$) were used to represent historical climate (averaging period 1950–2000). The same five variables were used to characterize potential future climate in 2030 and 2070, based on two Global Climate Models (GCMs), CSIRO-Mk3.0 [65] and MIROC-H (Centre for Climate Research, Japan) with the A1B and A2 SRES scenarios [66]. These were available as part of the CliMond dataset. The two GCMs were selected from 23 GCMs for the CliMond dataset based on three criteria [64]:

1. The temperature, precipitation, mean sea level pressure and specific humidity variables required for CLIMEX were available for these two GCMs.
2. The models have relatively small horizontal grid spacing.
3. They performed well compared to other GCMs in representing basic aspects of observed climate at a regional scale [67].

The A1B and A2 scenarios were selected to typify the range of possible climate suitability for *L. camara* in 2030 and 2070. No

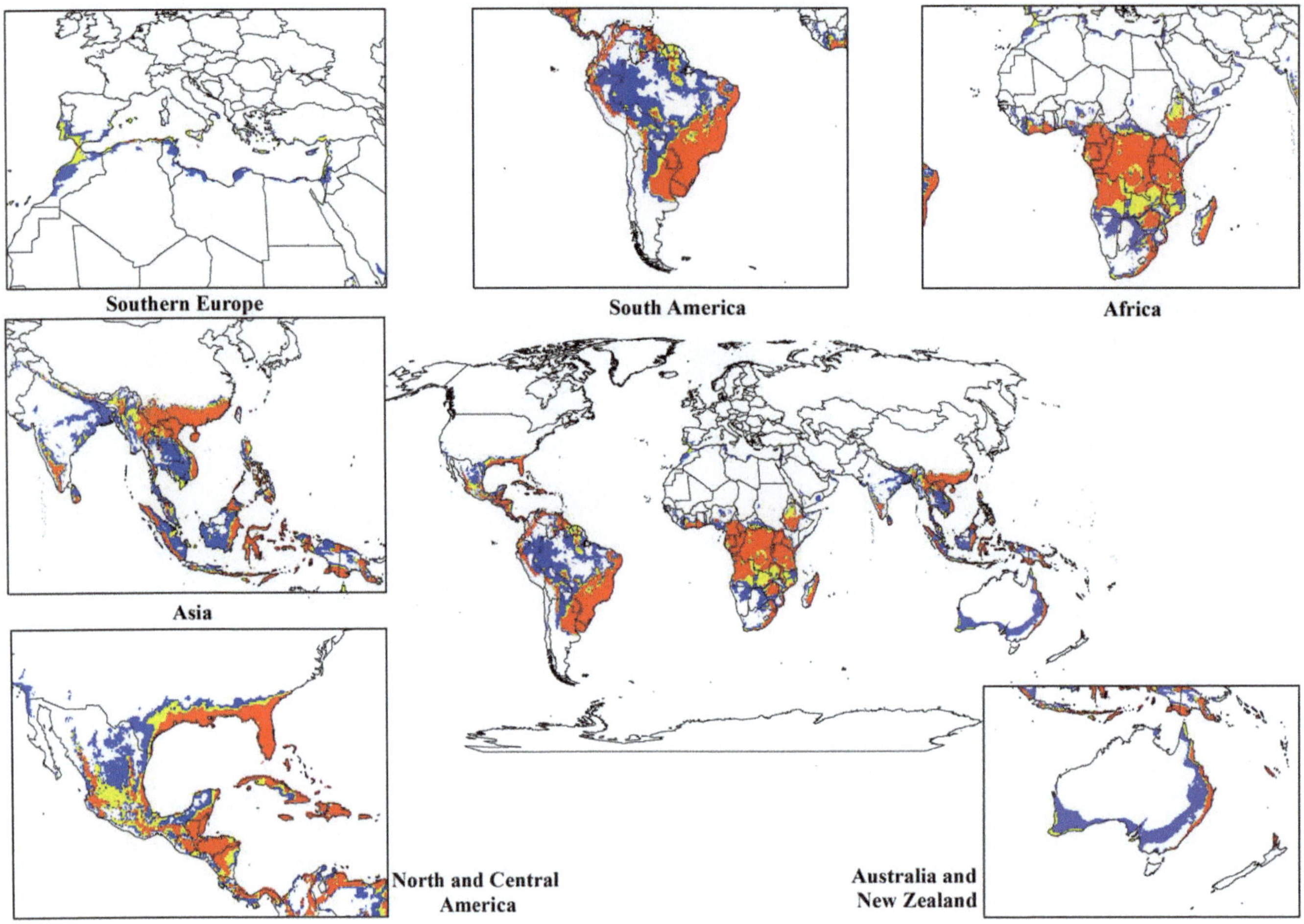

Figure 5. The climate (EI) for *L. camara* in the 2030s projected using CLIMEX under the CSIRO-Mk3.0 GCM running the SRES A2 scenario. White areas indicate unsuitable climate areas (EI = 0), blue areas indicate marginal climate areas (EI = 1–10), yellow areas indicate suitable climate areas (EI = 10–20) and red areas indicate highly suitable climate areas (EI>20).

scenarios from the B family of SRES scenarios were included in this paper because recent analyses of trends in factors such as global temperature and sea rise [68] showed that the observed increases were much higher than the hottest SRES scenario. The A1B scenario describes a balance between the use of fossil and non-fossil resources while A2 describes a varied world with high population growth but slow economic development and technological change. The projection dates of 2030 and 2070 were chosen because they provide a reasonable snapshot of two periods, one in the near future in 20 years' time and one much later in the future in 60 years' time.

Fitting CLIMEX Parameters

Sutherst [40] and Kriticos and Leriche [69] suggested that using both native and exotic distribution data in fitting CLIMEX parameters could produce a model that better approximates the potential distribution of the taxa being modelled than one that relies solely on native range data. They suggested that the constraints imposed by biotic influences in the species' native range may be absent in exotic locations, thus allowing it to expand its range beyond its Hutchinsonian realized niche [70]. Stress parameters were fitted to the known native (Central and South America) and naturalized (South Africa and Asia) distribution of the species while the phenology data from Brazil was used to fit growth parameters [62–63]. Each of the parameters was adjusted iteratively until a satisfactory agreement was reached between the potential and known distribution of lantana in these areas. The fitted parameters were checked to ensure that they were biologically reasonable. Australian distribution data was reserved for validation of the model.

Cold Stress

Two cold stress mechanisms were used to define the southern limits of lantana distribution in Argentina and northern limits in Nepal, Pakistan and China. Lantana seldom occurs where temperatures frequently fall below 5°C [71], and prolonged freezing temperatures kill aerial woody branches and cause defoliation [1]. Therefore, intolerance to frost was incorporated by accumulating stress when the average monthly minimum temperature fell below 5°C with the frost stress accumulation rate (THCS) set at -0.004 week^{-1}. This cold-stress mechanism allowed the species to survive in Kathmandu (27°42′N 85°18′E) [72]. The Cold-Stress Degree-day Threshold (DTCS) was set at 15°C days, with the stress accumulation rate (DHCS) set at -0.0022 week^{-1} so that the potential distribution was restricted to the known southern limits in Buenos Aires and northern limits in India, Nepal and China. This form of cold stress accounts for the need for the plant to grow at a minimal rate in order to offset

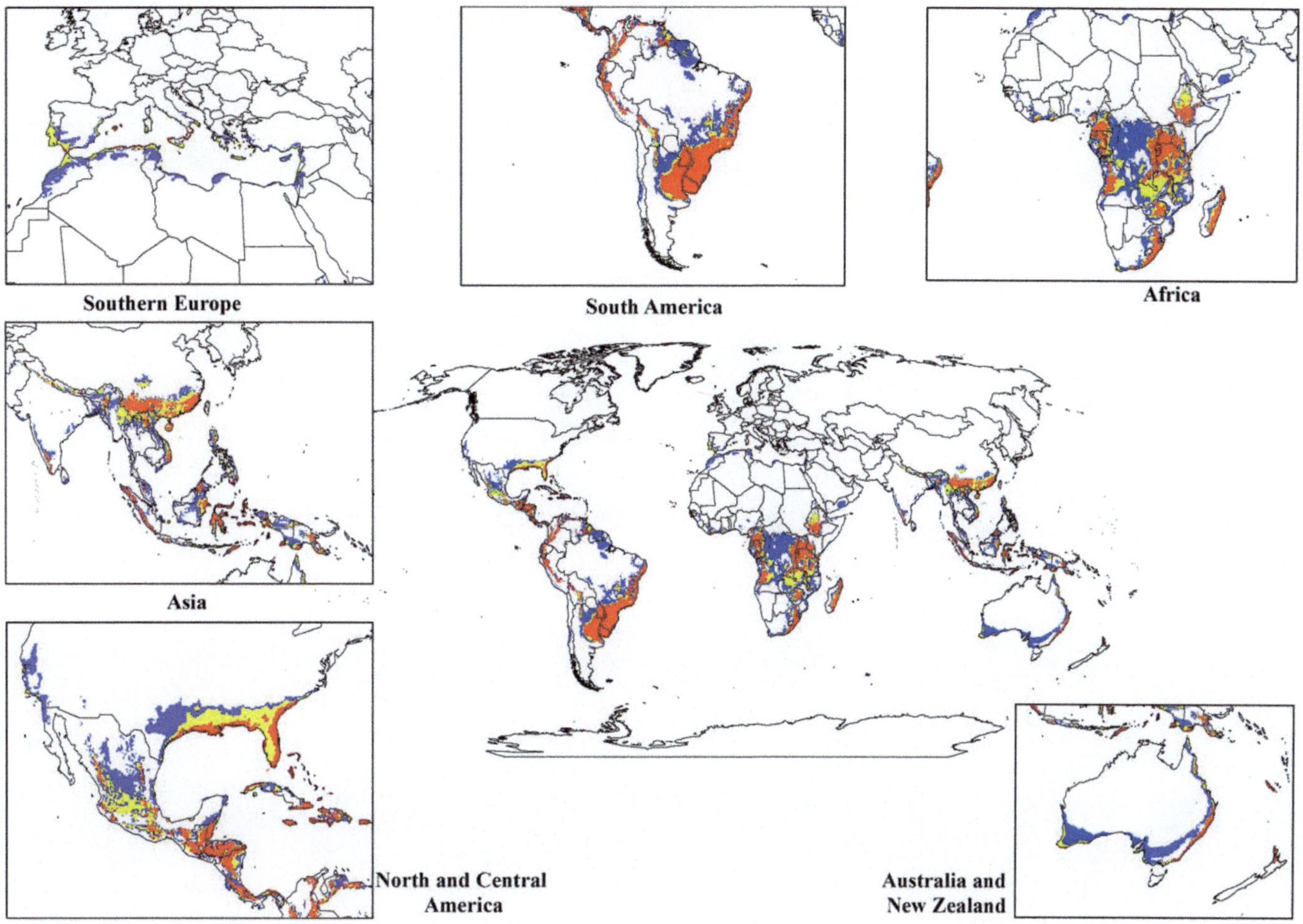

Figure 6. The climate (EI) for *L. camara* in the 2070s projected using CLIMEX under the CSIRO-Mk3.0 GCM running the SRES A1B scenario. White areas indicate unsuitable climate areas (EI = 0), blue areas indicate marginal climate areas (EI = 1–10), yellow areas indicate suitable climate areas (EI = 10–20) and red areas indicate highly suitable climate areas (EI>20).

respiration losses. If the temperatures are insufficient for the plant to grow this minimal amount, it needs to draw on photosynthate reserves.

Heat Stress

The heat stress parameter (TTHS) was set at 33°C, the same level as the limiting high temperature (DV3) with a stress accumulation rate (THHS) of 0.001 week^{-1}, which allowed lantana to persist along the Western Ghats [73] as well as in Bengal and Assam in India where it is reportedly common [57].

Dry Stress

The dry stress parameter was set at the same level (0.1) as the lower soil moisture threshold (SM0) because soil moisture related stresses probably begin at the same soil moisture levels where growth stops. The stress accumulation rate of −0.01 week^{-1} was set to exclude the species from the drier western parts of South Africa where it survives only as an ornamental plant [74].

Wet Stress

The wet stress threshold (SMWS) was set to 1.6 and the accumulation rate (HWS) set at 0.01 week^{-1} since lantana can tolerate up to 3000 mm of rainfall per year as long as the soil is not waterlogged for prolonged periods [1,75]. These settings allowed the species to grow well in Indonesia and the Philippines [53] as well as in central Burma, but excluded it from the wetter coastal areas [57].

Temperature Index

In South Africa, lantana is found in areas with a mean annual surface temperature greater than 12.5°C [76]. The seasonal phenology data for Iguazu (25°33′S, 54°34′W) in Brazil showed that 'cold winter temperatures caused cessation of growth with a substantial loss in leaves and side-branches' [63]. Winter temperatures in Iguazu can get as low as 8°C. Thus, the limiting low temperature (DV0) was set at 10°C to reduce growth appropriately during winter months in Iguazu. This value was chosen as a compromise between the South African distribution data and the phenology data from Iguazu. According to Day et al. [1], lantana does not appear to have an upper temperature limit. The summer temperatures in Iguazu rarely exceed 33°C and thus the limiting high temperature DV3 was set at 33°C, which allowed it to survive in Iguazu where it grows rapidly during summer [63]. The lower (DV1) and upper (DV2) optimal temperatures were set at 25°C and 30°C, respectively, based on seasonal phenology at Iguazu, and these provided a good fit to the observed distribution in South America, Asia and South Africa.

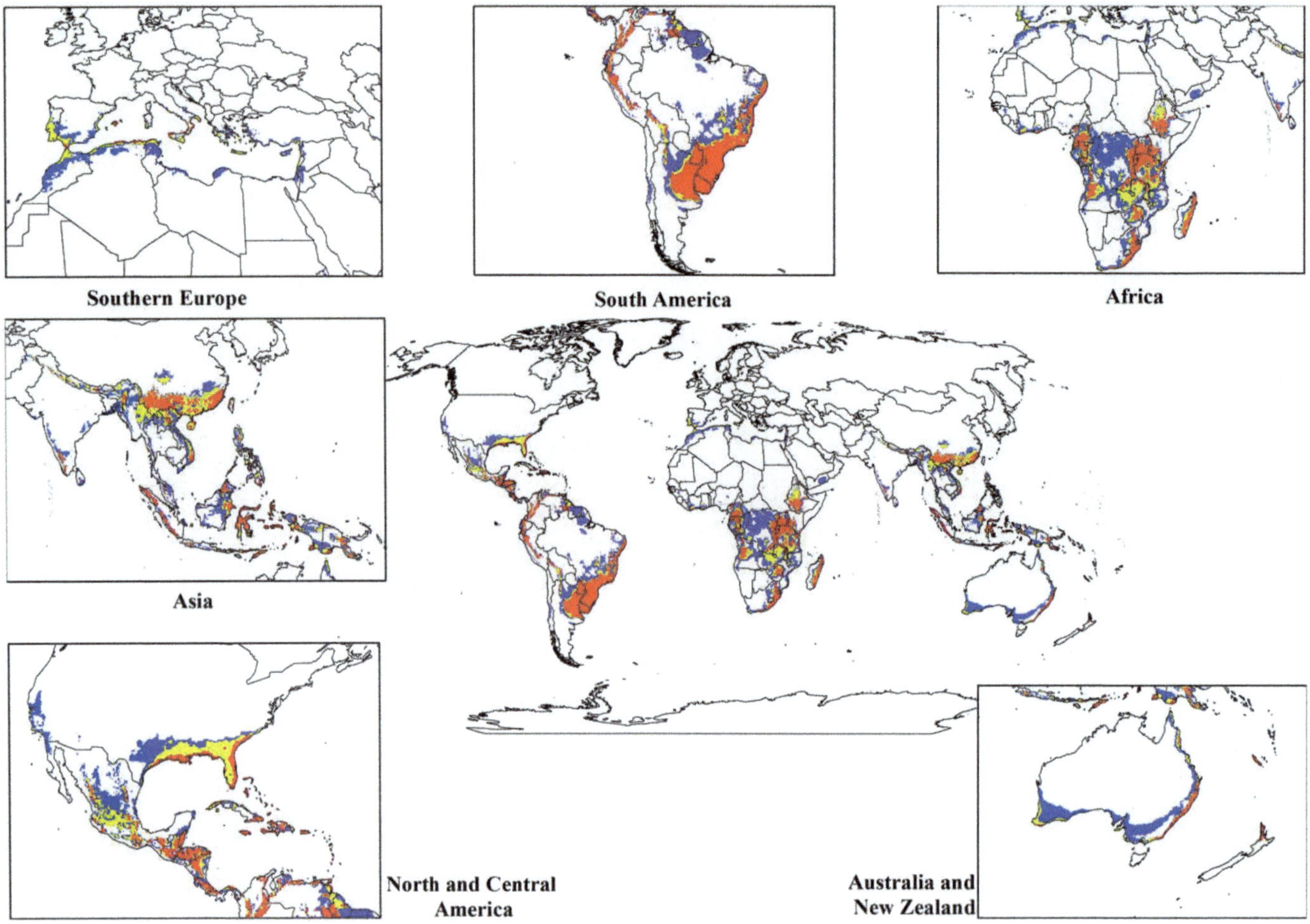

Figure 7. The climate (EI) for *L. camara* in the 2070s projected using CLIMEX under the CSIRO-Mk3.0 GCM running the SRES A2 scenario. White areas indicate unsuitable climate areas (EI = 0), blue areas indicate marginal climate areas (EI = 1–10), yellow areas indicate suitable climate areas (EI = 10–20) and red areas indicate highly suitable climate areas (EI>20).

Moisture Index

The lower moisture threshold (SM0) was set at 0.1, corresponding to the permanent wilting point for many plants [35]. This excluded lantana from the drier western parts of South Africa where it survives only as an ornamental [74] but allowed it to survive in Israel where Danin [77] reported lantana as 'a common component of the wasteland vegetation in the lowlands of the Mediterranean territories of Israel'. However, lantana may survive in certain areas of Israel due to irrigation since one of its other common habitats is irrigated cultivation such as date palm plantations and orchards [77]. The lower (SM1) and upper (SM2) optimum moisture thresholds were set at 0.5 and 1.2, respectively, to improve species growth during the months of January to March in Iguazu [63]. The upper soil moisture threshold (SM3) was set at 1.6 to allow it to grow in the Philippines and Indonesia where it has been reported as a troublesome weed [53].

Annual Heat Sum

The PDD thermal accumulation (number of degree days) mechanism did not appear to contribute to the definition of the South American or Asian distribution and so this parameter was not used.

The parameters are shown in Table 1. These parameters were used to model potential lantana distribution under the reference climate (averaging period 1950–2000) as well as climate change scenarios described above.

Results

Current Climate

The modelled global climate suitability for lantana (Figure 2) compares well with its known native distribution in South and Central America as well as its exotic range in South Africa and Asia (Figure 1). A comparison of Figures 1 and 2 showed that the present global distribution of lantana is consistent with the Ecoclimatic Index values resulting from the CLIMEX model. Much of the tropics and subtropics are projected to have suitable climatic conditions for lantana. Large areas of South and Central America, the southern states of the USA, Asia, sub-Saharan Africa, Madagascar and the high volcanic Pacific island groups such as Fiji, Vanuatu, Samoa and New Caledonia, among others, have highly suitable climate for the species. Warm temperate areas such as northern New Zealand and southern Mediterranean Europe including Portugal, Italy and Greece are predicted to have unsuitable climates.

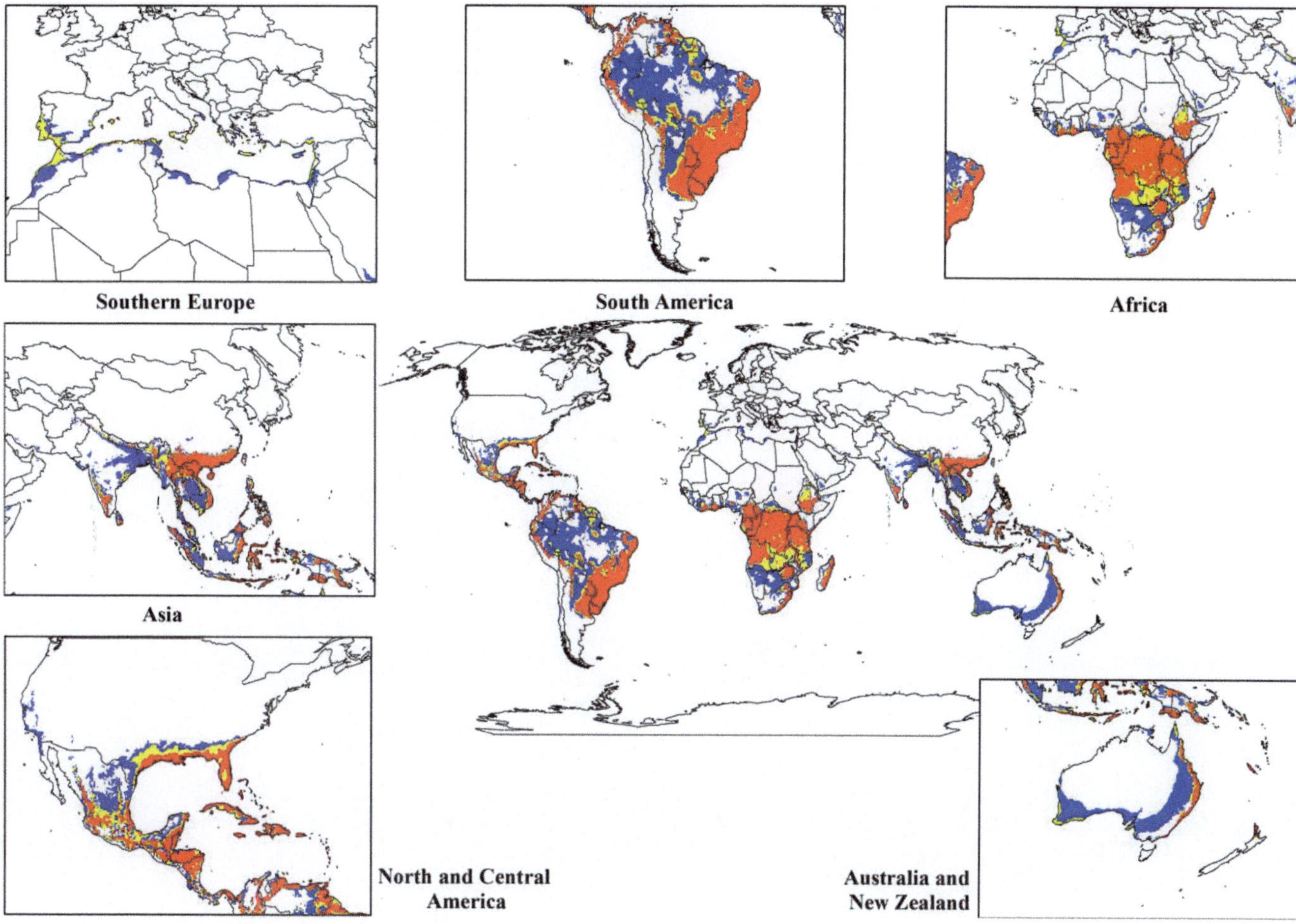

Figure 8. The climate (EI) for *L. camara* in the 2030s projected using CLIMEX under the MIROC-H GCM running the SRES A1B scenario. White areas indicate unsuitable climate areas (EI = 0), blue areas indicate marginal climate areas (EI = 1–10), yellow areas indicate suitable climate areas (EI = 10–20) and red areas indicate highly suitable climate areas (EI>20).

The current and potential distribution of lantana in Australia is shown in Figure 3. The occurrence records for Australia, which were reserved for model validation and not used for model fitting, accord well with the modelled climate suitability for the continent, and the present Australian distribution is consistent with the Ecoclimatic Index. Approximately 87% of the occurrence records fall within the suitable and highly suitable categories. In Australia, the model projects much of the eastern coast from Cape York in northern Queensland to southern New South Wales (NSW) to be climatically suitable (Figure 3). However, no occurrence records were found for Cape York Peninsula because despite a few isolated infestations in this region, lack of human disturbance limits the rate of spread [78]. Coastal areas in south-west Western Australia are projected to have suitable climate for lantana, conforming to the actual distribution since small infestations occur in these areas [1]. Central Australia is projected as being unsuitable, mainly due to dry stress.

Future Climate

For both the climate change models, a contraction in the suitable climate areas was observed worldwide (Figures 4, 5, 6, 7, 8, 9, 10, and 11) with this trend exacerbated in the 2070 scenario. The two GCMs showed moderately variable results but within each of the models, minimal sensitivity was seen between the two emission scenarios.

In South America, suitable climate areas for lantana are substantially reduced throughout northern Argentina, Uruguay, Bolivia, Peru, Paraguay, large parts of Brazil, French Guiana, Surinam, Guyana, coastal Venezuela and Colombia. A similar trend is seen in Central America with suitable climate areas for lantana contracting in Panama, Costa Rica, Nicaragua, Honduras and Guatemala. By 2030, a reduction in suitable climate for lantana is projected in all of these countries and this trend is exacerbated by 2070 (Figures 6, 7, 10, and 11). Warming under future climate scenarios is projected to lead to a substantial reduction in suitable climate for lantana in this region. In North America, some differences can be seen between the two GCMs in coastal areas of southern states such as Florida, Louisiana and Texas in North America. Under the CSIRO-Mk3.0 GCM, these areas are projected to remain climatically suitable until 2070 (Figures 6 and 7) while the same areas are projected as marginal to unsuitable with the MIROC-H GCM (Figures 10 and 11).

In Africa, suitable climatic areas for lantana are projected to contract substantially with only parts of Ethiopia, Uganda, Tanzania, Zambia, Angola, Gabon and Republic of Congo remaining suitable in 2070 under both GCMs and both SRES scenarios (Figures 6, 7, 10, and 11). Nevertheless, much of the

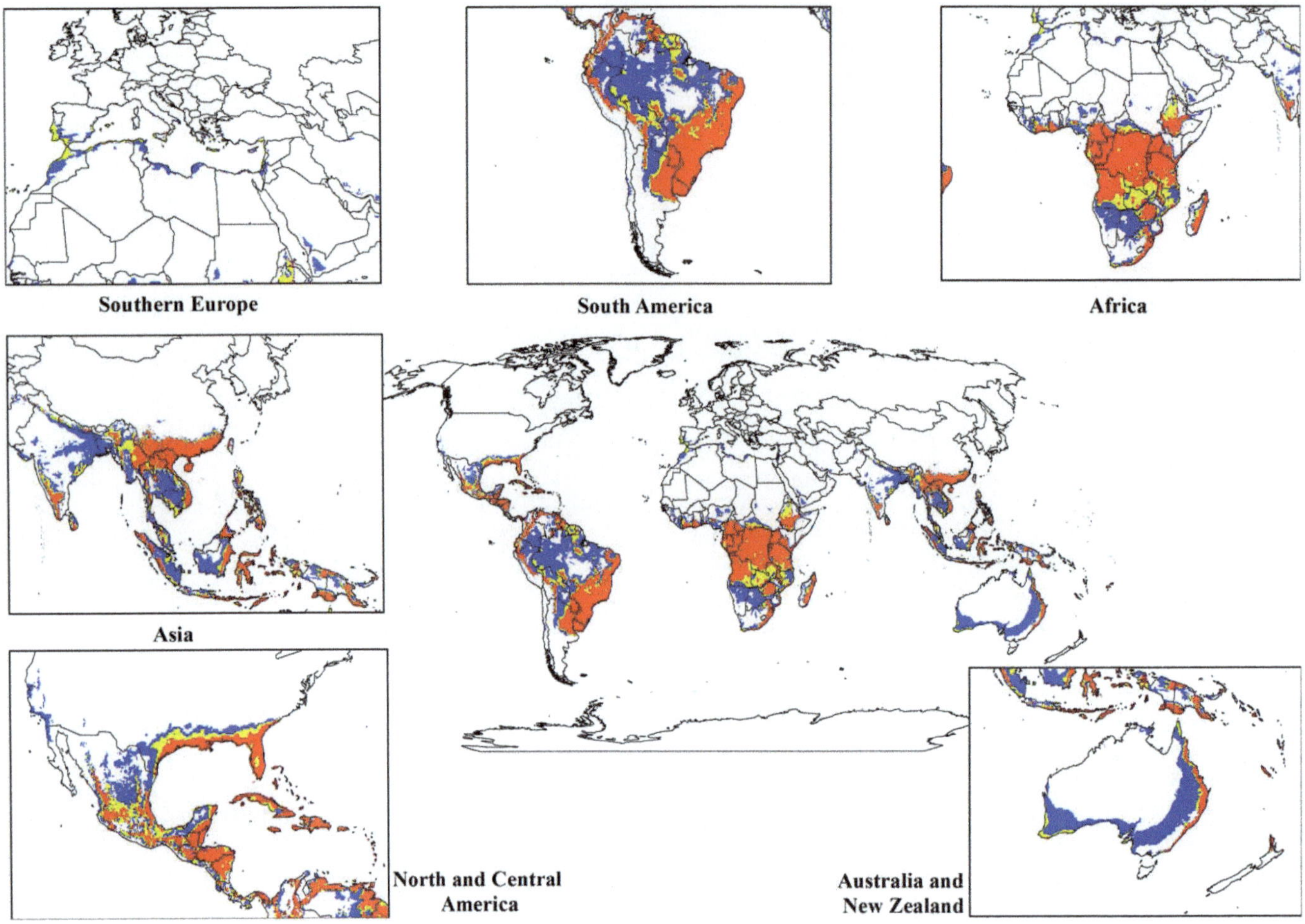

Figure 9. The climate (EI) for *L. camara* in the 2030s projected using CLIMEX under the MIROC-H GCM running the SRES A2 scenario. White areas indicate unsuitable climate areas (EI = 0), blue areas indicate marginal climate areas (EI = 1–10), yellow areas indicate suitable climate areas (EI = 10–20) and red areas indicate highly suitable climate areas (EI>20).

continent shows high climatic suitability for lantana until 2030 (Figures 4, 5, 8, and 9). In South Africa, lantana range appears to expand further inland, mainly in the Eastern Cape and Kwazulu-Natal provinces, west of the Swaziland border as well as into Lesotho and this is particularly apparent by 2070 under both GCMs.

In Asia, there is a considerable reduction in the projected potential range under climate change scenarios, especially for countries such as India, Sri Lanka, Myanmar, Thailand, Cambodia and Vietnam. However, in China the potential range shifts further inland and this is especially noticeable in the MIROC-H 2070 scenario (Figures 10 and 11). Lantana potential range shifts south into new areas in Australia (Victoria, South Australia and Tasmania) and a range expansion is seen in the south-west corner of Western Australia under both GCMs. Coastal areas in North Africa (Morocco and Algeria) and Southern Europe (Portugal, Spain, Italy and Greece) are projected to have suitable climate areas for lantana by 2070, particularly under the CSIRO-Mk3.0 GCM (Figures 6 and 7).

Discussion

This study has modelled the suitable climate area for *Lantana camara* under current climate and future climate scenarios using CLIMEX. The model provides a good fit to the current global distribution records as well as the current Australian distribution, which was reserved for model validation purposes. Under historical climate, much of the tropics and subtropics are modelled as having suitable climatic conditions for lantana. On the African continent, most of eastern and central Africa, parts of West Africa as well as eastern Madagascar are projected to have suitable to highly suitable climatic conditions for lantana greatly exceeding its current known distribution here. This could be a function of the lack of reporting from this region or invasion lag. Other non-climatic factors such as lack of dispersal opportunities or biotic interactions could also inhibit lantana from spreading in these regions. These results highlight areas where more detailed risk assessments on lantana invasion may be prudent. Capacity building, effective implementation of existing laws on movement and transport of lantana together with a public education campaign in this region may contribute towards more effective management. A similar case could be made for Asia where lantana's potential distribution exceeds its current distribution. Maps such as the ones produced in this study can be a useful tool in public awareness campaigns so as to enlist the help of local communities in the management of existing infestations and the prevention of further invasion.

Although cold stress appears to be the main factor limiting its distribution, dry stress prevents its establishment in the drier

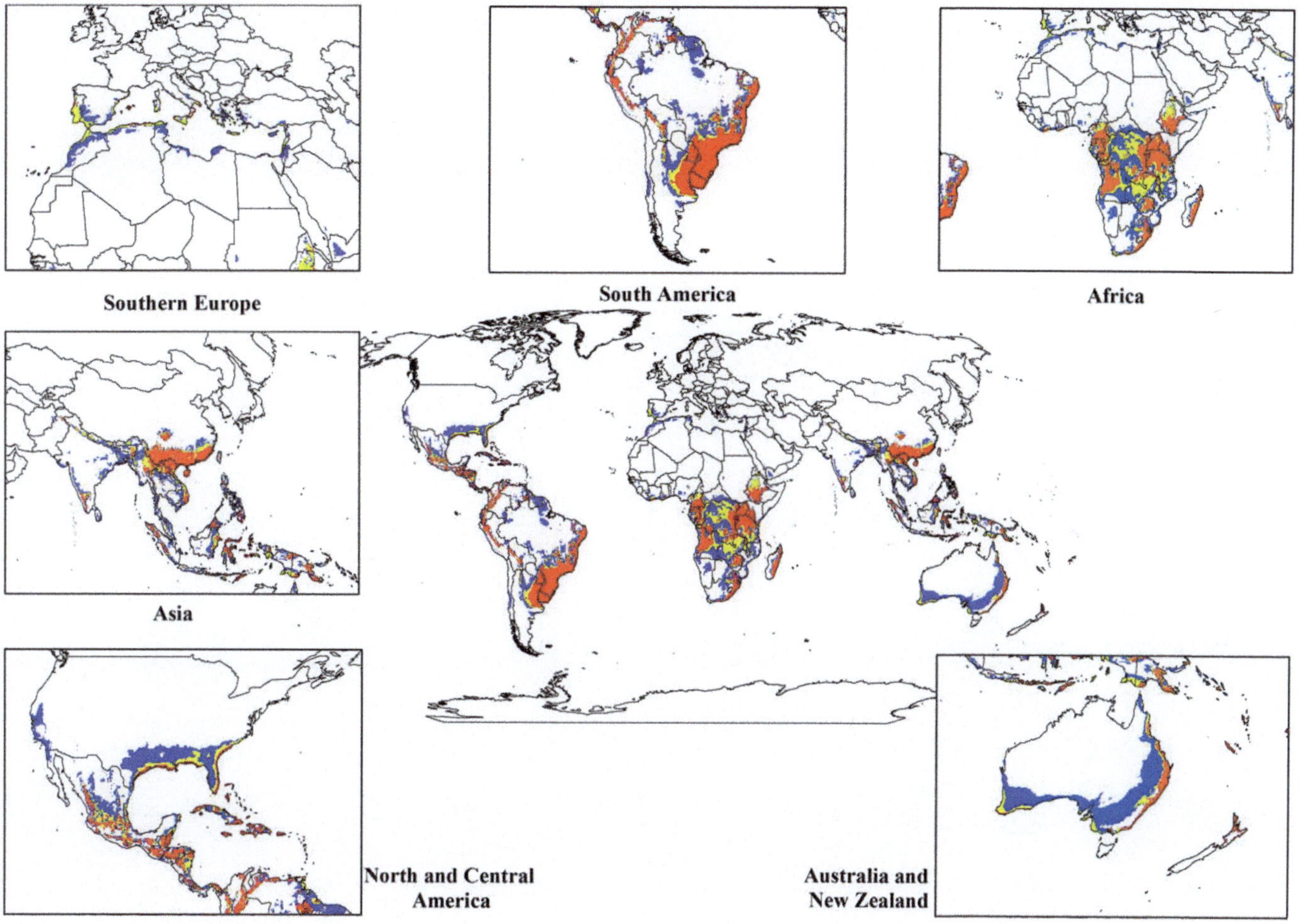

Figure 10. The climate (EI) for *L. camara* in the 2070s projected using CLIMEX under the MIROC-H GCM running the SRES A1B scenario. White areas indicate unsuitable climate areas (EI = 0), blue areas indicate marginal climate areas (EI = 1–10), yellow areas indicate suitable climate areas (EI = 10–20) and red areas indicate highly suitable climate areas (EI>20).

western parts of South Africa and inland Australia. Within the marginal areas identified in inland Australia, lantana would be patchily distributed and restricted to favourable microhabitats. Thus in these areas, it would pose limited threat and dispersal is also likely to be very slow in these regions.

The results of the climate change modelling give an indication of the possible changes in the potential distribution of lantana. As the climate changes, some areas where lantana currently occurs may become climatically unsuitable. All scenarios considered in this study indicated an overall contraction in the climatically suitable area for lantana in the future (Figures 4, 5, 6, 7, 8, 9, 10, and 11). Some of this reduced potential range for lantana covers important biodiversity hotspots of the world (e. g., coastal forests of Eastern Africa, lowland forests of west Africa, Indo-Burma region, Western Ghats of India and Sri Lanka). Whilst this result is likely to be encouraging for weed managers, it is probably paralleled by similar climatic threats to biodiversity [79]. Nonetheless, these results may be useful in making informed choices about the allocation of resources for weed control by highlighting areas where climate suitability is expected to decrease in the future.

The results identify new areas of the world that may be at risk of lantana invasion due to changes in climate, and which may warrant strategic control measures to prevent its spread. Although an overall reduction in the potential distribution is projected in the Americas, Africa and Asia under the future climate scenarios examined here, the potential for range expansion in North Africa, Europe, Australia and New Zealand was identified. In South Africa and China, lantana's potential distribution may expand further inland into new areas in the future. Such areas may require more detailed risk assessments on lantana invasion. The assessment and management of risks from weeds depends, to a large extent, on projections of habitat suitability so that threat levels can be assessed. The response of exotic species to changes in climate must form an integral part of such assessments [80,81]. In particular, areas that are currently at risk and that will continue to be at risk from lantana in the future are identified in this study. These areas could be important for biodiversity conservation, particularly in biodiversity hotspots such as southwest Australia, Atlantic Forests of South America and Caribbean Islands. These areas also include important agricultural areas worldwide such as southern states of North America, coastal areas of southern China, east coast of Australia and South Africa. Our results can be used in decision-making processes by land managers in prioritizing areas for eradication and in determining areas where containment would be cost-effective [82].

Under future climates, lantana may expand into areas that are currently too cool for it to survive and this can be seen in improved suitability in Europe. Biosecurity agencies in these countries should

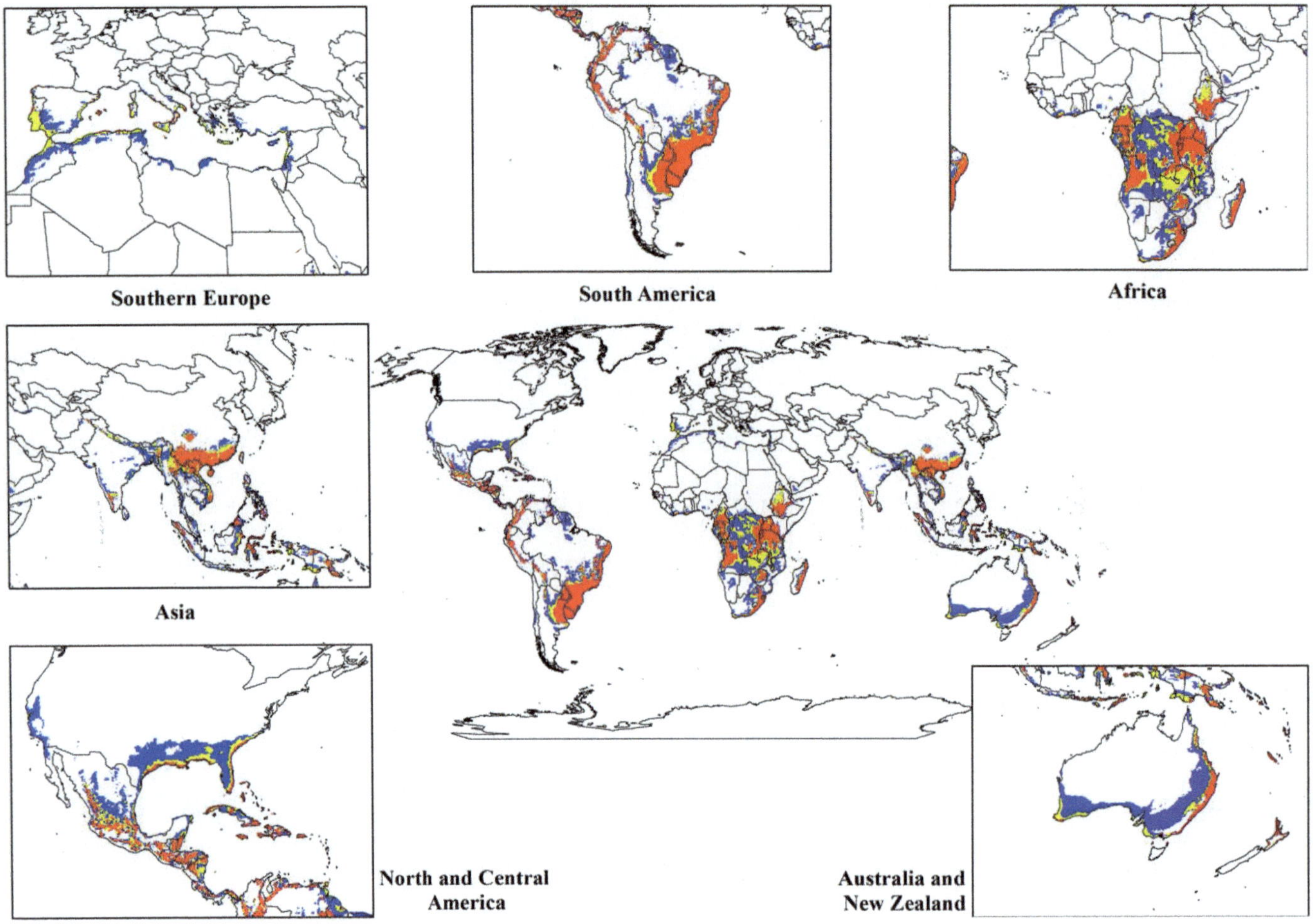

Figure 11. The climate (EI) for *L. camara* in the 2070s projected using CLIMEX under the MIROC-H GCM running the SRES A2 scenario. White areas indicate unsuitable climate areas (EI = 0), blue areas indicate marginal climate areas (EI = 1–10), yellow areas indicate suitable climate areas (EI = 10–20) and red areas indicate highly suitable climate areas (EI>20).

be aware of this potential threat and monitor areas that have been identified in this study for early signs of outlying lantana populations becoming invasive. Currently lantana is grown as an ornamental potted plant in northern Italy and is widely used in private and public gardens in central and southern Italy [83]. Areas of Portugal and Greece have also been identified as becoming climatically suitable for lantana to naturalize in the future. Simple and low-cost strategies such as weed alerts, identification and distribution of replacement garden ornamentals, low-cost surveillance and hygiene efforts to prevent lantana spreading to new areas may be a worthwhile investment on the part of biosecurity agencies in these countries. Climatic suitability for lantana may decrease, even leading to range contraction, in places where conditions become too warm and wet (e.g. northern Australia). Changes in climate may also have implications for the biological control of lantana since the distribution of biocontrol agents will also likely alter with climate change [84]. Lantana was one of the first weeds to be targeted for classical biological control at the turn of the century [53]. Since then 36 insect species have been released in 33 countries throughout its invaded range with disappointing results [48]. It will be important to establish ongoing monitoring of current biological control programs for lantana throughout its invaded range so that changes may be detected early and appropriate action taken.

Since CLIMEX is based on climate, non-climatic factors that affect species' distributions such as dispersal potential, biotic interactions and type of habitat are not included explicitly in the modelling process. However, the modelling method employed here should capture any effects from the release from natural enemies [70] that are apparent in lantana's exotic range, thus approximating its fundamental niche [14]. Moreover, the uncertainties associated with the state of climate modelling and uncertainty in future global greenhouse gas emission patterns [85] mean that models based on future climate scenarios should be treated as elaborate sensitivity analyses, indicative of the direction and magnitude of change that may be expected in the future. The climate suitability *projections* show areas of climatic suitability for lantana and are not *predicted* future distributions. Lantana's bird-dispersed berries make it an effective disperser [86], able to expand its range rapidly to occupy a broad range of environments within its climatic tolerance. Once established, lantana can survive long periods of drought [30,86]. It can also grow on poor soils and pure sand substrates if there is adequate soil moisture [63,75].

The climate models for lantana presented here may be useful for its management, particularly under future climate change. These models may be adapted for: informing decisions regarding allocation of resources for weed management towards areas where risk of invasion remains and away from areas where climatic suitability is likely to decrease under future climate, inform management decisions in preventing the spread of lantana into new areas, and prioritizing lantana management initiatives in areas which are currently at risk and will remain at risk of invasion in the future.

The modelling presented here ignored the direct effects of increasing concentrations of atmospheric CO_2 on factors that affect the ability of lantana to grow and persist (e.g. its water use efficiency). It would be useful to supplement our knowledge of the direct effects of increasing [CO_2] on invasive plants in general at both a physiological level and at the ecosystem level. Because of the expense and difficulties of running free-air carbon emission (FACE) studies, there have been very few studies of ecosystem water use efficiency. The differences in projected range changes based on assumptions drawn from studies of individual plants [37] are markedly different to those based on results from FACE experiments [37,87]. Cheaper "open top" CO_2 experiments may offer a compromise that would allow researchers to explore single plant and ecosystem water use efficiency changes for a broad range of species and climate types. This knowledge could have profound impacts on our ability to model likely range changes under future climates.

Those areas that have been identified as suitable or highly suitable for lantana are at greatest risk, and the projection of future climate scenarios provides useful insights into the potential distribution of this highly invasive weed. The identification of important biodiversity conservation and agricultural assets that may be affected by the anticipated changes in range of this species as well as new areas at risk of invasion under climate change should facilitate strategies to manage its spread.

Author Contributions

Conceived and designed the experiments: ST LK NR DJK. Performed the experiments: ST DJK. Analyzed the data: ST DJK. Contributed reagents/materials/analysis tools: ST DJK. Wrote the paper: ST LK NR DJK.

References

1. Day MD, Broughton S, Hannan-Jones MA (2003) Current distribution and status of *Lantana camara* and its biological control agents in Australia, with recommendations for further biocontrol introductions into other countries. BioControl 24: 63N–76N.
2. Henderson S, Dawson TP, Whittaker RJ (2006) Progress in invasive plants research. Progress in Physical Geography 30: 25–46.
3. Mack RN, Simberloff D, Mark Lonsdale W, Evans H, Clout M, et al. (2000) Biotic invasions: causes, epidemiology, global consequences, and control. Ecological Applications 10: 689–710.
4. Vitousek PM, D Antonio CM, Loope LL, Westbrooks R (1996) Biological invasions as global environmental change. American Scientist 84: 468–468.
5. Binggeli P (1996) A taxonomic, biogeographical and ecological overview of invasive woody plants. Journal of Vegetation Science 7: 121–124.
6. Guisan A, Zimmerman NE (2000) Predictive habitat distribution models in ecology. Ecological Modelling 135: 147–186.
7. Hutchinson GE (1957) Concluding remarks. In: Cold Spring Symposium on Quantitative Biology Yale University, New Haven, CT, USA. pp 415–427.
8. Araújo MB, Guisan A (2006) Five (or so) challenges for species distribution modelling. Journal of Biogeography 33: 1677–1688.
9. Soberon J, Peterson AT (2005) Interpretation of models of fundamental ecological niches and species' distributional areas. Biodiversity Informatics 2: 1–10.
10. Pearson RG, Dawson TP (2003) Predicting the impacts of climate change on the distribution of species: Are bioclimate envelope models useful? Global Ecology and Biogeography 12: 361–371.
11. Andrewartha HG, Birch LC (1984) The ecological web: more on the distribution and abundance of animals. Chicago: University of Chicago Press.
12. Woodward FI (1987) Climate and Plant Distribution. Cambridge: Cambridge University Press. 174 p.
13. Sutherst RW, Maywald G (1985) A computerized system for matching climates in ecology. Agriculture Ecosystems & Environment 13: 281–299.
14. Wharton TN, Kriticos DJ (2004) The fundamental and realized niche of the Monterey Pine aphid, *Essigella californica* (Essig) (Hemiptera: Aphididae): implications for managing softwood plantations in Australia. Diversity and Distributions 10: 253–262.
15. Chejara VK, Kriticos DJ, Kristiansen P, Sindel BM, Whalley RDB, et al. (2010) The current and future potential geographical distribution of *Hyparrhenia hirta*. Weed Research 50: 174–184.
16. Watt MS, Kriticos DJ, Manning LK (2009) The current and future potential distribution of *Melaleuca quinquenervia*. Weed Research 49: 381–390.
17. Poutsma J, Loomans AJM, Aukema B, Heijerman T (2008) Predicting the potential geographical distribution of the harlequin ladybird, *Harmonia axyridis*, using the CLIMEX model. BioControl 53: 103–125.
18. Sutherst RW, Bourne AS (2009) Modelling non-equilibrium distributions of invasive species: a tale of two modelling paradigms. Biological Invasions 11: 1231–1237.
19. Begon M, Harper JL, Townsend CR (1986) Ecology: individuals, population and communities. Oxford: Blackwell Scientific Publications.
20. Kriticos DJ, Yonow T, McFadyen RE (2005) The potential distribution of *Chromolaena odorata* (Siam weed) in relation to climate. Weed Research 45: 246–254.
21. Kriticos DJ, Randall RP (2001) A comparison of systems to analyze potential weed distributions. In: Groves RH, Panetta FD, Virtue JG, eds. Weed Risk Assessment. Collingwood: CSIRO Publishing. pp 61–79.
22. Webber BL, Yates CJ, Le Maitre DC, Scott JK, Kriticos DJ, et al. (2011) Modelling horses for novel climate courses: insights from projecting potential distributions of native and alien Australian acacias with correlative and mechanistic models. Diversity and Distributions 17: 978–1000.
23. Davis AJ, Jenkinson LS, Lawton JH, Shorrocks B, Wood S (1998) Making mistakes when predicting shifts in species range in response to global warming. Nature 391: 783–786.
24. Follak S, Strauss G (2010) Potential distribution and management of the invasive weed *Solanum carolinense* in Central Europe. Weed Research 50: 544–552.
25. IPCC (2007) Climate Change 2007: Synthesis Report. Summary for Policy-makers. Intergovernmental Panel on Climate Change, Cambridge University Press.
26. Dukes JS, Mooney HA (1999) Does global change increase the success of biological invaders? Trends in Ecology & Evolution 14: 135–139.
27. Sutherst RW, Baker RH, Coakley SM, Harrington R, Kriticos DJ, et al. (2007) Pests under global change – meeting your future landlords? . In: Candadell J, Pataki D, Pitelka L, eds. In Terrestrial ecosystems in a changing world. New York: Springer, Berlin Heidelberg. pp 211–223.
28. Walther G-R, Roques A, Hulme PE, Sykes MT, Pyaek P, et al. (2009) Alien species in a warmer world: risks and opportunities. Trends in Ecology & Evolution (Personal edition) 24: 686–693.
29. Achhireddy NR, Singh M (1984) Allelopathic Effects of Lantana (*Lantana camara*) on Milkweedvine (Morrenia odorata). Weed Science 32: 757–761.
30. Sharma GP, Raghubanshi AS, Singh JS (2005) Lantana invasion: An overview. Weed Biology and Management 5: 157–165.
31. Gentle CB, Duggin JA (1997) *Lantana camara* L. invasions in dry rainforest - open forest ecotones: The role of disturbances associated with fire and cattle grazing. Australian Journal of Ecology 22: 298–306.
32. Lamb D (1991) Forest regeneration research for reserve management: some questions deserving answers. In: Goudberg N, Bonell M, Benzaken D, eds. Tropical Rainforest Research in Australia: present status and future directions for the Institute for Tropical Rainforest Studies, Townsville. pp 177–181.
33. Fensham RJ, Fairfax RJ, Cannell RJ (1994) The invasion of *Lantana camara* L. in Forty Mile Scrub National Park, north Queensland. Australian Journal of Ecology 19: 297–305.
34. Hearne Scientific Software (2007) CLIMEX Software Version 3.0.2 Hearne Scientific Software Pty Ltd, Melbourne.
35. Sutherst RW, Maywald G, Kriticos DJ (2007) CLIMEX Version 3: User's Guide. Melbourne: Hearne Scientific Software Pty Ltd.
36. Andrewartha HG, Birch LC (1954) The distribution and abundance of animals. Chicago: University of Chicago Press. 782 p.
37. Kriticos DJ, Sutherst RW, Brown JR, Adkins SW, Maywald GF (2003) Climate change and the potential distribution of an invasive alien plant: *Acacia nilotica* ssp indica in Australia. Journal of Applied Ecology 40: 111–124.
38. van der Ploegg RR, Bohm W, Kirkham MB (1999) On the origin of the theory of mineral nutrition of plants and the law of the minimum. Soil Science Society of America Journal 63: 1055–1062.
39. Shelford VE (1963) The ecology of North America. Urbana, IL: University of Illinois Press.
40. Sutherst RW (2003) Prediction of species geographical ranges. Journal of Biogeography 30: 805–816.
41. Gujral GS, Vasudevan P (1983) *Lantana camara* L., a problem weed. Journal of Scientific and Industrial Research 42: 281–286.
42. Mabberley DJ (1997) The Plant Book. Cambridge: Cambridge University Press.
43. Munir AA (1996) A taxonomic review of *Lantana camara* L. and *L. montevidensis* (Spreng.) Briq. (Verbenaceae) in Australia. . Journal of the Adelaide Botanical Gardens 17: 1–27.

44. Sanders RW (1987) Taxonomic significance of chromosome observations in Caribbean species of Lantana (Verbenaceae). American Journal of Botany 74: 914–920.
45. Waterhouse DF, Norris KR (1987) Biological control: Pacific prospects. Melbourne: Inkata Press.
46. Moldenke HN (1944) The Known Geographic Distribution of the Members of the Verbenaceae and Avicenniaceae. Botanical Gazette 106: 158–164.
47. Kohli RK, Batish DR, Singh HP, Dogra KS (2006) Status, invasiveness and environmental threats of three tropical American invasive weed (*Parthenium hysterophorus* L., *Ageratinum conyzoides* L., *Lantana camara* L.) in India. Biological Invasions 8: 1501–1510.
48. Thomas SE, Ellison CA (2000) A century of biological control of *Lantana camara*: Can pathogens make a significant difference? In: Spencer NR, ed. Proceedings of the X International Symposium on Biological Control of Weeds Montana State University, Montana, USA.
49. Gooden B, French K, Turner PJ, Downey PO (2009) Impact threshold for an alien plant invader, *Lantana camara* L., on native plant communities. Biological Conservation 142: 2631–2641.
50. Dobhal PK, Kohli RK, Batish DR (2011) Impact of *Lantana camara* L. invasion on riparian vegetation of Nayar region in Garhwal Himalayas (Uttarakhand, India). Journal of Ecology and the Natural Environment 3: 11–22.
51. Sharma GP, Raghubanshi AS (2006) Effect of *Lantana camara* L. cover on local depletion of tree population in the Vindhyan tropical dry deciduous forest of India. Applied Ecology and Environmental Research 5: 109–121.
52. Humphries SE, Stanton JP (1992) Weed assessment in the Wet Tropics World Heritage area of north Queensland Wet Tropics Management Agency.
53. Holm LG, Plucknett DL, Pancho JV, Herberger JP (1991) The World's Worst Weeds: Distribution and Biology. Malabar: Krieger Publishing Company.
54. Smith LS, Smith DS (1982) The naturalised *Lantana camara* complex in eastern Australia. Brisbane: Botany Branch, Queensland Department of Primary Industries. 26 p.
55. GBIF (2010) GBIF Data Portal. Available at: http://www.gbif.org/.
56. SAPIA (2006) Monitoring the emergence and spread of invasive alien plants in southern Africa. Available at: http://www.agis.agric.za.
57. Biswas K (1934) Some foreign weeds and their distribution in India and Burma. Indian Forester 60: 861–865.
58. Chen S, Gilbert MG (1994) Flora of China. Available at: http://www.efloras.org.
59. Jafri S (1974) Flora of Pakistan. Available at: http://www.efloras.org.
60. Press JR, Shrestha KK, Sutton DA (2000) Annotated Checklist of the Flowering Plants of Nepal. Available at: http://www.efloras.org.
61. Thakur ML (1992) Lantana weed (*Lantana camara* var. aculeate Linn) and its possible management through natural insect pests in India. Indian Forester 118: 466–488.
62. Winder JA (1982) The effects of natural enemies on the growth of Lantana in Brazil. Bulletin of Entomological Research 27: 599–616.
63. Winder JA Factors affecting the growth of Lantana in Brazil Reading: University of Reading.
64. Kriticos DJ, Webber BL, Leriche A, Ota N, Macadam I, et al. (2011) CliMond: global high-resolution historical and future scenario climate surfaces for bioclimatic modelling. Methods in Ecology and Evolution. no-no p.
65. Gordon HB, Rotstayn LD, McGregor JL, Dix MR, Kowalczyk EA, et al. (2002) The CSIRO Mk3 Climate System Model, CSIRO Atmospheric Research Technical Paper No. 60, CSIRO Atmospheric Research, Aspendale.
66. IPCC (2000) Special Report on Emissions Scenarios: A Special Report of Working Group III of the Intergovernmental Panel on Climate Change. Cambridge, UK.
67. Hennessy KJ, Colman R (2007) Global Climate Change Projections. Climate Change in Australia – Technical Report 2007 KB. Pearce, PN. Holper, M. Hopkins, WJ. Bouma, PH. Whetton, KJ. Hennessy, SB. Power, eds. 148 p, CSIRO, Melbourne, Australia.
68. Rahmstorf S, Cazenave A, Church JA, Hansen JE, Keeling RF, et al. (2007) Recent climate observations compared to projections. Science 316: 709.
69. Kriticos DJ, Leriche A (2010) The effects of climate data precision on fitting and projecting species niche models. Ecography 33: 115–127.
70. Keane RM, Crawley MJ (2002) Exotic plant invasions and the enemy release hypothesis. Trends in Ecology & Evolution 17: 164–170.
71. Cilliers CJ (1983) The weed, *Lantana camara* L., and the insect natural enemies imported for its biological control into South Africa. Journal Of Entomological Society of Southern Africa 46: 131–138.
72. Maharjan SR, Bhuju DR, Khadka C (2006) Plant community structure and species diversity in Ranibari Forest, Kathmandu. Nepal Journal of Science and Technology 7: 35–44.
73. Murali KS, Sidappa Setty R (2001) Effect of weeds *Lantana camara* and *Chromelina odorata* growth on the species diversity, regeneration and stem density of tree and shrub layer in BRT sanctuary. Current Science 80: 675–678.
74. Cilliers CJ, Neser S (1991) Biological Control of *Lantana camara* (Verbenaceae) in South Africa. Agriculture Ecosystems & Environment 37: 57–75.
75. Thaman R (1974) *Lantana camara*: Its introduction, dispersal and impact on islands of the tropical Pacific ocean. Micronesica 10: 17–39.
76. Stirton CH (1977) Some thoughts on the polyploidy complex *Lantana camara* L. (Verbenaceae). In: Annecke DP, ed. Second National Weeds Conference of South Africa. Stellenbosch, South Africa: Balkema. pp 321–340.
77. Danin A (2000) The inclusion of adventive plants in the second edition of Flora Palaestina. Willdenowia 30: 305–314.
78. DERM (2010) Mapping Lantana using Landsat: A Remote Sensing Centre Report. Department of Environment and Resource Management, Brisbane.
79. Colwell RK, Brehm G, Cardelus CL, Gilman AC, Longino JT (2008) Global warming, elevational range shifts, and lowland biotic attrition in the wet tropics. Science 322: 258–261.
80. Crossman ND, Bass DA (2008) Application of common predictive habitat techniques for post-border weed risk management. Diversity and Distributions 14: 213–224.
81. Crossman ND, Bryan BA, Cooke DA (2011) An invasive plant and climate change threat index for weed risk management: Integrating habitat distribution pattern and dispersal process. . Ecological Indicators 11: 183–198.
82. Skurka Darin GM, Schoenig S, Barney JN, Panetta FD, DiTomaso JM (2011) WHIPPET: A novel tool for prioritizing invasive plant populations for regional eradication. Journal of Environmental Management 92: 131–139.
83. Garibaldi A, Pensa P, Minuto A, Gullino ML (2008) First report of *Sclerotinia sclerotiorum* on *Lantana camara* in Italy. The American Phytopathological Society 92: 1369.
84. Kriticos DJ, Crossman ND, Ota N, Scott JK (2009) Climate change and invasive plants in South Australia. Canberra: CSIRO Climate Adaptation Flagship. 97 p.
85. Kriticos DJ, Alexander NS, Kolomeitz S (2006) Predicting the potential geographic distribution of weeds in 2080. In: Preston C, Watts JH, Crossman ND, eds. 15th Australian Weeds Conference. Adelaide, Australia: Weed Management Society of South Australia. pp 27–34.
86. Stock D, Johnson K, Clark A, van Oosterhout E (2009) Lantana Best Practice Manual and Decision Support Tool. Yeerongpilly: The State of Queensland, Department of Employment, Economic Development and Innovation.
87. Kriticos DJ (2008) Weeds getting steamed up in the tropics: climate chnage effects on ecosystem water use efficiency and woody plant range. In: Van Klinken RD, Osten VA, Panetta FD, Scanlan JC, eds. 16th Australian Weeds Conference. Cairns: Queensland Weeds Society. pp 506–509.

4

Radial Growth of Qilian Juniper on the Northeast Tibetan Plateau and Potential Climate Associations

Chun Qin[1], Bao Yang[1]*, Thomas M. Melvin[2], Zexin Fan[3], Yan Zhao[4], Keith R. Briffa[2]

1 Key Laboratory of Desert and Desertification, Cold and Arid Regions Environmental and Engineering Research Institute, Chinese Academy of Sciences, Lanzhou, China, **2** Climatic Research Unit, School of Environmental Sciences, University of East Anglia, Norwich, United Kingdom, **3** Key Laboratory of Tropical Forest Ecology, Xishuangbanna Tropical Botanical Garden, Chinese Academy of Sciences, Kunming, China, **4** Institute of Geographic Sciences and Natural Resources Research, Chinese Academy of Sciences, Beijing, China

Abstract

There is controversy regarding the limiting climatic factor for tree radial growth at the alpine treeline on the northeastern Tibetan Plateau. In this study, we collected 594 increment cores from 331 trees, grouped within four altitude belts spanning the range 3550 to 4020 m.a.s.l. on a single hillside. We have developed four equivalent ring-width chronologies and shown that there are no significant differences in their growth-climate responses during 1956 to 2011 or in their longer-term growth patterns during the period AD 1110–2011. The main climate influence on radial growth is shown to be precipitation variability. Missing ring analysis shows that tree radial growth at the uppermost treeline location is more sensitive to climate variation than that at other elevations, and poor tree radial growth is particularly linked to the occurrence of serious drought events. Hence water limitation, rather than temperature stress, plays the pivotal role in controlling the radial growth of Sabina przewalskii Kom. at the treeline in this region. This finding contradicts any generalisation that tree-ring chronologies from high-elevation treeline environments are mostly indicators of temperature changes.

Editor: Paulo Cavalcanti Gomes Ferreira, Universidade Federal do Rio de Janeiro, Brazil

Funding: The study was jointly funded by the NSFC (Grant No. 41002050, 41071130), and Interdisciplinary Innovation Team project of the Chinese Academy of Sciences (29Y329B91). KRB and TMM acknowledge support from NERC (NE/G018863/1). Bao Yang gratefully acknowledges the support of K.C. Wong Education Foundation, Hong Kong. The funders had no role in study design, data collection and analysis, decision to publish, or preparation of the manuscript.

Competing Interests: The authors have declared that no competing interests exist.

* E-mail: yangbao@lzb.ac.cn

Introduction

Due to its exposure to extremely harsh environmental conditions, such as low air temperatures and severe water shortage, tree growth near the alpine treeline is potentially extremely sensitive to climate change [1]. Consequently, such environments have received increasing attention from dendroclimatologists. Based on the principle of limiting factors in dendroclimatology, many dendroclimatologists consider that the variability of annual tree growth at the lower treeline is likely controlled by variability in precipitation, while at the upper treeline it is more closely controlled by the variability of temperature, especially at high latitudes, or at high elevations in the generally humid Alps and in some semi-arid areas [2–6]. However, uniform growth pattern (where the tree growth is controlled by a common climate signal at upper and lower treelines) has also been found in the arid regions of western central Asia [7], in the northeastern region of the arid central Tianshan Mountains [8], in a subtropical monsoonal region of South America [9] and in a low latitude humid region of North America [10,11]. Therefore, further researches are required to clarify the association between elevation and climate influences on tree growth in the regions of various climate types.

The Northeast Tibetan Plateau, located at the convergence of the westerly and easterly summer monsoons, is highly sensitive to climate change. The altitude of the terrain in this region varies from 2000 m to 5050 m above sea level (a.s.l.). Qilian juniper and Qinghai spruce (*Picea crassifolia Kom.*) are endemic and widespread dominant species, on south-facing slopes and north-facing slopes, respectively, in the alpine forest of the northeast Tibetan Plateau. Both species are generally considered as good candidates for obtaining effective paleoclimate proxies [12–19]. There are a number of previous studies of tree growth and climate variability in the treeline ecotone in this area [20–29]. Some researchers consider that temperature is the main growth-limiting factor in the upper treeline ecotone: for example in the eastern Qilian Mountains [27], in the eastern mountainous regions of the Qaidam Basin [23,28], in the semi-humid climates of the southeastern [20] and eastern [21] of the Tibetan Plateau. Other researchers have found that the growth responses of tree-line forests to climate variability show different patterns, associated with different climate conditions in the Anyemaqen Mountains [24] and the southern Tibetan Plateau [29], and have shown that short-term growth-climate relationships are unstable in the Qilian Mountains [26]. In summary, the effect of climate change on the alpine treeline in our study area remains complex and a matter of debate. The focus of attention in this study concerns the main limiting factor for tree growth in the treeline ecotone: is it temperature, precipitation, or both? Thus, we start from the plant physiological/ecological viewpoint, exploring altitude and age-dependent characteristics of tree growth in this species, and combine this with dendroclimatology, exploring the statistical association between tree radial growth time series and climate data records to establish a case study on the northeastern Tibetan

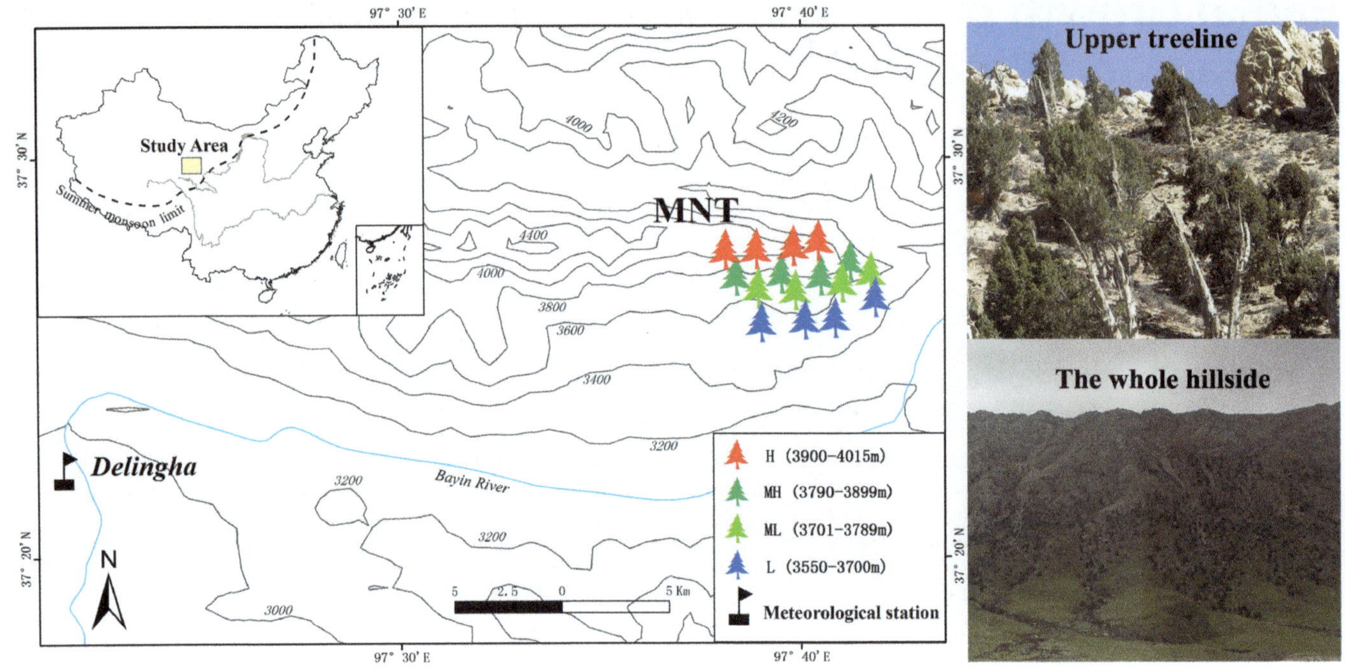

Figure 1. Locations of the tree-ring sampling sites and the meteorological station on the northeast Tibetan Plateau (black flag: Delingha meteorological station; red tree, olive green tree, green tree and blue tree indicates the ring-width collection areas at different elevations: H (3900 m–4015 m), MH (3790 m–3899 m), ML (3701 m–3789 m), and L (3550 m–3700 m); two photographs were taken of the upper treeline and the whole hillside, respectively).

plateau, China. This information will hopefully provide new insights that will improve our understanding of the influence of climate variability on the radial growth of Qilian juniper (*Sabina przewalskii Kom.*) at the alpine treeline in the semiarid region.

Materials and Methods

All necessary permits were obtained for the described field studies from the Administration of Delingha Forestry.

Study area

The study area is located in the Zongwulong Mountains, on the northeast Tibetan Plateau. The altitude of the study area varies between 3200 m and 5030 m a.s.l, with an average altitude of 4000 m. There is perennial snow cover above 4500 m.a.s.l. Qilian juniper (on south-facing slopes), Qinghai spruce (on north-facing slopes) and more prostrate shrub growth are found below 4100 m.a.s.l.. Qilian juniper is mostly distributed in the upper regions of the mountains, while the tops are mostly characterized by rock debris and rocky outcrops. The mountains are mainly composed of middle-Jurassic sandstone, conglomerate and limestone. Alpine forest gray–brown soils, subalpine shrub meadow soils, alpine meadow soils, and alpine chestnut soils are typical in this area.

The individual sampling sites are located in a sampling region called Manitu (MNT, 97.68°E, 37.46°N, Figure 1), on a south-facing (sunny) slope in the middle of the Zongwulong Mountains. The slope angle ranges from 15° to 65°. The alpine forest is pure Qilian juniper forest. The closest available meteorological records are from Delingha national weather station (37.37°N, 97.37°E, 2982.4 m.a.s.l.), which is 41 kilometers from MNT and between 500 m and 1000 m below the study sites. The mean, maximum and minimum annual temperatures at Delingha were 3.84°C, 11.38°C and −2.55°C in the period 1956–2011, respectively. Mean winter (from previous December to current January) and summer (from June to July) temperatures were −9.4°C and 15.7°C. The average annual precipitation was 173 mm, with 73.7% of the annual precipitation falling between May and August. The average relative humidity was 38.8%. In summary, our study area represents a typical mid-latitude arid to semi-arid alpine region (Figure 2).

Tree-ring width data

All the tree-ring width (TRW) samples were collected at breast height with an increment borer, on a single hillside. After air drying, the core surfaces were prepared with razor blades and the

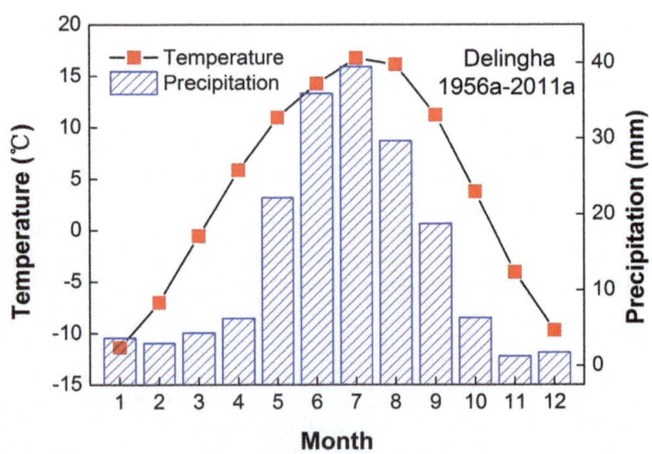

Figure 2. Climate diagrams of temperature and precipitation at Delingha meteorological station for the 1956–2011.

Table 1. Characteristics of raw (measured) tree-ring data and of the standard chronologies for each of four (High, Middle High, Middle Low and Low) elevations and for the overall composite chronology.

Site code	Altitude (m)	Slope(°)	Core/Tree	Time span(AD)	ML	Max/Mean/Min TRW (mm)
H	3900–4015/3965	40–65	165/93(45)	955–2011	452	0.667/0.331/0.127
MH	3790–3899/3818	40–65	147/85(37)	1005–2011	447	0.671/0.352/0.131
ML	3701–3789/3760	35–60	165/91(54)	940–2011	436	0.874/0.394/0.205
L	3550–3700/3603	15–45	117/62(31)	1110–2011	369	1.331/0.513/0.169
ALL	-	-	594/331(167)	940–2011	430	1.331/0.39/0.127
Site code	**MS**	**AC(1)**	**R/R1/R2***	**PC1*(%)**	**SNR***	**EPS***
H	0.33/0.22	0.70/0.33	0.57/0.89/0.57	44.9/59.4	26/50	0.98
MH	0.36/0.27	0.70/0.50	0.57/0.92/0.56	44.9/61.1	17/36	0.97
ML	0.32/0.25	0.73/0.54	0.4/0.78/0.39	30.8/42.5	14/28	0.96
L	0.34/0.25	0.74/0.62	0.52/0.87/0.49	43.4/56.8	6/11.8	0.92
ALL	0.34/0.25	0.72/0.49	0.48/0.85/0.47	37.8/66.5	42/54	0.99

Altitude: elevation range/median elevation; Slope: topographic slope angle; Core/Tree: number of cores and trees collected at this elevation, the value in round brackets is the number of trees for which pith offsets are available; Time span: the period when the EPS value is greater than 0.85,which means the time series is robust. ML: mean segment length; Max/Mean/Min TRW: maximum, mean and minimum tree-ring width; MS: mean sensitivity of raw ring data and standard chronology; AC1: first-order autocorrelation of raw ring data and standard chronology; R/R1/R2: mean inter-series/within-tree/between-tree correlation coefficient; PC1: percent variance explained by the first principal component of raw ring data and standard chronology; SNR: signal-to-noise ratio of raw ring data and standard chronology; EPS: expressed population signal, calculated for ARSTAN standard chronologies for 30-year intervals with 15-year overlaps.

*Calculated for the common period 1750–2000 (1700–2000 for the composite chronology).

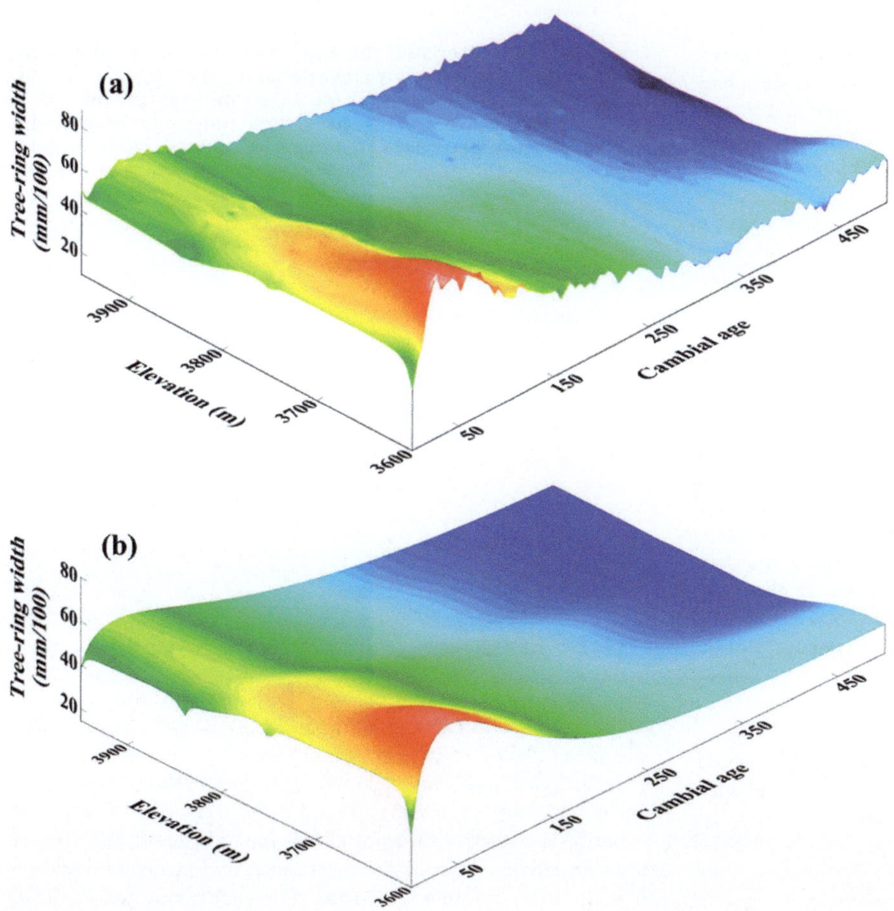

Figure 3. Trend surface for cambial age, elevation (m), tree-ring width (mm/100) (a) and radial ring-width growth trends (mm/100) (b) at MNT, Delingha, China. The surface is interpolated from four original unsmoothed curves of radial ring width (see Figure 5) using inverse distance weighting.

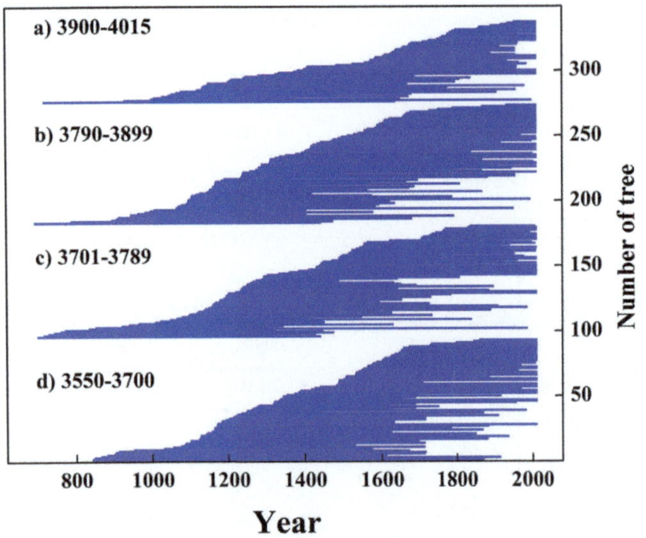

Figure 4. Comparison of sample depths over time for the four elevation zones.

surface contrast was enhanced with chalk. Ring widths were recorded with a LINTAB II measuring system at a resolution of 0.01 mm, and all cores were cross-dated by visual growth pattern matching, skeleton plotting and statistical tests available in the software package TSAP [30,31]. Cores displaying poor correlation with the master series were removed. Finally, 594 increment cores from 331 trees were retained. Detailed information on sampling sites and the TRW data at different elevations are listed in Table 1. Based on the elevation range and the characteristics of the sample distribution, we separated the data into four different altitude belts stretching from the lower tree line to the upper tree line, while ensuring that sufficient samples were available in each altitude belt. These four ranges were as follows: High (H: 3900–4015 m.a.s.l.), Middle High (MH: 3790–3899 m.a.s.l.), Middle Low (ML: 3701–3789 m.a.s.l.), and Low (L: 3550–3700 m.a.s.l.).

Climate data

Monthly mean/maximum/minimum temperature, total precipitation and mean relatively humidity data were computed from daily records for Delingha station obtained from the China Meteorological Data Sharing Service System (http://cdc.cma.gov.cn/). The instrumental records cover 1956–2011. The nearest Palmer Drought Severity Index (PDSI) gridpoint data (1850–2010, for a 2.5°×2.5° grid, centered on 36.25°N, 98.75°E) were extracted from the self-calibrated PDSI data set, in this case calculated using Penman-Monteith Potential Evapotranspiration (updated to 2010) [32]. For consistency with other climate data, we used the PDSI data for the period 1956 to 2010.

Radial tree growth analysis

Measured tree-ring radial growth series comprise age-related trends, climate-related growth variations, the effects of disturbance from within and outside the forest community, and residual variability associated with random disturbance effects [33,34]. Because the canopy density of Qilian juniper is less than 0.3, we adopted the simplifying assumption that trees were growing in a relatively competition-free, open-canopy forest. We assumed a situation where an idealized series of radial growth increment measurements contains only age-related effects manifested as a long-term radial growth trend with climate-related variations

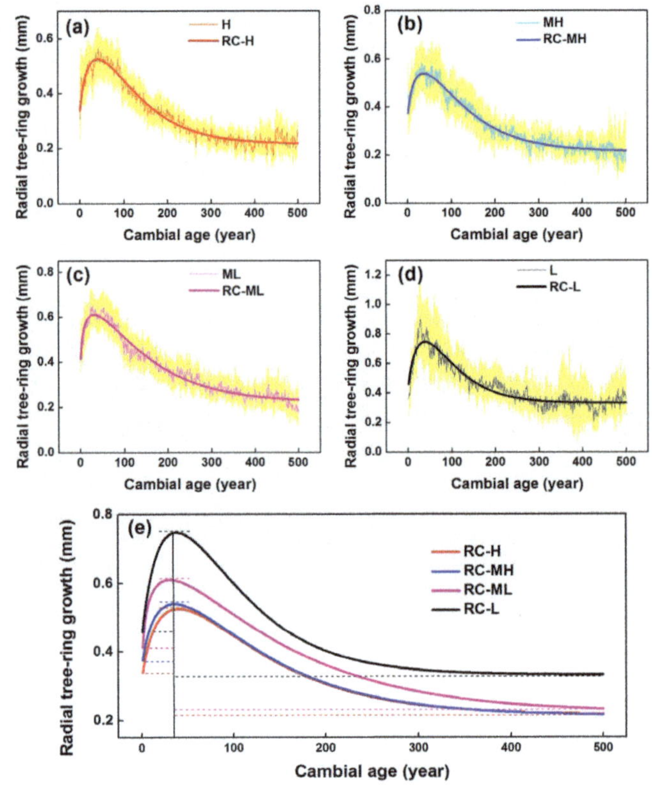

Figure 5. Characteristics of the radial tree-ring biological growth trends at different elevations: (a) H (3900 m–4015 m), (b) MH (3790 m–3899 m), (c) ML (3701 m–3789 m), and (d) L (3550 m–3700 m), based on Hugershoff functions fitted to the means measurement data. The thin line is a mean curve for radial tree-ring biological growth at different elevations. The thick line is a curve fitted to the Hugershoff function. The 95% confidence interval is shown by light yellow shading.

superimposed. We also note that "elevation is a site factor affecting mean annual precipitation and temperature regimes" [3]. Tree-ring width (TRW) variability is therefore attributable to a combination of altitude-related variability (including climate) and age-related trends, specifically:

$$TRW = E + A,$$

where E is the elevation-related variability, and A represent the age-related trends.

Chronology analysis

To investigate the climate-related variability, 'growth trends' were firstly removed from the series of ring width measurements. All processing was carried out using the ARSTAN program [34]. The raw TRW series were 'standardized' to remove biological growth trends and other low-frequency variations considered attributable to stand dynamics. Before standardization, the variance of each series was stabilized using a data-adaptive power transformation based on the local mean and standard deviation [35]. In order to remove age-related biological trends, a modified negative exponential curve was fitted to each raw series. The tree-ring indices were obtained by calculating residuals between raw measurements and fitted values. All detrended index series were then averaged using the biweight robust mean (rather than the

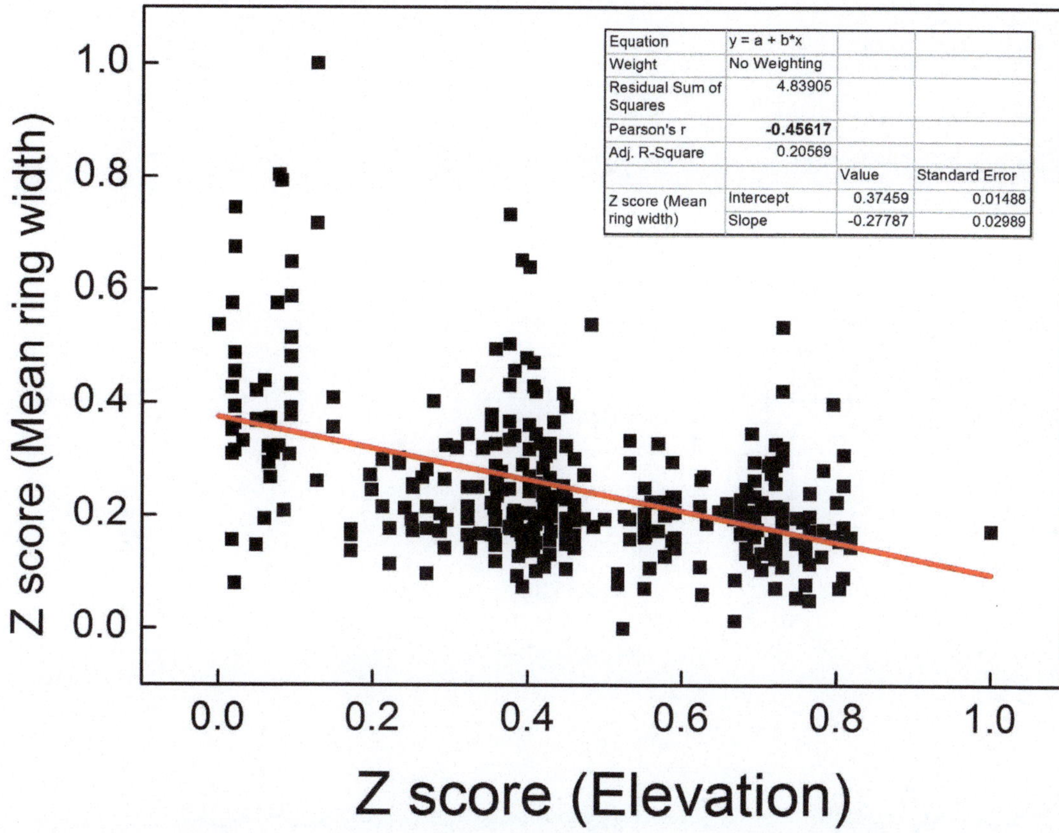

Figure 6. Scatter diagram showing relationships between the elevation and mean ring width after Z score transformation using 265 trees. The Pearson's correlation coefficient is −0.29, which is significant at the p = 0.01 level.

arithmetic mean) in order to reduce the influence of outliers [36]. Four TRW chronologies were developed in this way, each representing tree growth over time, on relatively short-to-medium timescales (years to centuries) in each of the four altitude belts.

Standard (STD) and residual (RES) are two types of tree chronology produced by ARSTAN, each representing complimentary information useful for climate reconstruction. However, there are essential differences in terms of trend persistence. Due to the removal of low-order persistence, the RES chronology only represents relatively 'high-frequency' information, while STD contains more low-frequency information originating from persistent variance in single trees. To investigate growth-climate relationships, we employed both types of TRW chronology. The relevant data will be deposited in PANGAEA (http://www.pangaea.de/).

Growth-climate response analysis

Due to the biological persistence in TRW series, TRW series often represent an effectively smoothed response to climate variability registered on a year-to-year basis. To explore the statistical associations between different monthly-mean climate factors and ring-width indices, we selected the period encompassing the previous May to the current October as the "potential effective climatic window". Individual monthly climate–growth relationships were analyzed using the software DENDROCLIM 2002 [37]. Based on the results, the optimum seasonal assemblies were selected for the growth-climate response analysis.

Unfortunately, there are only 56 years of instrumental climate records available, which is too short to identify the natural relationships between tree radial growth and climate change trends with strong confidence. Meanwhile, TRW index data show 1956–2011 to be a period of relatively high growth (mean value 1.11) relative to the long-term index average (the mean value is 0.99 from 1110 to 2011), limiting our ability to assess whether there are existing significant altitude-related impacts of climate on radial growth in our study area. We also note that the chronology construction approaches used here (i.e. the STD and RES chronologies from ARSTAN – see earlier) may not retain all of the long-timescale climate-related information. We selected six 'typical' 50-year periods (comprising three high-growth periods and three relatively low-growth periods): 1204–1253 (high); 1254–1303 (low), 1435–1484 (low), 1550–1599 (high), 1684–1733 (low), and 1961–2010 (high). We treated each TRW index series as a 50-dimensional vector and computed the Euclidean distance between each individual vector and the mean 50-dimensional vector (the mean value of each series on each year), using cluster analysis, to measure the difference between each single TRW index at different elevations.

The equation for Euclidean distance is $D(X_i, x_{i0}) = \sqrt{\sum_1^n ((X_i - X_{i0}) * (X_i - X_{i0}))}$, where X_i is the ith-dimension value of each vector and X_{i0} is the mean value of all the ith-dimension values. Here, $n = 50$ years.

Extreme climate events analysis

Missing rings are an excellent indicator of likely extreme climate events when the meteorological conditions are assumed to be beyond the physiological limit for tree growth. In general, we would expect the frequency of missing rings to be associated

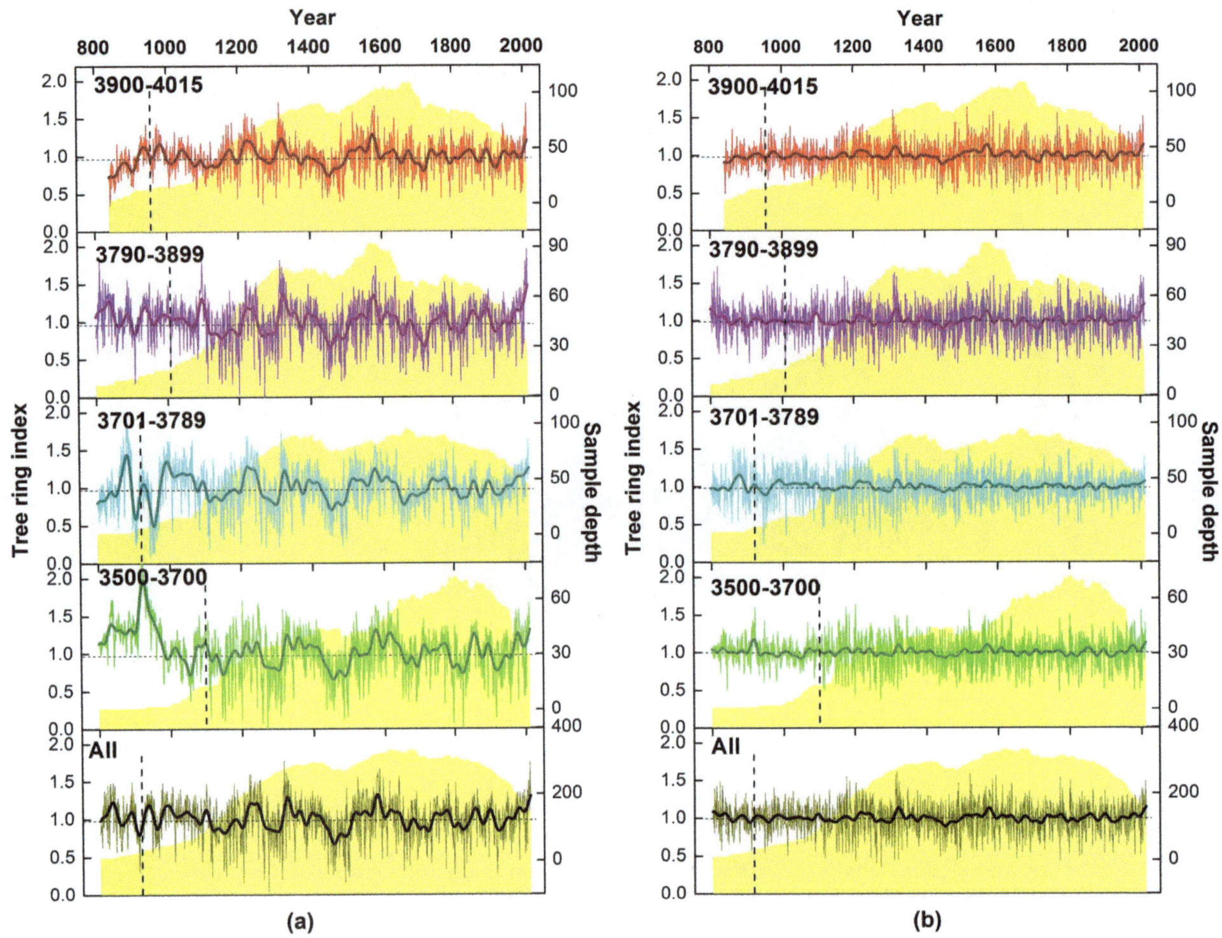

Figure 7. Four standardized chronologies (a), four residual chronologies (b) and the composite chronology (All) and their sample depth at different elevations from MNT. The regional chronology is derived by using all available ring-width series from the four sub-chronologies. The thick line is the 11-year Fast Fourier Transform (FFT) smoothed series. To the right of the vertical dotted line the chronology EPS value is above 0.85, which indicates a statistical reliable series. The light yellow shaded area shows the number of series. The numbers in the upper left corner of all subfigures are the elevation ranges of each zone.

mostly with the occurrence of extreme drought-related events in this semi-arid area. We wished to investigate the nature of the relationship between missing ring and climate variability and whether there is evidence of significant altitude-related effects on this relationship. In order to answer these questions, we computed the temporal variability of the relative missing ring ratio in each altitude belt. To reduce sample depth-related effects (i.e. differing replication), homogenization of sample depth was applied prior to computing the relative missing ring ratio. The Kruskal-Wallis H test, a type of K independent samples nonparametric test, was applied to test the difference.

Table 2. Correlation coefficient matrix between the four STD (upper right corner) and RES (lower left corner) chronologies and the composite chronology during 1956–2011 and 1110–2011.

R	H	MH	ML	L	All
H	1	0.91/0.92	0.88/0.87	0.88/0.81	0.96/0.95
MH	0.95/0.94	1	0.94/0.92	0.89/0.88	0.97/0.98
ML	0.92/0.91	0.92/0.95	1	0.89/0.90	0.96/0.96
L	0.93/0.87	0.93/0.93	0.9/0.93	1	0.95/0.92
All	0.98/0.97	0.98/0.99	0.95/0.97	0.97/0.95	1

All the values are significant at the $p = 0.01$ level.

Results

Characteristics of radial growth

In order to establish a robust picture of the physiological-ecological trend curve, we examined tree cambial age only up to 500 years, when the sample depths were 19 trees (for the high elevation group: H), 14 trees (MH), 22 trees (ML), and 13 trees (L), respectively. Figure 3 clearly shows that radial growth of Qilian juniper has a distinct age-related trend, regardless of whether the trees grow at high or low altitude, starting with a very rapid initial growth rate peaking at ages 25–45. Subsequently, the growth rate gradually decreases with increasing age. This pattern is consistent with our previous research on the same species [38] in the upstream Heihe river region (38°09′N, 99°57′E).

Meanwhile, altitude-related variations are evident during the different lifestages of Qilian juniper. Due to the similar distribution of young and old trees within each altitude belt (Figure 4), the four

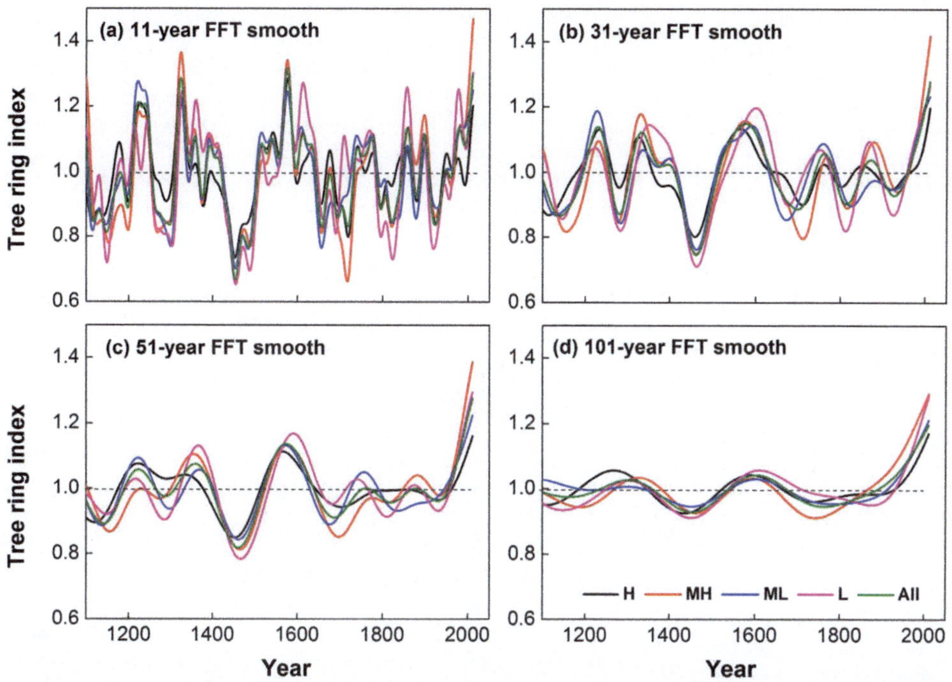

Figure 8. Comparison of low-frequency variability in four STD chronologies at different elevations and the composite chronology during 1110–2011.

Table 3. Comparisons of the three optimum seasonal assemblies of various climate elements for each of the STD chronologies.

	Temperature		Min.Tem.		Max.Tem.		Precipitation		PDSI	
	SA	R	SA	R	SA	R	SA	R	SA	R
	P9-C3	0.24	P9-C3	0.19	P9-C3	0.26	P7-C6	0.6	C3-C6	0.5
H	P1-P12	0.23	P1-P12	0.19	P1-P12	0.21	P12-C6	0.56	P9-C8	0.46
	P12-C2	0.24	P12-C2	0.2	P12-C2	0.26	P3-C6	0.56	P9-C2	0.33
	P9-C3	0.51	P9-C3	0.48	P9-C3	0.38	P7-C6	0.74	C3-C6	0.57
MH	P1-P12	0.49	P1-P12	0.48	P1-P12	0.3	P12-C6	0.67	P9-C8	0.48
	P12-C2	0.46	P12-C2	0.45	P12-C2	0.37	P3-C6	0.65	P9-C2	0.31
	P9-C3	0.46	P9-C3	0.45	P9-C3	0.28	P7-C6	0.68	C3-C6	0.61
ML	P1-P12	0.45	P1-P12	0.45	P1-P12	0.25	P12-C6	0.66	P9-C8	0.51
	P12-C2	0.41	P12-C2	0.43	P12-C2	0.25	P3-C6	0.65	P9-C2	0.34
	P9-C3	0.28	P9-C3	0.29	P9-C3	0.11	P7-C6	0.61	C3-C6	0.46
L	P1-P12	0.28	P1-P12	0.3	P1-P12	0.09	P12-C6	0.58	P9-C8	0.4
	P12-C2	0.25	P12-C2	0.27	P12-C2	0.13	P3-C6	0.58	P9-C2	0.27
	P9-C3	0.39	P9-C3	0.37	P9-C3	0.26	P7-C6	0.69	C3-C6	0.55
PC1	P1-P12	0.38	P1-P12	0.37	P1-P12	0.22	P12-C6	0.65	P9-C8	0.48
	P12-C2	0.35	P12-C2	0.35	P12-C2	0.26	P3-C6	0.64	P9-C2	0.32
	P9-C3	0.38	P9-C3	0.36	P9-C3	0.26	P7-C6	0.68	C3-C6	0.56
All	P1-P12	0.37	P1-P12	0.36	P1-P12	0.21	P12-C6	0.64	P9-C8	0.48
	P12-C2	0.34	P12-C2	0.34	P12-C2	0.26	P3-C6	0.63	P9-C2	0.32

Min. Tem.: monthly minimum temperature; Max. Tem.: monthly maximum temperature; PDSI: Palmer Drought Severity Index; SA: optimum seasonal assembly; R: correlation coefficient; H/MH/ML/L: the site code of the four different elevations (see Table 1); PC1: first principal component of the four chronologies; All: the composite chronology; P9-C3: previous September to Current March; 0.24/0.11: correlation coefficient with STD and RES chronologies, respectively.

Table 4. Pearson correlation coefficients (1956–2011) between STD chronologies and optimal seasonal precipitaiton and temperature and the same correlations using first-order differential series.

	Precipitation(P7-C6)		Temperature (P12-C2)	
	R	R_{ac1}	R	R_{ac1}
H	0.6	0.62	0.24	0.04
MH	0.74	0.69	0.46	0.02
ML	0.68	0.62	0.41	−0.03
L	0.61	0.66	0.25	−0.06
All	0.68	0.66	0.34	−0.01

R: correlation coefficient; R_{ac1}: correlation coefficient of the first-order differential series. H, MH, ML, L, All are site codes for different elevations (see Table 1). Precipitation (P7-C6) is the accumulated precipitation from previous July to current June. Temperature (P12-C2) is the mean temperature from previous December to current February.

ring-width groups display similar age-related "disturbances" in the age-related expectation of ring-width as a function of tree age [39]. In general, TRW decreases with increasing elevation (Figure 3, Table 1), regardless of whether a tree is old or young. However, the amplitude of the average 'cycle' of changing ring width throughout the lifespan of trees is markedly lower (from 0.29 mm to 0.16 mm in the beginning 50 years, and from 0.42 mm to 0.30 mm over the whole 500 years) with increasing altitude (Figure 5). By 500 years the annual growth rate has reduced from 0.33 mm to 0.21 mm. The age of maximum ring width attainment is similar at each of the four elevations (specifically, H: 41, MH: 37, ML: 31, L: 39; years of age). The radial ring-width decline with age after trees reach about 50 years old is steeper at lower elevation sites. This physiological phenomenon is directly related to the tree growth rate (i.e. increase of diameter with age): lower elevation trees exhibit a faster growth rate. Such differences in radial growth rates at the same biological age are the result of local environmental effects of temperature and precipitation due to different elevations.

Mean series length (i.e. the age of sample trees) increases with increasing elevation, which likely reflects more human disturbance (e.g., logging, grazing, man-made fire) at the lower boundary and less disturbance at the upper treeline. It is worth noting that the confidence associated with fitting trend surfaces in the 350–500 years range is low, because the minimum of the mean series length in four elevations (Table 1) is nearly 350 years; this necessarily restricts our discussion of the biological age 350–500 due to the effective "high-pass-filtering effect" on the variance spectra of tree indices and on the chronology series imposed by the standardization methods used here. This is the so-called "segment length effect" described in Cook et al. [40] and Esper et al. [41].

In order to quantify the relationships between TRW and elevation, we removed the fast growth period (about 300 years) at the beginning of each series for each tree and computed the Z-Score of TRW and elevation of 265 trees, and plotted a scatter diagram (Figure 6). The results show that TRW and elevation have a significant negative correlation ($R = -0.28, N = 265$) at the 0.01 level. Therefore, elevation significantly affects radial (physiological) tree growth of Qilian juniper species. These results provide a basis for the priority of growth rate classification or altitude classification pretreatment of TRW over the tree-ring growth trend removal when exploring the use of Regional Curve Standardisation.

Characteristics of the tree ring index

Four STD and RES chronologies were established at four elevations (Figure 7). The mean sensitivity and first-order autocorrelation of the tree ring indices are considerably reduced relative to the values for ring width data. However, the percent variance explained by the first principal component of the index data is increased by comparison. The high correlation coefficients between all the chronologies (listed in Table 2) show that there are no significant statistical differences between standardized tree ring chronologies at different altitudes for the periods 1110–2011 ($R>0.81, N=902, p=0.01$) and 1956–2011 ($R>0.88, N=56, p=0.01$). It is worth noting that RES showed better correlation than STD at the four elevations, due to removal of more of the lower order trends. That is to say, there is greater consistency in the high-frequency variability than in medium or lower-frequency trends, though there were no statistically significant differences apparent in the lower-frequency trends in the chronologies. The consistency between the STD chronologies is shown by Fast Fourier Transform (FFT) filter smoothing of the chronologies at different time scales during the last 902 years (Figure 8). Principal components analysis of these data shows that the first component (PC1) explains 91.3% variance, with very similar loadings of 0.97/0.97/0.94/0.94 corresponding to the H/MH/ML/L chronologies respectively. Therefore, all four chronologies from different elevations exhibit a very strong common pattern of variability during the past 902 years. Finally, STD and RES chronologies

Table 5. Response surface regression results and their F values at the four different elevations and the composite site.

Site	A_{00}	A_{01}	A_{02}	A_{10}	A_{11}	A_{20}	F	$F_{0.01}(5,49)/F_{0.01}(5,48)$
H	0.142/−0.1	−0.264/0.163	−0.216/0.149	0.866/0.6	0.573/0.022	−0.259/−0.047	8.8/7.4**	3.42/3.43
MH	0.116/−0.081	0.025/0.189	−0.12/0.129	0.863/0.67	0.496/0.022	−0.284/−0.046	19.1/10.7**	3.42/3.43
ML	0.128/−0.025	0.018/0.156	−0.121/0.097	0.761/0.608	0.315/0.004	−0.19/−0.071	10.2/7.3**	3.42/3.43
L	0.328/−0.115	−0.278/0.12	−0.227/0.163	0.917/0.626	0.351/0.081	−0.309/−0.045	8.9/9.3**	3.42/3.43
All	0.199/−0.09	−0.148/0.16	−0.181/0.141	0.901/0.638	0.461/0.037	−0.287/−0.049	11.8/9.2**	3.42/3.43

The response surface was defined by a quadratic surface function: $Z(X,Y) = A_{00} + A_{01}Y + A_{02}Y^2 + A_{10}X + A_{11}XY + A_{20}X^2$.
Here, Z = Z score value of standardized tree ring index at different elevations and the whole site; X = Z score value of mean precipitation (previous July to current June); Y = Z score value of mean temperature (previous September to current March). The F value is calculated by $F = \frac{SS_R/p}{SS_D/(n-p-1)}$, with notation as follows. SS_R: regression sum of squares, SS_D: residual sum of squares, p: degrees of freedom of the regression function, here $p = 5$, n: number of active data, here n = 55. If $F > F_\alpha$, the response surface regression is significant at the α level.
**F value is significant at p = 0.01 level.

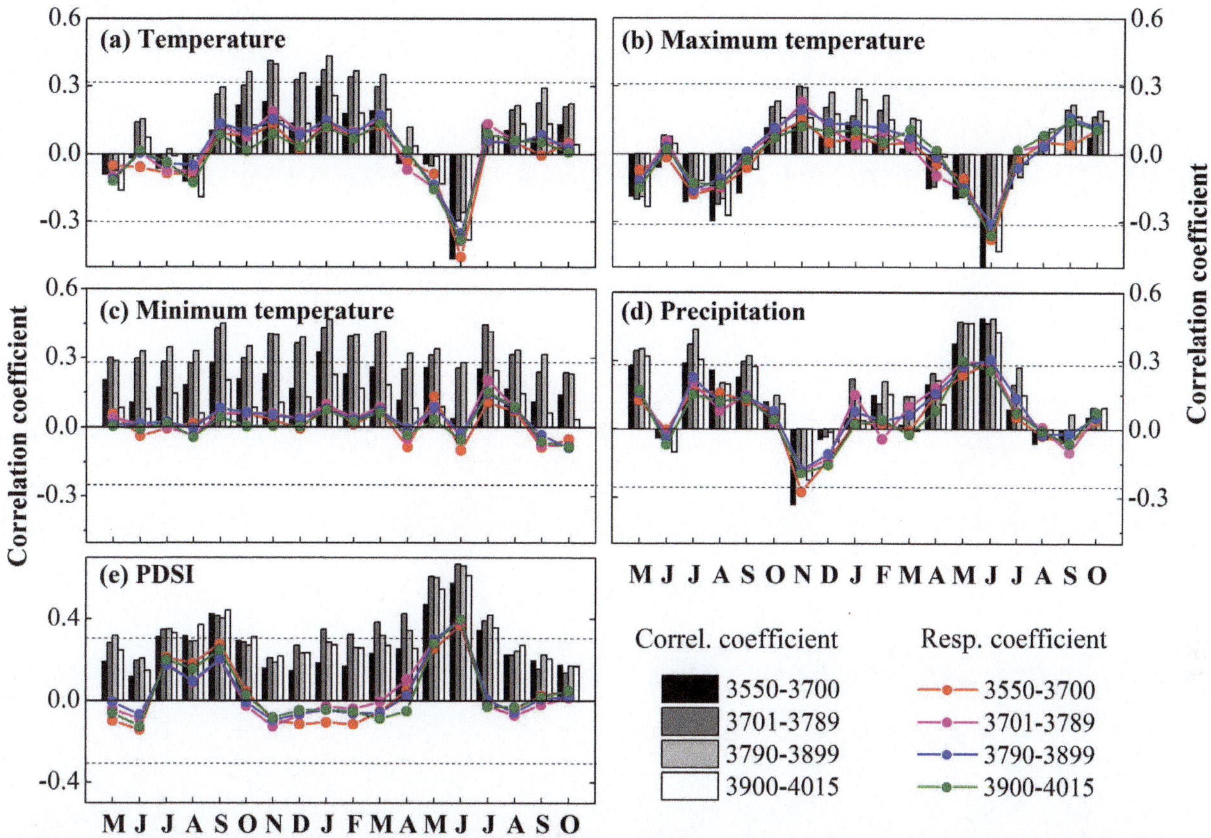

Figure 9. Correlation (histogram) and response function (line) between the four STD chronologies at different elevations and monthly mean values of temperature (a), maximum temperature (b), minimum temperature (c), precipitation (d), and PDSI (e). The horizontal dashed line indicates the 95% confidence level for the correlation.

were developed using all increment cores collected from the four altitude belts, and is referred to here as "All" chronology (Tabs. 1–5, 8; Figure 8).

A common period analysis for the regional chronology was carried out for the period 1750–2000, and some descriptive statistics are shown in Table 1. Signal strength of the standard chronology was assessed by the mean inter-series correlation (rbar), and the associated expressed population signal (EPS). The Rbar (>0.4), and EPS (>0.90) during 1750–2000 indicate that signal strength was sufficient for climate reconstruction of the medium to high-frequency climate variability.

Growth-climate response

Figure 9 shows that, as expected, there is a similar pattern in single-month correlation function analysis between climate variability and the different elevation STD chronologies over the common period 1956–2011. Correlation with TRW and mean temperature is significantly positively correlated during the winter half year (from previous September to current March) and is significantly negatively correlated with temperature during the current June. However, only the correlation with temperature in the current June is confirmed to be significant (at the 0.05 level) by the response function analyses. The relationship between maximum temperature and TRW was similar in general, except for slight differences in the magnitudes of changes. There was a different pattern in the minimum temperature, which appears to be positively correlated in every month. However, this was not verified by the response function analysis. Apparently significant positive correlations were found for monthly precipitation in May, July and September of the previous year and May and June of the current year, while a significant negative correlation was found for November of the previous year. Correlation with precipitation in the current May and June was confirmed by response function analysis, which is also validated by the process-based Vaganov-Shashkin models [42] in this region. The correlations between TRW and PDSI are positive in every month, as confirmed for the current May and June in the response function analysis.

Table 3 shows the apparent optimum seasonal associations are P7-C6, P9-C3, and C3-C6 for precipitation, temperature and PDSI, respectively. Here, we focus attention on precipitation and temperature because these strongly influence PDSI. We first focus on the winter temperature and compute the Pearson's correlation coefficient between STD chronologies and winter temperature along with its first-order differential series at different elevations (Table 4). This reveals that there is no correlation ($R = -0.06$–0.04, $N = 56$) on inter-annual timescales (i.e. with the high frequency signal) and that the raw correlation is based only on a correspondence in trend ($R = 0.24$–0.46, $N = 56$). Unfortunately, the longer-term climate association cannot be further explored with only 56 years of instrumental records. However, there is a distinct pattern between precipitation and tree radial growth (Table 4). There is a strongly positive correlation ($R = 0.62$–0.69,

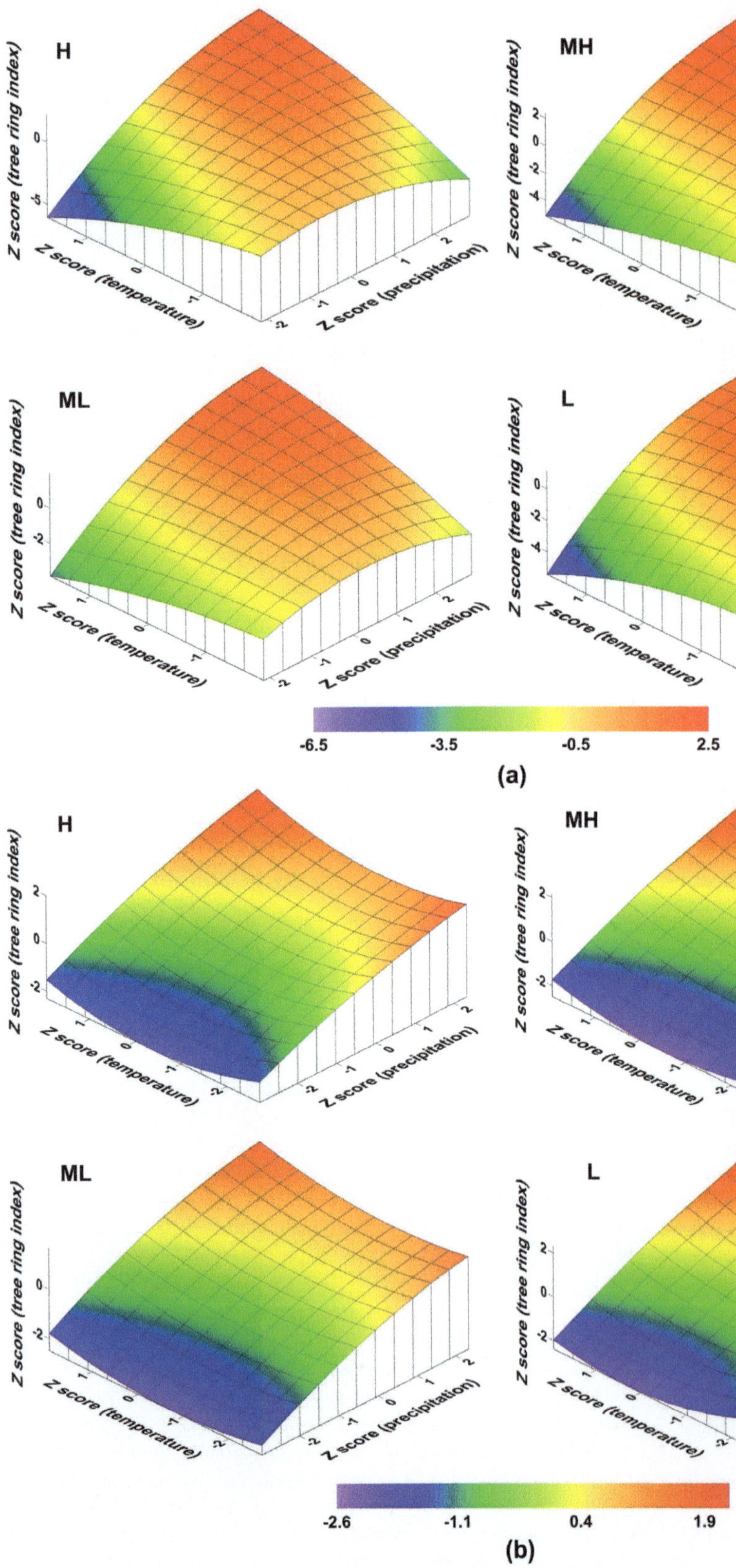

Figure 10. Response surfaces for standardized tree ring index (a) and first-order differential series of standardized tree ring index (b) for four different elevations: H, MH, ML, L. The height of each surface represents the magnitude of growth response to a given combination of annual (previous July to current June) precipitation and winter (previous September to current March) temperature. All data (including precipitation, temperature, and standardized tree ring index) are Z score transformed in order to eliminate the impact of different dimensional units.

Table 6. Comparisons of altitude-related characteristics for the mean tree biological age and corresponding sample size in six typical periods.

Elevation/m	1204–1253/AD	1254–1303/AD	1435–1484/AD	1550–1599/AD	1684–1733/AD	1961–2010/AD
3551–3600	63/6	104/9	229/10	338/11	284/15	265/12
3601–3650	69/1	119/1	51/3	155/4	73/15	355/3
3651–3700	176/9	217/10	369/11	490/10	582/4	183/4
3701–3750	108/10	156/11	146/13	232/14	305/15	576/8
3751–3800	108/36	155/41	236/43	270/43	249/51	351/27
3801–3850	284/11	245/13	186/10	274/12	351/11	417/5
3851–3900	55/12	93/16	241/19	260/25	215/14	517/7
3901–3950	78/14	121/16	254/19	303/24	199/16	407/8
3951–4000	113/12	111/16	233/21	284/25	153/33	380/16
4001–4050	149/6	199/6	274/7	288/8	312/5	589/3
>4050	150/1	150/1	150/1	150/1	150/1	

N = 56) between precipitation and tree growth on inter-annual time scales. This is more evident for the climate response surfaces calculated using the STD tree ring index (Figure 10a, Table 5) and its first-order differential series (Figure 10b, Table 5) in the four different elevations. These results lead us to conclude that any temperature-related inference based on these chronologies relies solely on a statistical association manifested only at low-frequency and it cannot be considered more than a coincident association in recent decades at these sites. Tree rings best represent precipitation variability, as might be expected given the severe lack of water at these sites.

Elevation effects on the chronology variability

To examine the natural role of altitude-related effects on chronology variability during different 'typical' periods (Table 6) in the last 902 years, we calculated the Euclidean distance for each of the tree-ring index series (i.e. for each single tree) in both the STD and residual data sets during each period. We first examined the altitude-related characteristics for both the mean tree biological age and sample size (Table 6) in order to assess whether a heterogeneous sample distribution might contribute to the tree biological age curve, or whether different sample sizes occur at different elevations. Most of the six periods have a relative even altitude distribution with respect to both the mean tree biological age and sample size, except for 1254–1303 and 1961–2010. So we considered the remaining four periods would most likely indicate any true altitude-related effect. Significant negative correlations ranging from −0.29 to −0.16 (with p values ranging from about 0 to 0.04, Figure 11a) were obtained for the relationships between elevation and standardized tree-ring series during most periods (the exception was 1254–1303: correlation $r = 0.02$, $p = 0.85$), which is anomalous with respect to the analysis results for the four chronologies at different elevations. Various "individual responses" of TRW index [26,43] are the most likely potential causes of such contradictory conclusions. However, there were no significant negative correlations with altitude in the residual tree ring series for any of the periods (Figure 11b) (correlation range $r = -0.05$ to 0.06, $p = 0.40$ to 0.82) with the exception of 1961–2010 ($r = -0.20$, $P = 0.05$). Therefore, there is an apparent altitude-related effect on standard tree ring index series, and no statistically discernable effect in the residual tree ring series. This suggests some systematic tendancy for greater growth at lower elevation associated with the greater low-frequency variability presented in the indices produced for standard chronologies. This finding is also confirmed using other methods for growth trend removal (e.g. Hugershoff; not shown in this article). It should be noted that while such effects are statistically significant on the scale of a single TRW index series level, they cannot be diagnosed using the four STD chronologies (i.e. after averaging individual trees index series). From the inter-annual, decadal, centennial to multi-centennial scale, there are no notable differences in the STD chronologies for the four elevations from 1110 to 2011 (Figure 12, Table 7). This finding is confirmed by the tree-ring study conducted at Qifeng site (~235 km far away) in the northeast Tibetan Plateau [13]. However, this situation contrasts with the view that climate responses are frequency dependent: e.g. where low-frequency variations from the upper forest border chronologies are likely related to warm-season temperature fluctuations, whereas high-frequency variations are related to precipitation, such as in the semiarid mountainous area of North America at similar latitudes [44,45]. Such differences in the frequency-dependent climate signal are due to different climate responses for the upper (temperature) and lower (precipitation) border chronologies, and may not be applicable to a region such as this or where the climate responses at the upper and lower forest borders are unknown.

Extreme climate events and their potential as climatic controls on tree growth

Relative missing ring ratios, at four altitude belts and for the whole hillside in MNT during 1200–2011 are plotted in Figure 13. In order to learn more about whether there are significant differences in missing ring responses to extreme climate variability at different altitudes, we used the Kruskal-Wallis H test. The results (Table 8) show that significant differences exist between high (>10%, $n = 13$, $p = 0.05$) and low (<2%, $n = 105$, $p = 0.015 < 0.05$) missing ring ratio events, although there was no significant difference ($n = 149$, $p = 0.397 > 0.05$) between the complete distributions of all the missing ring ratio events across the four altitude belts. As seen from Table 8, the trees near the upper treeline (mean rank H: 208.6, MH: 236.82) are more sensitive to climate change than those in the other altitude belts (mean rank ML: 202.18, L: 194.4), based on a frequency of relative missing ring ratio events of less than 2%. A similar

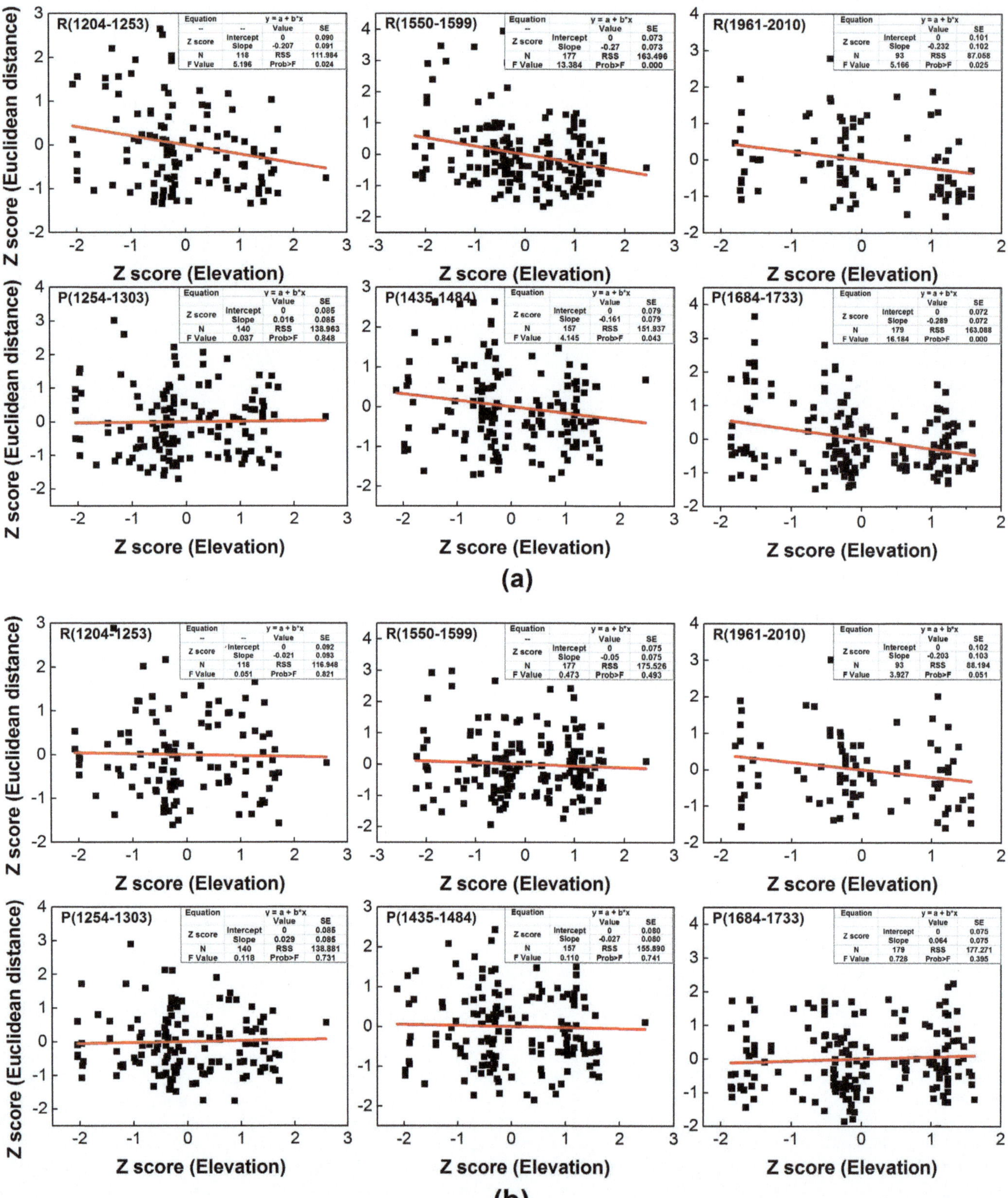

Figure 11. Comparison of altitude-related effects on standardized tree ring series (a) and residual tree ring series (b), which separately constitute the STD and RES chronologies during six typical periods (three high-value periods and three low-value periods). Euclidean distance is defined as the distance between each TRW series (50-dimensional vectors) and the mean TRW series (50-dimensional vector) of all the TRW series in the given periods. Both the elevation (m) and Euclidean distance are transformed to Z scores. R(1204–1253): high precipitation during 1204–1254; P(1254–1303):low precipitation during 1254–1303; SE: standard error; RSS: residual sum of squares; Prob>F: the probability that the null hypothesis for the full model is true (i.e., that all of the regression coefficients are zero).

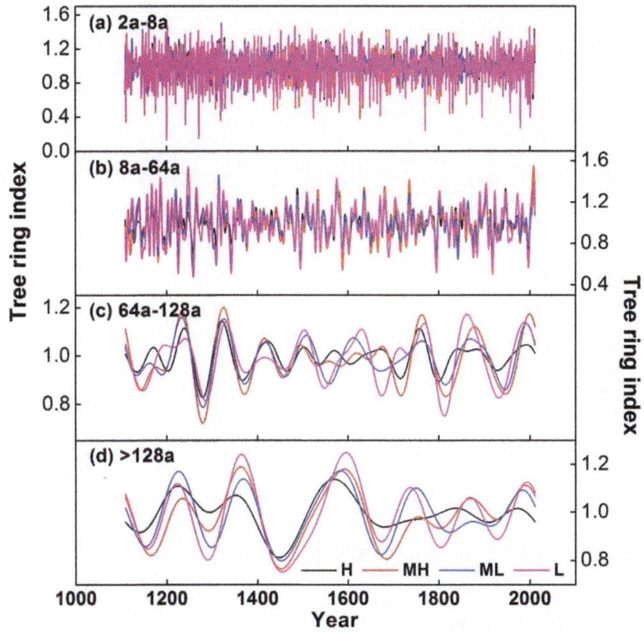

Figure 12. Comparisons of different frequency components of STD chronologies at four elevations during 1110–2011. The four frequency components are 1/8 to 1/2 cycles per year (cpy) (a), 1/64 to 1/8 cpy (b), 1/128 to 1/64 cpy (c), and less than 1/128 cpy (d), respectively.

Table 7. Comparisons of correlations of different frequency components for four elevation STD chronologies.

R*	H	MH	ML	L
H	1	0.96/0.92	0.93/0.85	0.91/0.75
MH	0.84/0.76	1	0.96/0.92	0.96/0.83
ML	0.74/0.78	0.87/0.82	1	0.95/0.87
L	0.56/0.68	0.70/0.86	0.81/0.86	1

*1/8 to 1/2 cpy and 1/64 to 1/8 cpy are on the upper right corner; 1/128 to 1/64 cpy, and less than 1/128 cpy are on the lower left corner.
All the values are significant at the p = 0.01 level.

interpretation applies to the data on extreme climate events (relative missing ring ratio greater than 10%, mean rank H: 34.77, L: 24.77). Our analysis suggests that researchers can reduce field work while still gaining valuable insights into the occurrence of past extreme climate events by focusing sample collection at the upper treeline.

What is the climatic cause of such missing ring events? Comparing their occurrence (i.e. years with more than 10% missing rings) with reconstructed PDSI data (Cook et al. [46], data for grid location 36.25°N, 98.75°E), reveals a mean PDSI value of −3.3 i.e. extreme dry years (Figure 13). This is confirmed by primary instrumental records during 1956–2011. Compared with temperature, precipitation displays a more robust and consistent relationship with missing ring events (Figure 14). Therefore, the missing ring events, especially those years with a relative missing ring ratio more than 10%, very likely indicate serious drought events. Figure 15 illustrates the composited climate pattern for the 9 missing ring event years that occurred within 1956–2011. Temperatures are "normal" for the year preceding the missing-ring year. There was some water shortage in the prior August (−16.15 mm), but this is likely not a key factor. However, serious water shortage in May (−17.19 mm) and June (−23.07 mm) in the tree growth year, which corresponds to the start period of radial growth are very likely the major cause. Simultaneously, high temperature (+1.32°C) in June probably reinforces the water stress. Extremely low precipitation, accompanied by abnormally high temperatures at the beginning of the radial growth season, is the likely cause of missing ring events.

Conclusions

Using samples collected from 331 trees at four elevations (3550 m–4020 m) on one single hillside, including 164 trees with exact pith offset information, we analyzed the characteristics of age-related effects and altitude-related effects on TRW in Qilian juniper on the northeastern Tibetan Plateau. Altitude impacts significantly on radial ring width: growth rates were lower at higher altitudes, regardless of age. Tree-ring width reaches a maximum between the ages 25–45.

There is a strong and significant correlation between the ring-width chronologies at four different altitudes. The chronology-climate associations at different elevations, as would be expected on this basis, show no significant differences, whether expressed as single-month correlations or by the use of response function analysis. Standardised tree growth is associated primarily with precipitation variability during previous July to current June. Despite coincident trends, winter temperatures prior to the growing season, have no influence on the high frequency tree growth signal. In the similarity analysis of standardized and residual tree ring series, some significant altitude-related effects were apparent but only on the scale of individual standardized tree

Table 8. Kruskal-Wallis H test on different missing ring ratio events during 1200–2011.

MRR(%)	N	Mean Rank				Chi-Square	Asymp.Sig.(2-tailed)
		H	MH	ML	L		
>10	13	34.77	28	18.46	24.77	7.826	0.05*
10-2	31	58.03	59.27	57.98	74.71	4.847	0.183
<2	105	208.6	236.82	202.18	194.4	10.435	0.015*
All	149	293.11	315.06	289.29	296.54	2.29	0.515

MRR: missing ring ratio; N: number of events; Mean Rank of H/MH/ML/L: mean ranks within each altitude belt; Chi-Square: Chi-Square coefficient; Asymp. Sig.: asymptotic significance, which means that the significance is very close to 0.
*the value is significant at the p = 0.05 level.

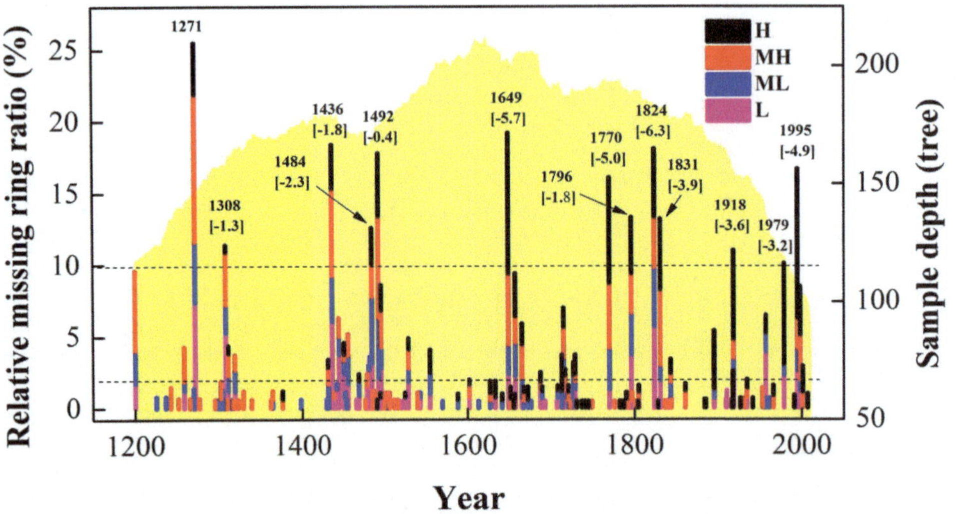

Figure 13. Comparison of relative missing ring ratios at four altitude belts and for the whole hillside in MNT, Delingha, northeast Tibetan Plateau during 1200-2011. Sample depth homogenization was applied to reduce sample-depth-related disturbance prior to computing the relative missing ring ratio. Bar heights indicate missing ring ratios for all samples each year, as visualized by different colored bars (the length of each bar corresponds to the relative missing ring ratio at each elevation). The light yellow shadow is the sample depth. The two dotted lines correspond to the relative missing ring ratios of 10% and 2%, respectively. The values on each bar are the corresponding year and PDSI value (in square brackets, Cook et al., 2010) for those years when the relative missing ring ratio exceeds 10%.

ring series, regardless of whether precipitation was relatively high or low. Elevation showed no significant influence on the variability of mean chronologies during the last 902 years or during any of the selected 50-year periods. Missing ring events are related to drought, and the uppermost treeline is more sensitive to climate change than that of low altitude forest. Serious water shortage in May and June of the growth year, which is strengthened by the high temperature in June of the growth year, are the main conditions needed to initiate missing ring events. There is no doubt that our interpretation of tree-growth/climate relationship in the cold and arid region could be helped by a better mechanistic understanding of tree-growth physiology. A constructive way forward may be to link small scale physiological tree growth models, driven by localised meteorological variables, with large-scale climate models and compare the results with radial growth data in this region.

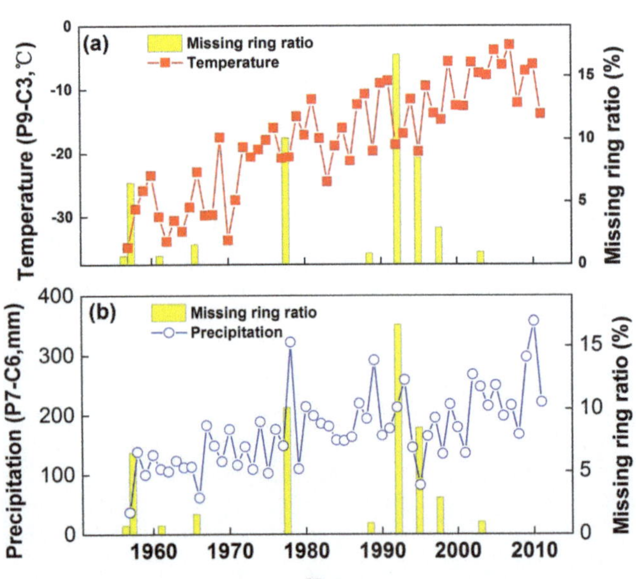

Figure 14. Comparison between missing ring ratio and the temperature (a) and precipitation (b) instrumental records at Delingha during 1956–2011. Temperature is the mean from previous September to current March. Precipitation is the total from previous July to current June.

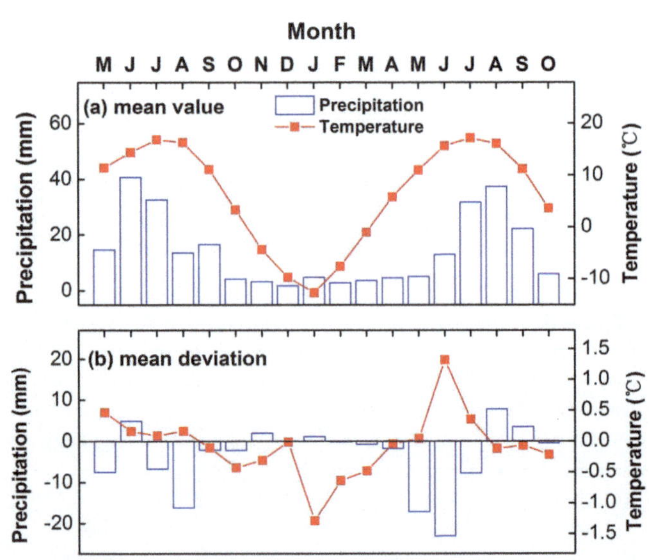

Figure 15. Average (a) and mean deviations (b) of precipitation and temperature from their long-term (1956–2011) mean values for all the missing ring events of the composite chronology.

Author Contributions

Conceived and designed the experiments: BY. Performed the experiments: CQ BY TM YZ KB. Analyzed the data: CQ BY. Contributed reagents/materials/analysis tools: CQ BY TM YZ KB ZF. Wrote the paper: CQ BY TM YZ KB ZF.

References

1. Carrer M, Nola P, Eduard JL, Motta R, Urbinati C (2007) Regional variability of climate-growth relationships in Pinus cembra high elevation forests in the Alps. Journal of Ecology, 95, 1072–1083.
2. LaMarche VC (1974) Paleoclimatic inferences from long tree-ring records. Science, 183, 1043–1048.
3. Fritts HC (1976) Tree Rings and Climate. Academic Press, London.
4. Splechtna BE, Dobry J, Klinka K (2000) Tree-ring characteristics of subalpine fir (Abies lasiocarpa (Hook.) Nutt.) in relation to elevation and climatic fluctuations. Ann For Sci 57:89–100.
5. Salzer MW, Hughes MK, Bunn AG, Kipfmueller KF (2009) Recent unprecedented tree-ring growth in bristlecone pine at the highest elevations and possible causes. Proc Natl Acad Sci 106 (48): 20348–20353.
6. Babst F, Poulter B, Trouet V, Tan K, Neuwirth B, et al. (2012) Site- and species-specific responses of forest growth to climate across the European continent. Global Ecol Biogeogr 22(6): 706–717.
7. Esper J, Frank DC, Wilson RJS, Büntgen U, Treydte K (2007) Uniform growth trends among central Asian low- and high-elevation juniper tree sites. Trees 21:141–150.
8. Wang T, Ren HB, Ma KP (2005) Climatic signals in tree ring of Picea schrenkiana along an altitudinal gradient in the central Tianshan Mountains, northwestern China. Trees 19: 735–741.
9. Morales MS, Villalba R, Grau HR, Paolini L (2004) Rainfall-controlled tree growth in high-elevation subtropical treelines. Ecology 85 (11): 3080–3089.
10. Biondi F (2001) A 400-year tree-ring chronology from the tropical treeline of North America. Ambio 30:162–166.
11. Biondi F (2002) Treeline dendroclimatology in the North American tropics. PAGES News 10:1.
12. Shao X, Xu Y, Yin ZY, Liang E, Zhu H, et al. (2010) Climatic implications of a 3585-year tree-ring width chronology from the northeastern Qinghai-Tibetan Plateau. Quaternary Science Reviews, 29, 2111–2122.
13. Yang B, Qin C, Huang K, Fan ZX, Liu JJ (2010) Spatial and temporal patterns of variations in tree growth over the northeastern Tibetan Plateau during the period 1450–2006. The Holocene, 20, 1235–1245.
14. Yang B, Qin C, Bräuning A, Burchardt I, Liu JJ (2011) Rainfall history for the Hexi Corridor in the arid northwest China during the past 620 years derived from tree rings. International Journal of Climatology, 31, 1166–1176.
15. Qin C, Yang B, Burchardt I, Hu XL, Kang XC (2010) Intensified pluvial conditions during the twentieth century in the inland Heihe River Basin in arid northwestern China over the past millennium. Global and Planetary Change, 72, 192–200.
16. Qin C, Yang B, Bräuning A, Sonechkin DM, Huang K (2011) Regional extreme climate events on the northeastern Tibetan Plateau since AD 1450 inferred from tree rings. Global and Planetary Change, 75, 143–154.
17. Kang SY, Yang B, Qin C, Wang JL, Shi F, et al. (2012) Extreme drought events in the years 1877e1878, and 1928, in the southeast Qilian Mountains and the air-sea coupling system. Quaternary International. doi:10.1016/j.quaint.2012.03.011.
18. Kang SY, Yang B, Qin C (2012) Recent tree-growth reduction in north central China as a combined result of a weakened monsoon and atmospheric oscillations. Climatic Change. DOI 10.1007/s10584-012-0440-6
19. Liang EY, Shao XM, Eckstein D, Huang L, Liu XH (2006) Topography- and species-dependent growth responses of Sabina przewalskii and Picea crassifolia to climate on the northeast Tibetan Plateau. Forest Ecology and Management, 236, 268–277.
20. Liang EY, Wang YF, Xu Y, Liu B, Shao XM (2010) Growth variation in Abies georgei var. smithii along altitudinal gradients in the Sygera Mountains, southeastern Tibetan Plateau. Trees 24: 363–373.
21. Li ZS, Liu GH, Fu BJ, Hu CJ, Luo SZ, et al. (2012) Anomalous temperature–growth response of Abies faxoniana to sustained freezing stress along elevational gradients in China's Western Sichuan Province. Trees (26): 1373–1388.
22. Yang B, He M, Melvin TM, Zhao Y, Briffa KR (2013) Climate Control on Tree Growth at the Upper and Lower Treelines: A Case Study in the Qilian Mountains, Tibetan Plateau. PLoS ONE 8(7): e69065. doi:10.1371/journal.-pone.0069065
23. Zheng YH, Zhu HF, Zhang YX, Shao XM (2009) Relationships between Sabina przewalskii radial growth and climatic factors at upper timberlines in eastern mountainous region of Qaidam Basin. Chinese Journal of Applied Ecology, 20, 507–512.
24. Fang KY, Gou XH, Chen FH, Peng JF, D'Arrigo R, et al. (2009) Response of regional tree-line forests to climate change: evidence from the northeastern Tibetan Plateau. Trees, 23, 1321–1329.
25. Peng JF, Gou XH, Chen FH, Li JB, Liu PX, et al. (2008) Altitudinal variability of climate-tree growth relationships along a consistent slope of Anyemaqen Mountains, northeastern Tibetan Plateau. Dendrochronologia, 26, 87–96.
26. Zhang YX, Shao XM, Wilmking M (2011) Dynamic relationships between Picea crassifolia growth and climate at upper treeline in the Qilian Mts., Northeast Tibetan Plateau, China. Dendrochronologia, 29, 185–199. doi:10.1016/j.dendro.2010.11.001.
27. Gou XH, Zhang F, Deng Y, Ettl GJ, Yang MX, et al. (2012) Patterns and dynamics of tree-line response to climate change in the eastern Qilian Mountains, northwestern China. Dendrochronologia, 30, 121–126.
28. Zhu HF, Zheng YH, Shao XM, Liu XH, Xu Y, et al. (2008) Millennial temperature reconstruction based on tree-ring widths of Qilian juniper from Wulan, Qinghai Province, China. Chinese Science Bulletin, 53(24), 3914–3920.
29. He MH, Yang B, Bräuning A (2013) Tree growth-climate relationships of Juniperus tibetica along an altitudinal gradient on the southern Tibetan Plateau. Trees 27:429–439.
30. Stokes MA, Smiley TL (1996) An Introduction to Tree-Ring Dating, pp. 1–73. The University of Chicago Press: Chicago, London.
31. Rinn F (2003) TSAP-Win: Time Series Analysis and Presentation for Dendrochronology and Related Applications. Version 0.55 User reference. Heidelberg, Germany (http://www.rimatech.com).
32. Dai A (2011) Characteristics and trends in various forms of the Palmer Drought Severity Index (PDSI) during 1900–2008. Journal of Geophysical Research, 116, D12115, doi:10.1029/2010JD015541.
33. Fritts HC, Mosimann JE, Bottorff CP (1969) A revised computer program for standardizing tree-ring series. Tree-Ring Bulletin, 29, 15–20.
34. Cook ER (1985) A time-series analysis approach to tree-ring standardization. PhD Thesis. The University of Arizona Press, Tucson.
35. Cook ER, Peters K (1997) Calculating unbiased tree-ring indices for the study of climatic and environmental change. The Holocene, 7, 361–370.
36. Cook ER, Kairiukstis A (Eds.) (1990) Methods of Dendrochronology: Applications in the Environmental Sciences. Kluwer Academic Publishers, Dordrecht.
37. Biondi F, Waikul K (2004) DENDROCLIM2002: a C++ program for statistical calibration of climate signals in tree-ring chronologies. Computers and Geosciences 30(3): 303–311.
38. Yang B, Qin C, Shi F, Sonechkin M (2012) Tree ring-based annual streamflow reconstruction for the heihe river in arid northwestern china from ad 575 and its implications for water resource management. The Holocene, DOI: 10.1177/0959683611430411.
39. Esper J, Niederer R, Bebi P, Frank D (2008) Climate signal age effects—Evidence from young and old trees in the Swiss Engadin. Forest Ecology and Management, 255, 3783–3789.
40. Cook ER, Briffa KR, Meko DM, Graybill DA, Funkhouser G (1995) The 'segment length curse' in long tree-ring chronology development for palaeoclimatic studies. The Holocene, 5, 229–237.
41. Esper J, Cook ER, Krusic PJ, Peters K, Schweingruber FH (2003) Tests of the RCS method for preserving low-frequency variability in long tree-ring chronologies. Tree-Ring Research, 59, 81–98.
42. Zhang YX, Shao XM, Xu Y, Wilmking M (2011) Process-based modeling analyses of Sabina przewalskii growth response to climate factors around the northeastern Qaidam Basin. Chinese Science Bulletin, 56, 1518–1525, doi: 10.1007/s11434-011-4456-5.
43. Carrer M (2011) Individualistic and time-varying tree-ring growth to climate sensitivity. PLoS ONE 6, e22813. doi:10.1371/journal.pone.0022813.
44. LaMarche VC (1974) Frequency-Dependent Relationships between Tree-Ring Series along an Ecological Gradient and Some Dendroclimatic Implications. Tree-Ring Bull, 34, 1–20.
45. Hughes MK, Funkhouser G (2003) Frequency-dependent climate signal in upper and lower forest border tree rings in the mountains of the great basin. Climatic Change 59, 233–244.
46. Cook ER, Anchukaitis KJ, Buckley BM, D'Arrigo RD, Jacoby GC, et al. (2010) Asian Monsoon Failure and Megadrought During the Last Millennium. Science, 328, 486–489.

5

Northern Hemisphere Glaciation during the Globally Warm Early Late Pliocene

Stijn De Schepper[1,2]*, Jeroen Groeneveld[3], B. David A Naafs[4¤], Cédéric Van Renterghem[5], Jan Hennissen[6], Martin J. Head[6,7], Stephen Louwye[5], Karl Fabian[8]

1 Department of Earth Science, University of Bergen, Bergen, Norway, **2** Geosciences Department, University of Bremen, Bremen, Germany, **3** MARUM – Center for Marine Environmental Sciences, University of Bremen, Bremen, Germany, **4** Alfred Wegener Institute for Polar and Marine Research, Bremerhaven, Germany, **5** Research Unit Palaeontology, Ghent University, Ghent, Belgium, **6** Department of Earth Sciences, University of Toronto, Toronto, Ontario, Canada, **7** Department of Earth Sciences, Brock University, St. Catharines, Ontario, Canada, **8** Norwegian Geological Survey, Trondheim, Norway

Abstract

The early Late Pliocene (3.6 to ~3.0 million years ago) is the last extended interval in Earth's history when atmospheric CO_2 concentrations were comparable to today's and global climate was warmer. Yet a severe global glaciation during marine isotope stage (MIS) M2 interrupted this phase of global warmth ~3.30 million years ago, and is seen as a premature attempt of the climate system to establish an ice-age world. Here we propose a conceptual model for the glaciation and deglaciation of MIS M2 based on geochemical and palynological records from five marine sediment cores along a Caribbean to eastern North Atlantic transect. Our records show that increased Pacific-to-Atlantic flow via the Central American Seaway weakened the North Atlantic Current and attendant northward heat transport prior to MIS M2. The consequent cooling of the northern high latitude oceans permitted expansion of the continental ice sheets during MIS M2, despite near-modern atmospheric CO_2 concentrations. Sea level drop during this glaciation halted the inflow of Pacific water to the Atlantic via the Central American Seaway, allowing the build-up of a Caribbean Warm Pool. Once this warm pool was large enough, the Gulf Stream–North Atlantic Current system was reinvigorated, leading to significant northward heat transport that terminated the glaciation. Before and after MIS M2, heat transport via the North Atlantic Current was crucial in maintaining warm climates comparable to those predicted for the end of this century.

Editor: Victoria C Smith, University of Oxford, United Kingdom

Funding: This research was supported by Deutsche Forschungsgemeinschaft grants SCHE 1665/2-1 and SCHE 1665/2-2 (SDS) and NA973/1-1 (BDAN), the University of Bergen (SDS), a MARUM Student Summer Fellowship (CVR), and a NSERC Canada Discovery Grant (MJH). The funders had no role in study design, data collection and analysis, decision to publish, or preparation of the manuscript.

Competing Interests: The authors have declared that no competing interests exist.

* E-mail: smad2@cantab.net

¤ Current address: Organic Geochemistry Unit, School of Chemistry, University of Bristol, Bristol, United Kingdom

Introduction

The early Late Pliocene (early Piacenzian) from 3.6 to ~3.0 million years ago (Ma) is the last sustained interval in Earth's history when global climate was warmer than today. The ~3.3–3.0 Ma time slab known as the mid-Piacenzian Warm Period (mPWP, Figure 1) has been studied intensively as a potential analogue for our future global climate [1]. The mPWP is characterised by ~3°C warmer global temperatures [2], 10–40 m higher sea-level [3], reduced continental ice sheets [4], and an Atlantic meridional overturning circulation (AMOC) comparable to [5] or stronger than [6] preindustrial levels. Atmospheric CO_2 concentrations were higher than preindustrial values, and likely as high as the modern anthropogenic values of ~400 ppm [7–9] (Figure 1C). The mPWP climate is a good approximation for the warm climatic conditions of the entire early Late Pliocene. This warm stable climate was nonetheless interrupted by a short-lived, intense global glaciation (3.305–3.285 Ma) during marine isotope stage (MIS) M2 [10,11] (Figure 1). In the LR04 Plio-Pleistocene benthic $\delta^{18}O$ stack [10], MIS M2 starts as a low-amplitude glaciation typical of the Pliocene, but deepens steeply between 3.305 and 3.285 Ma to reach values characteristic of early Quaternary glaciations. We distinguish this brief interval of intense glaciation (3.305–3.285 Ma) within the longer interval of MIS M2 (3.312–3.264 Ma) as defined in LR04 [10]. The associated glacio-eustatic sea level drop is reflected in a major depositional sequence boundary [12] with sea level estimated at 10 m±10–15 m, 40 m±10 m, or indeed up to 65 m±15–25 m below present [13–15] (Figure 1B). Given this large uncertainty in reconstructed sea level for MIS M2, it is difficult to quantify how the volume of the northern and southern hemisphere ice sheets changed. Using the Holocene-like, relatively cool and dry Arctic climate at Lake El'gygytgyn (northeast Arctic Russia) as an approximation of the broader Arctic climate, ice advance during MIS M2 is thought to have occurred in Alaska, Greenland, Svalbard and Antarctica, whereas substantial expansion in North America was less likely [16]. Estimates for ice volume increase in Antarctica correspond to a sea level drop of ~8 m [17] or even ~18 m [13], but cannot not fully explain the ~0.5‰ benthic foraminiferal $\delta^{18}O$ shift at this time [10]. Direct and indirect evidence of glaciation support expansion of the Antarctic ice sheet [18,19], a considerable ice advance of the Greenland and Svalbard/Barents Sea ice sheets

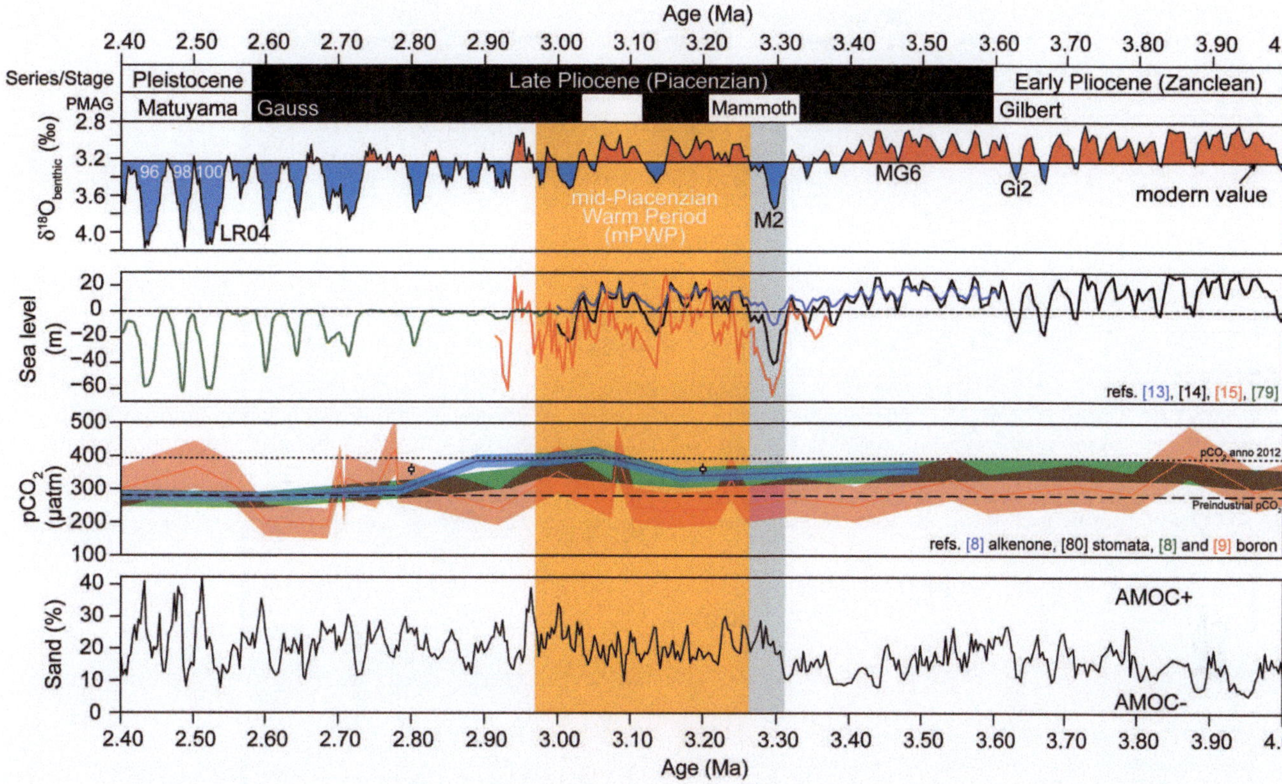

Figure 1. Marine isotope stage M2 in the long-term climate evolution of the Pliocene. (A) Time scale, including palaeomagnetic reversals (PMAG) and the LR04 benthic isotope stack [10], orange shading shows mid-Piacenzian Warm Period (= mid-Pliocene Warm Period), grey shading shows marine isotope stage MIS M2; (B) sea level estimates for the Pliocene to Pleistocene [13–15,79]; (C) Late Pliocene atmospheric carbon dioxide concentrations based on boron, alkenones and leaf stomata [8,9,80]; (D) long-term carbonate-sand record at ODP Site 999 as an indicator for Pacific water flow through the Central American Seaway into the Atlantic and AMOC [27].

[20–23], ice cap expansion in Iceland [24], and possibly in Alaska and the Canadian Rocky Mountains [25] (Figure 2).

Interrupting an interval of global warmth, MIS M2 has been proposed as an early, failed attempt by the Earth's climate to establish a pattern of intense and frequent Northern Hemisphere glaciations [26,27]. It was not until ~500,000 years later that this pattern emerged, likely due to decreasing atmospheric carbon dioxide concentrations during the Late Pliocene [8,28]. The decline in atmospheric carbon dioxide concentrations [7–9], increasing global ice volume [10,11], cooling of ocean surface waters [29–31], and tectonic closure of ocean gateways [27,32] since the Late Miocene may well have ultimately facilitated glaciation in the late Late Pliocene, but these long-term processes are an unlikely cause of the short-lived MIS M2 glaciation. Similarly, variations in astronomical forcing alone cannot explain the intense glaciation of MIS M2 because intervals with similar astronomical forcing occurred throughout the Late Pliocene without leading to intense glaciation. The isolated nature of the MIS M2 glaciation in the otherwise warm climate of the early Late Pliocene must be the result of a specific forcing, unique within this time period.

We established high-resolution palynological and geochemical records from five ocean drilling sites along a southwest–northeast transect in the North Atlantic covering the Caribbean Warm Pool, Gulf Stream, subtropical gyre and North Atlantic Current (NAC) over the interval 3.400–3.180 Ma to determine the role of ocean circulation in causing the extensive glaciation of MIS M2 (Figure 2). Our surface water mass, sea surface temperature (SST), relative salinity reconstructions, and carbonate-sand records provide direct evidence that the unique conditions responsible for glaciation during MIS M2 relate to an increased Pacific-to-Atlantic flow via the Central American Seaway (CAS) prior to MIS M2. This weakened northward heat transport due to a shift of the NAC. The conceptual model proposed here links an open CAS with glaciation in the Northern Hemisphere and contrasts with hypotheses that propose the closure of the CAS as a cause for the intensification of Northern Hemisphere glaciation around 2.6 Ma [33], or as a delaying factor [34] or a precondition for ice sheet expansion in the Northern Hemisphere [27].

Materials and Methods

Samples were collected at the IODP Bremen Core Repository (Germany) and Gulf Coast Repository (College Station, Texas, USA) from five sites constituting a transect between the Caribbean Sea (ODP Site 999), western North Atlantic (DSDP Site 603), and the eastern North Atlantic (DSDP Site 610, IODP Sites U1308 and U1313). The foraminiferal geochemistry data and palynomorph assemblages were acquired from the same samples at each of the five ocean drilling sites, and samples for biomarker (alkenone) analysis were taken from the same sample depths at three sites. All generated dinoflagellate cyst and geochemical proxy data are accessible through the database PANGAEA at http://doi.pangaea.de/10.1594/PANGAEA.804677. Previously published Mg/Ca and dinoflagellate cyst data [26,35] are also available at http://doi.pangaea.de/10.1594/PANGAEA.758710 and http://doi.pangaea.de/10.1594/PANGAEA.758711.

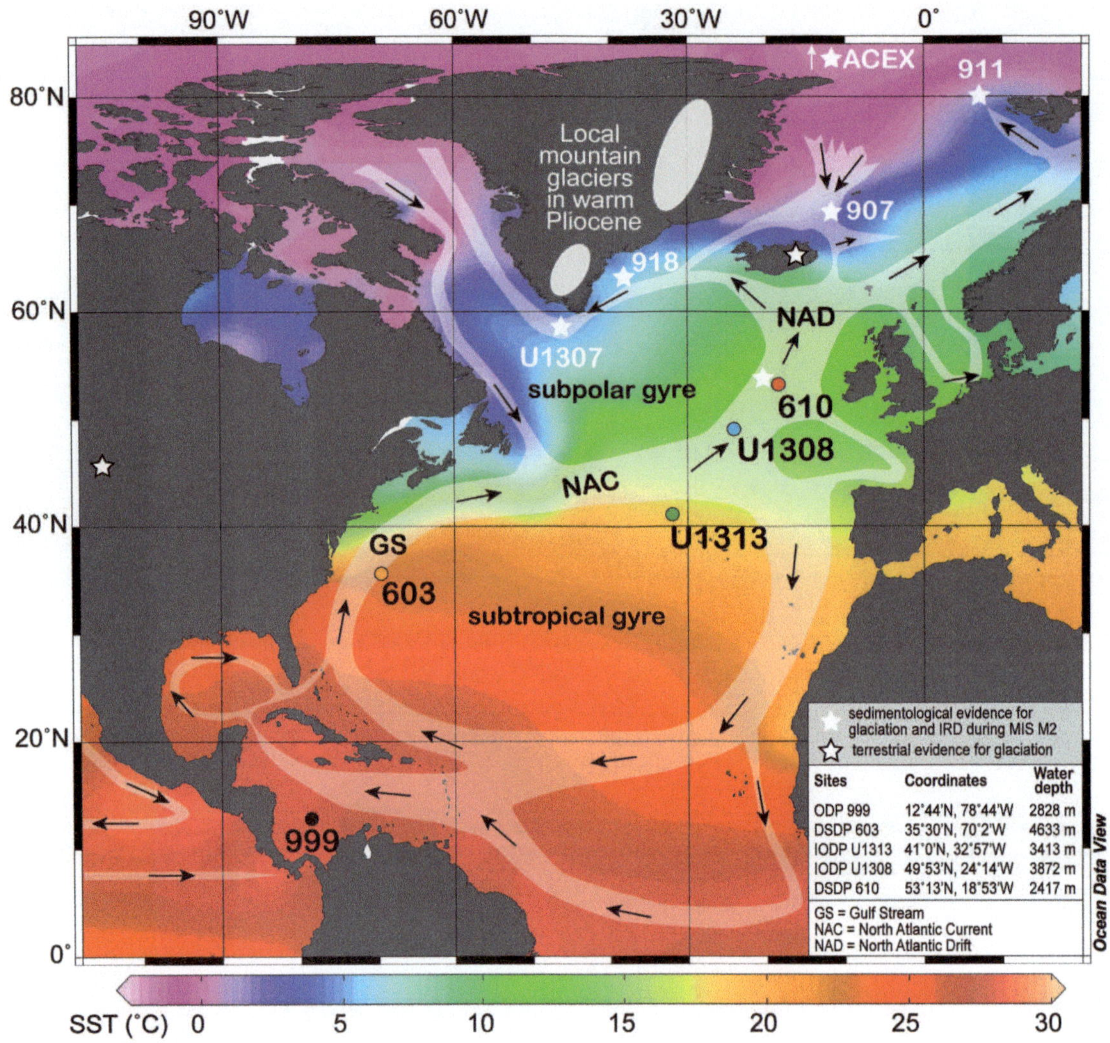

Figure 2. Modern North Atlantic surface circulation with modern sea surface temperatures (World Ocean Atlas 2005 [81]). Each studied site is indicated by the same colour in subsequent figures, and other sites discussed in the text are shown in white. Ice caps on Greenland are schematic representations of Pliocene reconstructions [4].

Dinoflagellate cyst preparation technique and assemblage interpretation

Our laboratory technique allows dinoflagellate cysts and foraminifera to be extracted from the same samples (full details in [26]). Each sample was first wet sieved at 125 µm to concentrate the foraminifera and ensure that the palynomorphs pass through the sieve for further processing. The fraction retained on the sieve (>125 µm) was dried and weighed before being picked for foraminifera. The sediment filtrate (<125 µm) was dried and weighed, and *Lycopodium clavatum* tablets were added before applying standard palynological preparation techniques involving cold HCl and HF acids [36]. No oxidation, alkali or ultrasonic treatments were used. Organic residues were sieved through a 10-µm nylon mesh and strew mounted onto microscope slides using glycerine jelly. Dinoflagellate cysts were counted under 400x magnification with counts varying between 44 and 527 (average 267) specimens per sample. In addition, acritarchs and terrestrial palynomorphs were also enumerated during the dinoflagellate cyst counts. Palynomorph concentrations and error estimates were then calculated based on the palynomorph and *Lycopodium clavatum* counts and the dry weight of the <125 µm fraction [37]. Only relative abundance variations that are statistically significant according to the procedure described in ref. [38] have been used for interpretation. Data presented in refs. [26] and [35] were used alongside our newly generated data (Figure 3).

The assemblage composition of dinoflagellate cysts in core-top samples is largely related to the present-day overlying water masses [39–42], and reflects the interplay between temperature, salinity, nutrients, sea ice cover and light availability. Present-day, last interglacial [43] and Pliocene [26] dinoflagellate cyst assemblages recovered from the eastern North Atlantic consisting of high abundances of *Operculodinium centrocarpum* sensu Wall & Dale (1966) (herein *O. centrocarpum*) all reflect the presence of the North Atlantic Current (NAC). At DSDP Site 610 and IODP Site U1308, *O. centrocarpum* concentrations are highest also when the relative abundances are high, independently corroborating the value of *O. centrocarpum* as a NAC indicator species.

Full authorial citations of species discussed in the text are given in Table 1.

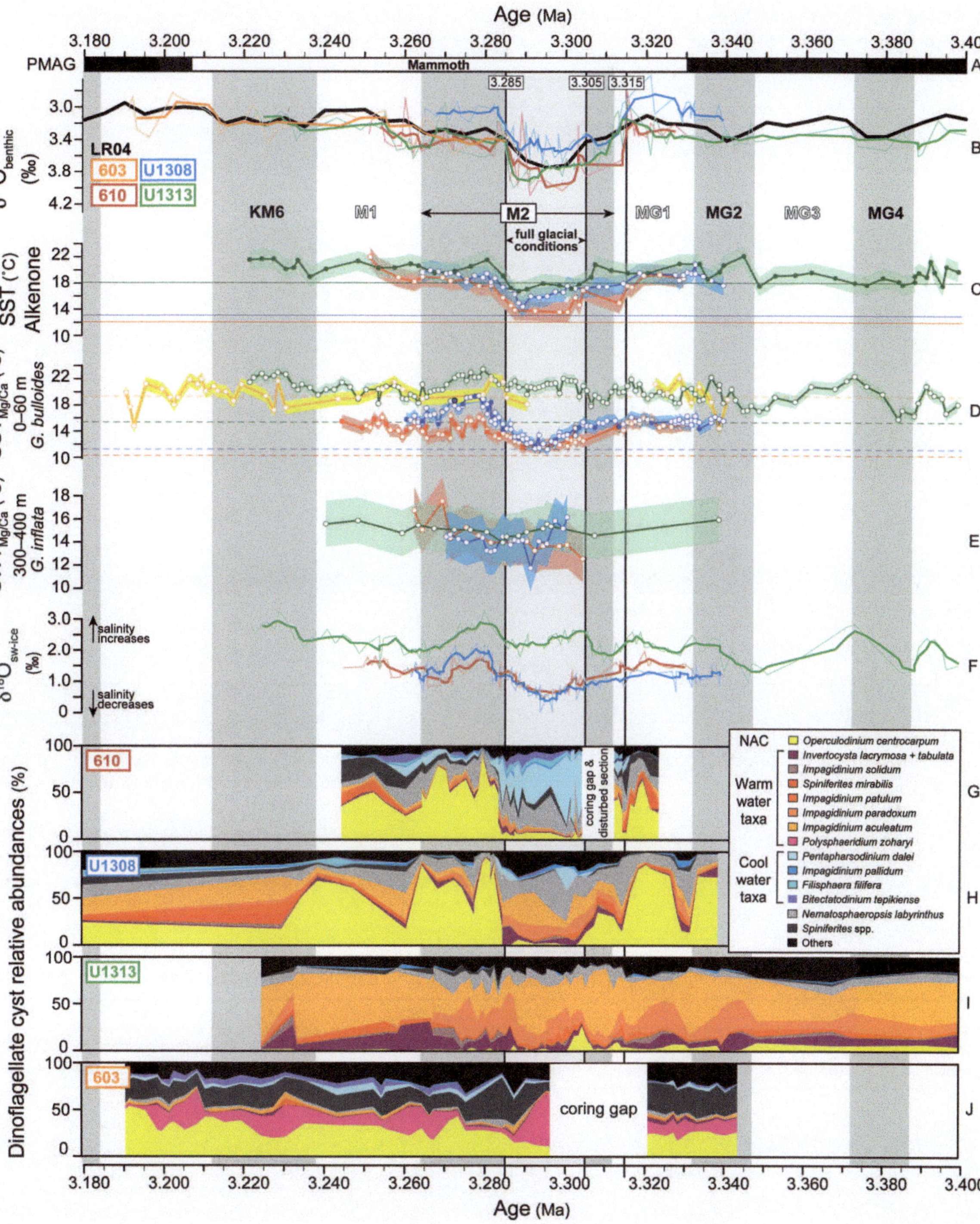

Figure 3. North Atlantic palaeoceanographic proxy records from DSDP Sites 603 and 610, and IODP Sites U1308 and U1313 between 3.400 and 3.180 Ma. Circles with white fill are data points from this study, circles with colour fill are from [30,55] (C) and [26,35] (D). Vertical grey bars represent glacials, white bars are interglacials. (A) palaeomagnetic reversals; (B) benthic isotope age models for each site tuned to the LR04 stack [10] (black line), thin coloured lines are data, thick coloured line is 4-point running mean; (C) alkenone SSTs (0–10 m water depth) including calibration related error (shading), horizontal lines represent modern average annual temperature at 0–50 m water depth for each site; (D) SSTs at 0–60 m water depth based on Mg/Ca of *G. bulloides*, including a 1°C error bar (shading), horizontal lines represent modern average spring (March-April-May) temperature at 0–50 m water depth for each site; (E) $SWT_{Mg/Ca}$ of *G. inflata* at 300–400 m water depth; (F) calculated $\delta^{18}O_{sw\text{-}ice}$ as estimate of salinity; thick coloured lines are a 4-point running mean; (G–J) dinoflagellate cyst assemblage composition. High abundances of *O. centrocarpum* (yellow) indicate an active NAC. Bluish colours = cool-water species, reddish colours = warm-water species; (G) and (H) contain data presented in [26] and [35]. (J) *B. tepikiense* and *F. filifera* are grouped together at DSDP Site 603 and are represented by the colour for *B. tepikiense* (purple); *Impagidinium* cf. *pallidum* is represented as *Impagidinium pallidum*.

Table 1. Dinoflagellate cyst species mentioned in the text and figures: abbreviation, full authorial citation and grouping.

Abbreviation	Full species name
B. tepikiense	*Bitectatodinium tepikiense* Wilson 1973
F. filifera	*Filisphaera filifera* Bujak 1984 emend. Head 1994
I. aculeatum	*Impagidinium aculeatum* (Wall 1967) Lentin & Williams 1981
I. paradoxum	*Impagidinium paradoxum* (Wall 1967) Stover & Evitt 1978
I. pallidum	*Impagidinium pallidum* Bujak 1984
I. patulum	*Impagidinium patulum* (Wall 1967) Stover & Evitt 1978
I. solidum	*Impagidinium solidum* Versteegh & Zevenboom in Versteegh 1995
I. lacrymosa	*Invertocysta lacrymosa* Edwards 1984
I. tabulata	*Invertocysta tabulata* Edwards 1984
N. labyrinthus	*Nematosphaeropsis labyrinthus* (Ostenfeld 1903) Reid 1974
O. centrocarpum	*Operculodinium centrocarpum* sensu Wall & Dale 1966
O. israelianum	*Operculodinium israelianum* (Rossignol 1962) Wall 1967
P. dalei	Cyst of *Pentapharsodinium dalei* Indelicato & Loeblich III 1986
P. zoharyi	*Polysphaeridium zoharyi* (Rossignol 1962) Bujak et al. 1980
RBC	Round brown cysts
S. mirabilis	*Spiniferites mirabilis* (Rossignol 1964) Sarjeant 1970, and
	Spiniferites hyperacanthus (Deflandre & Cookson 1955) Cookson & Eisenack 1974
Spiniferites/ Achomosphaera spp.	*Spiniferites* spp. Mantell 1850, and
	Achomosphaera spp. Evitt 1963
Others	Contains all other dinoflagellate cyst taxa counted.

Geochemistry: $\delta^{18}O$ and Mg/Ca of foraminifera, calculating and interpreting sea surface temperature and relative salinity ($\delta^{18}O_{sw\text{-}ice}$)

Foraminifera were picked from the >125 μm dry fraction of each sample. Planktonic foraminifera isotope data were measured using a Finnigan MAT 251 mass spectrometer at the Isotope Laboratory, Geosciences Department, University of Bremen using five specimens per sample of *Globigerina bulloides* (250–315 μm) for Sites 603, 610, U1308 and U1313 and five specimens per sample of *Globigerinoides sacculifer* (250–355 μm) for Site 999. Benthic foraminiferal isotope data are based on at least one >250 μm specimen of *Cibicidoides wuellerstorfi* or *Uvigerina perigrina* per sample. *Cibicidoides wuellerstorfi* $\delta^{18}O$ values have been corrected by adding 0.64 ‰ [44]. The standard deviation of the analyses is based on an

Table 2. Palaeomagnetic data for the reversal at the base of the Mammoth Subchron in DSDP Hole 603C (depth indicated in bold).

Site	Hole	Core, type		Section	Half	Top (cm)	Bottom (cm)	Depth (mbsf)	Sample name	Inclination (°)	Inc (low)	Inc (high)	Comments
603	C	17	X	1	W	91	93	136.51	C17S1091	−8.4	−20.8	4.8	Questionable reliability
603	C	17	X	2	W	47	49	137.57	C17S2047	−19.3	−43.1	13.2	Questionable reliability
603	C	17	X	3	W	11	13	138.71	C17S3011	−32.9	−35.0	−30.7	
603	C	17	X	3	W	42	44	139.02	C17S3042	−15.4	−18.1	−12.6	
603	C	17	X	3	W	102	104	139.62	C17S3102	−50.0	−51.7	−48.3	
603	C	17	X	3	W	107	109	139.67	C17S3107	−24.7	−29.3	−19.6	
603	C	17	X	4	W	119	121	141.29	C17S4119	−19.7	−34.4	−1.7	Questionable reliability
603	C	17	X	5	W	41	43	142.01	C17S5041	16.0	12.0	19.9	
603	C	17	X	6	W	42	44	143.52	C17S6042	14.8	11.1	18.3	
603	C	18	X	1	W	11	13	**145.31**	C18S1011	−22.9	−33.0	−11.0	
603	C	18	X	1	W	99	101	**146.19**	C18S1099	28.6	27.1	30.1	
603	C	18	X	1	W	117	119	146.37	C18S1117	39.6	36.5	42.5	

The reversal at the top of the Mammoth Subchron could not be assigned.

in-house Solnhofen carbonate standard with a value of 0.07‰. Values are reported relative to that of the Vienna Pee Dee Belemnite (VPDB) calibrated using National Bureau of Standards (NBS) 18, 19, and 20 standards.

For Mg/Ca measurements, we used 20–25 specimens per sample of *G. bulloides* (250–315 µm) or *G. sacculifer* (250–355 µm) and 20 specimens per sample of *Globorotalia inflata* (250–400 µm) from Sites 610, U1308 and U1313. The cleaning procedure for Mg/Ca measurements is described elsewhere [45]. After dissolution in 0.5 mL 0.075 M QD HNO_3, the samples were centrifuged and diluted for analysis on an ICP-OES (Perkin Elmer Optima 3300R) at the Geosciences Department, University of Bremen. The analytical precision of the Mg/Ca analyses for *G. bulloides*, *G. sacculifer*, and *G. inflata* combined was 0.17% (n = 459). Reproducibility based on replicate samples (n = 32) of both *G. bulloides* and *G. sacculifer* was ±0.11 mmol/mol (~3.3%). The validity of analyses was checked by analysing an artificial in-house standard to monitor drift of the ICP-OES (Mg/Ca = 2.93 mmol/mol) and the limestone standard ECRM752-1 (Mg/Ca = 3.75 mmol/mol) to allow inter-laboratory comparison [46]. Al/Ca, Fe/Ca, and Mn/Ca were simultaneously analysed with Mg/Ca to prevent contaminated samples from being included in the interpretation. We used the following calibration, established from core-top sediment samples in the North Atlantic, to transform the foraminiferal Mg/Ca ratios of *G. bulloides* into $SST_{Mg/Ca}$: Mg/Ca = 0.52 exp 0.10 T [47]. We interpret the $SST_{Mg/Ca}$ value of *G. bulloides* as spring to summer SSTs of the upper 60 m of the water column [26,48,49] because the oxygen isotope composition of *G. bulloides* reflects the northward-migrating phytoplankton spring bloom in the North Atlantic [47,50]. The $SST_{Mg/Ca}$ value of *G. sacculifer* represents the annual mixed-layer temperature of the upper 75 m of the water column for Caribbean Site 999 [51]. Mg/Ca values were transformed into palaeo-seawater temperatures using the following equation: Mg/Ca = 0.491 exp 0.033 T [52]. Although *G. inflata* calcifies throughout the water column, the $SWT_{Mg/Ca}$ based on mostly non-encrusted *G. inflata* represents the temperature of the permanent thermocline [53]. We used the following calibration to calculate temperatures: Mg/Ca = 0.72 exp 0.076 T. Combining analytical and calibration errors, we estimate the error on Mg/Ca palaeotemperature reconstruction for shallow-dwelling foraminifera as ±1.0–1.5°C [47], whereas for the deeper-dwelling *G. inflata* the error is estimated to be ±2–2.5°C [53].

The oxygen isotope composition of seawater ($\delta^{18}O_{sw}$) was calculated via a standard formula [54]. Since Mg/Ca and $\delta^{18}O$ were measured on the same planktonic foraminiferal species, the possible effects of seasonality and habitat differences are minimised. We used the LR04 global benthic foraminiferal $\delta^{18}O$ stack [10] as an approximation for changes in ice volume over the studied interval. After normalizing the LR04 record, we subtracted it from $\delta^{18}O_{sw}$, resulting in a $\delta^{18}O_{sw\text{-}ice}$ record that approximates local variations in salinity.

Alkenones

All alkenone data from Sites 610 and 1308 are new, whereas data from IODP Site U1313 have been published earlier [30,55] (Figure 3C). The modified alkenone unsaturation index $U_{37}^{k'}$ [56,57] was measured using a GC/TOF-MS system [58] on separate samples, taken from the same depths as those used for foraminiferal Mg/Ca and dinoflagellate cyst analyses. $U_{37}^{k'}$ in combination with a global core-top calibration was used to calculate annual mean SST (top 10 m) [59]. The analytical technique, calibration and reliability of alkenone-based SSTs for the Pliocene is detailed elsewhere [30,55]. The calibration error on the alkenone SSTs is ~1.5°C [59].

The global core-top calibration gives the highest correlation with annual mean SSTs, but locally alkenone-based SSTs could reflect the temperature of the growing season (spring in the North Atlantic) [60]. Although this affects the absolute SST estimates, it does not influence the relative trends in our records. The exception would be if the alkenone producers shifted their production season on a glacial/interglacial basis, but there is no evidence for such behaviour. However, if such shifts did occur during glacials the alkenone producers would have delayed their production towards summer to avoid the colder spring surface conditions. This implies that the cooling observed in the alkenone records during MIS M2 would actually underestimate the true cooling.

Carbonate sand fraction

We generated high-resolution carbonate sand fraction data from ODP Hole 999A over the study interval. In addition, we used the available low resolution, long-term carbonate sand fraction record of the same site [27]. The sand content (>63 µm) of deep-sea carbonates is considered [27] a sensitive indicator of changes in carbonate dissolution: sand content (foraminifer tests) decreases as dissolution progresses. A low-carbonate sand fraction was interpreted to reflect a poorly-ventilated deep Caribbean water mass. Carbonate dissolution at Site 999, caused by entry of Antarctic Intermediate Water (AAIW) into the Caribbean Basin in place of North Atlantic Deep Water (NADW), implies an open Central American Seaway and a weak overturning circulation [27].

Palaeomagnetic measurements

The positions of magnetic reversals for the Mammoth Subchron in DSDP Holes 603C and 610A [61], and IODP Hole 1308C [62] were re-measured in this study to increase precision by analysing discrete, oriented samples at 4–23 cm resolution (Tables 2–4). Oriented cubic polystyrene boxes (7.2 cm^3) were taken, avoiding visible mineral concretions and areas influenced by the coring process, from the working halves of Sections 603C-17X1 to 603C-18×1 (12 samples between 136.51 and 146.37 mbsf), Sections 610A-17H3 to 610A-17H4 (17 samples between 156.74 and 158.75 mbsf), Sections 610A-17H6 to 610A-18H1 (19 samples between 161.20 and 163.85 mbsf), and Sections U1308C-26H2 to U1308C-26H5 (20 samples between 230.48 and 239.45 mbsf). The discrete samples were measured at the Geosciences Department, University of Bremen on a cryogenic magnetometer (model 2G Enterprises 755 HR). The natural remanent magnetization (NRM) was demagnetized in nine steps (10–100 mT), and inclination and relative declination, and their confidence intervals were determined from line fits of straight-line segments in a Zijderveld diagram. Note that absolute declination depends on section and core orientations, whereas inclination values rely on the fact that the drill hole is to a very good approximation perpendicular. All ages of the magnetic reversals are according to the ATNTS 2004 [63].

In DSDP Hole 610A, the upper boundary of the Mammoth Subchron was found between 158.35 mbsf (positive inclination) and 158.75 mbsf (negative inclination). The reversal at the base of the Mammoth Subchron is more difficult to identify due to a coring gap between Cores 610A-17H and 610A-18H, and disturbed sediment in the upper 25 cm of Section 610A-18H1 [64], but must be located between 161.85 mbsf (negative inclination) and 163.11 mbsf (positive inclination). The reversals bounding the Mammoth Subchron in IODP Site U1308 are between 254.56 and 255.46 mcd at the top, and between 262.91

Table 3. Palaeomagnetic data for the reversals at the base and top of the Mammoth Subchron in DSDP Hole 610A.

Site	Hole	Core, type		Section	Half	Top (cm)	Bottom (cm)	Depth (mbsf)	Sample name	Inclination (°)	Inc (low)	Inc (high)	Comments
610	A	17	H	3	W	73	75	156.74	Mam-N1821	41.2	10.6	57.4	
610	A	17	H	3	W	88	90	156.89	Mam-N1822	48.9	36.4	57.2	
610	A	17	H	3	W	98	100	156.99	Mam-N1823	69.6	44.4	77.2	
610	A	17	H	3	W	108	110	157.09	Mam-N1824	62.9	53.9	68.5	
610	A	17	H	3	W	115	117	157.16	Mam-N1825	53	45.9	58.3	
610	A	17	H	3	W	127	129	157.28	Mam-N1826	56.6	50.7	61.1	
610	A	17	H	3	W	139	141	157.40	Mam-N1827	55.4	30.4	66.6	
610	A	17	H	4	W	6	8	157.57	Mam-N1828	51.9	39.2	60	
610	A	17	H	4	W	21	23	157.72	Mam-N1829	42.1	−42.5	69.8	
610	A	17	H	4	W	26	28	157.77	Mam-N1830	45.4	29	55.8	
610	A	17	H	4	W	36	38	157.87	Mam-N1831	49.8	34.3	59.4	
610	A	17	H	4	W	52	54	158.03	Mam-N1832	78.2	71.2	81.5	
610	A	17	H	4	W	69	71	158.20	Mam-N1833	48	40.1	54.1	
610	A	17	H	4	W	84	86	158.35	Mam-N1834	45.3	25	57.3	
610	A	17	H	4	W	92	94	**158.43**	Mam-N1835	22.7	2.5	38.4	
610	A	17	H	4	W	114	116	**158.65**	Mam-N1836	−17.2	−40	12.4	
610	A	17	H	4	W	124	126	158.75	Mam-N1837	−30.7	−41.8	−16.5	
610	A	17	H	6	W	69	71	161.20	Mam-N1838	−66.5	−70	−61.6	
610	A	17	H	6	W	88	90	161.39	Mam-N1839	−58.7	−61.6	−55.3	
610	A	17	H	6	W	93	95	161.44	Mam-N1840	−50.2	−53.7	−46.1	
610	A	17	H	6	W	101	103	161.52	Mam-N1841	−56.8	−58.9	−54.5	
610	A	17	H	6	W	117	119	**161.68**	Mam-N1842	64.1	62.3	65.7	
610	A	17	H	6	W	134	136	**161.85**	Mam-N1843	−39.2	−41	−37.3	
610	A	17	H	CC	W	11	13	**161.97**	Mam-N1844	20.8	10.9	29.6	
610	A	18	H	1	W	2	4	**162.63**	Mam-N1845	44.1	43	45.3	Disturbed?
610	A	18	H	1	W	8	10	**162.69**	Mam-N1846	−3.7	−6.5	−1	Disturbed?
610	A	18	H	1	W	14	16	**162.75**	Mam-N1847	43.4	41.8	45	Disturbed?
610	A	18	H	1	W	23	25	**162.84**	Mam-N1848	7.1	6.1	8.1	Disturbed?
610	A	18	H	1	W	27	29	**162.88**	Mam-N1849	−20.6	−23.9	-17.1	Disturbed?
610	A	18	H	1	W	31	33	**162.92**	Mam-N1850	−71.3	−74.9	−65.6	
610	A	18	H	1	W	41	43	**163.02**	Mam-N1851	−57.1	−59.4	−54.4	
610	A	18	H	1	W	50	52	**163.11**	Mam-N1852	34.2	31	37.3	
610	A	18	H	1	W	68	70	163.29	Mam-N1853	9.7	−8.9	26.6	
610	A	18	H	1	W	86	88	163.47	Mam-N1854	57.3	34.5	67.6	
610	A	18	H	1	W	101	103	163.62	Mam-N1855	37.4	−12.4	60.2	
610	A	18	H	1	W	124	126	163.85	Mam-N1856	17.5	4.7	28.7	

Depths in bold demonstrate the position of the reversals.
Note: The reversal at the base of the Mammoth Subchron is difficult to identify due to a core gap and potentially disturbed sediments in the upper 0–30 cm of Section 610A-18H1.

and 264.41 mcd at the bottom [62]. Our re-assessment places the top of the subchron between 254.59 and 254.67 mcd, and the bottom between 263.21 and 263.46 mcd. The offset between predicted and measured position in IODP Site U1308 is small and considered to be within the margin of accuracy of the methods. At IODP Site U1313, the reported palaeomagnetic reversal of the base of the Mammoth Subchron lies at 153.68 mcd ±0.1 m [65], within the glacial maximum of MIS M2. The offset may be the result of the field geometry and distance between the sites, variation in the magnetic lock-in time and depth in the sediment, and the sites used to determine the Mammoth Subchron in the Geomagnetic Polarity Time Scale.

Age models

An age model was established for each hole (Figures S1–S4) by tuning its benthic foraminiferal stable oxygen isotope record to the LR04 benthic foraminiferal isotope stack [10], with the palaeomagnetic reversals as guidelines only, using the software program AnalySeries 2.0.4.2 [66]. The accuracy of each age model depends on the accuracy of the LR04 benthic stack which is estimated at 15

Table 4. Palaeomagnetic data for the reversals at the base and top of the Mammoth Subchron in IODP Hole U1308C.

Site	Hole	Core, type		Section	Half	Top (cm)	Bottom (cm)	Depth (mbsf)	Sample name	Inclination (°)	Inc (low)	Inc (high)	Comments
1308	C	25	H	6	W	7	9	**230.48**	Mam-N1857	44.7	35.1	51.9	
1308	C	25	H	6	W	15	17	**230.56**	Mam-N1858	−38.1	−42.7	−32.8	
1308	C	25	H	6	W	25	27	230.66	Mam-N1859	−76.6	−79.9	−70.5	
1308	C	25	H	6	W	35	37	230.76	Mam-N1860	−62	−64.5	−59.1	
1308	C	25	H	6	W	42	44	230.83	Mam-N1861	−57	−58.6	−55.2	
1308	C	25	H	6	W	48	50	230.89	Mam-N1862	−51.8	−53.6	−49.8	
1308	C	25	H	6	W	56	58	230.97	Mam-N1863	−60.6	−61.5	−59.6	
1308	C	25	H	6	W	65	67	231.06	Mam-N1864	−69	−70.5	−67.3	
1308	C	25	H	6	W	77	79	231.18	Mam-N1865	−54.8	−56.4	−53.1	
1308	C	25	H	6	W	87	89	231.28	Mam-N1866	−72.4	−73.1	−71.6	
1308	C	25	H	6	W	94	96	231.35	Mam-N1867	−67	−67.7	−66.3	
1308	C	26	H	5	W	33	35	238.74	Mam-N1868	−51.7	−52.4	−51	
1308	C	26	H	5	W	40	42	238.81	Mam-N1869	−75.4	−76.1	−74.6	
1308	C	26	H	5	W	48	50	238.89	Mam-N1870	−54.8	−56.7	−52.8	
1308	C	26	H	5	W	56	58	238.97	Mam-N1871	−57.9	−61	−54.2	
1308	C	26	H	5	W	63	65	239.04	Mam-N1872	−66.5	−69	−63.5	
1308	C	26	H	5	W	72	74	239.13	Mam-N1873	13	−31.1	46.8	
1308	C	26	H	5	W	83	85	**239.24**	Mam-N1874	−42.9	−54.4	−24.8	
1308	C	26	H	5	W	94	96	**239.35**	Mam-N1875	61	0.6	74.4	
1308	C	26	H	5	W	104	106	239.45	Mam-N1876	65.6	62.9	67.9	

Depths in bold demonstrate the position of the reversals.

kyr between 3 and 4 Ma [10] and the accuracy of the graphic correlation of the benthic δ18O records with the LR04 tuning target. The tie-points used for the age models of each hole and their correlation coefficients are presented in Tables 5–9 and Figures S1–S5. The ages in the Results and Discussion section are reported with high precision (3 decimals) to demonstrate (1) the relative age difference between samples within one site, and (2) the relative age of events in relation to the onset (~3.315 Ma), full glaciation (~3.305–3.285 Ma), and termination of MIS M2 (~3.285 Ma). These ages should neither be considered as absolute ages, nor as evidence for suborbital age control.

The original age model of ODP Site 999 [27] is based on correlating the benthic $\delta^{18}O$ record to the astronomically dated benthic $\delta^{18}O$ records from equatorial East Pacific ODP Site 846 [67] and equatorial East Atlantic ODP Site 659 [68] for the time interval 5–2 Ma. The existing age model was updated [51] to the newly generated, orbitally-tuned age model of East Pacific IODP Site 1241 [69] (Figure S5). For this study, we used the LR04 benthic $\delta^{18}O$ stack [10] to fine-tune the glacial-interglacial transitions around MIS M2.

Table 5. Tie-points for the age model of DSDP Hole 610A.

Depth (mbsf)	Age (Ma)
158.60	3.207
159.00	3.233
159.32	3.237
159.90	3.253
160.20	3.263
161.27	3.285
161.56	3.290
161.75	3.301
163.05	3.315
163.89	3.326
164.09	3.332
167.08	3.596

Note: mbsf = metres below sea floor.

Table 6. Tie-points for the age model of IODP Hole U1308C.

Depth (mcd)	Age (Ma)
254.66	3.207
259.27	3.265
261.11	3.284
262.35	3.302
263.25	3.320
263.38	3.327
264.70	3.340

Note: mcd = metres composite depth.

Table 7. Tie-points for the age model of IODP Site U1313 (primary splice).

Depth (mcd)	Age (Ma)
149.92	3.224
150.83	3.237
153.07	3.285
154.39	3.311
157.23	3.372
159.72	3.419

Note: mcd = metres composite depth.

Table 8. Tie-points for the age model of DSDP Hole 603C.

Depth (mbsf)	Age (Ma)
136.70	3.194
138.25	3.214
139.79	3.237
140.27	3.252
142.85	3.286
145.71	3.330

Note: mbsf = metres below sea floor.

Results and Discussion

Events during interglacial MIS MG1 leading to glaciation

The early Late Pliocene was warmer than today, and prior to ~3.315 Ma our geochemical proxies and dinoflagellate cyst assemblages demonstrate a surface circulation comparable to today's but with elevated temperatures in the high-latitude North Atlantic. We record an active Gulf Stream over Site 603 as illustrated by the high SSTs (ca. 19.5°C) and the presence of *O. centrocarpum* and such warm water dinoflagellate cyst taxa as *Impagidinium aculeatum*, *I. paradoxum*, *I. patulum*, *I. solidum*, and *Polysphaeridium. zoharyi* which are also present there today [41]. Warm (ca. 20°C) and oligotrophic surface waters at the subtropical gyre Site U1313 are reflected in the dominance of *I. aculeatum*, *I. paradoxum*, *I. patulum*, and *Invertocysta* spp. An active NAC brought warm waters (15.2–18.6°C in the uppermost 60 m) northward over Sites U1308 and 610 (Figure 3C, 3D, 3G, 3H), expressed in the dinoflagellate cyst assemblages by the dominance of *O. centrocarpum* and the persistent presence of the warm water species *Spiniferites mirabilis*. The less steep meridional SST gradient compared to present, especially visible in the SST_{alk} and to lesser extent in the $SST_{Mg/Ca}$ (Figure 3C, 3D, 4), indicates generally warmer conditions in the higher latitudes compared to today.

Although deeper-water exchange via the CAS had been restricted since ~4.6 Ma [27], shallow Pacific-to-Atlantic exchange occurred well into the Late Pliocene [51]. This implies that Atlantic meridional overturning circulation (AMOC) [6] was able to function even when the CAS was partially open. Following a maximum in AMOC due to minimal Pacific-to-Atlantic through-flow around 3.6 Ma, a gradual increase in through-flow via an open CAS culminated immediately prior to MIS M2 [27,51]. At Caribbean Site 999, we record between ~3.320 and ~3.315 Ma a drop in SST and salinity ($\delta^{18}O_{sw\text{-}ice}$) (Figure 5D, 5E), a low carbonate sand-fraction (Figure 5F), and high productivity evidenced by high dinoflagellate cyst concentrations dominated by heterotrophic species (round brown cysts; Figure 5G, 5H). The low carbonate sand-fraction indicates a poorly ventilated deep Caribbean water mass and carbonate dissolution caused by entry of Antarctic Intermediate Water (AAIW) into the Caribbean Basin in favour of North Atlantic Deep Water (NADW) – interpreted as evidence of a weak overturning circulation [27]. The drop in SST and salinity point to an increased inflow of cooler, less saline Pacific waters to the Caribbean. Furthermore, the inferred high productivity is fully consistent with nutrient-rich waters from the Pacific entering the Caribbean [70]. Considered altogether, this evidence shows that Pacific-to-Atlantic through-flow via the CAS during interglacial MIS MG1 exceeded a critical threshold, thereby reducing the AMOC. This was likely aided by the high sea levels at that time [12,15] and a longer-term gradual weakening of the thermohaline circulation since 3.6 Ma [27] that brought the climate system closer to a tipping point. Prior to MIS MG1 (Figure 1D, 5C) high sea levels also occurred, but Pacific-to-Atlantic through-flow appears not to have weakened the NAC, and glaciation in the Northern Hemisphere remained restricted.

During maximal Pacific inflow via the open CAS during interglacial MIS MG1 (~3.315–3.320 Ma), contemporaneous changes occurred in the North Atlantic surface circulation. At ~3.315 Ma, a major reduction in northward flow of warm NAC waters is reflected at Sites 610 and U1308 by a major turnover of the dinoflagellate cyst assemblages within 1–2 kyrs, and an initial cooling of the surface waters is registered (Figure 3C, 3D, 3G, 3H). This corroborates modelling studies showing that an open CAS results in a weakened AMOC and hence northward heat transport [71,72]. Between ~3.315 and 3.305 Ma, a persistent NAC influence is recorded at Site U1308 when SSTs decreased further and *O. centrocarpum* (our NAC tracer) remained present in low abundance. At the same time we find a peak abundance of *O. centrocarpum* (Figure 3I) and surface-water cooling (Figure 3D) at subtropical gyre Site U1313. We interpret this sequence of events as an initial reduction in northward flowing warm water of the NAC at ~3.315 Ma, followed by a gradual southward deflection of the NAC between ~3.315 and 3.305 Ma. It is important to note that the initial reduction in northward transport of warm NAC water occurred at ~3.315 Ma within the interglacial MIS MG1, well before the MIS M2 glacial maximum at 3.295 Ma.

The southward shift of the NAC prior to MIS M2 led to the cessation of northward heat transport during the full glacial conditions of MIS M2 (3.305–3.285 Ma). During the glacial conditions, subtropical gyre circulation persisted as attested by a

Table 9. Tie-points for the age model of ODP Site 999.

Old age (Ma)	New age (Ma)
3.205	3.205
3.239	3.245
3.276	3.280
3.296	3.295
3.319	3.320
3.342	3.340
3.355	3.365
3.371	3.375

Note: old age from ref. [78], new age from this study.

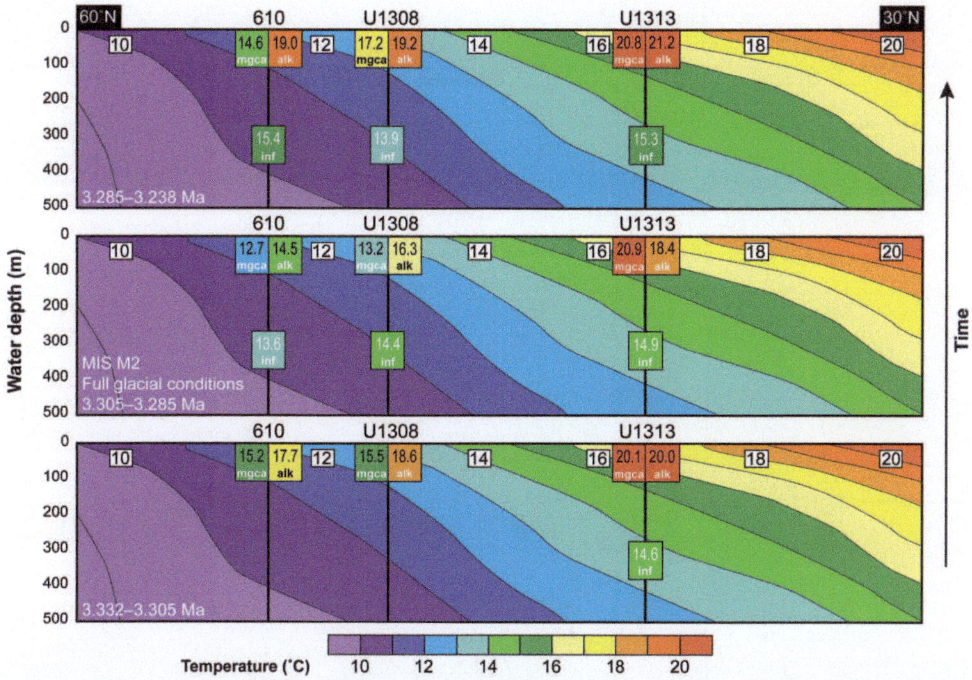

Figure 4. Reconstructed sea-surface temperatures for the studied intervals prior to, during and after the full glacial conditions of MIS M2 along an eastern North Atlantic transect through DSDP Site 610, and IODP Sites U1308 and U1313 from 60°N, 10°W to 30°N, 30°W. Background represents present day sea-water temperatures of the upper 500 m (from WOA2005, [81]). The insets below each site represent the surface water temperature based on alkenones (alk; surface water 0–10 m), Mg/Ca ratios of *Globigerina bulloides* (mgca; mixed layer 0–60 m) and sea-water temperature at 300–400 m based on Mg/Ca ratios of *Globorotalia inflata* (inf). The surface waters (SST_{alk}, $SST_{Mg/Ca}$ *G. bulloides*) show important cooling during MIS M2 with a steep N–S temperature gradient established. Low resolution sea-water temperature ($SWT_{Mg/Ca}$) reconstructions of *Globorotalia inflata* show that water at the permanent thermocline (300–400 m depth) remained stable and warm throughout the entire studied period. The glacial had no major effect on the deeper surface waters, except possibly at Site U1308 where one sample recorded temperatures as low as 11.8°C during MIS M2. The average values of 13.5, 14.2 and 15.1°C at Site 610, U1308 and U1313 respectively over the entire period illustrate that the entire upper water column during the Pliocene was warmer than today in the North Atlantic.

continuance of the Gulf Stream at Site 603 where no major changes in the dinoflagellate cyst assemblages were recorded (Figure 3J). The contrasting proxy evidence of cool surface SST_{alk} [30], warmer mixed-layer $SST_{Mg/Ca}$ and largely unchanged dinoflagellate cyst assemblages at subtropical gyre Site U1313 is difficult to interpret (Figure 3C, 3D, 3I). During the earlier glacials MIS MG4 and MG2, both geochemical proxies record a cooling, suggesting a fundamentally different oceanography for MIS M2. It is not known whether the divergence in $SST_{Mg/Ca}$ values during MIS M2 was caused by different a genotype of *G. bulloides* becoming dominant in a changed oceanographic setting [73]. Irrespective of the ultimate cause, we consider the contrasting SST proxy records in combination with the palynological data as evidence of a southward shift of the NAC that affected Site U1313 during MIS M2 but not during prior glacials.

Glaciation in the Northern Hemisphere during MIS M2

At the two northern sites, dinoflagellate cyst assemblages indicate subpolar conditions (*Bitectatodinium tepikiense*, *Filisphaera filifera*, *I. pallidum*, *Nematosphaeropsis labyrinthus*, *Pentapharsodinium dalei*) at Site 610 and oligotrophic conditions (*I. aculeatum*, *I. paradoxum*) at Site U1308 (Figure 3G, 3H), while surface waters at both sites cooled by 3–4°C to temperatures only just higher than today (Figure 3C, 3D). This cooling at the northern sites established a steep latitudinal SST gradient in the North Atlantic (Figure 4), causing the thermal isolation of Greenland from northward heat transport. As a comparison, a 3–4°C cooling of the Nordic Seas was necessary for the last glacial inception (~115,000 years ago) in Scandinavia [74]. The increased meridional SST gradient will have reduced air temperature and increased snowfall over most of North America, both factors favourable to ice sheet inception [29]. We demonstrate that sufficiently cool surface waters were present in the northern high latitude oceans, and propose that these were crucial for the glaciation in the Northern Hemisphere during MIS M2. The moisture required to build a large ice sheet in the Northern Hemisphere was presumably already present in the atmosphere, because Pliocene climates were generally wetter than today [75]. It is nevertheless likely that after the southward shift of the NAC and cooling of the northern high-latitude surface waters, carbon cycle (vegetation, CO_2) [28,71] and perhaps sea ice (albedo) feedbacks also contributed to the major glaciation during MIS M2. Nevertheless, the extent of Northern Hemisphere glaciation during MIS M2 remained smaller than a typical Quaternary glaciation, but may have been larger than at present. The higher $\delta^{18}O_{benthic}$ values during MIS M2 compared to today (Figure 1A; 3.74‰ vs. 3.23‰, [10]) indeed imply that MIS M2 ice sheets were larger than today. Indirect evidence of expanded ice sheets in the Northern Hemisphere is found in several sediment and ice-rafted debris records from the Arctic Ocean, Nordic Seas and northern North Atlantic [20–23] which indicate that the Greenland and Svalbard/Barents Sea ice sheets reached the coastline. Glacial deposits on Iceland [24] and possibly also in the Canadian Rocky Mountains and Alaska [25] demonstrate the presence of ice caps there. With SSTs approaching present day values (Figure 3C, 3D) and a Holocene-like Arctic climate prevailing during MIS M2 [16], the development of a significant

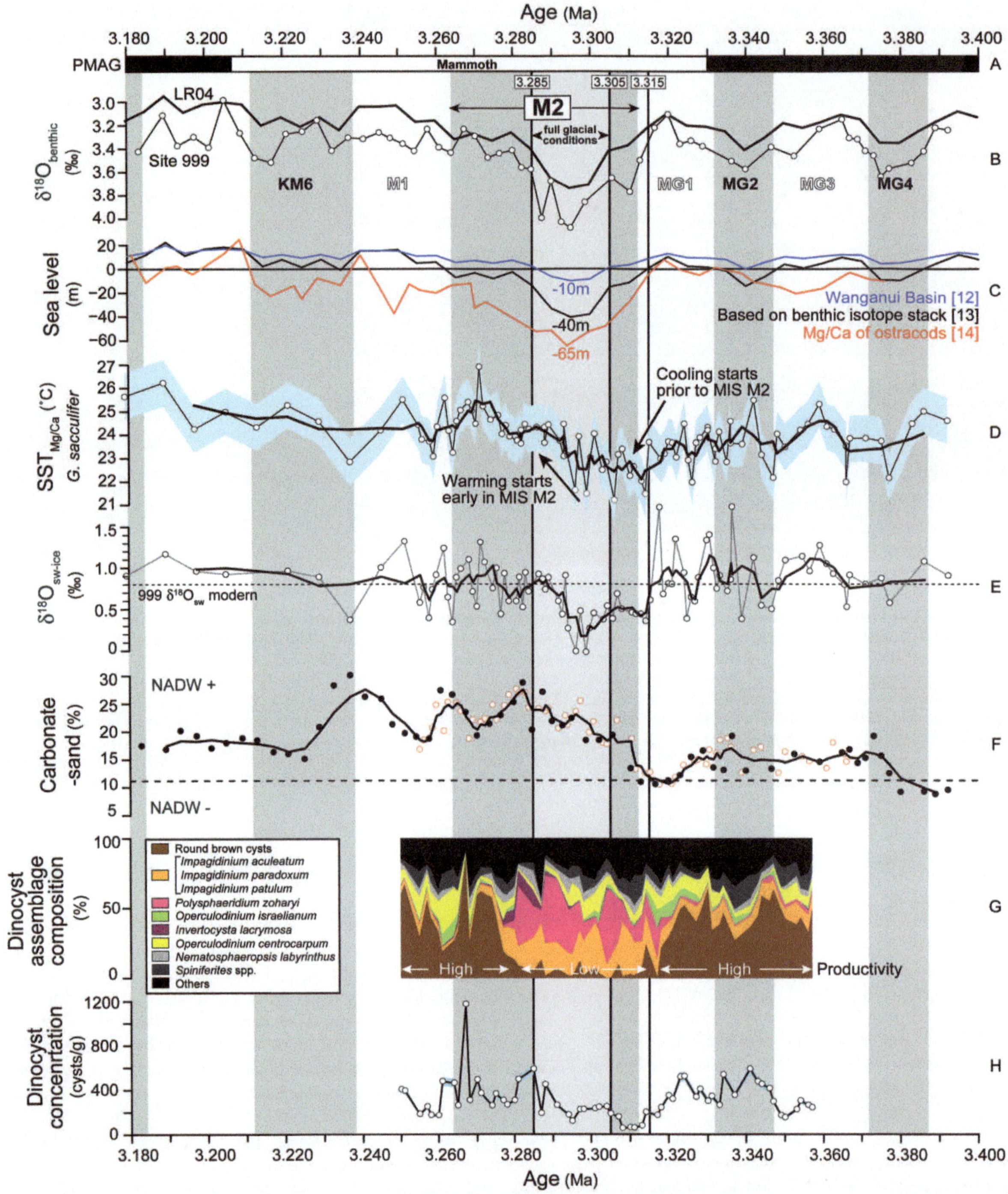

Figure 5. Caribbean Sea palaeoceanographic proxy records from ODP Site 999 between 3.400 and 3.180 Ma. Vertical grey bars represent glacials, vertical white bars are interglacials. All circles with white fill are data from this study. (A) Palaeomagnetic reversals; (B) benthic isotope age model for Site 999 tuned to the LR04 stack [10] (black line); (C) sea level estimates; (D) SST-$_{Mg/Ca}$ of *G. sacculifer*, thick black line represents 4-point running mean, shading represents calibration error; (E) $\delta^{18}O_{sw\text{-}ice}$ estimate of salinity; (F) carbonate-sand fraction, filled black circles are data from [27], thick line represents 4-point running mean. Lower (higher) values reflect decreased (in-creased) North Atlantic Deep Water (NADW) influence at the site. (G) dinoflagellate cyst assemblage composition; presence of round brown cysts indicate high productivity and inflow of Pacific water; (H) dinoflagellate cyst concentration, including error bar (light blue shading).

North American ice sheet is unlikely. Therefore, to explain the observed ~0.5‰ benthic isotope shift [10], a considerable expansion of the Antarctic ice sheet must have occurred also [16]. Nevertheless, the possibility of an ice cap in North America during MIS M2 should not be excluded given the evidence of an ice cap in the North American interior that did not reach the North Atlantic coastline at ~3.5 Ma [76], when glacials (e.g. MIS MG6) were less severe than during MIS M2 (Figure 1).

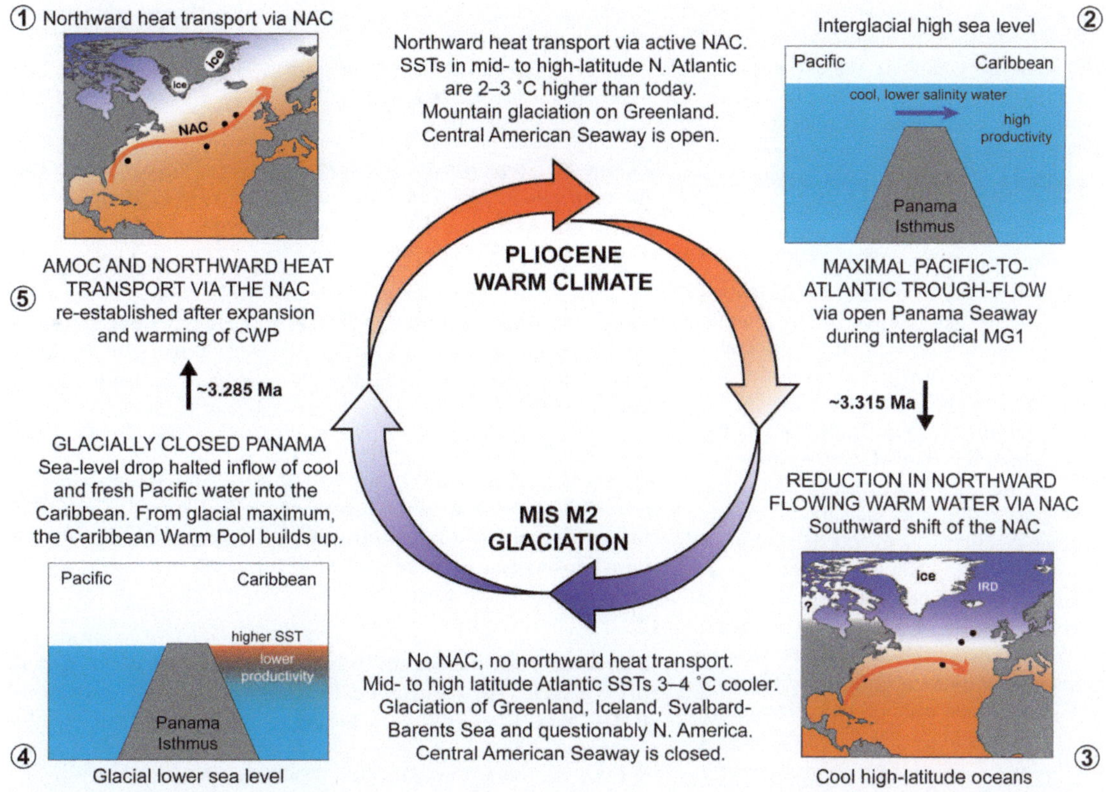

Figure 6. Conceptual model of glaciation and deglaciation of the Northern Hemisphere during MIS M2 in the otherwise globally warm early Late Pliocene. Numbers show sequence of events.

Glacial closure of the Central American Seaway led to deglaciation

Our results further demonstrate that the sea level drop at the full glaciation of MIS M2 [12,15] closed the CAS and effectively halted the inflow of Pacific water into the Atlantic realm. The oligotrophic conditions at Site 999 shown in the absence of heterotrophic dinoflagellate species and low cyst concentrations (Figure 5G, 5H) during MIS M2 suggest no inflow of nutrient-rich Pacific waters [70]. The increasing SST and salinity ($\delta^{18}O_{sw\text{-}ice}$) at Site 999 from the glacial maximum at ~3.295 Ma onwards show that Caribbean surface waters became warmer and more saline while remaining oligotrophic (Figure 5D, 5E, 5G, 5H). This is a reflection of the build-up of the Caribbean Warm Pool already from the glacial maximum onwards. The expansion and warming of the Caribbean Warm Pool are essential for re-establishing AMOC and northward heat transport, as observed for the last deglaciation [77]. At around 3.285 Ma, Site 999 is characterised by high SSTs and salinity, a return to biologically productive conditions, and increased carbonate preservation (high carbonate sand-fraction) (Figure 5F). This indicates a Caribbean Warm Pool sufficiently large and warm to re-invigorate the AMOC. The re-established northward heat transport and active NAC flowing along its modern pathway around 3.285 Ma is reflected by the rapid turnover within 1–2 kyrs of the dinoflagellate cyst assemblages in the eastern North Atlantic Sites 610 and U1308 where *O. centrocarpum* becomes dominant again (Figure 3G, 3H). By then, the warm climates of the mPWP [1] were established, with North Atlantic SSTs ~3°C above present values (Figure 3C, 3D), a modern-like AMOC [5] but with a reduced meridional sea-surface temperature gradient (Figure 4), and a Greenland ice sheet that was reduced to isolated mountain glaciers [4]. As such, the glaciation during MIS M2 appears responsible for its own demise.

Conclusions

Our study identifies links between CAS through-flow, NAC variability, high latitude sea-surface temperatures, and Northern Hemisphere glaciation during Late Pliocene MIS M2 (~3.30 Ma). We provide a conceptual model based on palynological and geochemical records in the North Atlantic and Caribbean for glacial expansion and consequent deglaciation during an otherwise globally warmer world (Figure 6).

A long-term global cooling trend (reflected in SST, CO_2, and ice volume records; Figure 1) preconditioned the Northern Hemisphere for glaciation during the early Late Pliocene. However, the ultimate tipping point for intense glaciation during MIS M2 was the through-flow of Pacific water via an open CAS into the Atlantic, ultimately resulting in a steep SST gradient in the North Atlantic and thermal isolation of the high latitudes. An open CAS as the trigger for Northern Hemisphere glaciation contrasts with the usually invoked CAS closure as either the cause, precondition, or delaying factor for the intensification of Northern Hemisphere glaciation which occurred 500,000 years later [27,33,34]. Recent modelling experiments indicate that the closure of the CAS actually had no effect on the Late Pliocene Greenland ice sheet, and demonstrate that declining atmospheric carbon dioxide concentrations were the driving factor behind the intensification of Northern Hemisphere glaciation at ~2.75 Ma [28,71].

Our records in fact demonstrate that the glacio-eustatic closure of the CAS during MIS M2 eventually re-established northward heat transport in the North Atlantic. Following the expansion of the Antarctic and Northern Hemisphere ice sheets (including Greenland, Iceland, Svalbard/Barents region and questionably the interior of North America) to a volume seemingly larger than present, sea level fell to more than 10 m and possibly as much as 65 m below present (Figure 1B). This closed the CAS and halted the flow of Pacific water into the North Atlantic, allowing the Caribbean Warm Pool to accumulate. In time, this re-invigorated the Gulf Stream/North Atlantic Current system and provided northward heat transport, leading to high-latitude North Atlantic surface waters that were 3°C warmer than present and consequent retreat of the Greenland ice sheet to mountainous areas in the east and southeast during the mPWP.

The transition from MIS M2 to the mPWP can be seen as the evolution of a world with comparable global temperatures to present and slightly larger ice sheets, to a world with global temperatures ~3°C higher than today and glaciation strongly diminished and localised in the Northern Hemisphere. Although operating on a longer time scale, this climate transition can provide valuable insights into the present anthropogenically-forced climate transition towards a globally warmer planet, being comparable to projections for the end of this century. In view of this projected climate warming, our results from the Late Pliocene show that high-latitude North Atlantic surface circulation and SSTs are a crucial factor in the expansion and contraction of Northern Hemisphere ice sheets.

Supporting Information

Figure S1 Age model for DSDP Hole 610A based on the correlation of oxygen isotope records from the studied intervals with the LR04 benthic oxygen isotope global stack [10]. *Left panel*: core-sections, polarity subchrons, including uncertainty interval for the exact position of each reversal of the Mammoth Subchron, and benthic isotope record against depth (mbsf). *Middle panel*: correlation of the benthic record (thin red line, raw data; thick red line, 4-point running mean) to the LR04 global stack of benthic isotope records [10] plotted against time. Grey shading represents the marine isotope stage boundaries from [10]: marine isotope stage M2 was defined between 3.264 and 3.312 Ma. We consider the full glaciation to occur between 3.305 and 3.385 Ma (light grey). Thin black lines between left and middle panel show the tie points used (listed in inset). *Right panel*: sedimentation rate based on our age model. *Inset* gives the tie points used, and correlation values of the benthic record running mean and raw data with the LR04 global stack. Note: Hole 610A shows a coring gap between Cores 610A-17H and 610A-18H, and sediment disturbance in the upper 25 cm of Section 610A-18H1.

Figure S2 Age model for IODP Site U1308 based on the correlation of oxygen isotope records from the studied intervals with the LR04 benthic oxygen isotope global stack [10]. *Left, middle and right panel and inset* as for Figure S1.

Figure S3 Age model for IODP Site U1313 based on the correlation of oxygen isotope records from the studied intervals with the LR04 benthic oxygen isotope global stack [10]. *Left, middle and right panel and inset* as for Figure S1.

Figure S4 Age model for DSDP Site 603 based on the correlation of oxygen isotope records from the studied intervals and palaeomagnetic reversals with the LR04 benthic oxygen isotope global stack [10]. *Left, middle and right panel and inset* as for Figure S1.

Figure S5 Shown on the left are the benthic $\delta^{18}O$ global LR04 stack [10] compared to the benthic $\delta^{18}O$ record of IODP Site 1241 [69]. On the right, the LR04 global stack is compared to the old [51] and new (this study) benthic $\delta^{18}O$ curve of ODP Site 999. The latter is a fine-tuning of the [51] record to the LR04 stack.

Acknowledgments

M. Segl is thanked for isotope measurements, S. Pape for assistance with the ICP-OES, and M. Hoins, S. Forke and J. Engelke for assistance in the lab (MARUM and University of Bremen). Samples were supplied by the Integrated Ocean Drilling Program. We thank Alan M. Haywood and one anonymous reviewer for the constructive comments.

Author Contributions

Conceived and designed the experiments: SDS JG BDAN. Performed the experiments: SDS JG BDAN CVR JH KF. Analyzed the data: SDS JG BDAN CVR JH MJH SL KF. Contributed reagents/materials/analysis tools: SDS JG BDAN MJH SL KF. Wrote the paper: SDS JG BDAN CVR JH MJH SL KF.

References

1. Dowsett HJ, Robinson MM, Haywood AM, Hill DJ, Dolan AM, et al. (2012) Assessing confidence in Pliocene sea surface temperatures to evaluate predictive models. Nat Clim Chang 2: 1–7. doi:10.1038/nclimate1455.
2. Haywood AM, Valdes PJ (2004) Modelling Pliocene warmth: contribution of atmosphere, oceans and cryosphere. Earth Planet Sci Lett 218: 363–377. doi:10.1016/S0012-821X(03)00685-X.
3. Raymo ME, Mitrovica JX, O'Leary MJ, Deconto RM, Hearty PJ (2011) Departures from eustasy in Pliocene sea-level records. Nat Geosci 4: 328–332. doi:10.1038/ngeo1118.
4. Dolan AM, Haywood AM, Hill DJ, Dowsett HJ, Hunter SJ, et al. (2011) Sensitivity of Pliocene ice sheets to orbital forcing. Palaeogeogr Palaeoclimatol Palaeoecol 309: 98–110. doi:10.1016/j.palaeo.2011.03.030.
5. Zhang ZS, Nisancioglu KH, Chandler MA, Haywood AM, Otto-Bliesner BL, et al. (2013) Mid-Pliocene Atlantic meridional overturning circulation not unlike modern. Clim Past 9: 1495–1504. doi:10.5194/cp-9-1495-2013.
6. Raymo ME, Grant B, Horowitz M, Rau GH (1996) Mid-Pliocene warmth: stronger greenhouse and stronger conveyor. Mar Micropaleontol 27: 313–326. doi:10.1016/0377-8398(95)00048-8.
7. Pagani M, Liu Z, LaRiviere J, Ravelo AC (2009) High Earth-system climate sensitivity determined from Pliocene carbon dioxide concentrations. Nat Geosci 3: 27–30. doi:10.1038/ngeo724.
8. Seki O, Foster GL, Schmidt DN, Mackensen A, Kawamura K, et al. (2010) Alkenone and boron-based Pliocene pCO_2 records. Earth Planet Sci Lett 292: 201–211. doi:10.1016/j.epsl.2010.01.037.
9. Bartoli G, Hönisch B, Zeebe RE (2011) Atmospheric CO_2 decline during the Pliocene intensification of Northern Hemisphere glaciations. Paleoceanography 26: PA4213. doi:10.1029/2010PA002055.
10. Lisiecki LE, Raymo ME (2005) A Pliocene-Pleistocene stack of 57 globally distributed benthic $\delta^{18}O$ records. Paleoceanography 20: PA1003. doi:10.1029/2004PA001071.
11. Mudelsee M, Raymo ME (2005) Slow dynamics of the Northern Hemisphere glaciation. Paleoceanography 20: PA4022. doi:10.1029/2005PA001153.
12. Miller KG, Kominz MA, Browning JV, Wright JD, Mountain GS, et al. (2005) The Phanerozoic record of global sea-level change. Science 310: 1293–1298. doi:10.1126/science.1116412.
13. Naish TR, Wilson GS (2009) Constraints on the amplitude of Mid-Pliocene (3.6–2.4 Ma) eustatic sea-level fluctuations from the New Zealand shallow-marine sediment record. Phil Trans A Math Phys Eng Sci 367: 169–187. doi:10.1098/rsta.2008.0223.
14. Miller KG, Wright JD, Browning JV, Kulpecz A, Kominz M, et al. (2012) High tide of the warm Pliocene: Implications of global sea level for Antarctic deglaciation. Geology 40: 407–410. doi:10.1130/G32869.1.

15. Dwyer GS, Chandler MA (2009) Mid-Pliocene sea level and continental ice volume based on coupled benthic Mg/Ca palaeotemperatures and oxygen isotopes. Phil Trans A Math Phys Eng Sci 367: 157–168. doi:10.1098/rsta.2008.0222.
16. Brigham-Grette J, Melles M, Minyuk P, Andreev A, Tarasov P, et al. (2013) Pliocene warmth, polar amplification, and stepped Pleistocene cooling recorded in NE Arctic Russia. Science 340, 1421–1427. doi:10.1126/science.1233137.
17. Pollard D, DeConto RM (2009) Modelling West Antarctic ice sheet growth and collapse through the past five million years. Nature 458: 329–333. doi:10.1038/nature07809.
18. Naish T, Powell R, Levy R, Wilson G, Scherer R, et al. (2009) Obliquity-paced Pliocene West Antarctic ice sheet oscillations. Nature 458: 322–328. doi:10.1038/nature07867.
19. Passchier S (2011) Linkages between East Antarctic Ice Sheet extent and Southern Ocean temperatures based on a Pliocene high-resolution record of ice-rafted debris off Prydz Bay, East Antarctica. Paleoceanography 26: PA4204. doi:10.1029/2010PA002061.
20. Kleiven HF, Jansen E, Fronval T, Smith TM (2002) Intensification of Northern Hemisphere glaciations in the circum Atlantic region (3.5–2.4 Ma) – ice-rafted detritus evidence. Palaeogeogr Palaeoclimatol Palaeoecol 184: 213–223. doi:10.1016/S0031-0182(01)00407-2
21. Knies J, Matthiessen J, Vogt C, Laberg JS (2009) The Plio-Pleistocene glaciation of the Barents Sea–Svalbard region: a new model based on revised chronostratigraphy. Quat Sci Rev 28: 812–829. doi:10.1016/j.quascirev.2008.12.002.
22. Moran K, Backman J, Brinkhuis H, Clemens SC, Cronin T, et al. (2006) The Cenozoic palaeoenvironment of the Arctic Ocean. Nature 441: 601–605. doi:10.1038/nature04800.
23. Sarnthein M, Bartoli G, Prange M, Schmittner A, Schneider B, et al. (2009) Mid-Pliocene shifts in ocean overturning circulation and the onset of Quaternary-style climates. Clim Past 5: 269–283. doi:10.5194/cp-5-269-2009.
24. Geirsdóttir Á (2011) Pliocene and Pleistocene glaciations of Iceland: a brief overview of the glacial history. In: Ehlers J, Gibbard PL, Hughes PD, editors. Developments in Quaternary Science, Vol. 15. Amsterdam: Elsevier. 199–210. doi:10.1016/B978-0-444-53447-7.00016-7.
25. Barendregt RW, Duk-Rodkin A (2011) Chronology and extent of Late Cenozoic ice sheets in North America: a magnetostratigraphical assessment. In: Ehlers J, Gibbard PL, Hughes PD, editors. Developments in Quaternary Science, Vol. 15. Amsterdam: Elsevier. 419–426. doi:10.1016/B978-0-444-53447-7.00032-5.
26. De Schepper S, Head MJ, Groeneveld J (2009) North Atlantic Current variability through marine isotope stage M2 (circa 3.3 Ma) during the mid-Pliocene. Paleoceanography 24: PA4206. doi:10.1029/2008PA001725.
27. Haug GH, Tiedemann R (1998) Effect of the formation of the Isthmus of Panama on Atlantic Ocean thermohaline circulation. Nature 393: 673–676. doi:10.1038/31447.
28. Lunt DJ, Foster G, Haywood AM, Stone E (2008) Late Pliocene Greenland glaciation controlled by a decline in atmospheric CO_2 levels. Nature 454: 1102–1106. doi:10.1038/nature07223.
29. Brierley CM, Fedorov AV (2010) Relative importance of meridional and zonal sea surface temperature gradients for the onset of the ice ages and Pliocene–Pleistocene climate evolution. Paleoceanography 25: PA2214. doi:10.1029/2009PA001809.
30. Naafs BDA, Stein R, Hefter J, Khélifi N, De Schepper S, et al. (2010) Late Pliocene changes in the North Atlantic Current. Earth Planet Sci Lett 298: 434–442. doi:10.1016/j.epsl.2010.08.023.
31. Lawrence KT, Sosdian S, White HE, Rosenthal Y (2010) North Atlantic climate evolution through the Plio-Pleistocene climate transitions. Earth Planet Sci Lett 300: 329–342. doi:10.1016/j.epsl.2010.10.013.
32. Cane MA, Molnar P (2001) Closing of the Indonesian seaway as a precursor to east African aridification around 3–4| million years ago. Nature 411: 157–162. doi:10.1038/35075500.
33. Driscoll NW, Haug GH (1998) A short circuit in thermohaline circulation: a cause for Northern Hemisphere glaciation? Science 282: 436–438. doi:10.1126/science.282.5388.436.
34. Berger W, Wefer G (1996) Expeditions into the Past: Paleoceanographic studies in the South Atlantic. In Wefer G, Berger WH, Siedler G, Webb DJ, editors. The South Atlantic: Present and Past Circulation. Berlin, Heidelberg: Springer-Verlag. 363–410. doi: 10.1007/978-3-642-80353-6_21.
35. De Schepper S, Fischer EI, Groeneveld J, Head MJ, Matthiessen J (2011) Deciphering the palaeoecology of Late Pliocene and Early Pleistocene dinoflagellate cysts. Palaeogeogr Palaeoclimatol Palaeoecol 309: 17–32. doi:10.1016/j.palaeo.2011.04.020.
36. De Schepper S, Head MJ (2008) New dinoflagellate cyst and acritarch taxa from the Pliocene and Pleistocene of the eastern North Atlantic (DSDP Site 610). J Syst Paleont 6: 101–117. doi:10.1017/S1477201907002167.
37. Stockmarr J (1971) Tablets with spores used in absolute pollen analysis. Pollen et Spores 13: 615–621.
38. Heslop D, De Schepper S, Proske U (2011) Diagnosing the uncertainty of taxa relative abundances derived from count data. Mar Micropaleontol 79: 114–120. doi:10.1016/j.marmicro.2011.01.007.
39. Harland R (1983) Distribution maps of Recent dinoflagellate cysts in bottom sediments from the North Atlantic Ocean and adjacent seas. Palaeontology 26: 321–387.
40. Rochon A, de Vernal A, Turon J-L, Matthiessen J, Head MJ (1999) Distribution of recent dinoflagellate cysts in surface sediments from the North Atlantic Ocean and adjacent seas in relation to sea-surface parameters. AASP Contributions Series 35. Dallas, Texas: AASP Foundation. 146.
41. Zonneveld KAF, Marret F, Versteegh GJM, Bogus K, Bonnet S, et al. (2012) Atlas of modern dinoflagellate cyst distribution based on 2405 datapoints. Rev Palaebot Palynol. doi:10.1016/j.revpalbo.2012.08.003.
42. Radi T, de Vernal A (2008) Dinocysts as proxy of primary productivity in mid–high latitudes of the Northern Hemisphere. Mar Micropaleontol 68: 84–114. doi:10.1016/j.marmicro.2008.01.012.
43. Eynaud F, Turon J-L, Duprat J (2004) Comparison of the Holocene and Eemian palaeoenvironments in the South Icelandic Basin: dinoflagellate cysts as proxies for the North Atlantic surface circulation. Rev Palaebot Palynol 128: 55–79. doi:10.1016/S0034-6667(03)00112-X.
44. Shackleton NJ, Hall MA (1984) Oxygen and carbon isotope stratigraphy of Deep Sea Drilling Project Hole 552A: Plio-Pleistocene glacial history. DSDP Init Rep 81: 599–609. doi:10.2973/dsdp.proc.81.116.1984.
45. Barker S, Greaves M, Elderfield H (2003) A study of cleaning procedures used for foraminiferal Mg/Ca paleothermometry. Geochem Geophys Geosyst 4: 8407, doi:10.1029/2003GC000559.
46. Greaves M, Caillon N, Rebaubier H, Bartoli G, Boharty S, et al. (2008) Interlaboratory comparison study of calibration standards for foraminiferal Mg/Ca thermometry. Geochem Geophys Geosyst 9: Q08010. doi:10.1029/2008GC001974.
47. Elderfield H, Ganssen G (2000) Past temperature and $\delta^{18}O$ of surface ocean waters inferred from foraminiferal Mg/Ca ratios. Nature 405: 442–445. doi:10.1038/35013033.
48. Schiebel R, Bijma J, Hemleben C (1997) Population dynamics of the planktic foraminifer *Globigerina bulloides* from the eastern North Atlantic. Deep Sea Res I 44(9–10): 1701–1713. doi:10.1016/S0967-0637(97)00036-8.
49. Chapman MR (2010) Seasonal production patterns of planktonic foraminifera in the NE Atlantic Ocean: implications for paleotemperature and hydrographic reconstructions. Paleoceanography 25: PA1101. doi:10.1029/2008PA001708.
50. Ganssen GM, Kroon D (2000) The isotopic signature of planktonic foraminifera from NE Atlantic surface sediments: implications for the reconstruction of past oceanic conditions. J Geol Soc Lond 157: 693–699. doi: 10.1144/jgs.157.3.693.
51. Steph S, Tiedemann R, Prange M, Groeneveld J, Nürnberg D, et al. (2006) Changes in Caribbean surface hydrography during the Pliocene shoaling of the Central American Seaway. Paleoceanography 21: PA4221. doi: 10.1029/2008PA001645
52. Nürnberg D, Müller A, Schneider RR (2010) Paleo-sea surface temperature calculations in the equatorial east Atlantic from Mg/Ca ratios in planktic foraminifera: A comparison to sea surface temperature estimates from $U^{k'}_{37}$, oxygen isotopes, and foraminiferal transfer function. Paleoceanography 15: 124–134. doi:10.1029/1999PA000370.
53. Groeneveld J, Chiessi CM (2011) Mg/Ca of *Globorotalia inflata* as a recorder of permanent thermocline temperatures in the South Atlantic. Paleoceanography 26: PA2203. doi:10.1029/2010PA001940.
54. Shackleton NJ (1974) Attainment of isotopic equilibrium between ocean water and benthonic foraminifera genus *Uvigerina*: Isotopic changes in the ocean during the last glacial. Colloq Int CNRS 219: 203–210.
55. Naafs BDA, Hefter J, Acton G, Haug GH, Martinez-Garcia A, et al. (2012) Strengthening of North American dust sources during the late Pliocene (2.7Ma). Earth Planet Sci Lett 317–318: 8–19. doi:10.1016/j.epsl.2011.11.026.
56. Brassell SC, Eglington G, Marlowe IT, Pflaumann U, Sarnthein M (1986) Molecular stratigraphy: a new tool for climatic assessment. Nature 320: 129–133. doi:10.1038/320129a0.
57. Prahl FG, Wakeham SG (1987) Calibration of unsaturation patterns in long-chain ketone compositions for paleotemperature assessment. Nature 330: 367–369. doi:10.1038/330367a0.
58. Hefter J (2008) Analysis of alkenone unsaturation indices with fast gas chromatography/time-of-flight mass spectrometry. Anal Chem 80: 2161–2170. doi:10.1021/ac702194m.
59. Müller PJ, Kirst G, Ruhland G, von Storch I, Rosell-Melé A (1998) Calibration of the alkenone paleotemperature index $U^{k'}_{37}$ based on core-tops from the eastern South Atlantic and the global ocean (60°N–60°S). Geochim Cosmochem Acta 62: 1757–1772. doi:10.1016/S0016-7037(98)00097-0.
60. Schiebel R, Brupbacher U, Schmidtko S, Nausch G, Waniek JJ, et al. (2011) Spring coccolithophore production and dispersion in the temperate eastern North Atlantic Ocean. J Geophys Res 116: C08030. doi:10.1029/2010JC006841.
61. Clement BM, Robinson F (1987) The magnetostratigraphy of Leg 94 sediments. DSDP Init Rep 94: 635–650. doi:10.2973/dsdp.proc.94.112.1987.
62. Expedition 303 Scientists (2006) Site U1308. Proc IODP 303/306: 1–98. doi:10.2204/iodp.proc.303306.108.2006.
63. Lourens L, Hilgen F, Shackleton NJ, Laskar J, Wilson D (2005) The Neogene. In: Gradstein FM, Ogg JG, Smith AG, editors. A Geological Time Scale 2004. Cambridge: Cambridge University Press. 409–430. doi: 10.1017/CBO9780511536045
64. Shipboard Scientific Party (1987) Site 610. DSDP Init Rep 94: 351–470. doi:10.2973/dsdp.proc.94.106.1987.
65. Expedition 306 Scientists (2006) Site U1313. Proc IODP 303/306: 1–124. doi:10.2204/iodp.proc.303306.112.2006.

66. Paillard D, Labeyrie L, Yiou P (1996) Macintosh Program performs time-series analysis. Eos Trans AGU 77: 379. doi:10.1029/96EO00259.
67. Shackleton NJ, Hall MA, Pate D, Pisias N, Mayer L, et al. (1995) Pliocene stable isotope stratigraphy of ODP Site 846. Proc ODP Sci Res 138: 337–356. doi:10.2973/odp.proc.sr.138.117.1995.
68. Tiedemann R, Sarnthein M, Shackleton NJ (1994) Astronomic timescale for the Pliocene Atlantic $\delta^{18}O$ and dust flux records of Ocean Drilling Program Site 659. Paleoceanography 9: 619–638. doi: 10.1029/94PA00208.
69. Tiedemann R, Sturm A, Steph S, Lund SP, Stoner JS, et al. (2006) Astronomically calibrated timescales from 6 to 2.5 Ma and benthic isotope stratigraphies, Sites 1236, 1237, 1239, and 1241. Proc ODP Sci Res 202: 1–69. doi:10.2973/odp.proc.sr.202.210.2007.
70. Schneider B, Schmittner A (2006) Simulating the impact of the Panamanian seaway closure on ocean circulation, marine productivity and nutrient cycling. Earth Planet Sci Lett 246: 367–380. doi:10.1016/j.epsl.2006.04.028.
71. Lunt DJ, Valdes PJ, Haywood AM, Rutt IC (2008) Closure of the Panama Seaway during the Pliocene: implications for climate and Northern Hemisphere glaciation. Clim Dyn 30: 1–18. doi:10.1007/s00382-007-0265-6.
72. Klocker A, Prange M, Schulz M (2005) Testing the influence of the Central American Seaway on orbitally forced Northern Hemisphere glaciation. Geophys Res Lett 32: L03703. doi:10.1029/2004gl021564.
73. Darling KF, Kucera M, Wade CM, von Langen P, Pak D (2003) Seasonal distribution of genetic types of planktonic foraminifer morphospecies in the Santa Barbara Channel and its paleoceanographic implications. Paleoceanography 18(2): 1032. doi:10.1029/2001PA000723.
74. Born A, Kageyama M, Nisancioglu KH (2010) Warm Nordic Seas delayed glacial inception in Scandinavia. Clim Past 6: 817–826. doi:10.5194/cp-6-817-2010.
75. Salzmann U, Haywood AM, Lunt DJ, Valdes PJ, Hill DJ (2008) A new global biome reconstruction and data-model comparison for the Middle Pliocene. Glob Ecol Biogeogr 17: 432–447. doi:10.1111/j.1466-8238.2008.00381.x.
76. Gao C, McAndrews JH, Wang X, Menzies J, Turton CL, et al. (2012) Glaciation of North America in the James Bay Lowland, Canada, 3.5 Ma. Geology 40: 975–978. doi:10.1130/G33092.1.
77. Carlson AE, Oppo DW, Came RE, Legrande AN, Keigwin LD, et al. (2008) Subtropical Atlantic salinity variability and Atlantic meridional circulation during the last deglaciation. Geology 36: 991–994. doi:10.1130/G25080A.1.
78. Steph S, Tiedemann R, Groeneveld J, Sturm A, Nürnberg D (2006) Pliocene changes in tropical East Pacific upper ocean stratification: response to tropical gateways? Proc ODP Sci Res 202: 1–51. doi:10.2973/odp.proc.sr.202.211.2006.
79. Bintanja R, van de Wal RSW (2008) North American ice-sheet dynamics and the onset of 100,000-year glacial cycles. Nature 454: 869–872. doi:10.1038/nature07158.
80. Kürschner W, van der Burgh J, Visscher H, Dilcher D (1996) Oak leaves as biosensors of late Neogene and early Pleistocene paleoatmospheric CO_2 concentrations. Mar Micropaleontol 27: 299–312.
81. Locarnini RA, Mishonov AV, Antonov JI, Boyer TP, Garcia HE (2006) World Ocean Atlas 2005, Volume 1: Temperature. In: Levitus S, editor. NOAA Atlas NESDIS 61. Washington, D.C.: U.S. Gov. Printing Office. p. 182. Available: http://www.nodc.noaa.gov/OC5/indprod.html.

6

Causes and Consequences of Past and Projected Scandinavian Summer Temperatures, 500–2100 AD

Ulf Büntgen[1,2]*, Christoph C. Raible[2,3], David Frank[1,2], Samuli Helama[4], Laura Cunningham[5], Dominik Hofer[2,3], Daniel Nievergelt[1], Anne Verstege[1], Mauri Timonen[6], Nils Chr. Stenseth[7], Jan Esper[8]

1 Swiss Federal Research Institute WSL, Birmensdorf, Switzerland, **2** Oeschger Centre for Climate Change Research, Bern, Switzerland, **3** Climate and Environmental Physics, Physics Institute University of Bern, Bern, Switzerland, **4** Arctic Centre, University of Lapland, Rovaniemi, Finland, **5** School of Geography and Geosciences, University of St Andrews, St Andrews, Scotland, **6** The Finnish Forest Research Institute, Rovaniemi, Finland, **7** Centre for Ecological and Evolutionary Synthesis CEES, University of Oslo, Blindern, Norway, **8** Department of Geography, Johannes Gutenberg University of Mainz, Mainz, Germany

Abstract

Tree rings dominate millennium-long temperature reconstructions and many records originate from Scandinavia, an area for which the relative roles of external forcing and internal variation on climatic changes are, however, not yet fully understood. Here we compile 1,179 series of maximum latewood density measurements from 25 conifer sites in northern Scandinavia, establish a suite of 36 subset chronologies, and analyse their climate signal. A new reconstruction for the 1483–2006 period correlates at 0.80 with June–August temperatures back to 1860. Summer cooling during the early 17th century and peak warming in the 1930s translate into a decadal amplitude of 2.9°C, which agrees with existing Scandinavian tree-ring proxies. Climate model simulations reveal similar amounts of mid to low frequency variability, suggesting that internal ocean-atmosphere feedbacks likely influenced Scandinavian temperatures more than external forcing. Projected 21st century warming under the SRES A2 scenario would, however, exceed the reconstructed temperature envelope of the past 1,500 years.

Editor: Jürgen Kurths, Humboldt University, Germany

Funding: Funding was provided by the Swiss Federal Research Institute WSL (www.wsl.ch). The funders had no role in study design, data collection and analysis, decision to publish, or preparation of the manuscript.

Competing Interests: The authors have declared that no competing interests exist.

* E-mail: buentgen@wsl.ch

Introduction

Annually resolved, large-scale temperature reconstructions of the last millennium rely on a handful of predictor records, which often reflect heterogeneous patterns of local- to regional-scale climate variability [1]. Due to the small number of highly resolved proxy data available prior to the Little Ice Age (LIA; ~1350–1850 AD), the few tree-ring width (TRW; mm/year) and maximum latewood density (MXD; g/cm^3) chronologies from northern North America and northern Eurasia that span the last millennium, become increasingly important further back in time. Long-term extra-tropical temperature estimates typically incorporate the same proxy records and use similar reconstruction methodologies [2]. For instance, millennium-long tree-ring composite data from the Torneträsk region in northern Sweden (e.g. [3] and references therein) contribute to most large-scale temperature reconstructions [4], thus clearly impact our understanding of the spatial extent, absolute timing and relative amplitude of the Medieval Climate Anomaly (MCA; ~900–1300 AD). At the same time, it appears particularly important to note that multi-decadal variations in surface air and sea surface temperature (SAT and SST) across the North Atlantic/Scandinavian sector are dominated by internal climate variability [5–9], which may does not always agree with short-term events and long-term trends observed in other areas or more generally larger spatial scales. Consequently, reconstructions of hemispheric to global average temperatures that partly rely on Scandinavian proxy data, likely contain climatic fingerprints of internal variability that might not be overly representative outside a limited region (e.g. [10]).

One source of internal climate variability across Scandinavia is the Atlantic Multidecadal Oscillation (AMO), which had a periodicity of ~65–70 years during the last 150 years [11]. Observational and model evidence indicates that the Atlantic Meridional Overturning Circulation (AMOC) plays an important role in driving the AMO [12], which in turn controls summer climate across the high northern latitudes (e.g. [7]). In addition to the AMO, atmospheric circulation modes, such as the North Atlantic Oscillation (NAO), the Arctic Oscillation (AO) or the East Atlantic Pattern (EAP) may influence Scandinavian climate variability on different timescales (e.g. [13–15]). Quantifying and distinguishing between such internally driven ocean-atmosphere-land dynamics and externally forced climate variations represents a major challenge for the detection and attribution of anthropogenic-induced climate change (e.g. [6,8,16]). The combined assessment of proxy-based climate reconstructions and model-based climate simulations therefore becomes essential for allowing studies to separate externally forced trends from internally driven variations, whenever the climatic signal is distinguishable from noise (e.g. [17]).

This study presents the largest update of regional MXD measurements undertaken to date, allowing modern growth-climate responses to be analysed into the 21st century. We aim to develop a new reconstruction of northern Scandinavian summer temperature variability over the past five centuries, compare this record with existing tree-ring evidence and model output, and disentangle the role of external climate controls from internal ocean-atmosphere oscillations.

Results

The 25 site chronologies show positive correlations with growing season temperatures between April and August (see supporting information (SI) Text S1 for details). The highest correlation is found with June–August (JJA) temperatures, corresponding to intense cell formation (lumen enlargement) and wood lignification (wall thickening). Significant ($p<0.001$) positive correlations with JJA temperatures (0.31–0.83) were obtained from the 36 MXD records including 25 site and 11 subset chronologies (Figure 1–3). Small differences in mean MXD (0.59–0.85 g/cm^3), but major differences in sample replication (18–1179 series), as well as mean tree age (39–324 years) are found (Figure S1). Less replicated site chronologies generally contain a lower temperature signal. Five site chronologies from the network's western margin, which are based on 22–29 samples only, show modest correlations (Table S1 and Figure S2), and thus were omitted as predictors in the final reconstruction.

Subset replication ranges from 181–1179 series, and is substantially larger than individual site replication (18–104 series). Some differences in the growth-climate response were expected within several subsets that combined physiological (species and age), ecological (shore/offshore), and methodological (sampling design) extremes. Nonetheless, all subset chronologies were strongly correlated with regional summer temperatures (>0.76), suggesting a considerably high degree of common variance within the MXD network. Lower, but still significant ($p<0.001$) correlations of 0.29–0.63 were found between the 36 MXD chronologies and JJA coastal Norwegian SSTs averaged over the 65–70°N and 5–15°E region.

The 36 MXD chronologies reflect similar growth trends over the past ~150 years (Figure 2A) and show strong cross-correlation with each other (r = 0.72 for 1860–1977). All records contain below average MXD values in the 1860s, ~1900s, and in the 1960s. Above average values are evident from ~1920–1950 and during the last ~15 years. The highest MXD values occur in 1937. The lowest MXD values are observed in 1902. A similar picture derives from the eight instrumental records of JJA temperatures and the corresponding CRUTEM3v grid-box mean (Figure 2B). Cross-correlation between the eight station measurements is 0.70 (1860–1977), which is even below the analogous value of the 36 MXD chronologies. Minimum and maximum JJA values of the individual station records represent a realistic range of local temperatures that can be utilized as a target envelope for the MXD chronologies (Figure 2C). This envelope remains relatively narrow throughout the past 150 years; however, increased spread between the station measurements is somewhat evident during the first three decades of the 20th century. This period coincides with decreased correlations among the individual MXD chronologies. The good fit observed between the proxy and target data also demonstrates that both high and low frequency coherence has been maintained (Figure 2C). The composite record tracks the recent and particularly the earlier warmth of the 1930s, remains within the target envelope, encompasses 70% of northern Scandinavian JJA temperature variations (CRUTEM3v; 1860–2006), and further reflects 36% of coastal Norwegian SST variability (SSTV2; 1860–2002). The correlation between instrumental SAT and Norwegian SST is 0.63 when calculated over the common period 1860–2002, suggesting that SST similarly

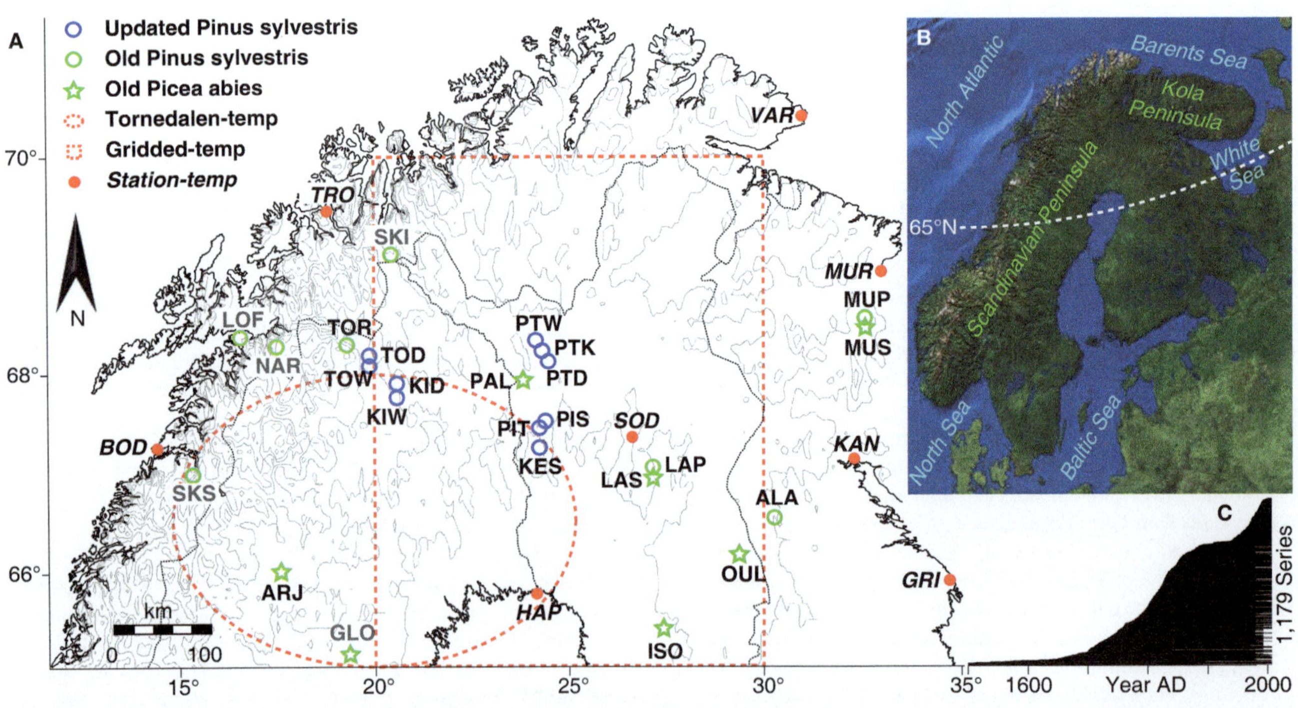

Figure 1. Spatial characteristics of the northern Scandinavian tree-ring network. (**A**) The 25 MXD site chronologies and the eight instrumental stations incorporated within this study. Grey labels refer to five MXD sites from the westerly portion of the network that showed a modest sensitivity to temperature and thus were not included in the final reconstruction. (**B**) Overview of the North Atlantic/Scandinavian sector. (**C**) Temporal distribution of the 1,179 MXD samples during the past five centuries.

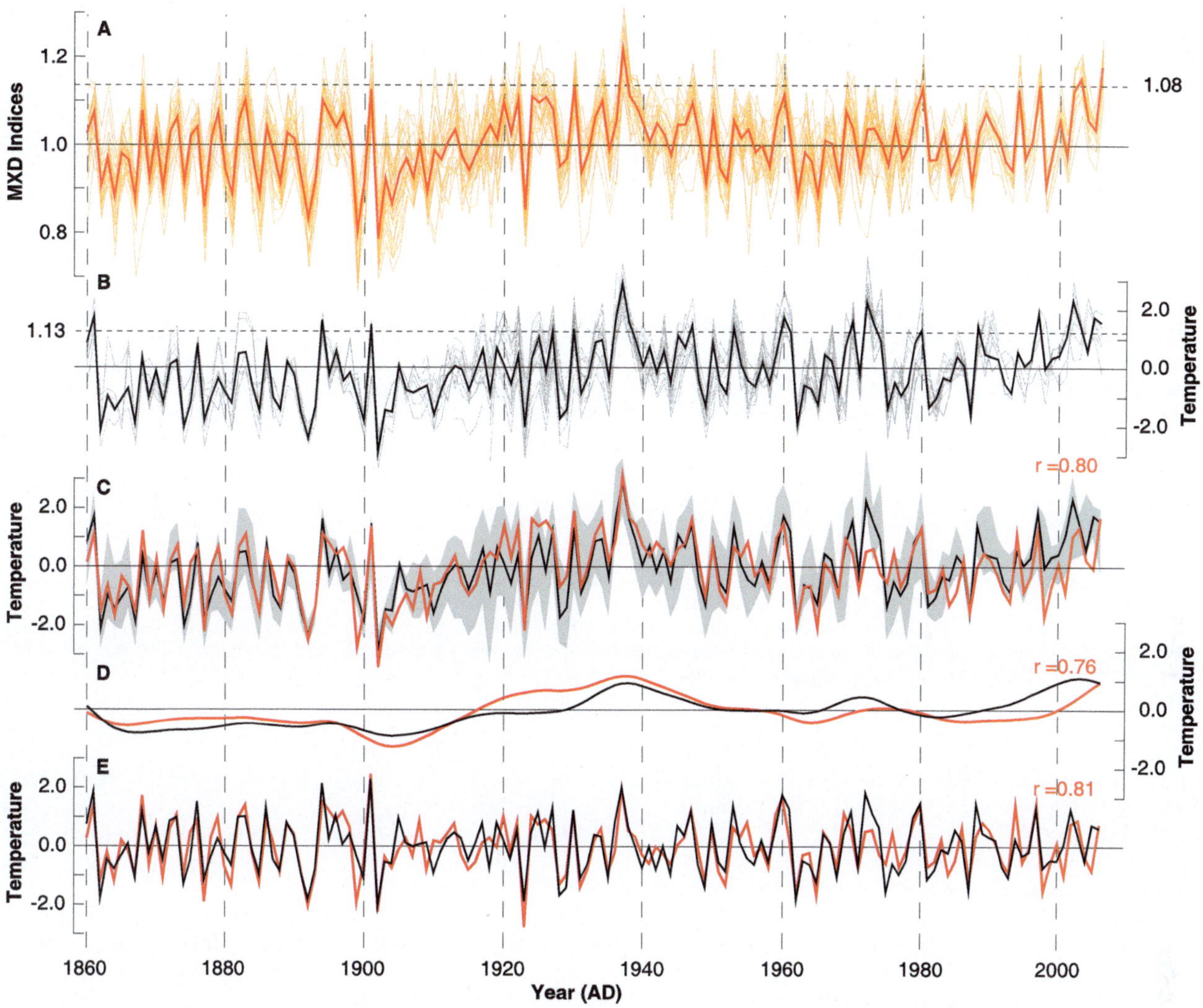

Figure 2. Agreement between high to low frequency variability of the tree-ring proxy and instrumental target data over the past 150 years. (**A**) Site and subsets RCS chronologies (orange) and their mean (red). (**B**) JJA temperatures of eight instrumental stations (grey), and the grid-box mean (black). Dashed horizontal lines and corresponding values indicate the mean of the last decade (1997–2006). (**C**) Instrumental gridded (black) and reconstructed proxy (red) temperatures superimposed on the min/max range of the station measurements (grey shading), expressed as anomalies from 1961–1990, and split into (**D**) 20-year low-pass and (**E**) 20-year high-pass components. Correlation (1860–2006) of the unfiltered and filtered timeseries is shown in red.

influences Scandinavian MXD formation and SAT variation. Detailed calibration/verification statistics are summarized in Table 1.

The proxy and instrumental data correlate at 0.80 over 1860–2006. Correlation coefficients remain approximately constant after 20-year high- and low-pass filtering, which is suggestive of near equal fidelity among inter-annual to multi-decadal variability. Nevertheless, a detailed examination of the proxy/target relationship over different frequency domains indicates some decadal-scale disagreements during the early and late 20th century (Figure 2D–E), as well as some discrepancy in the timing of positive extremes. The 20-year low-pass filtered timeseries exhibit slightly higher reconstructed values, relative to measured temperatures, from ~1915–1945 and slightly lower temperatures from ~1985–2006. The level, direction and timing of disagreement between the proxy and target records, however, do not indicate any systematic 'Divergence' during the 20th century (e.g. [18]).

The proxy reconstruction is found to be less sensitive to warm than to cold extremes (Table S2), a trait shared with other MXD-based temperature reconstructions, which typically better capture cold temperature events (e.g. [3,19–21]). Five of the ten coldest reconstructed and observed summers are common to both records, whereas only two out of the ten warmest years are indicated by both the instrumental and tree-ring data (Table S2). However, it should be noted that both the warmest (1937) and coldest (1902) instrumentally recorded summers are robustly retained by the MXD data.

The new summer temperature reconstruction spans the period 1483–2006 and reveals inter-annual to multi-decadal fluctuations without clear indication for cooler LIA and warmer recent conditions (Figure 4A), which contrasts evidence from other regions (e.g. [19,22]) and larger scales (e.g. [1,23]). Warming is most pronounced around 1500, 1660, 1780, and 1930. Cooling is most distinct during the first half of the 17th century. The ten warmest summers occur sporadically throughout the reconstruc-

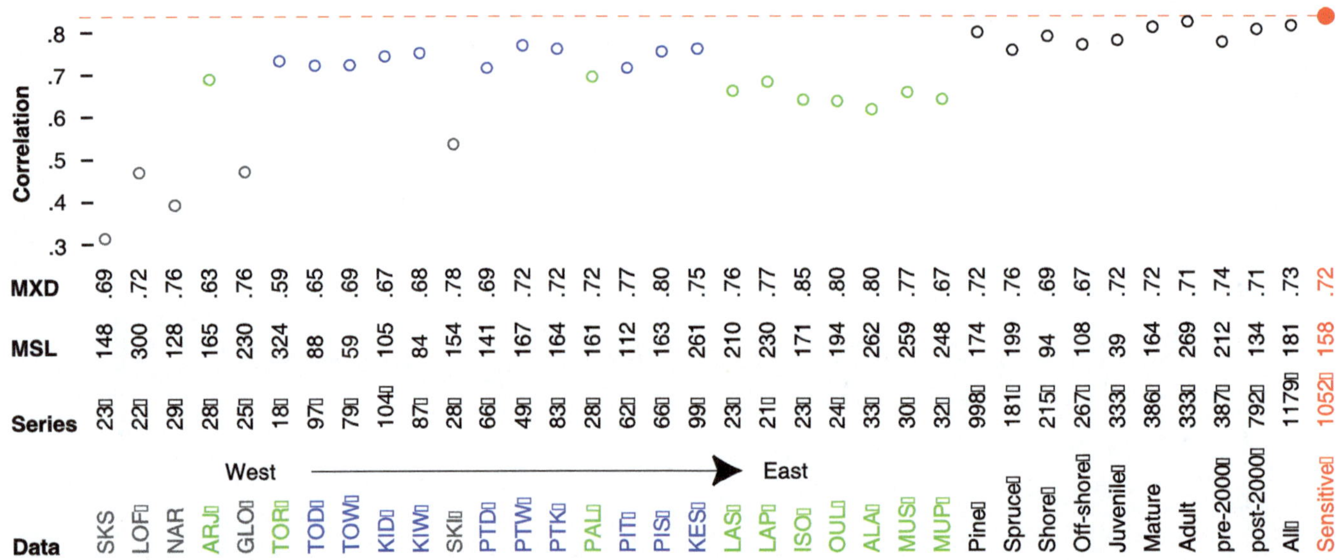

Figure 3. Climate response of the tree-ring network revels significant positive correlations between MXD chronologies and June–August temperatures. Correlation of the 36 MXD site and subset chronologies against Scandinavian JJA temperatures (1860–1977). Information on maximum latewood density (MXD; g/cm^3), mean segment length (MSL; years), and replication (series) is provided for each dataset. Grey [green/blue] site codes refer to less-sensitive (pre/post-2000) chronologies. The red dot refers to the sensitive subset herein used for reconstruction purpose, with the dashed line highlighting the corresponding maximum correlation with JJA temperature (0.83). Note that all correlations exceed the 99.9% significance level.

tion (Table S2). The warmest reconstructed JJA temperature occurred in 1937 (3.11°C), while the coldest reconstructed summer occurred in 1633 (−4.19°C).

Spatial significance of the new reconstruction is approximately restricted to the Scandinavian Peninsula and the Baltic Sea >55°N and <45°E (Figure S3A). This pattern is confirmed by non-significant (p>0.1) correlations between the new Scandinavian reconstruction and independent MXD-based summer temperature reconstructions from the Polar Urals, Pyrenees and Alps that range between 0.1 and −0.07 (1483–1989). Spatial correlations with SST are positive for the surrounding northeast Atlantic, as well as the North and Baltic Seas. Positive correlations with the western part of the North Atlantic around 60–70°W and 30–40°N are also observed (Figure S3B), however, probably reflect a dynamic seasonal structure of surface air pressure rather than a continual pattern of water circulation. This assumption is supported by spatial field correlations between our new reconstruction and gridded (5°×5°) June–August SLP (HadSLP2r; 1860–2006) (Figure S4).

Below-average pressure anomalies centred over the Baltic Sea are associated with the 20 coldest reconstructed summers that exhibited in northern Scandinavia between 1659 and 1999 (Figure S5A). These negative pressure anomalies are teleconnected with positive pressure anomalies over the Bay of Biscay, thus describing a European north/south dipole structure related to anomalous cold westerly winds directed towards Scandinavia. Anomalously high northern Scandinavian summer temperatures, on the other hand, match a well-defined pattern of positive pressure anomalies over the Baltic Sea, and negative anomalies west of Iceland (Figure S5B). The southern European counterpart of negative pressure extends over the Mediterranean basin.

Our northern Scandinavian JJA temperature history agrees well with a TRW-based reconstruction of summer AO indices [24] (r = 0.39 between 1650–1975; Figure S5C). Amplification of the

Table 1. Calibration and verification statistics indicate proxy/target agreement on all frequency domains.

	Full Calibration	Late Calibration	Early Calibration	Extra Calibration
	1860–2006	**1934–2006**	**1860–1932**	**1816–1859**
r	0.80	0.76	0.83	0.89
(low/high)	(0.76/0.81)	(0.70/0.78)	(0.89/0.84)	(0.96/0.88)
DW	1.55	2.10	1.42	1.73
	Extra Verification	Early Verification	Late Verification	Full Verification
RE	0.64	0.64	0.63	0.45
CE	0.61	0.39	0.38	0.41

Correlation coefficients of 12 proxy/target pairings, independently computed over different split periods and frequency domains, and for unfiltered and 20-year low- and high-pass filtered data range from 0.70–0.96 with a mean of 0.83 (p<0.001). Corresponding Durbin and Watson (DW) [48] statistics range from 1.42–2.10 (Table S3), and RE/CE statistics range from 0.45–0.64/0.38–0.61.

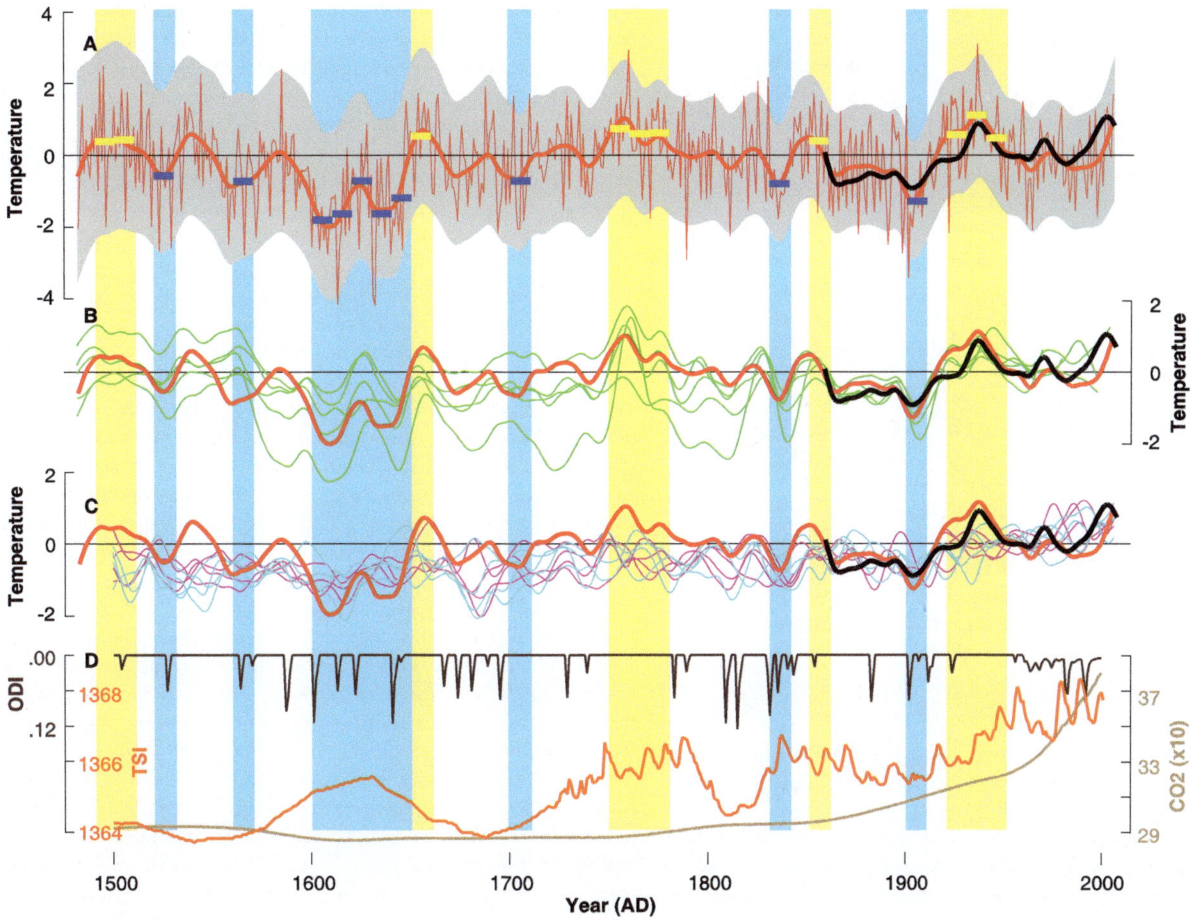

Figure 4. Reconstructed and modelled northern Scandinavian summer temperature variations over the past five centuries. (A) Actual (black) and reconstructed (red) JJA temperature anomalies (°C) with error estimates (grey) and the ten warmest and coldest decades superimposed (colour boxes). **(B)** Comparison of the actual and reconstructed temperatures with five existing (green) reconstructions (see Table S3 for details). **(C)** Comparison of the actual and reconstructed temperatures with CCSM3 SAT (pink) and SST (blue) model simulations. Mean and variance of the data are scaled against JJA temperature (1860–2006), expressed as anomalies (1961–1990) and 20-year low-pass filtered. **(D)** Solar (TSI; Total Solar Irradiance), volcanic (ODI; Optical Depth change in the visible band), and CO_2 (ppm) forcing as used in the model simulation.

strong positive agreement, however, may possibly result from overlapping data within this study and the AO estimate, and the fact that the AO study considered overall temperature sensitive proxy data [24]. Correlations between the temperature and AO reconstructions increase to 0.61, 0.70 and 0.73 after 10-year, 20-year and 30-year low-pass filtering, respectively. This frequency-dependence supports the strong multi-decadal coherency between variations in northern European summer temperature and high latitudinal pressure fields.

Comparison between our new record and five existing northern Scandinavian summer temperature reconstructions reveals a sound picture over the past two centuries (Figure 4B), but describes some offset prior to ~1800. It should be noted that these records are not methodologically independent and include some common data (Table S3). The five existing records show slightly lower correlations ($r = 0.52–0.80$) with gridded JJA temperatures (CRUTEM3v; averaged over 65–70°N and 20–30°E; 1860–1970) than the new reconstruction ($r = 0.85$). The five existing records express an average cross-correlation of 0.59 for the 802–1970 common period. Our new reconstruction correlates with the individual existing reconstructions between 0.42 and 0.76 (1483–1970), and at 0.74 with their average. After 20-year low-pass filtering, this value slightly decreases to 0.67. The five individual records fall within the uncertainty range of our new reconstruction and portray a similar sequence of warm and cold periods.

Comparison of our reconstruction with the SAT and SST of four ensemble simulations reveals that the model output falls within the proxy uncertainty but contains marginally less amplitude (Figure 4C). Some offset between cooler reconstructed and warmer simulated temperatures is found during the last half of the 20th century with the proxy values better matching the instrumental target. Comparison of the reconstructed and simulated temperature variations and trends with external solar, volcanic and CO_2 forcing (as used in the model) retains a rather weak link over the past 500 years (Figure 4D). Since systematic perturbations of external forcing are not evident in the reconstructed and simulated Scandinavian temperature history, internal climate dynamics must have played an important role in determining past summer temperatures within this region. It should be noted that the four SAT (SST) model ensembles correlate on average at 0.50 (0.48) with each other (1500–2100). Significant differences in coherency are found before and after 1850 for the SAT (0.04 and 0.56) and SST (0.09 and 0.51) ensemble members. A similar pattern with overall higher correlations is obtained after 20-year low-pass filtering. Early/late correlation averages are 0.21/0.92 for SAT and 0.17/0.90 for

SST, indicating less externally forced agreement prior to the industrial era, for instance. Correlations between our new reconstruction and the four SAT (SST) ensemble members (after 20-year low-pass filtering) average at 0.37 (0.24) over the full 1483–2006 period of overlap, with only small differences before and after 1850.

The combined proxy/model approach allows variations in pre- and post-industrial northern Scandinavian summer temperature to be assessed over the period 500–2100 (Figure 5). The tree-ring data indicate lower temperatures before ~700 and around ~800, and then again in the early 17th century commonly known to be part of the LIA. The first warm spell is recorded at ~760, a prolonged interval of generally warm summers occurred between ~1000 and 1100, warm summers also appeared ~1430, ~1770, and again in the 1930s and the 1970s. The early values, however, should be interpreted cautiously, because the amount of available TRW and MXD measurement series used in five northern Scandinavian proxy records decreases back in time. Nevertheless, our synthesis clearly demonstrates that summer temperatures have previously been as markedly warm as the observed 20th century conditions including the high temperatures of the 1930s and 1970s. Despite this, the model output further suggests that, under the SRES A2 scenario, temperatures will exceed the range of past variability from ~2030 onwards (Figure 5).

Discussion

Our MXD compilation represents a major update in one of the key regions in a circumpolar network [20,22], enabling the relationship between forest productivity and recent climate variability to be evaluated into the 21st century. The ability of MXD chronologies from northern Scandinavian conifers to capture the timing, amplitude and geographical distribution of recent temperature fluctuations is evidently confirmed. There is no indication for any systematic 'Divergence Issue' (e.g. [18]), namely the failure of tree growth to track high and low frequency temperature changes. Furthermore, our results reliably reproduced not only inter-annual temperature variations, but also long-term trends over a well-defined spatial domain (e.g. [20,22]). Such findings are consistent with updated tree-ring compilations from northern Siberia [25] and the European Alps [26], although the actual reconstructions themselves reflect different signals of regional climate fluctuations. Constraints in capturing low frequency temperature variability that could potentially result from either the tree-ring data used or the detrending methodology applied, have been addressed by the inclusion of young trees resulting in an even distribution of series start dates throughout time and by the performance of RCS detrending (for details see below). Uncertainty ranges associated with the resulting temperature reconstruction have been estimated via cumulative error estimates (see below), which likely resulted in a rather conservative uncertainty range. The effects of site location, shore position, species-specific response, tree age, sampling design, growth standardization, and proxy calibration have been considered, and are expressed by the temporal variation of error estimates.

A slight offset between cooler proxy reconstructions and warmer model simulations is observed during the latter half of the 20th century (Figure 4). This discrepancy is potentially due to slightly overestimated temperatures within the ensemble members, as anthropogenic sulphate aerosols have not been included in the model forcing. A sensitivity experiment, including this forcing, though resulted in less warming during the 20th century [15]. It is further noted that both reconstructed and simulated Scandinavian SAT show only weak responses to natural external forcing, which obviously indicates the importance of internal climate dynamics within this region. As shown by [15], variations in SST, which in turn are connected to the North Atlantic Thermohaline Circulation, must be considered as important drivers of multi-decadal climate variability across the North Atlantic/European

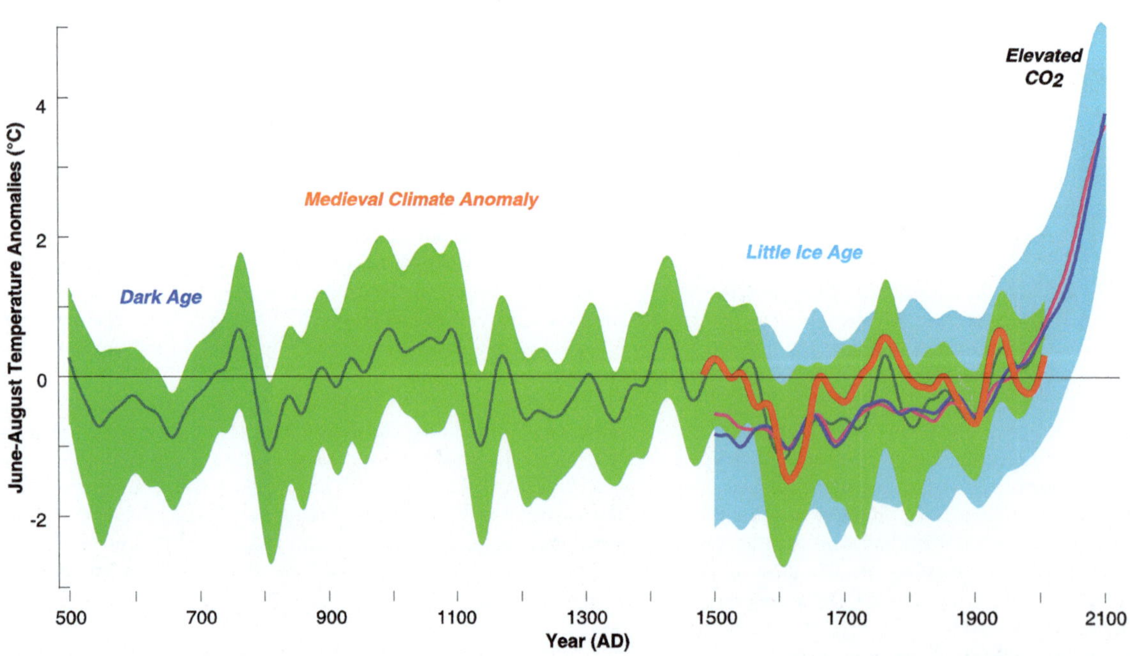

Figure 5. Proxy and model evidence of northern Scandinavian summer temperature variability from 1500–2100 AD. Reconstructed (red this study and green mean of five existing records) and projected (pink surface air and blue sea surface) temperature anomalies of the 1961–1990 reference period and 60-year low-pass filtered. Green and blue shadings indicate the uncertainty envelopes derived from the annual minimum and maximum values of the various proxy and model data, respectively.

sector and particularly over the Scandinavian study region. These internal atmosphere-ocean variations might also be strong enough to largely mask solar, volcanic, and even anthropogenic signals as suggested by the herein reconstructed and simulated temperatures of the past 500 years. Our results highlight the strength of a combined proxy/model approach to examining climatic variability, which extends the analytical skill of either using proxy reconstructions or model simulations only. The joint consideration of backward reconstructing and forward modelling further offers longer timescales to be evaluated. The Scandinavian records, however, differ from the situation commonly observed in Central Europe where impacts of solar and volcanic forcing on past temperature change have been undoubtedly identified (e.g. [19,27]). Based upon the available findings, it appears likely that the ocean may play a role in generating low frequency temperature variability.

Another important source of internal low frequency variability are atmosphere dynamics that comprise dominant pressure patterns, such as the AO, the NAO and the EAP, which in turn are often coupled with the underlying climate system (e.g. [5,28]). These pressure patterns may possibly exhibit linkages to temperature and precipitation fluctuations at decadal to multi-decadal timescales north of ~20°N. This suggestion is supported by the agreement between our new Scandinavian temperature history and a previous reconstruction of the summer AO [24].

Nevertheless, it is premature to conclude that globally observed effects of recent anthropogenic warming on ecology [29] may be less intense over Scandinavia, where internal climate variability likely moderated and modulated some of the latest temperature increase observed in other regions and at larger scales. Despite their dominance in large-scale temperature reconstructions, initial tests suggest that such hemispheric approaches are not particularly sensitive to the inclusion or exclusion of Scandinavian tree-ring data especially after ~1300 AD (Figure S6). Caution is advised as more detailed analyses at the hemispheric-scale are necessary to better differentiate between regional and lager scale climate change and even disentangle the relevant forcing agents at play. If Scandinavian summer temperatures indeed exhibit less post 1980s warming relative to earlier warm periods than many other regions across the Northern Hemisphere, however, still needs to be quantified.

The unprecedented degree of projected anthropogenic warming under the SRES A2 scenario as described by the modelling component of this study, would though imply severe alterations to ecosystem functioning and productivity, as well as trophic amplifications across a broad range of marine and terrestrial habitats and taxa [30,31]. Long-term stability of boreal population dynamics, sustainable food webs and biodiversity conservations are thus at risk [32,33]. Temperature proxies derived from oxygen isotopic analyses of foraminifera located along the Norwegian margin offer relatively high-resolution evidence on comparable timescales and thus have been used to link terrestrial and marine ecosystem responses to climate changes [34]. Previous temperature reconstructions have relied on physical-chemical relationships, namely that a 0.26‰ shift in the oxygen isotopes reflects a 1°C change in water temperature [34]. More recent endeavours have demonstrated the potential to develop shell-growth increments of *Artica islandica* in Norwegian coastal waters, which can contain well-defined climate signals [35–37]. Currently, these records only span several centuries. Nevertheless, it may be possible to extend them further, potentially even over the entire Holocene. If this would be achieved, the combination of such archives and the new MXD reconstruction presented here would enable further elucidation of the influence of ocean dynamics and internal forcing on temperature variability within this region.

Increased efforts to obtain new high-resolution proxy archives in tandem with an ensemble of model simulations is recommended not only to disentangle effects of external climate forcing from internal climate variability, but also to improve our understanding of associated spatiotemporal shifts in ecosystem responses to climatic variations. The expected outcome of such work would be relevant not only for climatologists, but also for ecologists, epidemiologists, oceanographers and even economists throughout the multiple effects of climate change on biological resources.

Materials and Methods

No additional sampling permits beside those organized by Mauri Timonen (MT; among the co-authors) were required to develop the herein presented MXD dataset (Dr. Yrjö Norokorpi, Area Manager, Natural Heritage Services of Metsähallitus, granted 15.09.2006 permission for tree-ring sampling to MT and the WSL research team of UB, DF, DN and JE). A total of 1,179 tree-ring core or disc samples were collected between 1978 and 2006 at 25 Scandinavian conifer sites located north of 65°N (Figure 1). Eighteen site chronologies consisted of pine (*Pinus sylvestris*), while seven were based on spruce (*Picea abies*). All of the 1,179 MXD measurement series were processed at the Swiss Federal Research Institute WSL (see [38] for methodological details). Pith-offset – the amount of years between the innermost ring and the expected germination age – was estimated for each core sample to range between 0 and 170 years. Non-climatic, i.e. biological-induced age trends were removed from the raw MXD measurement series using the Regional Curve Standardization (RCS) method on a site-by-site basis [39]. Series were first aligned by cambial age; a mean of the age-aligned series was then calculated and smoothed with a cubic spline of 10% the series length [40]. The resulting timeseries is termed the Regional Curve (RC) and used for detrending, i.e. deviations of the individual measurements from the RC were calculated as ratios. Dimensionless indices were transformed back to calendar years, and averaged per site using a robust bi-weighted mean. The variance of the subsequent mean chronologies was adjusted for changes in sample replication and inter-series correlation [41], with only periods whose replication exceeded 10 series were analysed further.

Gridded CRUTEM3v data (1860–2006) [42] averaged over 65–70°N and 20–30°E were employed as predictor variables for the reconstruction of Scandinavian summer temperatures (SI). Variability resulting from differences in site location, species-specific responses, lakeshore position (i.e. the distance between trees and the nearest lake), tree age, sampling design, standardization methodology, and model calibration was integrated in the estimated reconstruction error (see SI for details).

Eight homogenized station records (also included in the CRUTEM3v data) with continuous temperature measurements spanning >70 years were selected from the Global Historical Climatology Network (GHCN-adjusted; Figure 1 and Table S1), to estimate a likely range of northern Scandinavian temperature variations. Differences between local station records were used to calculate a data target envelope used for comparison with the MXD chronologies. The combined Tornedalen record was considered for extra-verification back to 1816 [43], and gridded SSTs from coastal Norway (65–70°N and 5–15°E, 1892–2002) were utilized for comparison (SSTV2) [44]. Correlations with gridded SAT and SST data, as well as comparison with MXD-based summer temperature reconstructions from the Polar Ural [25], the Pyrenees [45], and the Alps [19], were performed to evaluate the spatial signature of Scandinavian temperature variability. A composite analysis of gridded 500-hPa geopotential height field data [46] was used for the reconstructed 20 coldest and warmest Scandinavian summers of the period 1659–1999. This analysis allowed the assessment of possible impacts of large-scale atmospheric circulation patterns on more regional northern Scandinavian summer temperature extremes. A suite of five published tree ring-based summer temperature records from northern Scandinavian was gathered for comparison with our new record (Table S3). The five chronologies preserve inter-annual to multi-centennial variability and were rescaled against the same instrumental target record (CRUTEM3v) as the newly developed temperature reconstruction presented in this study. Finally, we used an ensemble of four transient simulations from the coupled Community Climate System Model version 3 (CCSM3) [47], covering the 1500–2100 period to derive proxy-independent temperature variations and run proxy/model comparisons (see SI for details).

Supporting Information

Text S1 Supporting information.

Figure S1 Temporal coverage of the 1,179 MXD series, of which 387 were developed before 2000 AD with 792 added in 2006, covers the entire Scandinavian peninsula north of 65°N and has a mean segment length (MSL) of 161 years with an maximum latewood density (MXD) of 0.72 g/cm3 (inset denotes the relationship between tree age and MXD value). Site chronologies start between 1400 and 1844, and their autocorrelation ranges from 0.23–0.50.

Figure S2 Correlation of the 25 MXD site chronologies after RCS detrending (circles) with monthly temperature means (CRUTEM3v) computed over the common 100 yr interval (1860–1977). Results are classified into 20 sensitive (green) and 5 less sensitive (grey; GLO, LOF, NAR, SKI, SKS) sites, with the horizontal lines referring to their average response.

Figure S3 (**A**) Correlation (r>0.3) of the reconstruction against gridded JJA SAT (1901–2006) and four MXD-based summer temperature records: 1 = Polar Ural (0.06), 4 = Tyrol (−0.07), 3 = Lötschental (0.13) and 2 = Pyrenees (0.00). (**B**) Correlation of the reconstruction against gridded JJA SST (1901–2003).

Figure S4 Spatial field correlation (r>0.3) between the new MXD-based reconstruction of northern Scandinavian summer temperature and gridded (5°×5°) SLP [HadSLP2r; 20] calculated for the June–August season and 1860–2006 period.

Figure S5 Spatial composite analysis at 500-hPa geopotential height of (**A**) the 20 coldest and (**B**) the 20 warmest reconstructed summers (JJA; 1659–1999), and (**C**) temporal comparison (1650–1975) of the reconstructed JJA temperatures (orange) with JJA estimates of the AO (light blue). Smoothed lines are 30-year low-pass filters and stars refer to the annual extremes composed.

Figure S6 Temperature reconstructions from Scandinavia (green) and the Northern Hemisphere that were 60-year low-pass filtered and standardized over their common period 831–1992. The two Northern Hemisphere versions either include (brown) or exclude (ochre) Scandinavian proxies.

Table S1 Characteristics of the (**A**) 25 MXD chronologies and (**B**) eight instrumental stations used. Grey site codes refer to less-sensitive chronologies, blue (green) refers to updated (pre-2000) site. The mean inter-series correlation (Rbar) and the Expressed Population Signal (EPS) were computed over 30-year windows lagged by 15 years [14]. Setting refers to the number of inhabitants if a station is located in urban terrain.

Table S2 (**A**) Temperature extremes of the target and proxy data computed over the common period 1860–2006. Italic letters refer to common extremes. (**B**) Reconstructed summer temperature extremes and associated error estimates computed over the full period 1483–2007.

Table S3 Summary information of the 5 Scandinavian tree-ring chronologies used for comparison with this study. 'Season' refers to the originally indicated response window, whereas 'Response' refers to correlations computed against Scandinavian JJA mean temperatures over the 1860–1970 common period.

Acknowledgments

We are particularly thankful to those colleagues that made their published tree-ring data available. CCSM3 simulations were performed at the Swiss National Supercomputing Centre (CSCS).

Author Contributions

Conceived and designed the experiments: UB DF JE. Performed the experiments: UB DF JE CR DH. Analyzed the data: UB DF JE CR DH DN AV. Contributed reagents/materials/analysis tools: LC SH MT. Wrote the paper: UB CR JE LC NS DH SH DF.

References

1. D'Arrigo R, Wilson R, Jacoby G (2006) On the long-term context for late twentieth century warming. J Geophys Res 111: doi:10.1029/2005JD006352.
2. Esper J, Wilson RJS, Frank DC, Moberg A, Wanner H, et al. (2005) Climate: Past Ranges and Future Changes. Quat Sci Rev 24: 2164–2166.
3. Grudd H (2008) Torneträsk tree-ring width and density AD 500–2004: a test of climatic sensitivity and a new 1500-year reconstruction of north Fennoscandian summers. Clim Dyn 31: 843–857.
4. Frank D, Esper J, Zorita E, Wilson R (2010) A noodle, hockey stick, and spaghetti plate: a perspective on high-resolution paleoclimatology. Wiley Interdiscip Rev Clim Change 1: 507–516.
5. Raible CC, Luksch U, Fraedrich K, Voss R (2001) North Atlantic decadal regimes in a coupled GCM simulation. Clim Dyn 18: 321–330.
6. Yoshimori M, Stocker TF, Raible CC, Renold M (2005) Externally-forced and internal variability in ensemble climate simulations of the Maunder Minimum. J Climate 18: 4253–4270.
7. Sutton RT, Hodson DLR (2005) Atlantic ocean forcing of North American and European summer climate. Science 309: 115–118.
8. Keenlyside NS, Latif M, Jungclaus J, Kornblueh L, Roeckner E (2008) Advancing decadal-scale climate prediction in the North Atlantic sector. Nature 453: 84–88.
9. Helama S, Timonen M, Holopainen J, Ogurtsov MG, Mielikäinen K, et al. (2009) Summer temperature variations in Lapland during the Medieval Warm Period and the Little Ice Age relative to natural instability of the thermohaline circulation on multi-decadal and multicentennial scales. J Quat Sci 24: 450–456.
10. Jones PD, Osborn TJ, Briffa KR (1997) Estimating sampling errors in large-scale temperature averages. J Clim 10: 2548–2568.
11. Schlesinger ME, Ramankutty N (1994) An oscillation in the global climate system of period 65–70 years. Nature 367: 723–726.
12. Delworth TL, Mann ME (2000) Observed and simulated multi-decadal variability in the Northern Hemisphere. Clim Dyn 16: 661–676.
13. Thompson DWJ, Wallace JM (1998) The Arctic Oscillation signature in the wintertime geopotential height and temperature fields. Geophys Res Let 25: 1297–1300.

14. Raible CC, Stocker TF, Yoshimori M, Renold M, Beyerle U, et al. (2005) Northern Hemispheric trends of pressure indices and atmospheric circulation patterns in observations, reconstructions, and coupled GCM simulations. J Clim 18: 3968–3982.
15. Hofer D, Raible CC, Stocker TF (2011) Variations of the Atlantic meridional overturning circulation in control and transient simulations of the last millennium. Climate of the Past 7: 133–150.
16. Hegerl GC, Crowley TJ, Hyde WT, Frame DJ (2006) Climate sensitivity constrained by temperature reconstructions over the past seven centuries. Nature 440: 1029–1032.
17. Zorita E, Moberg A, Leijonhufvud L, Wilson R, Brázdil R, et al. (2010) European temperature records of the past five centuries based on documentary information compared to climate simulations. Clim Change 101: 143–168.
18. D'Arrigo R, Wilson R, Liepert B, Cherubini P (2008) On the 'Divergence Problem' in Northern Forests: A Review of the Tree-Ring Evidence and Possible Causes. Glob Planet Change 60: 289–305.
19. Büntgen U, Frank DC, Nievergelt D, Esper J (2006) Summer temperature variations in the European Alps, AD 755–2004. J Clim 19: 5606–5623.
20. Briffa KR, Osborn TJ, Schweingruber FH, Jones PD, Shiyatov SG, et al. (2002) Tree-ring width and density around the Northern Hemisphere: Part 1, local and regional climate signals. Holocene 12: 737–757.
21. Battipaglia G, Frank DC, Büntgen U, Dobrovolny P, Brázdil R, et al. (2010) Five centuries of Central European temperature extremes reconstructed from tree-ring density and documentary evidence. Glob Planet Change 72: 182–191.
22. Briffa KR, Osborn TJ, Schweingruber FH, Jones PD, Shiyatov SG, et al. (2002) Tree-ring width and density around the Northern Hemisphere: Part 2, spatio-temporal variability and associated climate patterns. Holocene 12: 759–789.
23. Kaufmann DS, Schneider DP, McKay NP, Ammann CM, Bradle RS, et al. (2009) Recent Warming Reverses Long-Term Arctic Cooling. Science 325: 1236–1239.
24. D'Arrigo DD, Cook ER, Jacoby GC, Mann ME (2003) Tree-ring reconstructions of temperature and sea-level pressure variability associated with the warm-season Arctic Oscillation since AD 1650. Geophys Res Lett 30: 1549.
25. Esper J, Frank D, Büntgen U, Verstege A, Hantemirov RM, et al. (2010) Trends and uncertainties in Siberian indicators of 20th century warming. Glob Change Biol 16: 386–398.
26. Büntgen U, Frank DC, Wilson R, Career M, Urbinati C, et al. (2008) Testing for tree-ring divergence in the European Alps. Glob Change Biol 14: 2433–2453.
27. Fischer EM, Luterbacher J, Zorita E, Tett SFB, Casty C, et al. (2007) European climate response to tropical volcanic eruptions over the last half millennium. Geophys Res Lett 34: doi:10.1029/2006GL027992.
28. Gray ST, Graumlich LJ, Betancourt JL, Pederson GT (2004) A tree-ring based reconstruction of the Atlantic multidecadal oscillation since 1567 AD. Geophys Res Lett 31: L12205.
29. Parmesan C, Yohe G (2003) A globally coherent fingerprint of climate change impacts across natural systems. Nature 421: 37–42.
30. Stenseth NC, Mysterud A, Ottersen G, Hurrell JW, Chan K, et al. (2002) Ecological effects of climate fluctuations. Science 297: 1292–1296.
31. Edwards M, Richardson AJ (2004) Impact of climate change on marine pelagic phenology and trophic mismatch. Nature 430: 881–884.
32. Ims RA, Henden JA, Killengreen ST (2008) Collapsing population cycles. Trends in Ecol Evo 23: 79–86.
33. Kausrud KL, Mysterud A, Steen H, Vik JO, Østbye E, et al. (2008) Linking climate change to lemming cycles. Nature 456: 93–97.
34. Klitgaard Kristensen D, Sejrup HP, Haflidason H, Berstad IM, Mikalsen G (2004) Eight-hundred years temperature variability from the Norwegian continental margin and the North Atlantic thermohaline circulation. Paleoceanography 19: 10.1029/2003PA000960.
35. Schöne BR, Oschmann W, Rössler J, Freyre Castro AD, Houk SD, et al. (2003) North Atlantic Oscillation dynamics recorded in shells of a long-lived bivalve mollusk. Geology 31: 1237–1240.
36. Helama S, Schöne BR, Kirchhefer AJ, Nielsen JK, Rodland DL, et al. (2007) Compound response of marine and terrestrial ecosystems to varying climate: pre-anthropogenic perspective from bivalve shell growth increments and tree-rings. Marine Environmental Res 63: 185–199.
37. Helama S, Hood BC (2011) Stone Age midden deposition assessed by bivalve sclerochronology and radiocarbon wiggle-matching of Arctica islandica shell increments. J Archaeol Sci 38: 452–460.
38. Eschbach W, Nogler P, Schär E, Schweingruber FH (1995) Technical advances in the radiodensitometrical determination of wood density. Dendrochronologia 13: 155–168.
39. Esper J, Cook ER, Krusic PJ, Peters K, Schweingruber FH (2003) Tests of the RCS method for preserving low-frequency variability in long tree-ring chronologies. Tree-Ring Res 59: 81–98.
40. Cook ER, Peters K (1981) The smoothing spline: A new approach to standardizing forest interior tree-ring width series for dendroclimatic studies. Tree-Ring Bulle 41: 45–53.
41. Frank D, Esper J, Cook ER (2007) Adjustment for proxy number and coherence in a large-scale temperature reconstruction. Geophys Res Lett 34: doi10.1029/2007GL030571.
42. Brohan P, Kennedy JJ, Harris I, Tett SFB, Jones PD (2006) Uncertainty estimates in regional and global observed temperature changes: a new dataset from 1850. J Geophys Res 111: doi:10.1029/2005JD006548.
43. Klingbjer P, Moberg A (2003) A composite monthly temperature record from Tornedalen in Northern Sweden, 1802–2002. Int J Climatol 23: 1465–1494.
44. Kaplan A, Cane M, Kushnir Y, Clement A, Blumenthal M, et al. (1998) Analyses of global sea surface temperature 1856–1991. J Geophys Res 103: 567–589.
45. Büntgen U, Frank D, Trouet V, Esper J (2010) Diverse climate sensitivity of Mediterranean tree-ring width and density. Trees 24: 261–273.
46. Luterbacher J, Xoplaki E, Dietrich D, Rickli R, Jacobeit J, et al. (2002) Reconstruction of sea level pressure fields over the Eastern North Atlantic and Europe back to 1500. Clim Dyn 18: 545–561.
47. Yoshimori M, Raible CC, Stocker TF, Renold M (2010) Simulated decadal oscillations of the Atlantic meridional overturning circulation in a cold climate state. Clim Dyn 34: 101–121.
48. Durbin J, Watson GS (1951) Testing for serial correlation in least squares regression. Biometrika 38: 159–78.

7

3800 Years of Quantitative Precipitation Reconstruction from the Northwest Yucatan Peninsula

Alicia Carrillo-Bastos, Gerald A. Islebe*, Nuria Torrescano-Valle

El Colegio de la Frontera Sur, Unidad Chetumal, Herbario, Chetumal, Quintana Roo, México

Abstract

Precipitation over the last 3800 years has been reconstructed using modern pollen calibration and precipitation data. A transfer function was then performed via the linear method of partial least squares. By calculating precipitation anomalies, it is estimated that precipitation deficits were greater than surpluses, reaching 21% and <9%, respectively. The period from 50 BC to 800 AD was the driest of the record. The drought related to the abandonment of the Maya Preclassic period featured a 21% reduction in precipitation, while the drought of the Maya collapse (800 to 860 AD) featured a reduction of 18%. The Medieval Climatic Anomaly was a period of positive phases (3.8–7.6%). The Little Ice Age was a period of climatic variability, with reductions in precipitation but without deficits.

Editor: Gil Bohrer, The Ohio State University, United States of America

Funding: This work received funding from: Conacyt funding project 127613. www.conacyt.mx. The funders had no role in study design, data collection and analysis, decision to publish, or preparation of the manuscript.

Competing Interests: The authors have declared that no competing interests exist.

* E-mail: gislebe@ecosur.mx

Introduction

The understanding of climate dynamics and the prediction of future changes require the identification of patterns of climatic change at different temporal scales [1]. To find these patterns, it is necessary to take into account continuous readings that go beyond instrumental measurements, which provide a temporally limited perspective.

Fossil pollen is a natural source of climatic information. Vegetation is subject to environmental conditions that favor certain species depending on their climatic preference. By calibrating modern pollen samples with climate variables, it is possible to reconstruct quantitative climatic parameters [2].

Precipitation is one of the most important and sensitive parameters of the tropical climate [3]. The reconstruction of precipitation is of great interest because it permits the validation of climatic change models by examining the patterns of the past, present, and future [4]. On the Yucatan Peninsula, the fossil pollen record has been analyzed in terms of vegetation change. However, the inferences concerning climate have been expressed only in qualitative terms (wetter/drier). In the Yucatan region, the variability in precipitation had an important influence on the cultural history of the Mayan civilization [5]. Previous studies on the peninsula have indicated that the droughts suffered during the Classic period contributed to the collapse of the Mayan civilization [6–9].

A quantitative estimate of the precipitation over the last 1500 years is available for the peninsula. The record is based on measurements of $\delta^{18}O$ values in stalagmites from the Tzabnabh cave [8]. In that study, it was determined that precipitation decreased by 36–52% relative to modern precipitation during the multiple droughts at the end of the Classic period. Recently, based on quantitative analysis of high-resolution climate records, it was concluded that the droughts of the Terminal Classic period were due to the reduction in the frequency and intensity of summer storms [9].

The Yucatan Peninsula is influenced by a number of different components of the climate system, creating a zone of mixed climates. For this reason, it is necessary to investigate whether the changes that occurred in this region were similar in magnitude, frequency and time in order to understand climatic variability. These changes are of particular importance because studying the events of the recent past will permit us to foresee what may occur in the near future, as climatic forcing that operated during the late Holocene did not significantly change [10]. With the purpose of contributing to the knowledge of climatic variability of the Yucatan Peninsula, this study presents a reconstruction of the vegetation from the northwest corner of the peninsula and the first quantitative reconstruction of precipitation for the last 3800 years.

Study Area

The study area is located in the northwest Yucatan Peninsula between the coordinates 21.5661N and 88.0865W in the biosphere reserve of Ria Lagartos (Figure 1). Because of its geographical position and proximity to the sea, the site is influenced by several different patterns of atmospheric circulation, such as the trade winds, polar air masses and the convection currents that provide precipitation [11]. The climate type is BSo(h')w(x)iw, warm-semiarid and evaporation exceeds precipitation. The largest portion of the annual precipitation is received during the summer (62%), when the Intertropical Convergence Zone (ITCZ) and Bermuda Azores high-pressure cell move to the north and the trade winds intensify. The remaining precipitation (38%) occurs during the dry season from November to May. The influence of polar air masses occurs during the months of

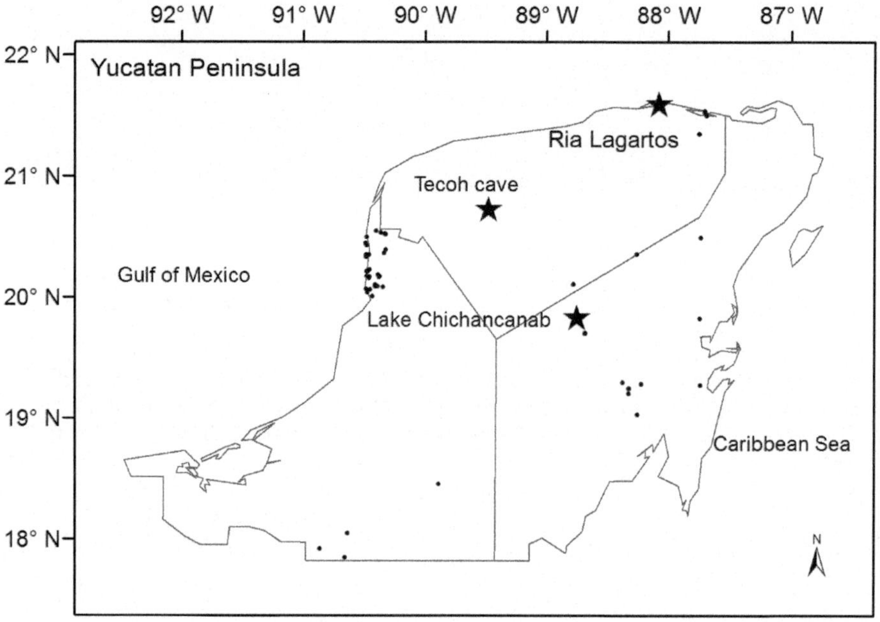

Figure 1. Map of area studied showing the sites mentioned in the discussion and the location of the modern pollen samples.

September to April [12]. The average temperature is 26°C [12], and the annual precipitation is 760 mm.

Several different vegetation regimes are found in the reserve: medium forest, dry forest, low forest, mangroves, coastal dunes and wetlands [11]. Vegetation near the coring site includes *Metopium brownei*, *Bursera simaruba*, *Haematoxylum campechianum*, *Conocarpus erectus*, *Plumeria sp.*, and *Bravaisia sp.*

Methods

Calibration of Pollen Rain Data and Precipitation

Pollen data. Samples of modern pollen rain were collected from different points of the Yucatan Peninsula (Figure 2), from a previous study [13] and present work. All are superficial sediment samples that were taken at a depth between 3 and 5 cm. At each site, 2–3 samples were taken and homogenized into one sample to avoid the effect of over-representation of local vegetation. A total of 57 samples were obtained.

The samples were treated with the standard techniques used for extraction and concentration: KOH, HCl, acetolysis [14], suspended glycerine gelatin (first study [13]) and an ultra-kit at 100% in this study. Tablets of *Lycopodium* were added to all the samples at the beginning of the treatments as an exotic marker. The pollen was identified with the help of the atlas developed by Palacios-Chavez et al. [15] and the palynological collection of Herbario-Ecosur (CIQR).

A minimum of 300 pollen of taxa belonging to pollen sum taxa were counted with an optical microscope with objectives of 400× and 1000×. Aquatic taxa and fungal spores were excluded from the pollen sum.

A total of 78 taxa were identified. To establish criteria of standardization and consistency in the data and to reduce bias, only taxa with percentages ≥1% and present in at least 5% of the samples were included [16]. Following this procedure, 30 species (Figure 2) were selected and the percentages were recalculated accordingly.

Precipitation data. Precipitation values were determined based on 65 meteorological stations on the Yucatan Peninsula from national weather services. At present, there are a total of 157 stations on the peninsula. However, only the 65 selected had at least 25 years of records during the period 1981 to 2010. The precipitation values for each sample of pollen rain were calculated by ordinary Kriging using the ArcGis 10 extension Geostatistical Analyst.

Modern Pollen–precipitation Relationship

The first step in the quantitative reconstruction of precipitation is to determine if the relationship between the plant species and the environmental variable follows a linear or a unimodal model [2]. For this, it is necessary to estimate the gradient length of the environmental variable expressed in standard deviation (SD) units of biological turnover [2]. To estimate this, detrended canonical correspondence analysis (DCCA) [17] was conducted, using the software CANOCO 4.5 [18]. The following parameters were used: precipitation as the only predictor variable, detrending by segment, non-linear rescaling and pollen data transformed to the square root. During this analysis, four samples exhibited extreme values and as a result were discarded when constructing the transfer functions.

Model of the Transfer Function

The length of the gradient of the environmental variable in SD units is an estimate of the behavior of the species along the length of this gradient [2]. If the gradient is small (2 SD units), the taxa behave monotonically through the gradient and the use of linear models of regression and calibration are appropriate [2]. In the DCCA, the length of the gradient obtained was <2 SD. Therefore, the transfer function was estimated using the method of partial least squares (PLS) regression [19], using software C2 1.7.2 [20].

PLS regression is a linear method that, by using inverse regression, calculates the transfer function that relates the modern pollen data to the environmental variable. This function can then be applied to fossil pollen data to reconstruct precipitation [2]. Evaluation of the prediction and selection of the best model with

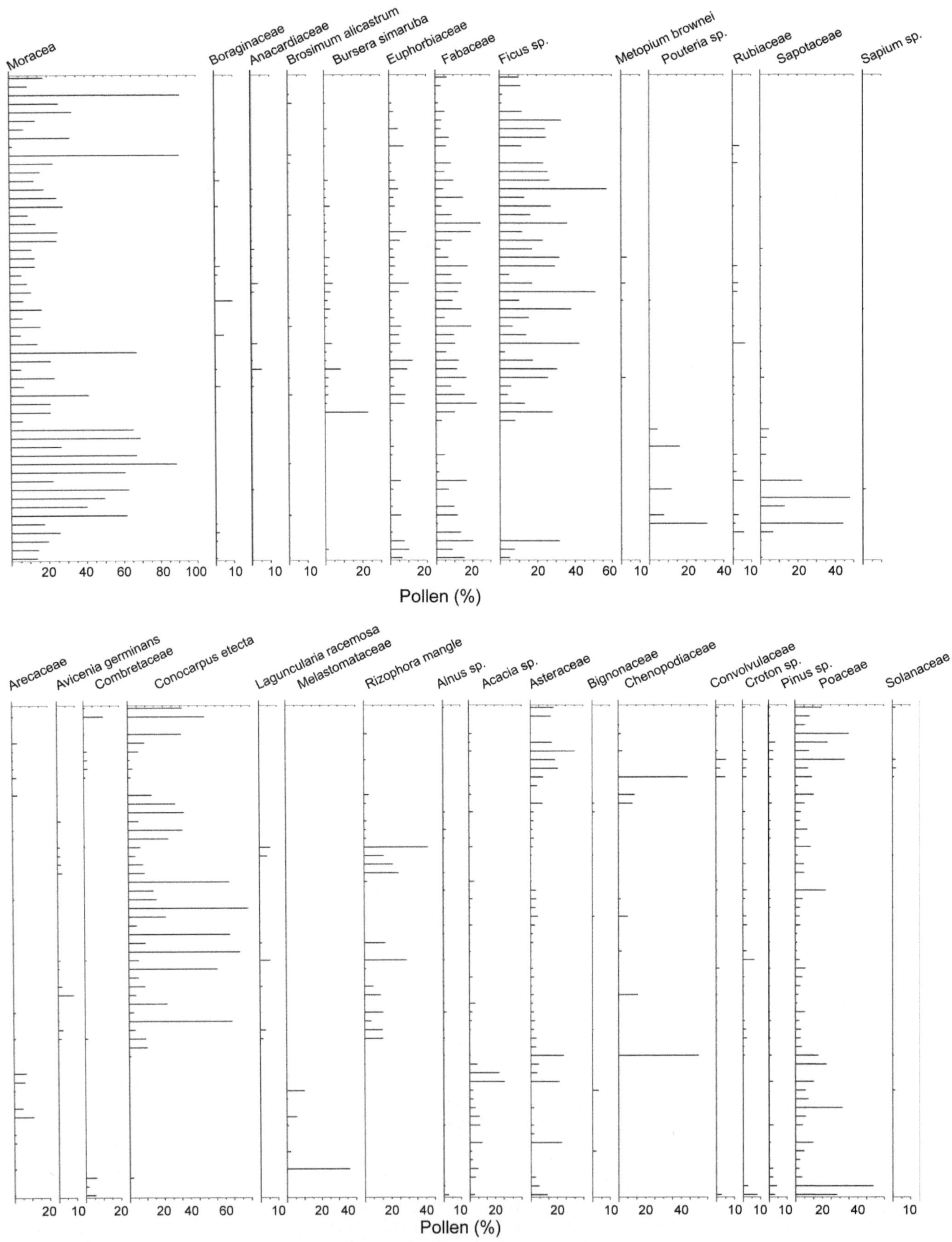

Figure 2. Percentage of taxa used in the calibration. The samples are sorted in ascending order with respect to the precipitation.

the smallest number of components are based on the combination of the values of maximum bias, root mean squared error (RMSEP), the coefficient of determination (r^2) and the residuals, all four of which are based on the method of cross-validation by bootstrapping 1000 interactions. The model selected was a two component PLS regression.

The precipitation anomalies are calculated using the following formula: $A = (R_i - Rm)*100/Rm$, where A is the anomaly of every period, R_i is the estimated precipitation in the same period and R_m is the average of all recorded precipitation.

Fossil Data

With a Russian corer, 50 cm sections of fossil sediment were obtained in a lagoon situated 5 to 7 km from the coastline. The total length of the core was 200 cm. The sediment was packed and transported to the palynology laboratory of Ecosur-Chetumal. The sediment was processed by physical and chemical treatments using the following protocol: treatment with HCl, filtration, digestion by KOH, acetolysis and mounted in Ultra-Kit at 100%. During the treatment with HCl, tablets of *Lycopodium* were added as an exotic marker. A total of 74 samples of fossilized pollen were obtained at an interval of 3 cm.

Taxa were separated by groups of ecological preference: tropical forest, disturbed taxa and mangrove. The pollen data are available in the Neotoma Database (http://www.neotomadb.org/uploads/data_subs/LAGARTOS.zip). The results obtained were graphed in a pollen diagram using the software TILIA 1.7.16 [21]. Pollen zones were determined by classification analysis based on the index of similarity using the tool CONISS, integrated into the same software package. The chronology was estimated by a linear regression applied to three radiocarbon dates (AMS) (Table 1) measured in the organic matter of the samples. The samples were analyzed in the laboratory of Beta Analytic Inc. in Miami, Florida. The radiocarbon dates were calibrated with the program CALIB 6.1.1 [22] using the IntCal09 calibration curve [23], rounded to the nearest decade and converted to calendar ages using 1950 AD as 0 cal yr BP.

Ethics Statement

The samples are from locations where no specific permissions were required, and field studies did not involve endangered or protected species.

Results

Modern Pollen – precipitation Relationship

In the DCCA, the length of the gradient was 1.69 SD. The correlation between the first ordination axis of the taxa and precipitation was 0.803, while the percentage of variance accumulated between taxa and the precipitation was 95.6%. The eigenvalue of the first ordination axis was 0.201. The ordination diagram showed separation of taxa in the first axis, which generated the correlation with the precipitation (Figure 3).

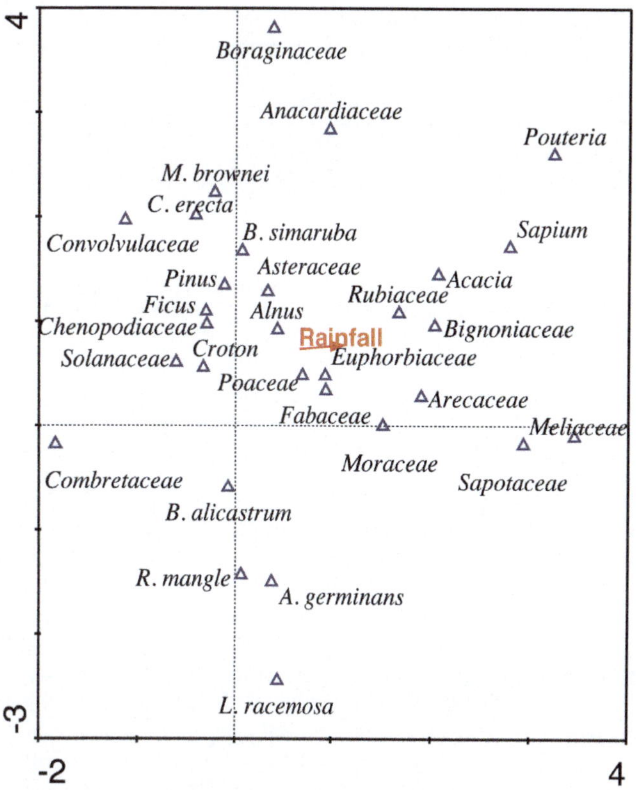

Figure 3. Ordination diagram results of DCCA.

Model of the Transfer Function

In the construction of the transfer function, a model with two components is used (Table 2). This model showed the lowest value of the RMSEP (131 mm) and a medium–high value for the coefficient of determination (0.703). The value of maximum bias (280.612) was not the smallest, but with this number of model components, the percentage of reduction in the RMSEP was 7.041% and the value of significance was 0.04.

Chronology and Reconstruction of the Vegetation

The resulting equation of the linear regression applied in conjunction with the calibrated ages (Table 1) produced, as a result, an age of 1850 BC at the bottom of the record.

Table 1. Radiocarbon ages (AMS) and sedimentation rates.

Sample code	Depth (cm)	Radiocarbon age (yr BP)	Calibrated age (cal yr BP)	Calendar age (BC/AD)	Relative area under probability distribution (2σ)	Sedimentation rate (mm/yr)
Beta-317982	84.1	2160±30	2056–2185	106–235 BC	0.54807	0.40
Beta-325222	171.36	3020±30	3141–3338	1191–1388 BC	0.974839	0.78
Beta-325223	193.8	3130±30	3319–3409	1369–1459 BC	0.847747	1.80

Regression equation for chronology *Age = (depth+9.1809)/0.055* $r^2 = 0.95987$.

Table 2. Model performance.

PLS component	RMSEP	r^2	Maximum bias	% Change	t-test significance
1	140.97	0.486	348.176	...	...
2	**131.05**	**0.703**	**280.612**	**7.041**	**0.044**
3	133.21	0.735	255.696	−1.653	0.579
4	142.485	0.761	253.917	−6.956	0.984

Performance statistics of the four components of the PLS pollen-rainfall transfer function. Root mean squares error of prediction (RMSEP), coefficient of determination (r^2), and maximum bias and reduction percentage in RMSEP given based on a bootstrapping cross-validation method. The select two-component model is shown in bold.

The average calculated rate of sedimentation is 0.99 mm/yr. In the fossil pollen count, we identified 53 pollen taxa that belonged to 37 families.

Only those taxa composing ≥1% in at least 5% of the samples were used for the quantitative analysis (Figure 4). The curve of *Zea mays* is also represented because it gives valuable information about the presence of human activity. After applying the classification analysis to the percentages of pollen, four pollen zones were established, which represent the principal phases of vegetation change.

Zone I (200-130 cm, 1850-580 BC). In this zone, the main change is between taxa of the tropical forest and the disturbance taxa. The tropical forest oscillates between 50 and 90%, dominated by Moraceae (15–80%), followed by *Ficus sp.*, Fabaceae and Euphorbiaceae. In the disturbed vegetation, the dominant taxa is Chenopodiaceae (40%) and, in lesser proportions, *Croton sp.*, Asteraceae and Poaceae. Mangrove vegetation is scarce and does not reach 5%. The first record of *Z. mays* appeared around 1840 BC.

Zone II (130-97 cm, 580 BC-20 AD). The change during this zone is due to the decrease in disturbance elements (percentages ≤20%) and the increase in forest taxa (70–89%). *Pouteria sp.* and Rubiaceae are present. *Brosimum alicastrum* is present almost continuously, and Moraceae increases (values ≥58%). There is also a slight increase of mangrove (values ≤3.3%), especially *Conocarpus erectus*. *Z. mays* appears discontinuously.

Zone III (97-57 cm, 20–750 AD). This zone is marked by an increase of disturbance vegetation and mangroves and decrease of forest taxa. *B. alicastrum* is not present at the start of this zone, and *Bursera simaruba* appears, while Moraceae drops to 10%. *Z. mays*, Asteraceae and *Pinus sp.* are continuous throughout and in greater percentage than in the previously mentioned zones. Chenopodiaceae and Poaceae predominate in this vegetation zone. The presence of *C. erectus* is continuous and in percentages between 2 and 12%.

Zone IV (57-0 cm, 750 AD - present). Percentages of disturbed taxa are low, such as Chenopodiaceae (≤2%), Poaceae (≤8%), Asteraceae (≤1%), and *Croton sp.* (≤1%). *Z.mays* does not appear, and *Pinus sp.* are found. Moraceae is the dominant element

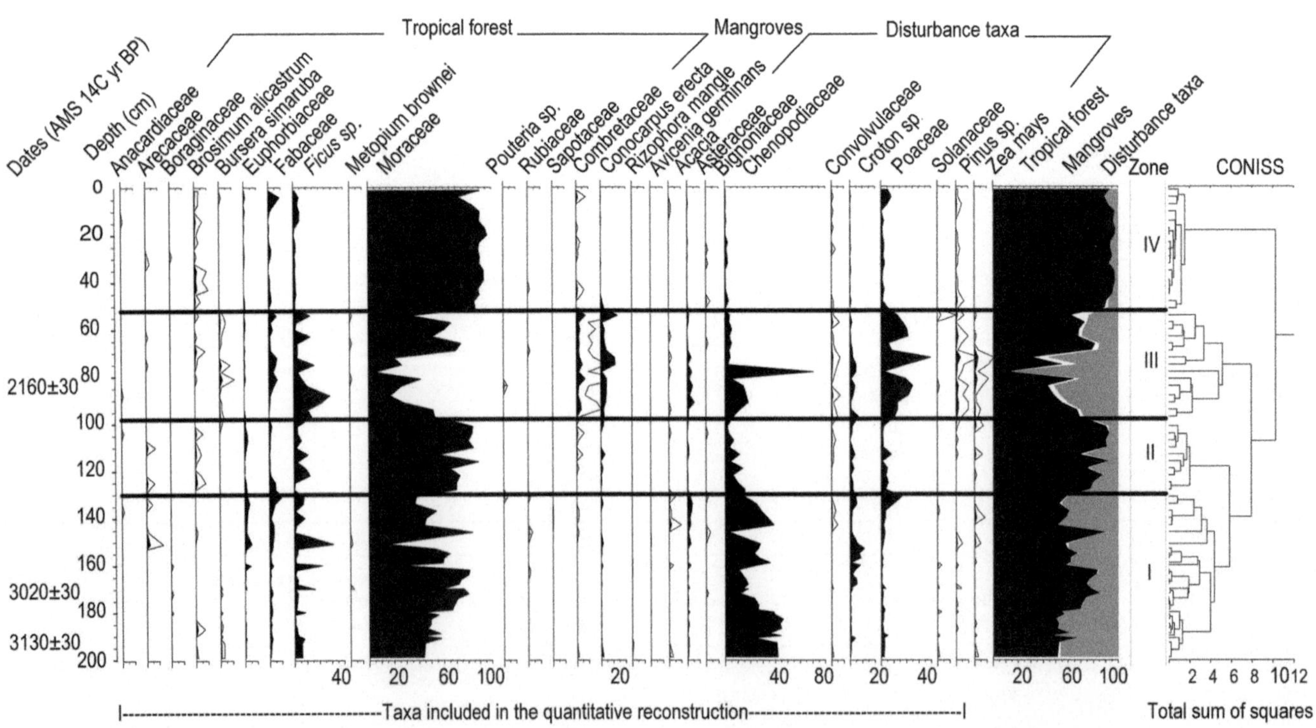

Figure 4. Diagram of fossil pollen.

(≥70%). *Ficus sp.* are lower. *B. simaruba* is not present, while *B. alicastrum* is present continuously. The curve of *C. erectus* is discontinuous.

Quantitative Precipitation Reconstruction

The application of the transfer function to the pollen data of Ria Lagartos permits the reconstruction of precipitation variability during the last 3800 years. The average precipitation for this period is 850 mm with a SD of 66 mm. Four phases of change are detected: the first from 1850-1000 BC with an average of 883 mm, the second from 1000-50 BC with an average of 870 mm, the third from 50 BC-500 AD with an average of 748 mm and the fourth from 500–1770 AD with an average of 898 mm (Figure 5). Upon calculating the anomalies of precipitation, it appears that the greatest deficit occurred during the third phase. The years 180, 420 and 800 AD feature negative anomalies of 21% in the first extreme drought and 18% in the other two droughts.

However, the surplus did not even reach 9%. There were two surpluses in the second phase of change (820 and 540 BC) and two in the fourth phase (1420 and 1600 AD).

Discussion

Modern Pollen – precipitation Relationship

The longitude of the gradient obtained with the DCCA was low (<2 SD). Therefore, the relationship between taxa and precipitation has a monotonic relationship. From ecological studies, it is observed that species abundance exhibits a unimodal behavior with respect to an environmental variable [2]. Each taxon grows best at an optimum value of the environmental variable and is unable to fully develop at higher or lower values [24]. The range captured by the values of precipitation is most likely only a fraction of the unimodal curve. This explains why the taxa in this study showed monotonic and unimodal behavior. For the same reason, it is probable that the ordination diagram will show a short vector for precipitation. However, the diagram can show that the first axis of ordination separates the majority of the groups and their correlation with precipitation was 0.803.

Along this axis, the referred taxa act as indicators of humidity. Conserved vegetation in the paleoecological studies of the region [25–28] are ordered in the direction of the precipitation vector (e.g., Moraceae, Meliaceae, Sapotaceae, *Pouteria sp.*, Rubiaceae and Sapindaceae). Taxa associated with less humid conditions and with disturbance vegetation are grouped on the opposite side of the graph (e.g., Chenopodiaceae, *Croton sp.*, Asteraceae, Solanaceae, Convolvulaceae and Poaceae).

It is important to note that the second axis contributes to the ordering of the taxa. Although it is not possible to determine the responsible variable, it is most likely related to the condition of the soil. Species such as *Rhizophora mangle*, *Laguncularia racemosa* and *Avicennia germinans* are all clearly separated. *C. erectus* is associated with mangroves (but only away from the influence of the tide) and is found grouped together with *Metopium brownie* and *B. simaruba*, which prefer well-drained soils but can withstand periodic floods typical of low areas of the peninsula [29].

Model of the Transfer Function

For the Yucatan Peninsula, no other quantitative reconstructions that use pollen as a proxy exist. This absence rules out a comparison between the statistical values obtained in the construction of the transfer function.

However, the evaluation and selection of a model can be based on a combination of the different parameters. One of these parameters is RMSEP, which in our chosen model took the lowest value (Table 2).

This value indicates the prediction ability of the modern dataset, and therefore, we assert that the calibration model can function as a tool of prediction [2].

It is also necessary to look for a low value for the maximum bias, as it measures systematic differences in the prediction [30]. It is not the lowest value of all the models, but it is much lower in relation to the model of a single component. The medium-high r^2 value indicates a strong relationship between the inferred values and those observed [2]. The maximum bias and r^2 statistics were neither the lowest nor the highest, respectively, but the lower RMSEP was certainly important. For a model to be useful, it is necessary that the RMSEP is reduced to at least 5% in relation to the model with the least number of components [31].

Furthermore, it is important to consider that the p value must be <0.05 to ensure that the estimates of the model are not just due to

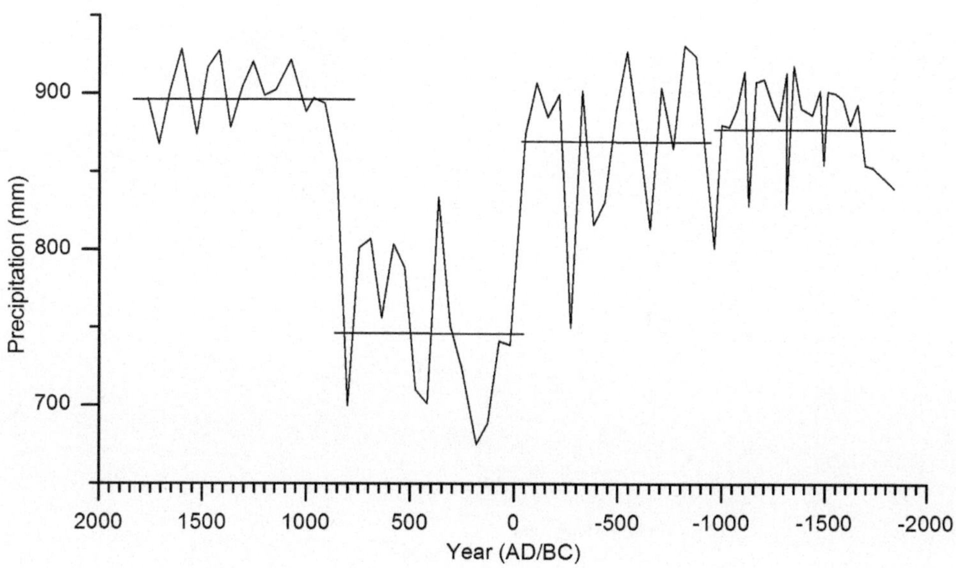

Figure 5. Inferred precipitation from fossil pollen. The horizontal lines indicate the averages for every phase.

chance. In this work, the p value of the reconstruction was 0.044. Therefore, we can consider that this value is statistically significant.

The graphs that compare observed values to estimated values (Figure 6a) and the observed values to the residuals (estimated minus observed) (Figure 6b) show that the model predicts values greater than 800 mm relatively well but is less reliable for medium values. However, the low values tend to be overestimated, with differences of at least 160 mm. Overall, the model is generally better with high values than with low values.

Chronology and Reconstruction of the Vegetation

Research has revealed that the vegetation of the region has been subject to the impact of human activities since early times. The first evidence of maize occurs around 1840 BC. The presence of taxa such as Malvaceae, *Acacia sp.*, Burseraceae, *Ficus* sp. and Fabaceae indicates that, during this stage, the vegetation consisted of secondary forest with Chenopodiaceae, Poaceae, and some cultivated plants. Data from other sites on the Yucatan Peninsula show conditions of low humidity during this stage [6], [26]. For this reason, we may infer that the secondary character of the vegetation is a reaction to the climatic conditions. The presence of Rubiaceae and the increase in abundance of Moraceae indicate that the forest recovered in the late phase of the middle Holocene (1550-1130 BC).

At the beginning of the late Holocene, a reduction in Moraceae and an increase of *Ficus sp.* and Chenopodiaceae are recorded, suggesting disturbed conditions in the region. This disturbance was not related to agricultural activity, as neither *Z. mays* nor elements associated with cultivation, such as *Croton sp.*, are present. Circa 950 BC, the vegetation acquired a more open character. This is inferred by the increase in the percentage of *Pinus sp.*, which do not form part of the regional vegetation. The presence of this pollen indicates that the arboreal cover was sparse enough to permit the deposition of allochthonous pollen [32]. During this time, the vegetation was a mosaic of low tropical forest, savannas and crops. From 550 BC to 50 AD, the increase in Moraceae; the almost continuous presence of *B. alicastrum*; and the presence of *Pouteria sp.*, Rubiaceae and Sapotaceae, indicate the recovery of the vegetation.

From 50 to 750 AD, the tropical forest suffered the greatest transformation of the studied period. The pollen record shows that tree cover decreased circa 190 AD. This is demonstrated by the reduction of tropical forest taxa (<50%) such as Moraceae (<20%) and the increase of *Pinus sp.* and *Ficus sp.* This change in the vegetation coincides temporally with the Preclassic drought reported in the Maya region [7], [25–26], [33–35]. In 370 AD, the forest changed to savanna. Pollen from arboreal elements decreased by 15%, while the taxa of disturbance (Poaceae, Chenopodiaceae, Convolvulaceae, *Pinus sp.*, among others) reached a total of 84%. During both transformations, agricultural activity was present, but maize pollen also decreased notably during these two events. This suggests that the climate was one of the triggers of the vegetation transformation.

The highest intensity of agricultural activity occurred in 470 AD based on the increase in pollen from maize as well as an increase in Poaceae and *Pinus sp.*, among other elements of disturbance. These taxa reach values of 56%, while forest taxa contribute only 29%. This suggests an expansion of savannas and crops but with more arboreal vegetation present than the period of land clearing discussed earlier.

The forest later recovered, and taxa suggest a decrease in the expansion of cultivated and grassland areas, although grasslands continued to be an important component of the vegetation (between 14 and 29%).

The vegetation mosaic of this time was composed of low tropical forest, savannas, crops and mangrove. Evidence of maize was last recorded around 800 AD and was absent in the next analyzed date (860 AD). For this reason, we infer a decrease of agricultural activities and the cultural collapse between 800 and 860 AD.

Circa 1779 AD, the youngest estimated age of this study, forest taxa percentages are the highest of the core. *B. alicastrum* is present continuously and, together with the high percentages of Moraceae and the presence of Anacardiaceae, Rubiaceae and Boraginaceae, suggests more conserved vegetation. There are, however, slight variations which coincide with small oscillations in the curve of *Pinus sp.* and Poaceae, which suggest phases of reduced arboreal cover.

Quantitative Precipitation Reconstruction

The statistical values in the construction of the transfer model indicate that the estimated precipitation from the Ria Lagartos core corresponds to changes in vegetation. However, the

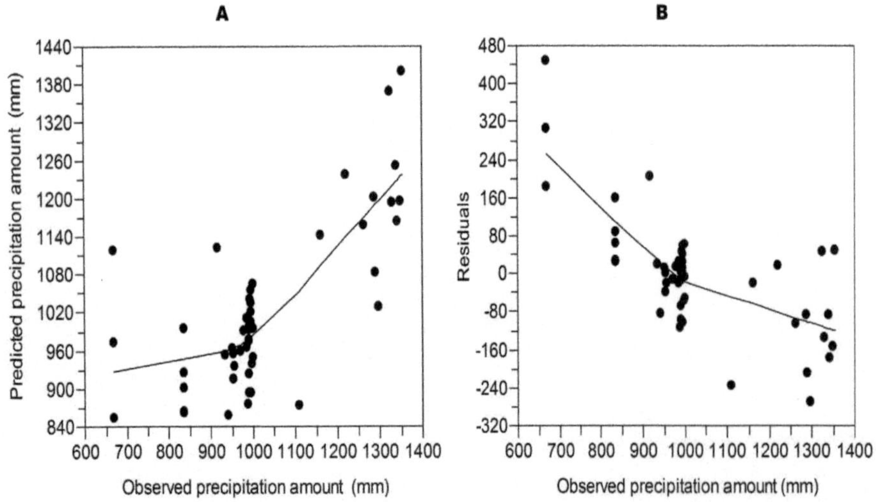

Figure 6. Model diagnostic plots. (A) Relation between estimated (with the transfer function) vs. observed precipitation. (B) Residual relations (observed-estimated) vs. observed precipitation.

vegetation of the region has been impacted since early times by human activity, which may interfere the reliability of the precipitation reconstruction. Nevertheless, as suggested by Leyden et al. [30], even under human influence, the pollen signal maintains a relationship to climatic changes. Therefore, it is indeed useful to compare pollen to the record of precipitation anomalies at Ria Largatos, to the record of *Z. mays* and to the $\delta^{18}O$ record of Lake Chichancanab [6] (Figure 7). With these comparisons, it is clear that precipitation deficits (droughts) and surpluses (wet phases) do not always correlate with rises and falls in the percentage of *Z. mays*.

In the first phase of the precipitation record, surpluses were more frequent than droughts and longer in duration (Figure 5). Precipitation was 3% higher than the average of the last 3800 years. During this time (1850-1000 BC), isotope values of Lake Chichancanab [6] are lower than the mean, suggesting wet conditions. Changes in the maize record do not match changes in the precipitation record, which suggest that variations in precipitation correspond to climate variability. This phase can be divided into two stages. During the first (1850-1250 BC), precipitation is above average and reaches surpluses of 6.7%. This wet stage can also be observed in a previous study of Ria Largatos [33] and in the Cariaco Basin [34]. During the second stage, droughts occur (precipitation 3.4% below average), but the surpluses are of similar magnitudes to those of the first stage. The beginning (950 BC) of the second phase is marked by a drought where the precipitation decreases by 6.4%. This drought coincides with an increase in the $\delta^{18}O$ values of Lake Chichancanab [6], Lake Tzib [26] and Lake Punta Laguna [36], along with the decrease in the percentage of forest taxa in the previous study of Ria Largatos [33]. We detect an increase in the maize record. It is probable then that the decrease in precipitation is less than the reconstruction suggests.

The largest surplus (8.75%) is recorded around 320 BC, which agrees with lower $\delta^{18}O$ values of Lake Chichancanab [6] and with one of the highest percentages of Ti in the last 4000 years in the Cariaco Basin [34]. The recorded increase in precipitation also coincides with an increase in solar radiation [37]. We find that the most intense drought of this phase occurred in 250 BC. The precipitation decreased 12.4%. This drought is apparently a result of the modification of the vegetation by agricultural activities. In Lake Chichancanab, higher $\delta^{18}O$ values are also observed but do

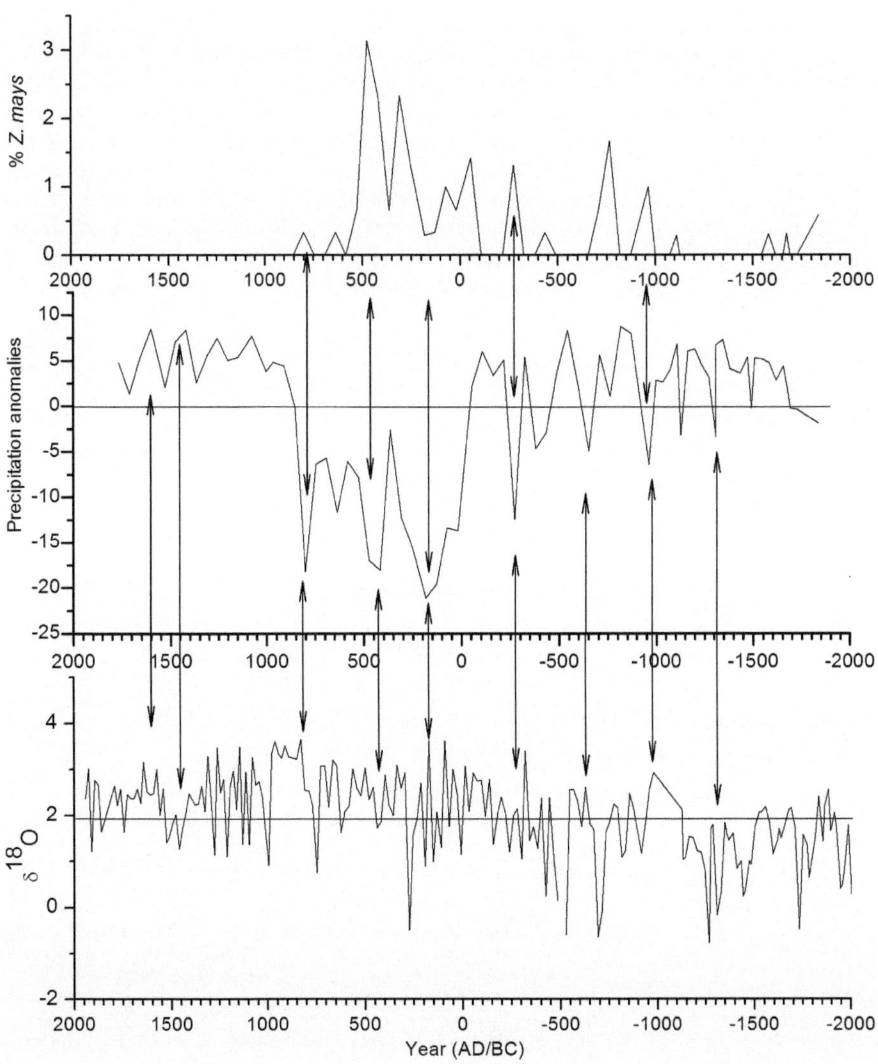

Figure 7. Comparison of records. (A) Record of *Z. mays* of this same study. (B) Anomalies of the inferred precipitation. (C) Record of $\delta^{18}O$ Lake Chichancanab [6].

not hold up a reduction of precipitation, nor do we observe such a reduction in Cariaco. However, changes are detected in the records of maize and other elements associated with cultivation, such as *Croton sp.* The second phase ended with an increase in the precipitation (5–6%).

The third phase is clearly differentiated by repeated droughts. During this time, precipitation was below average by an average of 12.6% but at times more than 18%. The first drought lasted from around 130 to 236 AD. During this drought, precipitation dropped by 21%. This phase coincided with what is known as the Preclassic drought reported in various works in the region [26–27], [33–34]. During this event, important changes in the environment and in Mayan cities occurred, including the collapse of El Mirador (Guatemala), a city that flourished during the Preclassic period [38]. It is important to note that, during this drought, there was no increase in the percentage of maize. The precipitation record confirms that this was a result of the climate and not an artifact of agricultural activity. However, the drought that occurred around 420 AD does coincide with an increase in evidence of maize. In fact, the drought coincides with the highest percentage of maize in the record. Therefore, it is probable that the 18% decrease in precipitation is overestimated. The third drought occurred in 800 AD when precipitation declined 18.2% and, in contrast to the previous drought, no high values of *Z. mays* were observed. This drought coincides with the high $\delta^{18}O$ values in the study of Lake Chichancanab [6] and decreased precipitation in the Cariaco Basin [28]. The drought also coincides temporally with the second phase of the disintegration of Mayan lands, during which various cities were abandoned. Among these cities were Yaxchilan, Calakmul, Piedras Negras, Copan and Naranjo [38]. Following this drought, maize disappeared from the record and is likely related to the collapse around this time. The collapse is estimated to have occurred between 800 and 860 AD.

Medina-Elizalde proposed that the reduction in precipitation during the final Classic Period was 36–52% based on quantitative precipitation reconstruction using isotopes in stalagmites. The values obtained in this study were more conservative. Nevertheless, it is important to consider that the model constructed tended to overestimate the low values (Figure 6). Therefore, it is likely that droughts were more intense than the values suggest here.

The fourth phase of change (800–1760 AD) is characterized as a wet period. The increase in precipitation was rapid, and the anomaly exceeded an average of 5% with no deficits recorded, despite variable precipitation. Medina-Elizalde [8] suggests that the revitalization of the Puuc region in the north of the peninsula was favored by the precipitation increase, which increased the production of food and could subsequently support a larger population. This event coincides temporally (860–900 AD) with the beginning of this fourth phase, during which the precipitation increased considerably. The period from 900 AD to 1300 AD experienced precipitation surpluses of more than 7.5%. This period coincides with the global event known as the Medieval Climatic Anomaly [39].

Between 1300–1760 AD, precipitation was high overall but also more variable. This period falls into the Little Ice Age (LIA), which was drier than the Medieval Climatic Anomaly but still wetter than the rest of the record. The precipitation reconstruction from the Tzabnabh cave [8] suggests that the LIA was relatively more humid, which agrees with the precipitation record of the present study, yet contrary to earlier findings [34]. However, other studies on fossil pollen on the peninsula have found droughts in this phase [27], [33].

Conclusions

The calibration of modern fossil pollen with observed precipitation led to the construction of a predictive and reliable model. The model works best with medium values, while low values tend to be overestimated.

The application of a transfer function permitted the reconstruction of precipitation for the last 3800 years. In the area of Ria Lagartos, precipitation anomalies show that the magnitudes of deficits (21%) exceed the surpluses (<9%). Four phases of change have been detected; the third (50 BC – 800 AD) was the driest. The drought related to the abandonment of the Preclassic Period featured an estimated 21% reduction in precipitation, while the drought of the Mayan collapse (800–860 AD) saw a decrease of 18%. The Medieval Climatic Anomaly was a period of positive anomalies (3.8–7.6%), while the Little Ice Age was a more variable period, with reductions but without droughts. Fossil pollen indicates agricultural activity since the early phases (1840), and those activities provoked important changes in the vegetation during the Classic Period. Moreover, data indicate the vulnerability of the vegetation to the combined effect of agriculture and climate. When both agents operate in conjunction, it is possible to cause major transformations in vegetation, such as occurred in 370 AD.

Acknowledgments

Valuable comments by the reviewers are acknowledged.

Author Contributions

Conceived and designed the experiments: ACB GAI NTV. Performed the experiments: ACB GAI NTV. Analyzed the data: ACB GAI NTV. Contributed reagents/materials/analysis tools: GAI. Wrote the paper: ACB GAI NTV.

References

1. Labeyrie L, Cole J, Alverson K, Stocker T (2003) The History of Climate Dynamics in the Late Quaternary. In Alverson KD, Bradley RS, Pedersen TF, editors. Paleoclimate, global change and the future. Berlin: Springer. 33–61.
2. Birks HJ (1995) Quantitative palaeoenvironmental reconstructions. In Maddy D, Brew JS, editors. Statistical modelling of Quaternary science data. Technical Guide.
3. Riehl H (1979) Climate and Weather in the Tropics. London: Academic Press. 611p.
4. Birks HJ (2005) Quantitative palaeoenvironmental reconstructions from Holocene biological data. In Mackay A, Battarbee R, Birks J, Oldfield F, editors. Global change in the Holocene. Great Britain: Hodder education. 107–123.
5. Hodell DA, Brenner M, Curtis J (2000) Climate change in the northern American tropics and subtropics since the last ice age. In Lentz DL, editor. Imperfect balance, landscape transformations in the pre-Columbian Americas. New York: Columbia University Press. 13–38.
6. Hodell DA, Curtis JH, Brenner M (1995) Possible role of climate in the collapse of classic Maya civilization. Nature 375: 391–394.
7. Curtis JH, Hodell DA, Brenner M (1996) Climate variability on the Yucatan peninsula (Mexico) during the past 3500 years, and implication for Maya cultural evolution. Quatern Res 16: 37–47.
8. Medina-Elizalde M, Burns SJ, Lea DW, Asmerom Y, van Gunten L, et al. (2010) High resolution stalagmite climate record from the Yucatán peninsula spanning the Maya terminal classic period. Earth Planet Sci Lett. 298: 255–262.
9. Medina-Elizalde M, Rohling EJ (2012) Collapse of Classic Maya Civilization related to modest reduction in precipitation. Sci. 335: 956–959.
10. Mann ME (2007) Climate over the past two millennia. Annu. Rev. Earth Planet. Sci. 35: 111–136.
11. CONANP (2007) Programa de conservación y manejo reserva de la biosfera Ría Lagartos. Secretaría de Medio Ambiente y Recursos Naturales. 266p.

12. CNA (2006) Jefatura de Proyecto de Aguas Superficiales. Subgerencia Regional Técnica. Gerencia Regional Península de Yucatán. Comisión Nacional del Agua. México.
13. Torrescano-Valle N (2007) Reconstrucción paleoambiental del Holoceno Medio-Tardio en la parte centro-sur de la península de Yucatán, México. Doctoral Thesis. El Colegio de la Frontera Sur.
14. Erdtman G (1969) Handbook of palynology. An Introduction to the study of Pollen grains and spores. Hafner Publishing Co. 486p.
15. Palacios-Chavez R, Ludlow-Wiechers B, Villanueva-Gutiérrez R (1991) Flora palinológica de la reserva de la biosfera de Sian Ka'an, Quintana Roo, Mexico. CICRO. 321p.
16. Correa-Metrio A, Cabrera KR, Bush MB (2010) Quantifying ecological change through discriminant analysis: a paleoecological example from the Peruvian Amazon. J Veg Sci 21: 695–704.
17. ter Braak CJF (1986) Canonical correspondence analysis: a new eigenvector technique for multivariate direct gradient analysis. Ecology 67: 1167–1179.
18. ter Braak CJF, Šmilauer P (2002) CANOCO Reference Manual and CANODRAW User's Guide: Software for Canonical Community Ordination (Version 4.5). Microcomputer Power (Ithaca, New York). 500 p.
19. Wold S, Ruhe A, Wold H, Dunn WJ (1984) The collinearity problem in linear regression: the partial least squares (PLS) approach to generalized inverses. SIAM J Sci Comput 5: 735–743.
20. Juggins S (2007) C2 version 1.5 user guide. Software for ecological and palaeoecological data analysis and visualization. Newcastle University, Newcastle upon Tyne, UK. 73p.
21. Grimm EC (2011) TILIA software version 1.7.16. Illinois State Museum, Research and Collection Center. Springfield USA. Available: http://intra.museum.state.il.us/pub/grimm/tilia/.
22. Stuiver M, Reimer PJ, Reimer RW (2005) CALIB 5.0 Available: http://calib.qub.ac.uk/calib/.
23. Reimer PJ, Hughen KA, Guilderson TP, McCormac FG, Baillie MGL, et al. (2002) Preliminary Report of the first workshop of the IntCal04 Radiocarbon Calibration/Comparison Working Group. Radiocarbon 44: 653–661.
24. ter Braak CJF (1987) Unimodal models to relate species to environment. DLO-Agricultural Mathematics Group. 266p.
25. Torrescano-Valle N, Islebe GA (2006) Tropical forest and mangrove history from south eastern Mexico: a 5000 yr pollen record and implications for sea level rise. Veg Hist Archaeobot 15: 191–195.
26. Carrillo-Bastos A, Islebe GA, Torrescano-Valle N, González NE (2010) Holocene vegetation and climate history of central Quintana Roo, Yucatan Peninsula, Mexico. Rev Palaeobot Palyno 160: 189–196.
27. Gutiérrez-Ayala LV, Torrescano Valle N, Islebe GA (2012) Reconstrucción paleoambiental del Holoceno Tardío de la reserva Los Petenes, Península de Yucatán, México. Rev Mex Cienc Geol 29: 749–763.
28. Leyden BW (2002) Pollen evidence for climatic variability and cultural disturbance in the Maya Lowlands. Anciente Mesoam 13: 85–101.
29. Pennington DT, Sarukhán J (2005) Árboles tropicales de México: manual para la identificación de las principales especies. Mexico: Universidad Autónoma de México, Fondo de cultura económica. 523 p.
30. ter Braak CJF, Juggins S (1993). Weigthed averaging partial least squares regression (WA-PLS): an improve method for reconstructing environmental variables from species assemblages. Hydrobiologia 269/270: 485–502.
31. Birks HJB (1998) Numerical tools in quantitative palaeolimnology-progress, potentialities, and problems. J. Paleolim 20: 307–332.
32. Islebe GA, Hooghiemstra H, Brenner M, Curtis JH, Hodell DA (1996) A Holocene vegetation history from lowland Guatemala. The Holocene 6: 265–271.
33. Aragón-Moreno AA, Islebe GA, Torrescano-Valle N (2012) A ~3800-yr, high-resolution record of vegetation and climate change on the north coast of the Yucatan Peninsula. Rev Palaeobot Palynol 178: 35–42.
34. Haug GH, Gunther D, Peterson LC, Sigman DM, Hughen KA, et al. (2003) Climate and the Collapse of Maya Civilization. Sci 299: 1731–1735.
35. Hodell DA, Brenner M, Curtis JH, Guilderson T (2001) Solar Forcing of Drought Frequency in the Maya Lowlands. Sci 292: 367–1370.
36. Hodell DA, Brenner M, Curtis JH (2007) Climate and cultural history of the Northeastern Yucatan Peninsula, Quintana Roo, Mexico. Climatic Change 83; 215–240.
37. Steinhilber JAA, Beer J (2008) Solar modulation during the Holocene. Astra 4: 1–6.
38. Gill RB (2008) Las grandes sequías mayas, agua, vida y muerte. México Distrito Federal: Fondo de cultura económica. 561 p.
39. Mann ME (2002) Medieval climatic optimum. In: MacCracken MC, Perry JS, editors. Encyclopedia of global environmental change. Chichester: Jonh Wiley and Sons 514–516.

8

Molecular Phylogeography and Evolutionary History of *Poropuntius huangchuchieni* (Cyprinidae) in Southwest China

Xiaoyun Wu[1,3], Jing Luo[1], Song Huang[4], Ziming Chen[1], Heng Xiao[1]*, Yaping Zhang[1,2]*

1 Laboratory for Conservation and Utilization of Bio-resource, Key Laboratory for Animal Genetic Diversity and Evolution of High Education in Yunnan Province, School of Life Sciences, Yunnan University, Kunming, Yunnan, China, **2** State Key Laboratory of Genetic Resources and Evolution, Kunming Institute of Zoology, Chinese Academy of Sciences, Kunming, Yunnan, China, **3** Key Laboratory of Animal Models and Human Disease Mechanisms, Chinese Academy of Sciences and Yunnan Province, Kunming Institute of Zoology, Chinese Academy of Sciences, Kunming, Yunnan, China, **4** Kunming Zoo, Kunming, Yunnan, China

Abstract

Background: The evolution of the Yunnan Plateau's drainages network during the Pleistocene was dominated by the intense uplifts of the Qinghai-Tibetan Plateau. In the present study, we investigated the association between the evolutionary histories of three main drainage systems and the geographic patterns of genetic differentiation of *Poropuntius huangchuchieni*.

Methodology/Principal Findings: We sequenced the complete sequences of mitochondrial control region for 304 specimens and the sequences of Cytochrome *b* gene for 15 specimens of the species *P. huangchuchieni* and 5 specimens of *Poropuntius opisthoptera*. Phylogenetic analysis identified five major lineages, of which lineages MK-A and MK-B constrained to the Mekong River System, lineages RL and LX to the Red River System, and lineage SW to the Salween River System. The genetic distance and network analysis detected significant divergences among these lineages. Mismatch distribution analysis implied that the population of *P. huangchuchieni* underwent demographic stability and the lineage MK-B, sublineages MK-A1 and LX-1 underwent a recent population expansion. The divergence of the 5 major lineages was dated back to 0.73–1.57 MYA.

Conclusions/Significance: Our results suggest that *P. opisthoptera* was a paraphyletic group of *P. huangchuchieni*. The phylogenetic pattern of *P. huangchuchieni* was mostly associated with the drainage's structures and the geomorphological history of the Southwest Yunnan Plateau. Also the differentiation of the major lineages among the three drainages systems coincides with the Kunlun-Yellow River Movement (1.10–0.60 MYA). The genetic differentiation within river basins and recent demographical expansions that occurred in some lineages and sublineages are consistent with the palaeoclimatic oscillations during the Pleistocene. Additionally, our results also suggest that the populations of *P. huangchuchieni* had keep long term large effective population sizes and demographic stability in the recent evolutionary history, which may be responsible for the high genetic diversity and incomplete lineages sorting of *Poropuntius huangchuchieni*.

Editor: Senjie Lin, University of Connecticut, United States of America

Funding: This work was supported by National Basic Research Program of China,National Natural Science Foundation of China, Bureau of Science and Technology of Yunnan Province.The funders had no role in study design, data collection and analysis, decision to publish, or preparation of the manuscript.

Competing Interests: We have no completing interests.

* E-mail: xiaoheng@ynu.edu.cn (HX); zhangyp@mail.kiz.ac.cn (YZ)

Introduction

Past tectonic movements and climatic changes have greatly shaped the structure of hydrographic systems [1,2,3]. Likewise, since freshwater fish species are strictly confined to freshwater drainages, the historical connections, capture, reversal and separation of rivers are a driving force behind their diversification and speciation [1,4,5,6,7]. Consequently, the genetic structure and the dispersal of freshwater fish species is also strongly connected to historical and ecological changes to the aquatic environment [1,4,5,6,7], so by examining the historical biogeography of freshwater fishes could provide a natural link in understanding concurrent geographical and biotic evolution of a given region [8].

As the southeastern neighbor of Qinghai–Tibetan Plateau, the Yunnan Plateau has consistently responded to each period of the Qinghai–Tibetan Plateau's uplifting, which reached its present elevation some 3.4 MYA, during the Pliocene. The dramatic uplifting of Yunnan Plateau at the same time not only dramatically re-shaped the composition and configurations of geographic and aquatic environments, but also greatly changed the region's climate [9,10]. The three big rivers of the Yunnan Plateau–the Mekong, Salween, and Red Rivers–also experienced remarkable changes to their drainage networks during these geologic movements [11]. These movements had a markedly different influence in shaping the geomorphologic configurations of each different region of the Yunnan Plateau. As such, the configuration of the modern rivers differs significantly from their earlier paleo-configuration [9]. The resulting complex evolutionary patterns of the river drainages in Yunnan Plateau make it an especially interesting region for exploring the phylogeography of aquatic species.

Chu [12] and Chu and Chen [13] illustrated the relationship of the six river systems in Yunnan Plateau, based on similarity of genera and suggested that the six river systems can be divided into two groups: Jinsha-Nanpan-Red and Mekong-Salween-Irrawaddy (Figure S1). In addition, Several molecular phylogenetic studies demonstrated that the evolutionary history of some fish species (e.g., glyptosternoid fishes, subfamily Schizothoracinae) are associated with the uplifts of the Tibetan plateau and the concomitant river rearrangements (mostly river capture) [14,15,16]. These studies also suggest that the Mekong River had a close relationship with the Salween River, while the Red River had closely related with Jinsa River, Nanpan River and Beipan River. However, because these studies were performed on the fish species that mainly inhabit the cold water of the upper river systems, it is difficult to know whether the evolutionary patterns are actually fitted to the basins of the Salween, Mekong and Red Rivers in the southwest of the Yunnan Plateau.

A recent molecular study on the phylogeography of Yunnan spiny frog (*Nanorana yunnanensis*) suggested that the basins of the Salween, Mekong and Red Rivers in the southwest of the Yunnan Plateau have experienced little change since the early Pliocene, excepting in their upper portions near southeastern Tibet [17]. Previous geological studies on Quaternary sediments of these rivers, however, suggest that the mainstreams of the Salween, Mekong and Red Rivers roughly formed or joined up during the Kunlun-Yellow River Movement (KYM) (1.1–0.6 MYA) in the mid-Pleistocene [9,10,18]. On the surface, the results of the phylogeography of the Yunnan spiny frog seem inconsistent with the results of the geological studies of this region, suggesting the historical rearrangement of these rivers is still poorly understood.

To date, studies examining the phylogeography at intraspecific level or between closely related fish species inhabiting these drainages that can be used to evaluate the change to the drainage configurations in this region are lacking. In the present study, we examined the maternal phylogeographical pattern of *Poropuntius huangchuchieni* (Tchang, 1967) in the basins of the Mekong and Red Rivers and *P. opisthoptera* (Wu, 1977) in Salween River basin to investigate the association between the changes among the three river drainages and the genetic differentiation of freshwater fish species. *P. huangchuchieni*, a medium sized (about 130–300 mm) economic freshwater cyprinid fish, is widely distributed in the subtropical and tropical waters of Mekong and Red River basins [19,20]. The individuals of *P. huangchuchieni* dwell where the river flow becomes retarded, subsisting on a diet of plankton and mosses attached to stones or rotted wood.

In this study, we used the sequences of the mtDNA control region to estimate the genetic differentiation and phylogeographical patterns of *P. huangchuchieni*. To examine the taxonomic status of *P. opisthoptera*, we also determined the phylogenetic relationships between *P. huangchuchieni* and *P. opisthoptera*. Accordingly, the main objectives of this study are (1) to examine the population genetic structure and the demographic history of *P. huangchuchieni*; (2) to propose a historical biogeography and climate hypothesis to accommodate the phylogeographic patterns of lineages within *P. huangchuchieni*; and (3) to determine the taxonomic implications for *P. opisthoptera*.

Results

Sequence Variation and Genetic Diversity

After alignment, 965 bp DNA sequences of the complete control region were obtained from all *P. huangchuchieni* specimens. The control region sequences yielded 126 variable sites of which 100 were parsimony informative, defining 126 haplotypes (GenBank Access Number: KC567019-KC567146). The overall nucleotide diversity (π) and haplotype diversity (h) of all in-group sequences was 0.0286 ± 0.0139 and 0.9856 ± 0.0020, respectively. In total, 1140 bp DNA sequences of the complete *Cyt b* gene were determined from 20 in-group individuals, and these 20 *Cyt b* sequences contained 117 variable sites of which 87 were parsimony informative, further defining 18 haplotypes (GenBank Access Number: KC567001–KC567018).

Phylogenetic Analysis

ML and Bayesian phylogenetic analysis of the 126 haplotypes identified two major phylogroups, containing five highly independent lineages with strong statistical support (Fig. 1A). One group, named lineage SW, clustered all of the *P. opisthoptera* individuals from the Salween River, which was at the most basal position of the tree. The other group contained all *P. huangchuchieni* individuals, which formed four major evolutionary lineages: MK-A, MK-B, RL and LX (Fig. 1A).

Lineages of MK-A and MK-B were represented by nearly all of the haplotypes (except of Hap 81) from the Mekong River System, and co-occurred in most sample sites in the Mekong River System (Table 1). The haplotypes from Luosuo River (M3), Nanla River (M5) and Puwen River (M7) in the Mekong River system were only found in lineage MK-A, and the haplotypes from Menghan (M2) and Dazhong River (M8) were only found in MK-B (Table 1). The RL lineage contained the haplotypes from Amo River (R1), Tengtiao River (R3), Lvzhi River (R4) and Hedi River (R5) in the Red River System. Interestingly, two haplotypes–Hap48 from the Lvzhi River within the Red River System and Hap81 from Nanlei River within the Mekong River System–fell outside the main cluster of lineage RL and located at the basal position within lineage RL (Table 1). Lineage LX was comprised of all of the haplotypes from Lixian River (R2) and two haplotypes from Amo River (R1) (Table 1). Within MK-A, there were two sublineages, which were named as MK-A1 and MK-A2. MK-A1 was distributed in most of the sample sites (10 of 12 sample sites) of Mekong River System, whereas MK-A2 was distributed in only 4 sample sites (Figure 2). Lineage LX likewise contained two sublineages, which we named LX-1 and LX-2. LX-1 was only located in the Lixian River, while LX-2, which contained three haplotypes, was found in both the Lixian and Amo River (Figure 2).

In the phylogenetic trees based on the combined sequences of control region and the *Cyt b* gene, the five major lineages could be clearly detected (Fig. 1C), corresponding to those defined in the control region trees. Compared with the control region trees, the placement of Lineages SW and LX were slightly different; LX was placed at the most basal position and SW at the second basal position in the combined sequence trees. Based on the combined sequences, the positions of other three lineages were recovered with strong supports.

Population Genetic Structure

In total, among the 5 lineages, 126 haplotypes formed 9 unconnected haplotype networks at the 95% connection limit (Fig. 1B). MK-A was divided into two unconnected sublineages, MK-A1 and MK-A2. LX was also unconnected and formed two

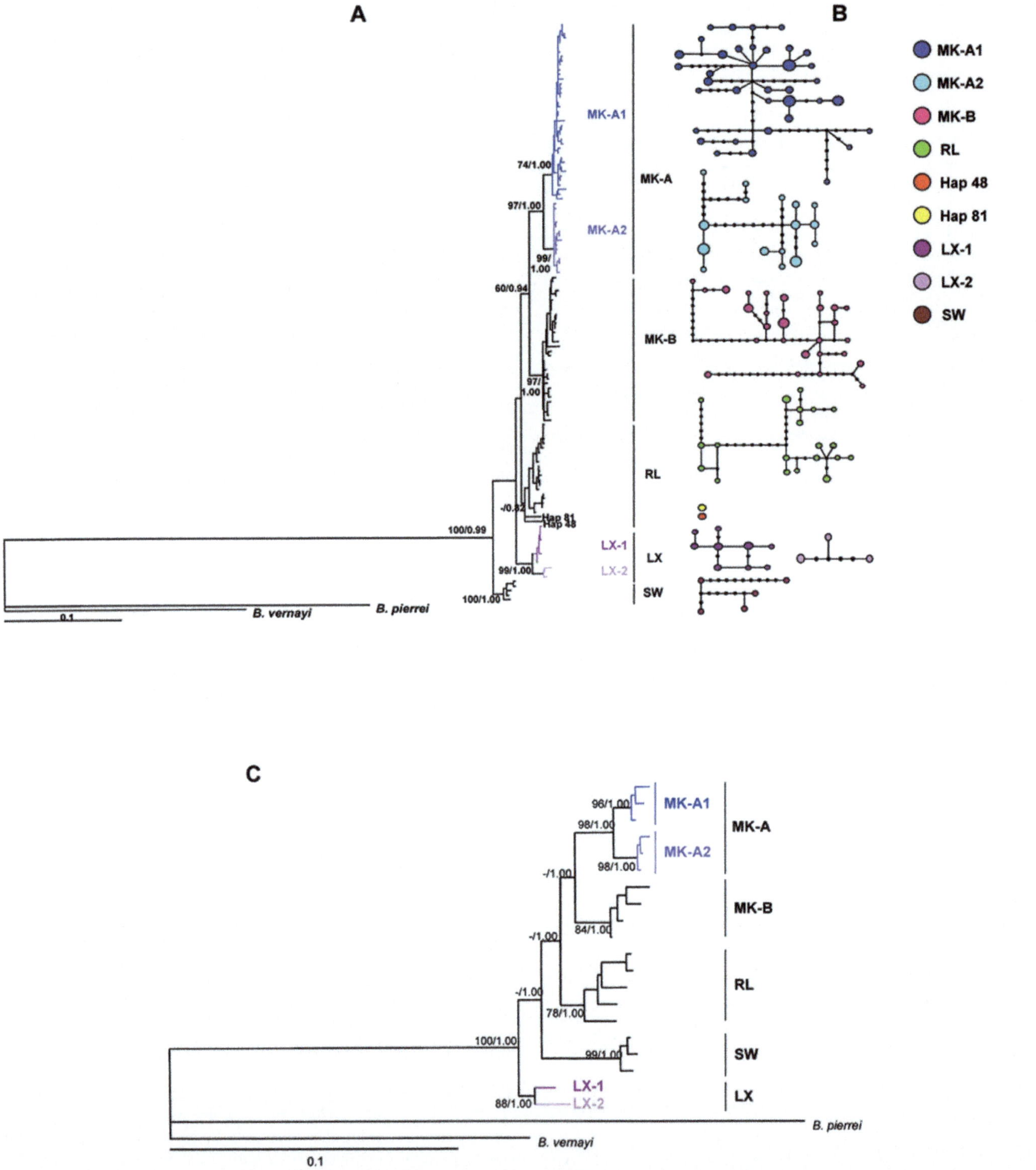

Figure 1. Phylogenetic trees reconstructed based on mitochondrial control region sequences of all haplotypes (A), the combined sequences of 20 haplotypes (C), and the haplotypes network analysis of the 126 haplotypes (B). (A) ML tree reconstructed based on mitochondrial control region sequences of all haplotypes under HKY+I+G model. Numbers on major nodes represents bootstrap values after 1,000 replications. If bootstrap values were less than 50%, they were defaulted. Trees were rooted by *H. pierrei* and one *H. vernayi*. (B) Haplotypes networks conducted based on the 126 haplotypes. Circle size is proportional to haplotype frequency. The number of black dots on line connected haplotypes represents mutation steps between haplotypes; when the mutation step is 1, it was defaulted. (C) 50% majority-role consensus tree inferred from ML and Bayesian analysis of combined sequences of 20 haplotypes under GTR+I+G model. Numbers at nodes represent the posterior probability for Bayesian analysis and bootstrap value for maximum likelihood (ML) analysis. If the bootstrap values were less than 50%, they were defaulted. Trees were rooted by *H. pierrei* and one *H. vernayi*.

Table 1. Sampling information and size in each lineage, as well as haplotype, and nucleotide diversity of populations based on mtDNA D-loop sequences.

Site code	Site description	Longitude	Latitude	Number of individuals in each lineage						*h*	π
				Total	MK-A	MK-B	RL	LX	SW		
M1	Mekong River; Jinghong	100.805 E	22.011 N	4	1	3				–	–
M2	Mekong River; Menghan	100.925 E	21.845 N	11		11				0.8333±0.1002	0.0062±0.0036
M3	Luosuo River, a tributary of Mekong River	101.262 E	21.927 N	28	28					0.8492±0.0538	0.0045±0.0026
M4	Mekong river; Guanlei	101.123 E	21.688 N	28	24	4				0.9200±0.0339	0.0088±0.0047
M5	Nanla River, a tributary of Mekong River	101.560 E	21.437 N	29	29					0.8941±0.0382	0.0154±0.0079
M6	Nanlei River, a tributary of Mekong River	99.608 E	22.369 N	34	19	13	2*			0.8699±0.0442	0.0271±0.0135
M7	Puwen River, a tributary of Luosuo River	101.067 E	22.505 N	24	24					0.7935±0.0566	0.0084±0.0045
M8	Dazhong River, a tributary of Mekong River	100.574 E	22.557 N	3		3				–	–
M9	Hei River, a tributary of Mekong River	100.251 E	22.636 N	11	3	8				0.7091±0.0827	0.0191±0.0104
M10	Weiyuan River, a tributary of Mekong River	100.709 E	23.485 N	21	20	1				0.8857±0.0473	0.0076±0.0041
M11	Mengjia River, a tributary of Mekong River	100.310 E	23.594 N	41	17	24				0.9207±0.0258	0.0220±0.0110
M12	Mekong River, Manwan	100.496 E	24.525 N	10	1	9				0.8889±0.0754	0.0147±0.0081
R1	Amo River, a tributary of Lixian River	101.753 E	23.114 N	27			25	2		0.9316±0.0245	0.0118±0.0062
R2	Lixian River, a tributary of Red River	102.293 E	22.574 N	21				21		0.9053±0.0441	0.0058±0.0032
R3	Tengtiao River, a tributary of Lixian River	103.155 E	22.604 N	6			6			–	–
R4	Lvzhi River, a tributary of Red River	101.582 E	24.243 N	2			2#			–	–
R5	Hedi River, a tributary of Red River	102.329 E	23.425 N	4			4			–	–
S1	Nanka River, a tributary of Salween River	99.385E	22.574N	5					5	–	–

*Hap 81;
#one of the individuals represented Hap 48.

independent networks, LX-1 and LX-2. RL was divided into three unconnected networks, where Hap 48 and Hap 81 formed two independent networks, while all the other haplotypes formed an integrate network.

The geographical structure at the river systems levels was determined using AMOVA analysis. Once all populations were grouped into the three river systems (Fst = 0.650, Fct = 0.331, P<0.0001), 33.09% of the variation was between the different river systems and 31.91% of the variation was between populations within the same river systems.

Divergence Time Estimates

In conducting the relative rate test, the null hypothesis of rate constancy could not be rejected for all lineages pairs. Accordingly, mtDNA divergence could instead be used to calculate the divergence times of the inferred mtDNA lineages. We calculated the net between–lineage mean distances using 20 *Cyt b* sequences, with values ranging from 0.016 to 0.033 among the inferred lineages (Table 2); the net between–lineage mean distances estimated using all the control region sequences were between 0.017–0.030 (Table 2). We also calculated the net between–lineage mean distances based on the same 20 specimens of control region sequences with values from 0.013 to 0.038 (results not shown) among the inferred lineages. Accordingly, it could be concluded that the evolutionary rate of the control region sequences was similar to that of the *Cyt b* gene among the five major lineages of *P. huangchuchieni.*

The most recent common ancestor (TMRCA) of all the ingroup sequences was dated back to around 1.57 MYA (95% CI = 1.03–2.21). Meanwhile, the estimated divergence time of the lineage LX from lineages MKs and RL was estimated at 1.28 MYA (95% CI = 0.89–1.70); the diversification time of lineage RL from lineages MKs was 1.11 MYA (95% CI = 0.79–1.46). The TMRCA of lineages MKs and MK-A were dated 0.96 MYA (95% CI = 0.67–1.27) and 0.73 MYA (95% CI = 0.49–0.99), respectively.

Population Historical Demography

The mismatch distribution analysis of Lineages MK-A, RL and LX were multimodal distribution of pairwise differences, and the Fu's Fs values were also not significant, indicating a long term demographic stability in these lineages (Fig. 3). Lineage MK-B displayed a unimodal mismatch distribution and a significant Fs value, suggesting a recent population expansion. Lineage SW was excluded from this analysis because of its small sample size. Two sublineages (MK-A1 and LX-1) fitted the expected distributions under the expansion model (Fig. 3). Mismatch distributions for sublineages MK-A1 and LX-1 were unimodal, with Fs values of −16.48 (P<0.01) and −3.48 (P<0.05), respectively (Table 3). Neither mismatch distribution nor neutrality tests, however,

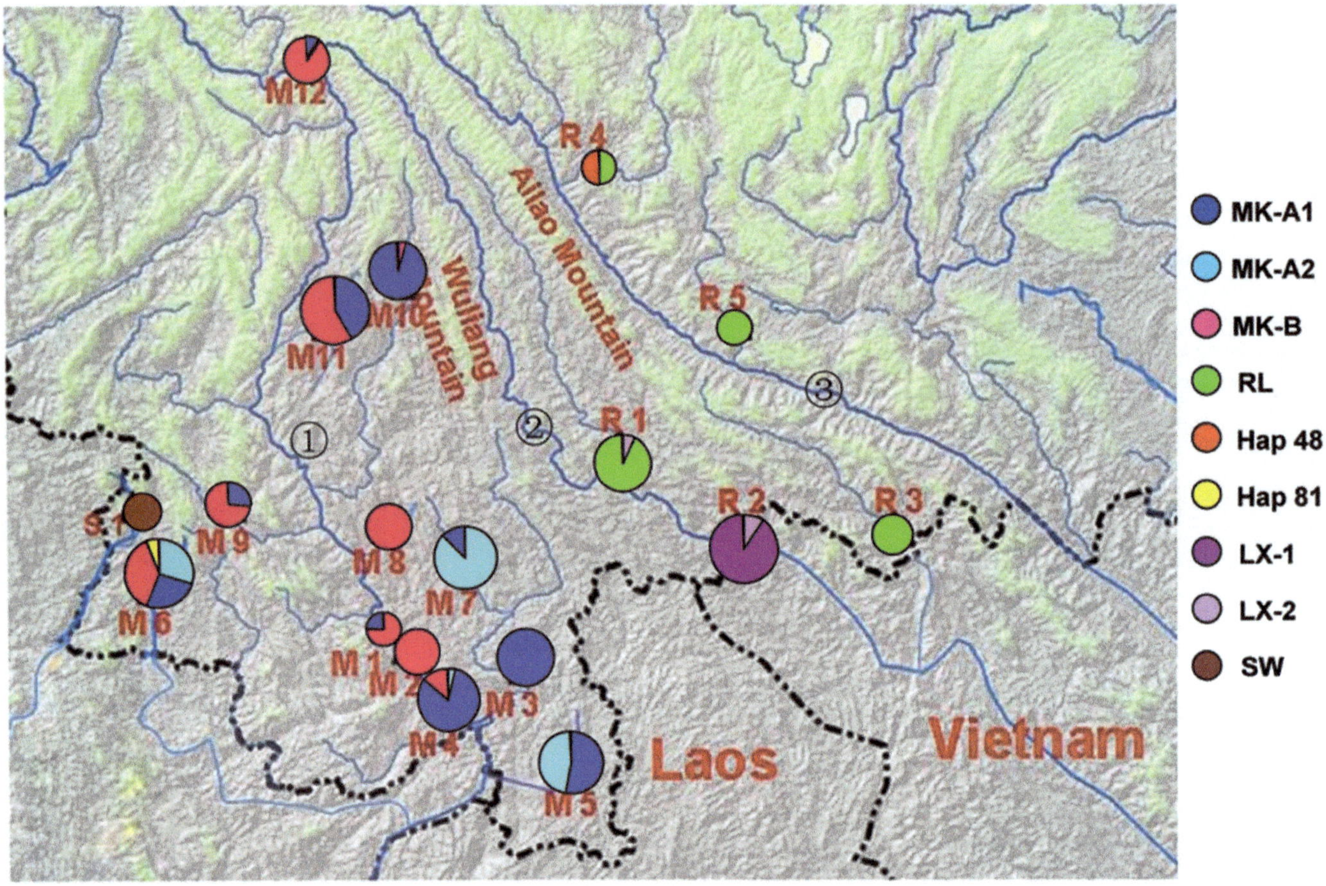

Figure 2. Geographic distribution of sample sites and the frequency of lineages and sublineages at each site. Circle area is proportional to the sample size. ①: Mekong River; ②: Lixian River; ③: Red River.

supported the expectation of population expansion for sublineage MK-A2. Given the small sample size of sublineage LX-2, the detection of population expansion was not performed. With the estimated tau value (τ) and a generation time of one year, the estimated times since population growth for lineage MK-B, and sublineages MK-A1 and LX-1 were respectively dated to around 140 ka, 100 ka, and 60 ka years ago (Table 3).

Table 2. Genetic distances among 5 major lineages based on *p* distance. Net genetic distances estimated based on the *Cyt b* data set are above the diagonal, net genetic distances estimated based on the D-loop data set are below the diagonal.

	MK-A	MK-B	RL	LX	SW
MK-A		0.016±0.003	0.019±0.004	0.024±0.004	0.033±0.005
MK-B	0.018±0.003		0.017±0.003	0.023±0.004	0.029±0.004
RL	0.019±0.004	0.016±0.003		0.024±0.004	0.028±0.004
LX	0.025±0.004	0.022±0.004	0.021±0.004		0.024±0.004
SW	0.035±0.005	0.037±0.006	0.026±0.005	0.031±0.006	

Morphometric Analysis

A total of 234 specimens were included in morphological analysis (lineage MK-A: n = 124; lineage MK-B, n = 55; lineage RJ, n = 25; lineage LX, n = 13; lineage SW, n = 18). The Canonical Variates Analysis (CVA) produced a scatter of specimens along the two first canonical axes (Fig. 4). Plotting canonical variables 1 (CV1) and canonical variables 2 (CV2) showed a clear morphometric space among groups. Together, these first two canonical variables collectively explain 83.64% of the total variability between groups (CV1 accounted for 65.09% and CV2 accounted for 18.55%). Meanwhile, 95% frequency ellipses showed an overlap in the scatter of data among the 5 groups, with the exception of lineages MK-A and LX not overlapping with lineage SW (*P. opisthoptera*) (Fig. 4). The CV1 sets lineage SW (*P. opisthoptera*) as having diverged from the other 4 lineages, being then associated with the relative position of the dorsal fin (Table 4). The CV2 sets lineage LX from lineages MK-A and MK-B, and corresponded to the relative heights of body and caudal peduncle (Table 4).

Discussion

Taxonomic Implications

The mitochondrial control region phylogenetic tree revealed the basal position of *P. opisthoptera*, which suggested a close relationship with *P. huangchuchieni*. However, the phylogenetic trees based on both the *Cyt b* gene and combined genes all showed the basal

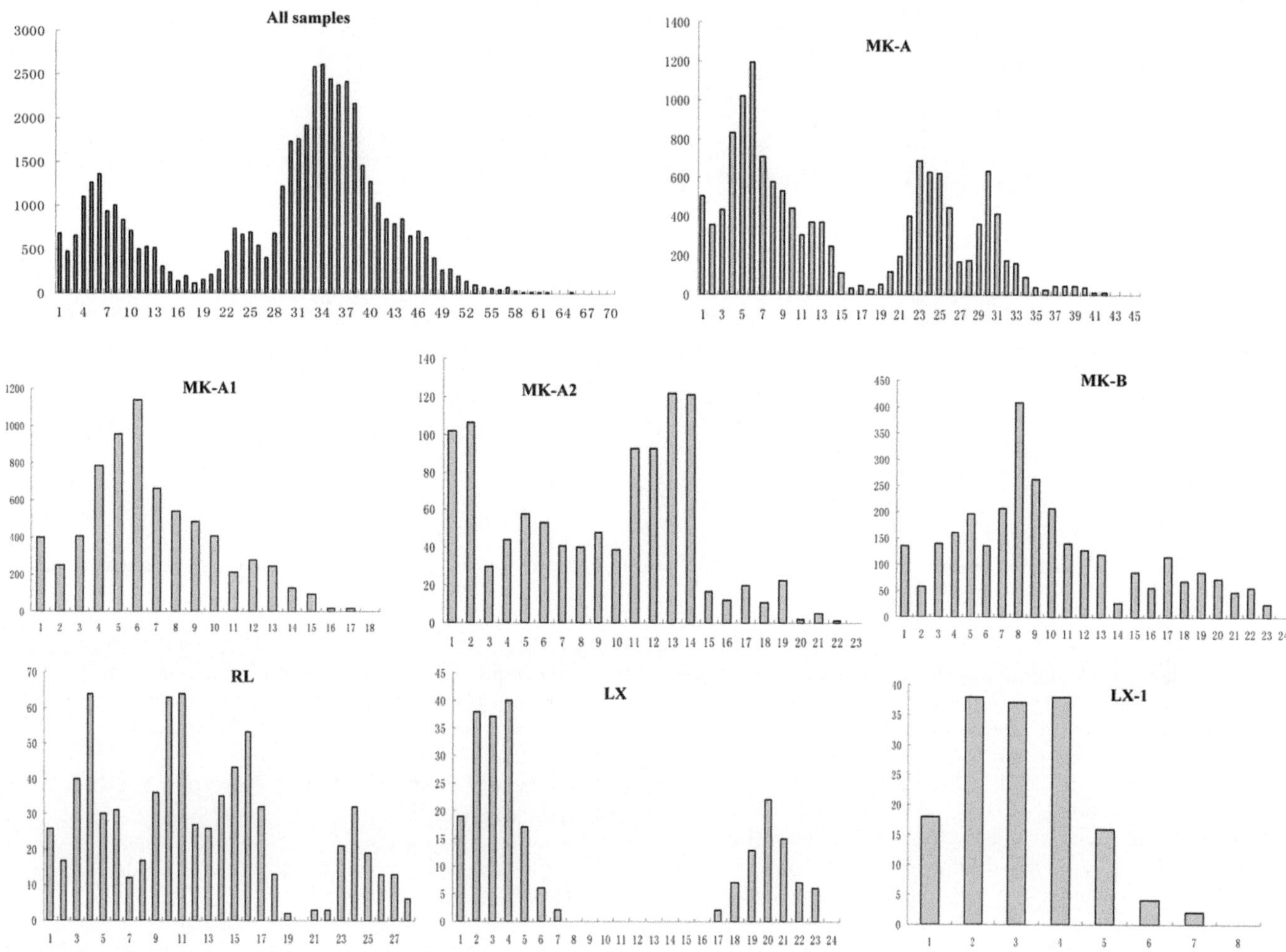

Figure 3. Mismatch distribution analysis of major lineages or sub-lineages of *P. huangchuchieni* based on control region sequences.

position of Lineage LX, and the second basal position of *P. opisthoptera*, indicating that *P. opisthoptera* may instead be a paraphyletic group of *P. huangchuchieni*. While the phylogenetic trees based on different data sets provided somewhat ambiguous results for the taxonomic inference of *P. opisthoptera*, together they firmly supported the notion that *P. opisthoptera* and the four major lineages of *P. huangchuchieni* all originated from a common ancestral stock.

The differentiation of *P. opisthoptera* from the four lineages of *P. huangchuchieni* may have resulted from the separation of the Salween River and Mekong River systems (see the next section). At the morphological character levels, the unique evidence that illustrated *P. opisthoptera* as a separate species from *P. huangchuchieni*

Table 3. Statistics of genetic diversity, tests of neutrality, and demographic parameters estimated for the major lineages and sublineages. These data were defaulted when number of individuals in a population was less than 10.

	No. of individuals	No. of haplotypes	Haplotype diversity	Nucleotide diversity	Tajima'D	Fu's *Fs*	Tau	T(KY)
MK-A	166	54	0.9593±0.0068	0.0110±0.0056	−0.0762	−8.804	–	–
MK-B	77	28	0.9482±0.0097	0.0143±0.0072	0.112	−6.842*	5.449	140
RL	39	26	0.9771±0.0109	0.0183±0.0092	−0.342	−2.828	–	–
LX	22	13	0.9328±0.0299	0.0171±0.0088	−0.040	−0.664	–	–
SW	5	5	–	–	–	–	–	–
MK-A1	119	39	0.9427±0.011	0.0060±0.0032	-1.256	-16.48*	3.906	100
MK-A2	47	16	0.9056±0.020	0.0083±0.0044	-0.536	0.112	–	–
LX-1	17	9	0.8824±0.0513	0.0022±0.0015	-0.912	-3.48*	2.229	60

**P*<0.05.

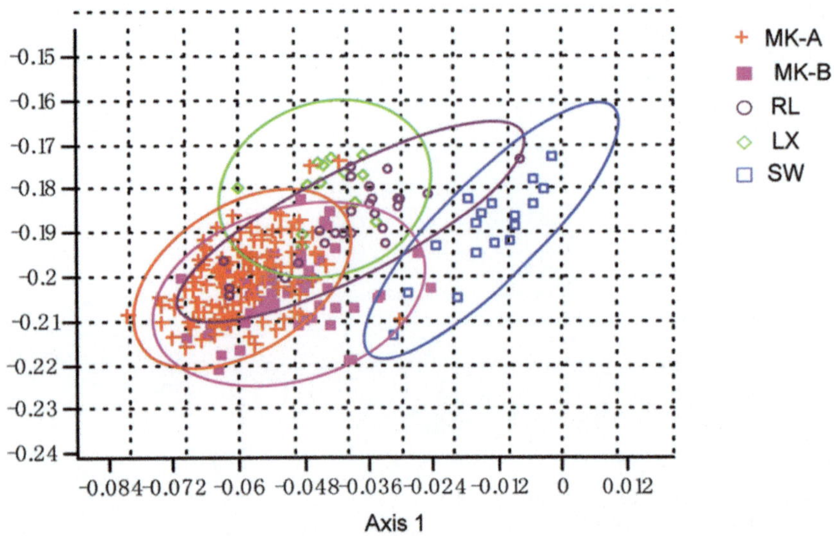

Figure 4. Canonical variate analysis (CVA) based on morphological characters of inferred mitochondrial lineages of *P. huangchuchieni*.

was the position of the dorsal fin [19,20]. In *P. opisthoptera*, the distance between the origin of the dorsal fin and the base of tail fin was shorter than that between the origin of the dorsal fin and the posterior end of preopercular bone. In *P. huangchuchieni*, the distance between the origin of the dorsal fin and the base of tail fin was no less than the distance between the origin of the dorsal fin and the posterior margin of eyes (Fig. 5). In our Canonical Variates Analysis (CVA), the position of the dorsal fin greatly contributed to the separation of *P. opisthoptera* from the four major lineages of *P. huangchuchieni*, in which the dorsal fin was placed more anteriorly in the four lineages of *P. huangchuchieni* but more posteriorly in *P. opisthoptera*. However, we detected several individuals in lineages MK-A, MK-B and RL that shared the posterior dorsal fin characteristic with individuals of *P. opisthoptera* (Fig. 4). This finding indicated that the position of dorsal fin was a continuous character and as such could not be used as diagnostic in order to discriminate the two species. Furthermore, this finding suggested that *P. opisthoptera* is not an effective species, but a paraphyletic group of *P. huangchuchieni*-a supposition consistent with our molecular analyses. Accordingly, our molecular and morphometric results both support that *P. opisthoptera* should be considered one of the major lineages of *P. huangchuchieni*.

Table 4. Factor loadings for the 10 highest measured variables and canonical correlations of the differentiation analysis for the 5 main lineages.

Variable	Axis 1	Variable	Axis 2
X3	**0.5450**	Y4	**0.5534**
X4	**0.5177**	Y7	**0.3923**
X9	−0.3194	Y10	−0.3603
Y1	0.2499	X8	−0.2847
Y7	−0.2415	X1	0.2244
Y10	−0.1909	Y6	0.2212
Y5	−0.1738	X4	0.1940
Y11	0.1628	Y5	0.1864
X2	0.1652	X9	−0.1802
X7	−0.1441	Y1	0.1759

Phylogeny and Phylogeographic Pattern in Southwestern Yunnan Plateau

Both our phylogenetic and population structure analyses provided clear evidence for five highly differentiated lineages within *P. huangchuchieni* (Fig. 1). Together, the five major lineages were almost constrained to the three different river systems, in which MK-A and MK-B were in Mekong, RL (except Hap81 which is from the Mekong River System) and LX in the Red River, and SW in the Salween River System. The highly structured phylogeographic pattern of *B. huangchuchieni* is consistent with the current drainage patterns in the Yunnan Plateau. The long-term geographic isolation was clearly an important factor in shaping the current genetic structure and distribution of the major lineages of *P. huangchuchieni* in the three river systems.

Though highly differentiated among the major lineages, the basal topology of our phylogenetic reconstruction is poorly supported. These results might then suggest that the differentiation events among the five major lineages occurred quite rapidly; and the splits of some lineages (i.e. Lineage LX, Lineage RL, Lineage MK-A and Lineage MK-B) likely took place within a comparatively very short period. Our analysis dated the divergence times among the five lineages of *P. huangchuchieni* to 0.96–1.57 MYA. The dated splits times among the major lineages of *P. huangchuchieni* in three river systems therefore can be tentatively associated to the Kunlun-Yellow River Movement (KYM) some 1.10–0.60 MYA [18], which is approximately consistent with the deformation events of the Yunnan Plateau [9] and the continuingly complex development of watersheds and watercourses during the intense Pleistocene geologic movements [9,10,18,21,22,23].

The differentiation of LX in the Red River System deserves particular attention, because it is located at the basal position of the molecular phylogenetic trees (Fig. 1C). The Lixian River is an intermediate drainage that had linked the Mekong and Red Rivers in recent bio-geographic history [24]. This pattern was further supported by phylogeographic analysis of another freshwater fish species *Opsarius pulchellus* (our currently unpublished data).

Figure 5. Locations of landmarks on the body surface of *P. huangchuchieni.* Note: 1: Tip of the mouth, junction of premaxilary and ethmoid; 2: Center of eyes; 3: Former basal point of dorsal fin; 4: Posterior basal point of dorsal fin; 5: Upper basal point of tail fin; 6: Centre basal point of tail fin; 7: Abdominal basal point of tail fin; 8: Posterior basal point of anal fin; 9: Former basal point of anal fin; 10: Former basal point of ventral fin; 11: Former basal point of pectoral fin.

However, the phylogenetic analyses based on both the control region and the combined sequences taken together provided consistent evidence supporting LX's divergence before the separation of the Mekong and Red River systems (Fig. 1A, 1B and 1C). We then conjectured that this discrepancy may be caused by the special geologic or climate changes in the recent evolutionary history, which might fragment the population of lineage LX. However, this explanation is only a hypothesis due to the relatively poor understanding the paleo-geologic and paleo-climatic changes in this region. The driving factors of lineage LX's differentiation are still unclear. Clearly, further investigation on the geologic research and phylogeography of other aquatic species might help to elucidate this question more fully.

Clearly, the particular evolutionary history of the three river systems revealed by our results is partially consistent with the known phylogeography of the fish species (e.g., glyptosternoid fishes, subfamily Schizothoracinae) inhabiting in the upper portions of these drainages near southeastern Tibet [14,15,16], which all suggest a close relationship between Salween River and Mekong River. Additionally, our results are inconsistent with the phylogeography of *Nanorana yunnanensis*, which proposed that Salween, Mekong and Red River basins in the western part of Yunnan Plateau experienced little change since the early Pliocene [17]. However, the deep river valleys were shown to have most likely acted as barriers for the montane mammal *Apodemus ilex* [26]. This discrepancy may potentially be due to the more complex geographic patterns that resulted from the evolution of the drainage systems, all of which continued to play different roles in regards to the genetic structure of different animals.

Population Differentiation within River Basins in Southwestern Yunnan Plateau

The population differentiated in Mekong and Red River Systems after the KYM vicariant event. For example, the lineage MK-B and MK-A which split into two sublineages, MK-A1 and MK-A2, in Mekong River System; the lineages RL and LX which split into two sublineages, LX-1 and LX-2, in the Red River System (Fig. 1). Independent geological or climatic evidence often serves to suggest that distinctive mtDNA phylogroups may have diverged in allopatry [25]. In addition, most patterns in mtDNA surveys also reveal the secondary admixture between allopatrically evolved phylogroups [25].

The estimated split time between the two lineages in Mekong River System was dated back to about 0.96 MYA. Lineages MK-A and LX diverged into two highly differentiated sublineages: MK-A1 and MK-A2 within the lineage MK-A, and LX-1 and LX-2 within the lineage LX (Fig. 1A, 1B and 1C), implying a long term genetic isolation. Additionally, the unimodal mismatch distributions and the neutrality tests (Fu's *Fs*) both suggest the population expansion event in lineage MK-B, sublineages MK-A1 and LX-1, estimated approximately 0.06–0.14 MYA and thus consistent with the largest glacial retreat that occurred 0.175 MYA [30,31,32]. Additionally, MK-A and MK-B have an overlapped distribution in most localities (Fig. 2; Table 1). Also, a few sublineages were detected co-occurring in sympatry within both the Mekong and Red River systems: i.e., haplotypes of sublineage MK-A1 and MK-A2 coexisted in some sample sites (e.g. M5, Mengla River; M7, Puwen River) in the Mekong River System; lineages LX and RL co-occurred at the Amo River within Red River System (Fig. 2). This pattern suggests secondary contact occurred after the initial divergence in the divergent lineages and sublineages within each drainage system. These demographic patterns might suggest that the sudden global cooling could then potentially explain a restricted gene flow and make the ancestral populations of lineages or sublineages restricted into different refugia, which followed a population expansion event and secondary admixture after the long term isolation between the differentiated lineages/sublineages in each drainage system. The differentiation, recent population expansions and secondary contact of lineages/sublineages within the Mekong and Red River Systems was also found in the phylogeographic analysis of another freshwater fish species *Opsarius pulchellus* (our currently unpublished data). These comparative findings suggest that the long term isolation between lineages caused by climate factors is likely the underlying mechanism of the observed population differentiation between the lineages of *P. huangchuchieni* in the Mekong and Red River Systems. The pattern of population expansion and secondary contact might well be attributable to the climatic oscillations during the extensive glacial period in the Pleistocene and to postglacial population expansions. The diverse environments in the southwestern Yunnan Plateau may have created refugia during glaciations which accordingly acted as barriers to subsequent expansion. However, in some lineages/sublineages (i.e. lineage RL, sublineage MK-A2), we detected a population equilibrium. This pattern could also be attributed to the diverse environments in the southwestern Yunnan Plateau and climatic oscillations during glacial periods, allowing some lineages to expand while others remained stable [33]. In addition, phylogeographic studies on *Opsarius pulchellus* (our currently unpublished data), some frogs which must live near rivers, like the montane frog [33], Yunnan spiny frog [17], and even the montane mammal *Apodemus ilex* [26], have similar population expansion and secondary contact patterns within the Mekong and Red River basins as well as within the Hengduan Mountains. On the whole, these evidences further suggest that the Pleistocene climatic changes affected the distribution and genetic structure of animals inhabiting the region.

During our analysis, we also recognized a multimodal mismatch distribution within all samples of *P. huangchuchieni* in this study (Fig. 3). This could potentially be attributed to a pattern of long term demographic stability and a large effective population size in the recent evolutionary history. These results reveal that population of *P. huangchuchieni* might have been widely distributed in the drainage and keep a high gene flow in a large population for a long period of time, even when the three river systems still interconnect as a single drainage. This pattern was further supported by the demographic stability and equivalent genetic diversity of three major lineages (lineages MK-A, RL and LX) in Mekong and Red River Systems (Table 3, Fig. 3). Because a small population colonized a new region, a recent population expansion event, characterized by a lower genetic diversity and a Poisson-like mismatch distribution, would be likely be found in the colonized region [27]. In this study, we found no evidence of a recent expansion event in the major lineages (except lineage MK-B) of *P. huangchuchieni* in these drainages. As such, it can be inferred that the population of *P. huangchuchieni* dispersed in the three river systems with a large population sizes and differentiated rapidly (Figs. 1, 3).

The large population sizes of these lineages in Mekong and Red River Systems might play a critical role in maintaining the higher levels of diversity and the evolution of lineages within *P. huangchuchieni* (Table 3), such as the prevention of the complete lineage sorting among drainages [25]. Therefore, we found two highly differentiated haplotypes (Hap 81 and Hap 48) in this study (Fig. 1), which were from peripheral populations of Mekong (M6) and Red (R4) River Systems, fall outside the main cluster of lineage RL and located at the basal position within lineage RL (Fig. 1A). The haplotype in the Mekong River System clustered into the lineage RL may have resulted from incomplete lineage sorting, an event that is more likely to occur within large effective populations which can accommodate more polymorphisms than a small sized population [28]. This situation likewise mainly occurs when population divergence is recent [29]. Accordingly, the large effective population of *P. huangchuchieni* and the recent Pleistocene differentiation of lineages are responsible to the incomplete lineage sorting between the Mekong and the Red River Systems.

Conclusion

The results of our study showed that the populations of *P. huangchuchieni* in three river systems (the Salween, Mekong and Red River Systems) were highly differentiated into five main lineages. Likewise, our results revealed that *P. opisthoptera* (lineage SW) was not a separate species but a paraphyletic group of *P. huangchuchieni.* The accumulated evidence suggests that the differentiation of the major lineages within the three river systems may be associated with the drainages' division of the intense uplifting of Qinghai-Tibetan Plateau during the KYM event. The genetic differentiation within river basins and recent demographical expansions occurred in some lineages and sublineages are consistent with the Pleistocene climatic oscillations. Additionally, our results also suggest that the population of *P. huangchuchieni* may have been widely distributed in the southwest region of Yunnan Plateau, maintaining a high gene flow in a large, stable population in the recent evolutionary history, which might be responsible to the incomplete lineage sorting. Additional studies on other aquatic species would likely help to clarify some of our preliminary findings and greatly add to our understanding on how geographic events and paleo-climate changes shape the genetic structure of aquatic populations in areas like the Yunnan Plateau.

Materials and Methods

Ethics Statement

P. huangchuchieni is a common economic fish. All specimens and muscle tissues were bought from local fish dealers. The specimens used in this study were dead. Samples were obtained following the regulations for the implementation of China on the Protection of Wild Animals (Presidential Decree [2004] No. 9).

Sample Collection

A total of 304 specimens of *P. huangchuchieni* and 5 individuals of *P. opisthoptera* were obtained from 18 geographic localities, covering most of the distribution area of *P. huangchuchieni* in China (Table 1; Fig. 2). As outgroups, we included one individual of each *Hypsibarbus pierrei* and *Hypsibarbus vernayi*, collected from a tributary of the Mekong River. Their species identities were confirmed using the morphological features outlined in previous reports by [19]. All specimens were preserved in 95% ethanol. Voucher specimens were deposited in the Museum of Yunnan University.

DNA Extraction, PCR Amplification, and Sequencing

Total genomic DNA was extracted from muscle tissues using the standard phenol-chloroform extraction method. The complete sequence of the mitochondrial control region was amplified with the primers used by Gilles *et al.* [34]. Afterward, the complete sequence of the mitochondrial cytochrome *b* gene (*Cyt b*) was amplified for a subset of samples, selected based on the topology of the control region sequences, using the primers outlines by Xiao *et al.* [35]. PCR amplifications were carried out in 50 ul reaction mixture containing 5 ul 10×PCR buffer (TaKaRa, Dalian), 0.2 mM dNTPs, 0.2 uM each primer, with 1.5 U Taq DNA polymerase (TaKaRa) and approximately 50 ng genomic DNA. Reaction condition were 3 min at 95°C, followed by 35 cycles of 1 min at 94°C, 1 min at 57°C (for the control region) or 52°C (for *Cyt b*), and 1 min at 72°C, and 7 min at 72°C. PCR products were purified using the Gel Extraction Mini Kit (Waston BioTechnologies, Shanghai).

PCR products were sequenced in an ABI Prism 3730 (Applied Biosystems) automatic sequencer. The complete sequences of control region were sequenced directly with PCR primers; the 19 complete sequences of cytochrome b were sequenced using PCR primers and two internal primers (L15286 and H15374) employed by Xiao *et al.* [36]. The sequencing reaction conditions were 25 cycles of 96°C for 30 s, 50°C 15 s and 60°C 4 min.

Statistical Analysis

All nucleotide sequences of the D-loop region, *Cyt b* and combined sequences (both *Cyt b* and D-loop) were aligned using MegAlign in DNAStar 6 (DNASTAR, Madison, USA), and further alignment was confirmed visually in BioEdit 7.0.9 [37]. Protein-coding nucleotide sequences were translated to amino acids to confirm alignment. The sequence polymorphic analysis was performed in MEGA 4 [38].

The control region phylogenetic relationships were reconstructed using Maximum Likelihood (ML) analysis and Bayesian phylogeny reconstruction to identify the evolutionary lineages of haplotypes. The hierarchical likelihood ratio tests criterion (hlRTs) implemented in Modeltest 3.7 [39] was used to determine the evolutionary model that best fit the empirical data. The HKY+I+G [40] model of sequence evolution was used for reconstructing the phylogeny. ML analysis for the control region haplotypes was evaluated using PhyML [41]. Support for nodes among branches was evaluated using non-parametric bootstrapping [42] with 1,000 replicates. Bayesian inference (BI) for both data sets was performed with MrBayes 3.1.2 [43] with four MCMC chains for 10,000,000 generations, saving trees every 100 generations. The first 20% of the generations were discarded as burn-in and posterior probabilities were determined by constructing a 50% majority rule consensus for the remaining trees.

In order to examine whether the combined sequences could be combined into a larger data matrix, we performed a partition-homogeneity (PH) test in Paup 4.0b10 with 1,000 replicates [44]. The PH test suggested that phylogenetic congruence between the *Cyt b* gene and D-loop region could not be rejected (P = 0.34). The phylogenetic relationships based on haplotypes of combined sequences were reconstructed using Maximum Likelihood (ML) analysis and Bayesian inference (BI). The best fitting model according to Modeltest 3.7 was GTR+I+G [40] for the combined sequences. ML reconstruction was performed in Paup 4.0b10 [44], involving a heuristic search, 10 random additional sequences and TBR branch swapping. Estimates of each lineage' robustness was obtained using 1,000 non-parametric bootstraps. Bayesian inference (BI) was performed with MrBayes 3.1.2 [43] with four MCMC chains for 10,000,000 generations, saving trees every 100 generations. The first 20% of the generations were discarded as burn-in and posterior probabilities were determined by constructing a 50% majority rule consensus for the remaining trees.

A haplotype network was constructed using TCS 1.2.1 [45] to estimate gene genealogies for the control region. The connection limit was fixed at 95% and gaps were treated as a 5th state. When ambiguities (closed loops or 'stranded' clades) occurred in the networks, they were resolved using published rules and predictions based on coalescence theory [46,47].

Genetic structure analysis was performed based on the control region using Arlequin 3.1 [48]. The haplotype diversity (H) and nucleotide diversity (π) [49] for each sampled population with a sample size $\geq$10 and each lineage/sublineage were estimated. A hierarchical analysis of molecular variance (AMOVA) was implemented to assess the significant population structure on different levels. The fixation index *F*st, *Fct* and *Fsc* [50] was used to estimate genetic differentiation at the three river systems levels. Both AMOVA and *F*st analysis were performed using pairwise difference with gamma correction; the significance was assessed by 10,000 permutations with Arlequin 3.1.1.1 [48].

A molecular clock relative rate test [51] was performed in RRTree [52] using all control region sequences. The relative rate test result was not significant (P>0.05), indicating the control region sequences fit the molecular clock hypothesis. To relate the genetic differentiation of the major lineages that were defined by genetic structure analysis to tectonic events of Tibet Plateau and adjacent regions, we estimated the divergence times between the major lineages using published nucleotide substitution rates, as there are no specific rates or fossil records available. To estimate the nucleotide substitution rate of the control region sequences in *P. huangchuchieni*, using Mega 4 we calculated the net between-lineage mean distances based on all of the control region sequences, 19 *Cyt b* sequences and 19 control region sequences [38]. The calculated net between-lineage mean distances were very similar between *Cyt b* sequences and the control region sequences (see Results). Generally, a rate of 2% per million years has been calibrated for *Cyt b* genes in multiple bony fishes [53,54]. Subsequently, the evolutionary rate of 2% per million years, which may potentially be suitable for the control region sequences of *P. huangchuchieni*, was applied to estimate the divergence time among lineages. The times to the most recent common ancestor (TMRCA) of the major lineages and the whole population were estimated using relax-clock molecular dating estimation implemented in the BEAST 1.5.2 [55]. Analyses using the HKY model of nucleotide substitution with gamma distributed rate variation among sites were performed. The Yule speciation method was assumed and the nucleotide substitution rate of 2% was used. Chains were run for 50 million generations, with the first 20% discarded as burn-in. The results were summarized through TRACER 1.5 [56].

The historical demographic expansion events of each evolutionary lineages or sublineages were examined using mismatch distribution, as implemented in Arlequin 3.1 [48]. We compared the observed frequency distribution of pairwise nucleotide differences among individuals. A multimodal distribution of pairwise differences was displayed in populations at a demographic equilibrium or in decline, whereas a unimodal distribution was in populations that had experienced a sudden demographic expansion. Additionally, values of Tajima' D [57] and Fu's *F*s [58] were also calculated to seek evidence of demographic expansions in lineages and sublineages. The statistical significance was tested with 5,000 permutations. The population expansion times of some lineages or sublineages were estimated using the equation $\tau = 2$ ut [59], where τ is the mismatch distribution age expansion parameter, u is the mutation rate per generation for the whole sequences, and t is the expansion time in number of generations. The value u was calculated using the formula $u = 2\ \mu k$, where μ is the mutation rate per nucleotide and k is the number of nucleotides assayed. We assumed a mutation rate per nucleotide of 2% for control region sequences of in-group individuals.

Geometric Morphometric Analysis

All specimens were placed on their right side, with their body posture and fins teased into a natural position. Images of 254 fish specimens (including 237 specimens of *P. huangchuchieni* and 17 specimens of *P. opisthoptera*) were collected using a digital camera at 11 landmarks (see Fig. 5) were digitized to describe the body shape of the fish using TpsDig2 [60]. Morphometric and statistical analyses were conducted in PAlaeontological STatistics (PAST) 1.81 [61]. The first step of each analysis was a two-dimensional procrustes to separate shape from size and eliminate variation in the position and orientation of the specimens. For comparison with results obtained from the mitochondrial marker, all groups were defined according to the lineages of mtDNA. The standardized data underwent canonical variate analysis (CVA) to estimate the discrimination among lineages. Population centroids with 95% confidence ellipses derived from the CVA were used to visualize the relationships among individuals of the lineages.

Supporting Information

Figure S1 The six main rivers distribute in Yunnan Plateau, China. The six rivers are divided into two groups, Jinsha-Nanpan-Red and Mekong-Salween-Irrawaddy, by the red lines.

Author Contributions

Conceived and designed the experiments: HX, YPZ. Performed the experiments: XYW. Analyzed the data: XYW,JL. Contributed reagents/materials/analysis tools: JL, SH, ZMC. Wrote the paper: XYW. Revised the manuscript: JL, YPZ.

Reference

1. Bermingham E, Avise JC (1986) Molecular zoogeography of freshwater fishes in the southeastern United States. Genetics 113: 939–965.
2. Durand JD, Persat H, Bouvet Y (1999) Phylogeography and postglacial dispersion of the chub (Leuciscus cephalus) in Europe. Mol Ecol 8: 989–997.
3. Dominguez-Dominguez O, Alda F, de Leon GP, Garcia-Garitagoitia JL, Doadrio I (2008) Evolutionary history of the endangered fish Zoogoneticus quitzeoensis (Bean, 1898) (Cyprinodontiformes: Goodeidae) using a sequential approach to phylogeography based on mitochondrial and nuclear DNA data. BMC Evol Biol 8: 161.
4. Zemlak TS, Walde SJ, Habit EM, Ruzzante DE (2011) Climate-induced changes to the ancestral population size of two Patagonian galaxiids: the influence of glacial cycling. Mol Ecol 20: 5280–5294.
5. Waters JM, Rowe DL, Apte S, King TM, Wallis GP, et al. (2007) Geological dates and molecular rates: rapid divergence of rivers and their biotas. Syst Biol 56: 271–282.
6. Bernatchez L, Wilson C (1998) Comparative phylogeography of nearctic and palearctic fishes. Mol Ecol 7: 431–452.
7. Dubut V, Fouquet A, Voisin A, Costedoat C, Chappaz R, et al. (2012) From Late Miocene to Holocene: processes of differentiation within the Telestes genus (Actinopterygii: Cyprinidae). PLoS One 7: e34423.
8. Wang JP, Lin HD, Huang S, Pan CH, Chen XL, et al. (2004) Phylogeography of Varicorhinus barbatulus (Cyprinidae) in Taiwan based on nucleotide variation of mtDNA and allozymes. Mol Phylogenet Evol 31: 1143–1156.
9. Cheng J, Liu X, Gao Z, Tang D, Yue J (2001) Effect of the Tibetan plateau uplifting on geographical environment of the Yunnan plateau. Geoscience 15: 290–296.
10. Ming Q, Shi Z (2006) The tentative inquiry on the formation time in the region of Three Parallel Rivers. Yunnan Geographic Environment Research 18: 1–4.
11. Clark MK, Schoenbohm LM, Royden LH, Whipple KX, Burchfiel BC, et al. (2004) Surface uplift, tectonics, and erosion of eastern Tibet from large-scale drainage pattern. Tectonics 23: doi:10.1029/2002TC001402.
12. Chu XL (1986) Ichthyofauna and its geographical subdivision in Yunnan, China. Indo-Pacific Fish Biology Proceedings of the Second International Conference on Indo-Pacific Fish. 471–476.
13. Chu XL, Chen YR (1989) The fishes of Yunnan, Part I Cyprinidae. Beijing: Science Press. 354–362.
14. Peng Z, Ho SY, Zhang Y, He S (2006) Uplift of the Tibetan plateau: evidence from divergence times of glyptosternoid catfishes. Mol Phylogenet Evol 39: 568–572.
15. Yang J, Yang JX, Chen XY (2012) A re-examination of the molecular phylogeny and biogeography of the genus Schizothorax (Teleostei: Cyprinidae) through enhanced sampling, with emphasis on the species in the Yunnan–Guizhou Plateau, China. J Zool Syst Evol Res 50: 184–191.
16. He D, Chen Y (2006) Biogeography and molecular phylogeny of the genus Schizothorax (Teleostei: Cyprinidae) in China inferred from cytochrome b sequences. J Biogeogr 33: 1448–1460.
17. Zhang DR, Chen MY, Murphy RW, Che J, Pang JF, et al. (2010) Genealogy and palaeodrainage basins in Yunnan Province: phylogeography of the Yunnan spiny frog, Nanorana yunnanensis (Dicroglossidae). Mol Ecol 19: 3406–3420.
18. Cui Z, Wu Y, G L (1998) Discovery and character of the Kunlun-Yellow River Movement. Chinese Science Bulletin 43: 833–836.
19. Chu X, Cui G (1989) Barbinae. In: Chu XL, Chen YR, editors. The fishes of Yunnan, China. Part I Cyprinidae. Beijing: Science Press. pp. 201–203.
20. Shan X, Lin R, Yue P, Chu X (2000) Barbinae. In: Yue PQ, chief editor. Fauna Sinica, Osteichthyes Cypriniformes III. Beijing: Science Press. pp. 3–170.
21. Li J (1999) Studies on the geomorphological evolution of the Qinghai-Xizang (Tibetan) Plateau and Asian monsoon. Marian Geology & Quaternary Geology 19: 1–11.
22. Li J, Li B, Wang F (1981) The process of uplift of the Qinghai-Xizang Plateau. Geological and Ecological Studies of Qinghai-Xizang Plateau. Beijing: Science Press. pp. 111–118.
23. Cheng J, Chen H, Luo S, Peng H (1994) On valley quaternary in the Jinsha River, Nujiang River, Lancang River and Yuanjiang River, Yunnan province. Geoscience 8 11–19.
24. Li F, Li X, Fu Q, Zhou W (2006) Fish diversity in Simao district, Yunnan. Journal of Guangxi Normal University 24: 77–80.
25. Avise J (2000) Phylogeography. The History and Formation of Species. Cambridge, Massachusetts: Harvard University Press.
26. Liu Q, Chen P, He K, Kilpatrick CW, Liu SY, et al. (2012) Phylogeographic study of Apodemus ilex (Rodentia: Muridae) in Southwest China. PLoS One 7: e31453.
27. Zink RM (2002) Methods in comparative phylogeography, and their application to studying evolution in the north american aridlands. Integr Comp Biol 42: 953–959.
28. Pollard DA, Iyer VN, Moses AM, Eisen MB (2006) Widespread discordance of gene trees with species tree in Drosophila: evidence for incomplete lineage sorting. PLoS Genet 2: e173.
29. April J, Mayden RL, Hanner RH, Bernatchez L (2011) Genetic calibration of species diversity among North America's freshwater fishes. Proc Natl Acad Sci U S A 108: 10602–10607.
30. Zhang D, Liu F, Bing J (2000) Eco-environmental effects of the Qinghai-Tibet plateau uplift during the Quternary in China. Environmental Geology 39: 1352–1358.
31. Shi Y (2002) Characteristics of late Quaternary monsoonal glaciation on the Tibetan Plateau and in East Asia. Quaternary International 97–98: 79–91.
32. Zheng B, Xu Q, Shen Y (2002) The relationship between climate change and Quternary glacial cycles on the Qinghai-Tibetan plateau: review and speculation. Quaternary International 97–98: 93–101.
33. Zhang M, Rao D, Yang J, Yu G, Wilkinson JA (2010) Molecular phylogeography and population structure of a mid-elevation montane frog Leptobrachium ailaonicum in a fragmented habitat of southwest China. Mol Phylogenet Evol 54: 47–58.
34. Gilles A, Lecointre G, Miquelis A, Loerstcher M, Chappaz R, et al. (2001) Partial combination applied to phylogeny of European cyprinids using the mitochondrial control region. Mol Phylogenet Evol 19: 22–33.
35. Xiao W, Zhang Y, Liu H (2001) Molecular systematics of Xenocyprinae (teleostei: cyprinidae): taxonomy, biogeography, and coevolution of a special group restricted in East Asia. Mol Phylogenet Evol 18: 163–173.
36. Xiao H, Chen SY, Liu ZM, Zhang RD, Li WX, et al. (2005) Molecular phylogeny of Sinocyclocheilus (Cypriniformes: Cyprinidae) inferred from mitochondrial DNA sequences. Mol Phylogenet Evol 36: 67–77.
37. Hall T (1999) BioEdit: a user-friendly biological sequence alignment editor and analysis program for Windows 95/98/NT. Nucleic acids symposium series 41: 95–98.
38. Tamura K, Dudley J, Nei M, Kumar S (2007) MEGA4: Molecular Evolutionary Genetics Analysis (MEGA) software version 4.0. Mol Biol Evol 24: 1596–1599.
39. Posada D, Crandall KA (1998) MODELTEST: testing the model of DNA substitution. Bioinformatics 14: 817–818.
40. Hasegawa M, Kishino H, Yano T (1985) Dating of the human-ape splitting by a molecular clock of mitochondrial DNA. J Mol Evol 22: 160–174.
41. Guindon S, Gascuel O (2003) A simple, fast, and accurate algorithm to estimate large phylogenies by maximum likelihood. Syst Biol 52: 696–704.
42. Felsenstein J (1985) Confidence limits on phylogenies: an approach using bootstrap. Evolution 39: 783–791.
43. Ronquist F, Huelsenbeck JP (2003) MrBayes 3: Bayesian phylogenetic inference under mixed models. Bioinformatics 19: 1572–1574.
44. Swofford D (2003) PAUP*: Phylogenetic Analysis Using Parsimony (* and other methods). Version 4.0b10. Sinauer Associates Sunderland, MA.
45. Clement M, Posada D, Crandall KA (2000) TCS: a computer program to estimate gene genealogies. Mol Ecol 9: 1657–1659.
46. Crandall K, Templeton A (1993) Empirical texts of some predictions from coalescent theory with applications to intraspecific phylogeny reconstruction. Genetics 134: 959–969.
47. Templeton AR, Sing CF (1993) A cladistic analysis of phenotypic associations with haplotypes inferred from restriction endonuclease mapping. IV. Nested analyses with cladogram uncertainty and recombination. Genetics 134: 659–669.
48. Excoffier L, Laval G, Schneider S (2005) Arlequin (version 3.0): an integrated software package for population genetics data analysis. Evol Bioinform Online 1: 47–50.
49. Nei M (1987) Molecular Evolutionary Genetics. New York: Columbia University Press. 512 p.
50. Excoffier L, Smouse PE, Quattro JM (1992) Analysis of molecular variance inferred from metric distances among DNA haplotypes: application to human mitochondrial DNA restriction data. Genetics 131: 479–491.
51. Takezaki N, Rzhetsky A, Nei M (1995) Phylogenetic test of the molecular clock and linearized trees. Mol Biol Evol 12: 823–833.

52. Robinson-Rechavi M, Huchon D (2000) RRTree: relative-rate tests between groups of sequences on a phylogenetic tree. Bioinformatics 16: 296–297.
53. Bermingham E, McCafferty A, Martin P (1997) Fish biogeography and molecular clocks: perspectives from the Panamanian Isthmus. In: Kocher T, Stepien C, editors. New York: Molecular Systematics of Fishes Academic Press. pp. 113–126.
54. Bowen W, Bass AL, Rocha LA, Grant WS, Robertson DR (2001) Phylogeography of the trumpetfishes (Aulostomus): ring species complex on a global scale. Evolution 55: 1029–1039.
55. Drummond AJ, Rambaut A (2007) BEAST: Bayesian evolutionary analysis by sampling trees. BMC Evol Biol 7: 214.
56. Rambaut A, Drummond A (2007) *Tracer1.5*. Available: http://beastbioedacuk/.
57. Tajima F (1989) Statistical methods for testing the neutral mutation hypothesis by DNA polymorphism. Genetics 123: 585–595.
58. Fu YX (1997) Statistical tests of neutrality of mutations against population growth, hitchhiking and background selection. Genetics 147: 915–925.
59. Rogers AR, Harpending H (1992) Population growth makes waves in the distribution of pairwise genetic differences. Mol Biol Evol 9: 552–569.
60. Rohlf F (2004) TpsDig. Department of Ecology and Evolution. SUNY, New York.
61. Hammer Ø, Harper D, Ryan P (2001) PAST: Paleontological Statistics Software Package for Education and Data Analysis. Palaeontologia Electronica 4: 9.

9

UV Irradiance and Albedo at Union Glacier Camp (Antarctica)

Raul R. Cordero[1]*, Alessandro Damiani[1], Jorge Ferrer[1], Jose Jorquera[1,2], Mario Tobar[1], Fernando Labbe[3], Jorge Carrasco[4], David Laroze[5]

1 Departamento de Física, Universidad de Santiago de Chile, Santiago, Chile, **2** Escuela Superior Politécnica del Litoral, Guayaquil, Ecuador, **3** Departamento de Ingeniería Mecánica, Universidad Técnica Federico Santa María, Valparaíso, Chile, **4** Dirección Meteorológica de Chile, Santiago, Chile, **5** Instituto de Alta Investigación, Universidad de Tarapacá, Arica, Chile

Abstract

We report on the first spectral measurements of ultraviolet (UV) irradiance and the albedo at a Camp located in the southern Ellsworth Mountains on the broad expanse of Union Glacier (700 m altitude, 79° 46′ S; 82° 52′W); about 1,000 km from the South Pole. The measurements were carried out by using a double monochromator-based spectroradiometer during a campaign (in December 2012) meant to weight up the effect of the local albedo on the UV irradiance. We found that the albedo measured at noon was about 0.95 in the UV and the visible part of the spectrum. This high surface reflectivity led to enhancements in the UV index under cloudless conditions of about 50% in comparison with snow free surfaces. Spectral measurements carried out elsewhere as well as estimates retrieved from the Ozone Monitoring Instrument (OMI) were used for further comparisons.

Editor: Juan A. Añel, University of Oxford, United Kingdom

Funding: Although the support of the Escuela Superior Politécnica del Litoral (Guayaquil, ECUADOR), CONICYT-REDES (Preis 130047), CONICYT-BMBF, FONDEF (IT13I10034), FONDECYT (Preis 1120639, 1140239 and Preis 1120764), Millennium Scientific Initiative (P10-061-F), CEDENNA, UTA-Project 8750-12, USACH-DICYT ASOCIATIVO, and UTFSM-DGIP, is gratefully acknowledged, the funders had no role in study design, data collection and analysis, decision to publish, or preparation of the manuscript.

Competing Interests: The authors have declared that no competing interests exist.

* E-mail: raul.cordero@usach.cl

Introduction

In general, surface UV climate is determined by the total ozone column, cloudiness, ground reflectivity (i.e. the albedo), and local aerosols. Although the latter may significantly modulate the surface UV, in Antarctica the load of aerosols is extremely low [1]. Moreover, while heavily overcast conditions can reduce surface UV irradiance up to 90% in the Antarctic Peninsula [2,3], the role of clouds is less important on the Antarctic plateau. Therefore, surface UV in Antarctica is driven by ozone and by albedo. The UV climatology at several Antarctic sites has been already studied [2–6].

On the Antarctic plateau, large seasonal ozone losses (which occur every year in the late August to early October period [7]) lead to a significant increase in surface UV radiation. Ground-based measurements have shown that the average spring erythemal irradiance for 1990–2006 is up to 85% greater than the modeled irradiance for 1963–1980 [8].

The high albedo of a snow-covered surface has a large effect on the global UV radiation, due to multiple reflections between the ground and the scattering atmosphere [9]. Model calculations of UV irradiance for cloudless sky show enhancements in irradiance levels of nearly 50% at 320 nm for a snow-covered surface, in comparison with snow-free surfaces [10]. Snow albedo depends on snow characteristics (grain size, age of snow, snow height, and soot on the snow) and on other factors such as the solar zenith angle (SZA), atmospheric parameters, geometric pattern of the snow surface, and morphology of the area surrounding the measurement site [11].

So far, the ozone depletion in Antarctica has been the dominant factor in the alterations observed in the UV irradiance. However, as a consequence of climate change, the albedo (that is the other major parameter determining the surface UV) is likely to change in its absolute amount, in its qualitative structure or in its temporal pattern.

The consequences of changes in the Antarctic albedo are beyond local variations in the UV. The surface energy budget of the Antarctic continent is significantly dependent on the surface albedo. Therefore a change in prevailing climatic conditions (triggered for example by a change in temperature) can enhance feedback mechanisms. The so-called albedo feedback is a mechanism that enhances climate change, particularly in regions with snow and ice cover [12]. As the Earth warms, the surface reflects less shortwave radiation to space due to changes in the coverage and reflectivity of the surface ice cover, which leads to additional warming [13]. If temperature increases in Polar Regions we can expect both a decrease of the surface albedo in some areas (due to ice melting) and an enhancement in other areas (due to the freshly fallen snow caused by increased precipitations) [14]. More snow and ice melting, more water vapor and in turn more clouds may lead to changes in the Earth's albedo and alterations in the global energy budget. Indeed, general circulation models (GCMs) [15–17] have shown a high sensitivity to changes in the Earth's albedo [16].

Since albedo significantly affects the UV climatology in Polar Regions (and also plays an important role in the Earth's climate warming pattern), its characterization is required. However, quality-controlled measurements of the albedo in Polar Regions are sparse and few spectral measurements have been reported [9,18–23].

Prior measurements at Amundsen-Scott (South Pole) Station and at Vostok Station found that the albedo was nearly independent of snow grain size and ranged from 0.96 to 0.98 in the UV and the visible part of the spectrum [18]; it has been also observed that the albedo slightly increased with the SZA (which agrees with prior results [24]).

Measurements in Sodankylä (67°22′N, 26°39′E, 179 m altitude) [21], and at Neumayer station (70°39′S, 8°15′W) [9], detected that the albedo decreased (up to 10%) as the SZA decreased through the day. The reductions range from 0.77 to 0.67 in the Arctic, and from 0.96 to 0.86 in the Antarctica [25]. The changes in the Antarctic albedo essentially agree with measurements carried out by using broadband instruments at Hells Gate Station (74°51′S, 163°48′E, 20 m altitude), at Neumayer Station, and at Dome Concordia Station (75°09′S, 123°06′E, 3232 m altitude) [11]. Snow metamorphism, sublimation during the day, and refreezing and/or crystal formation/precipitation during the night can explain the observed trends [11]. The spectral albedo of melting snow has been also measured near Barrow (Alaska) [19], near Davis Station (68°34′S, 77°58′E) [20] and in Sodankylä [22].

In this paper, we report on the first quality-controlled spectral measurements of the UV irradiance and the albedo at Union Glacier Camp, located in the southern Ellsworth Mountains on the broad expanse of Union Glacier; 3,030 km from the southern tip of Chile and about 1,140 km from the South Pole (700 m altitude, 79° 46′ S; 82° 52′W). The mean ice thickness of Union Glacier is 1450 m (with a deep subglacial topography; about 900 m below sea level [23]). The measurements were carried out by using the so-called USACH spectroradiometer from the Universidad de Santiago de Chile (USACH, Chile) during a campaign (in December 2012) meant to weight up the effect of the local albedo on the UV irradiance.

Materials and Methods

1. Ground-based Measurements

Quality-controlled measurements of the surface UV require double monochromator-based spectroradiometers. Instruments developed according to the specifications defined by the World Meteorological Organization (WMO) [26] and the Network for the Detection of Atmospheric Composition Change (NDACC) [27] can produce a radiometric stability better than 1% [28].

The accuracy of NDACC-certified instruments has been tested by intercomparison Campaigns [29,30,31]. The differences between NDACC-certified instruments are normally within the bounds defined by the involved uncertainties; up to 4% for UVA wavelengths and up to 10% for UVB wavelengths [32]. By comparison, for broadband instruments (calibrated by using spectroradiometers) uncertainties in the range 7%–16% have been reported [33,34]).

The UV Index (UVI) can be retrieved from spectral measurements by integrating the UV spectra (weighted by the Erythema action spectrum [35]). The uncertainty of UVI values computed from measurements by NDACC-certified instruments has been estimated to be about 5% [36]. Uncertainties of erythemal daily doses (computed also from spectral measurements) are expected to be similar.

Quality-controlled UV time series are limited. In Antarctica, the spectral UV irradiance has been monitored for almost two decades by the U.S. National Science Foundation [8]. Based on those time series, the UV climatology at several Antarctic locations (Palmer Station, McMurdo station and at South Pole station) has been reported [2,4,5].

Spectral measurements reported below were carried out by using the USACH spectroradiometer, which is based on a double monochromator Bentham DMc150F-U, 150 mm focal length, and 1800 lines/mm gratings, fitted with a photomultiplier (PMT) as detector. The Full With at Half Maximum (FWHM) of the USACH spectroradiometer is 2 nm. The system is operated within a temperature-controlled weatherproof box. An input optics for global irradiance fitted with a flat diffuser was used. Although it is not a NDACC-certified instrument, our spectroradiometer complies with NDACC specifications [27] and the WMO recommendations [26]. The USACH spectroradiometer sampled the irradiance every 1 nm (in the range 280–400 nm); scans were carried at a 30 min interval. The absolute calibration of the system was achieved by using a field calibrator fitted with a baffled 100 W quartz halogen lamp. In addition to in situ calibration, quality assurance included corrections for dark signal and for cosine error of the diffuser (see [26] for details). Based on the certificate of the lamp and the transfer of calibrations, we estimated the uncertainty involved in the absolute calibration to be up to 4% for UVA wavelengths and up to 10% for UVB wavelengths (see [32] for details).

2. Satellite Estimates

Satellite-derived estimates of the total ozone column (TOC) data have been compared with ground-based measurements [37,38]. However, under specific conditions (e.g., high satellite SZA, high surface albedo) and/or for Polar Regions (e.g., at high latitudes), a poorer agreement has been reported [39].

Satellite-derived estimates of the surface UV are normally derived from other products (e.g., ozone, albedo, aerosols and cloud cover) by using radiative models and then uncertainties tend to be somewhat higher especially for partly cloudy and overcast conditions [40,41], and over snow-covered surfaces [42,43].

Over polluted areas, validations of OMI-derived data have generally reported that the UV estimates are biased high [44–52]. Among the factors that explain the overestimation have been pointed out the limited spatial resolution [46,51], the lack of sensitivity of OMI to the boundary layer [39,41,44,45,48,52], and the effect of high SZA on the nadir-viewing instruments [39]. However, over snow-covered surfaces the OMI-derived dose is generally lower than the ground-based measurement because the OMI algorithm uses climatological surface albedo that may then be lower than the actual effective surface albedo [42,43]. Part of the problem is that a portion of the observed reflectivity may be incorrectly interpreted as cloud cover, which reduces the estimated irradiance [3].

As shown below, in this study our ground-based measurements were compared with OMI data. The UV estimates were retrieved from OMUVB products (version 1.3). We selected overpass OMI data minimizing the distance between the ground station and the center of the satellite pixel and, due to the presence of the mountains, discarding pixels at altitudes greater than 1000 m asl. The ozone estimates shown below over the period 2004–2012 were retrieved from OMTO3 products.

3. Radiative Transfer Models

Surface UV can also be computed by using radiative transfer models from a set of parameters that include the albedo, the total ozone column, the SZA, as well as the characteristics of aerosols and clouds. Under cloudless conditions and for UVB wavelengths the uncertainties of model products can be and up to 20% for sites with very large aerosol load up, and up to 9% for unpolluted sites [53,54].

The reflectivity of the areas with spatial variation in the albedo [55,56] is often characterized by using the "effective" albedo (the albedo that when used as input into a model reproduces the measured spectrum) [57,58].

Comparisons with ground-based measurements have validated model products [59,60] under cloudless conditions. Although the characterization of the cloud effect is still difficult, models are useful for checking the consistency of surface measurements [2]. In this work, as radiative transfer model we used UVSPEC [61]. The selected radiative transfer solver was the DISORT solver [62]; the extraterrestrial spectrum was quoted from Gueymard [63].

4. Meteorology at Union Glacier Camp

In general, the meteorology at the Union Glacier Camp is defined by the high pressure centered in the interior of the continent, the orography of the region, and the fact that it is located a distance of ~750 km from the edge of Ronne Ice Shelf and the coast line of Ellsworth Land. An automatic Weather Station (AWS) deployed by the Antarctic Logistic & Expeditions (ALE) Company, nearby the Union Glacier runway (79°46 S, 83°16 W) in early February 2010, provides meteorological data to characterize the area. According with the almost 3 years of data, the near-surface air temperature shows a "winter coreless" behavior reaching around −26 to −28°C from April to September and around −6°C in January. The average relative humidity after temperature correction for freezing environment [64] is ~77%; while, the prevailing wind direction year-round is from the South-South west (190–200° magnetic direction) with an average wind speed of ~20 ms^{-1}, revealing the katabatic origin of the wind field [65] affecting the area. On the other hand, the annual cloud cover fraction, according with recent results of the joint CloudSat-CALIPSO satellite data [66] over the southern Ellsworth Mountains region, can range between 30–40% in summer and 60–70% in winter.

Results and Discussion

1. Irradiance Measurements

Despite the fact that no specific permissions are required for the locations/activities reported in our paper, the Chilean Antarctic Institute (INACH, www.inach.cl) issued a permission after our request for measurements that we carried out at an installation in Antarctica (at Union Glacier Camp) under their supervision. Moreover, we confirm that the field studies did not involve endangered or protected species.

Figure 1a shows the spectral measurements carried out on 04.12.2012 (day.month.year) under cloudless conditions. Figure 1b depicts the spectra measured on 11.12.2012, also under cloudless conditions at noon (SZA = 57°) and at midnight (SZA = 77°). Figure 1c shows the UVI computed from our spectral measurements for 4-13.12.2012. The predominant cloudless conditions led to the smooth curves shown in figure 1c.

Figures 1b and 1c also show some satellite estimates. Crosses in figure 1b stand for OMI readings at 305 nm, at 310 nm, at 324 nm and 380 nm. OMI-derived estimates of UVI are also depicted by crosses in figure 1c. OMI measures the solar reflected and backscattered radiation in the wavelength range from 270 nm to 500 nm, that are used to retrieve the TOC, aerosol and cloud cover characteristics, surface UV irradiance and gas traces. The ground pixel size at nadir position is 13×24 km (along × across track) for total ozone column.

Despite the different spectral resolution (OMI resolution is 0.55 nm in the ultraviolet and 0.63 nm in the visible while the FWHM of our instrument was 2 nm), figure 1b shows a reasonable good agreement between our measurements and the OMI readings at 305 nm, at 310 nm, at 324 nm and 380 nm. Without correcting the effect of the resolution, the differences were within the range ±10% at 305 nm, at 324 nm and 380 nm; differences up to 30% were found at 310 nm.

In figure 1c, it can be observed that the UVI peaked at 3.8 on 13.12.2012; this value is about 10% lower than the corresponding UVI estimate retrieved from OMI. Also on 13.12.2012, the difference between the Erythemal Daily Dose (EDD) computed from our ground-based measurements (4113 J/m2), and the corresponding OMI-derived estimate (4448 J/m2), was about 8%. The OMI estimates of the EDD during the Campaign ranged from 4000 J/m2 on 11.12.2012 to 4448 J/m2 on 13.12.2012. When comparing ground-based estimates and satellite-derived data, we found during the campaign differences that ranged from 2% to 10% (in the case of the UVI) and from 0.1% to 8% (in the case of the EDD). These differences were anyway always within the expanded uncertainty bounds of our measurements [36].

Figure 2a shows the time series of the OMI-derived estimates of UVI at noon at Union Glacier Camp. Due to the high SZA, OMI readings are not available for the period May–August. As shown in figure 2a, annually in spring, the noon-time UVI is typically lower than 6 (although UVI estimates higher than 8 have been retrieved due to the seasonal ozone depletion). As shown in figure 2a, the average of UVI estimates computed for November 2008 was about 20% greater than the average for the same month over the 4 preceding years.

Figure 2b shows the ozone estimates over the period 2004–2012 retrieved from OMTO3 products. Large seasonal ozone losses are apparent in figure 2b and values close to 100 DU have been recorded every year in October [67]. By comparing figures 2a and 2b, becomes clear that the seasonal ozone depletion leads to significant increments in the UVI and in its variability. In this work, the monthly average and the corresponding variability were taken as average and the standard deviation, respectively, of the daily estimates. Table 1 shows the monthly average and the corresponding variability of OMI products. The average of the daily ozone estimates for September (158 DU) is typically about 40% lower than for January (275 DU). The variability of ozone data ranges between 5% in January and 26% in November (see Table 1). In the case of UV products, the variability of the "clear-sky UVI" computed for October (34%) is roughly 3 times greater than for January (11%).

We compared the ozone estimates retrieved from OMI, with the ozone computed from our measurements of the UV spectra carried out in December 2012. The ozone was retrieved from our measurements of the global irradiance by applying a method that implied comparing the ratio (between irradiances measured at different wavelengths) with a synthetic chart of this ratio computed for a variety of ozone values [68]. Figure 3 shows the total ozone column progression for 4-13.12.2012 computed from our ground-based measurements. As shown in the figure, our daily measurements agree with OMI-derived estimates of ozone (see crosses in Figure 3). Although the detected differences were slight (ranging from 0.1% to 3% (i.e. up to about 10 DU)). due to the high sensitivity of the UVI to the ozone, they can explain most of the differences in the UVI estimates shown in figure 1c. Estimations of the total ozone column retrieved from spectral measurements at

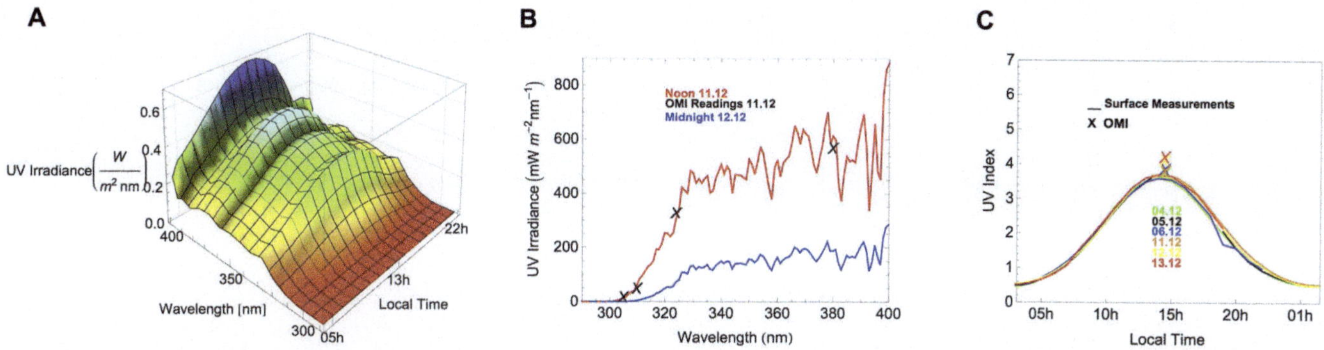

Figure 1. UV Irradiance at Union Glacier Camp. a) Ground-based measurements of the UV irradiance on 04.12.2012 (cloudless conditions). b) Spectra measured at noon on 11.12.2012 (cloudless conditions, SZA = 57°, red line) and at midnight on 12.12.2012 (cloudless conditions, SZA = 77°, blue line). Crosses stand for OMI readings at 305 nm, at 310 nm, at 324 nm and at 380 nm. c) Lines stand for UVI computed from ground-based spectral measurements on 04.12.2012 (green line), on 05.12.2012 (black line), on 06.12.2012 (blue line), on 11.12.2012 (orange line), on 12.12.12 (yellow line), on 13.12.2012 (red line). Crosses stand for OMI estimates of UVI.

other Antarctic locations have been also found to agree with TOMS data [2,5,69].

We further exploited our ground-based measurements studying the possibility of an unlikely aerosol modulation. The sky in the Antarctic interior is breathtakingly clear: there is no source of dust or other particulate matter but no measurements of the properties of aerosol loading at Union glacier Camp have been reported. We carried a limited number of spectral measurements of both the global and the diffuse irradiance (by using a shadow ring) at Union Glacier Camp around noon on 09.12.2012. Both, the AOD and the single scattering albedo (SSA) were retrieved from these quality-controlled spectral measurements by applying methods based on the comparison of the measured spectral irradiance with UV spectra computed by using a radiative transfer model [70–72]. According to these methods, the retrieved values of the AOD and SSA are those leading to the best match between the measured

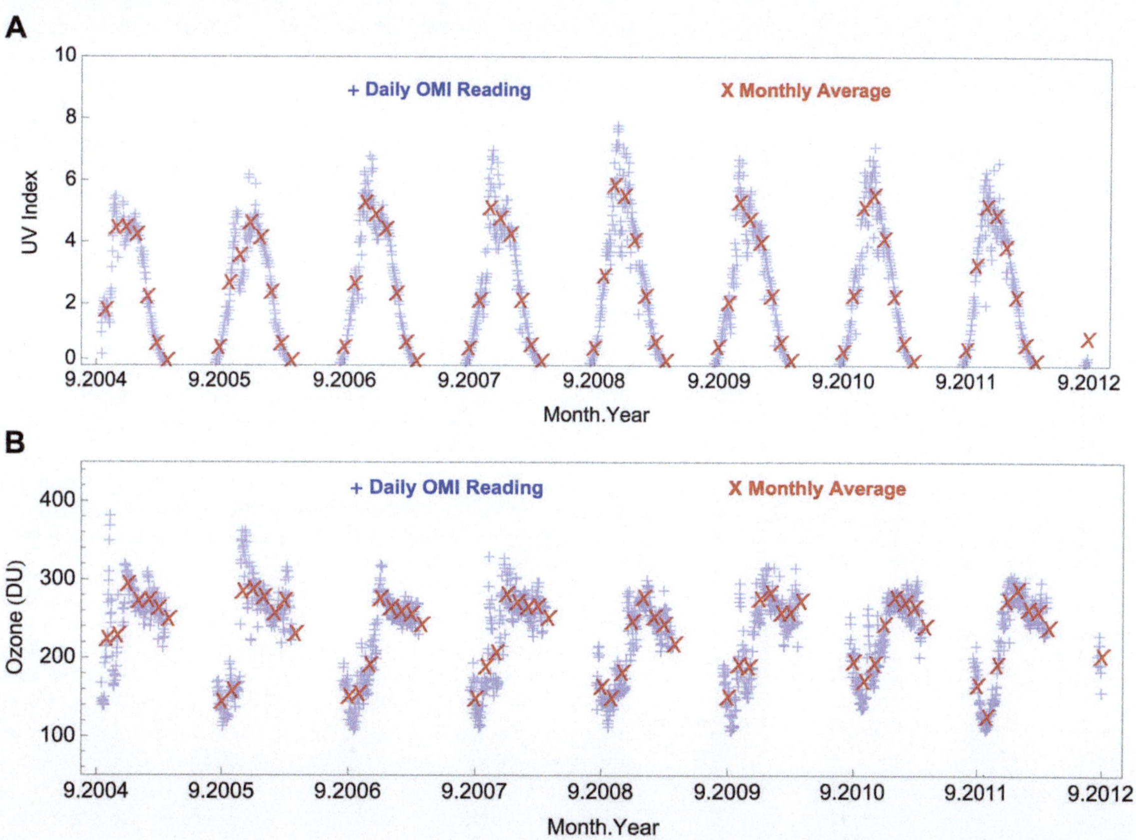

Figure 2. Time series of OMI-derived data at Union Glacier Camp. The number "9" in the abscise tick labels means "September". The blue crosses stand for the daily estimates; the red crosses stand for the monthly averages of the OMI-derived data. (a) Noontime UV index (UVI). (b) Total ozone column.

Table 1. UV Climatology at Union Glacier Camp computed by using the OMI-derived data over the period 2004–2012.

		Jan	Feb	March	April	Sept	Oct	Nov	Dec
Ozone [DU]	Average	275	261	259	241	158	161	203	267
	Variability	13	16	20	18	28	40	53	30
	Variability [%]	5	6	8	7	18	25	26	11
UVI (Clear-Sky)	Average	4.14	2.27	0.70	0.11	0.53	2.59	5.17	5.03
	Variability	0.46	0.61	0.34	0.03	0.40	0.89	1.17	0.74
	Variability [%]	11	27	49	30	77	34	22	14
UVI	Average	4.09	2.22	0.68	0.11	0.53	2.52	5.02	4.95
	Variability	0.53	0.64	0.35	0.03	0.40	0.90	1.22	0.78
	Variability [%]	13	29	51	30	77	36	24	16
Erythemal	Average	3825	1691	425	38	278	1767	4058	4849
Daily Dose	Variability	756	657	249	23	245	799	889	782
(J/m̂2)	Variability [%]	20	39	58	59	88	45	22	16
Surface Albedo	Average	0.89	0.87	0.83	0.78	0.76	0.85	0.88	0.89

The variability was taken as being equal to the standard deviation of the OMI-derived values.

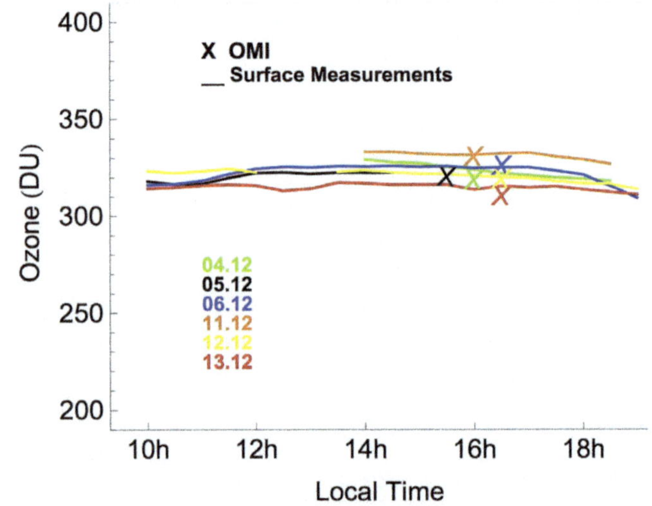

Figure 3. Ozone Column at Union Glacier Camp. Lines stand for total ozone column retrieved from ground-based measurements. Crosses stand for ozone estimates retrieved from OMI. Color indicates the dates: 04.12.2012 (Green); 05.12.2012 (black); 06.12.2012 (blue); 11.12.2012 (orange), 12.12.2012 (yellow), 13.12.2012 (red).

and the computed spectra [70–72]. By using the methods described above, we estimated that the AOD and the SSA at Union Glacier Camp are about 0.02 and 0.99, respectively. These values stand for an extremely low aerosol loading. Actually, due to the uncertainty associated with our aerosol estimates, assuming the total lack of aerosols in the area is also reasonable [9]. At Neumayer, extremely low aerosol loading has been recently measured [1].

Since our measurements were carried our under cloudless conditions, they did not allow us to weigh up the effect of local clouds on the UV irradiance. However, the effect of the high albedo of the surrounding snow-covered surface was assessed (see sections 4 and 5).

2. Albedo Measurements

The albedo is normally taken as being equal to the ratio between the up- and downwelling global radiation. Although measurements of the broadband albedo are normally carried out by using two radiometers [26], our spectral measurements were carried out (at 300–800 nm wavelength range) by using a single spectroradiometer and a rotating input optics that was set up at about 2.5 m above the snow surface. This means that the albedo assessments required measuring sequentially the up- and downwelling global radiation. Due to the time needed to accomplish a scan, the up- and downwelling spectra were separated in time about 5 min. Pairs of scans were carried out at 30 min intervals. A similar method based on the rotation of the input optics has been previously used in Antarctica [9]. Since the measurements of the up- and downwelling radiation were not carried out at the same time (i.e. at the same SZA), the spectra measured at 30 min intervals were linearly interpolated. Then, we estimated the up- and downwelling radiation for certain common SZAs.

Figure 4 shows the spectral albedo measured around noon. No significant differences were observed between measurements carried out on different days around noon under cloudless conditions. As shown in figure 4, the spectral albedo around noon was nearly constant in the UV and visible range but decreased at wavelengths longer than 600 nm; the lowest albedo (about 0.86) was measured at 800 nm. The diminishment of the albedo with

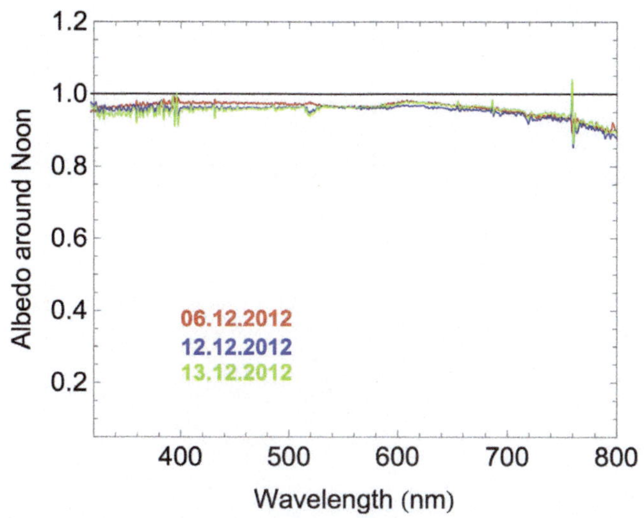

Figure 4. Spectral albedo measured under cloudless conditions around noon at Union Glacier Camp. Color indicates the dates: 06.12.2012 (Red); 12.12.2012 (Blue); 13.12.2012 (Green). Note that the narrow spectral variations in figure 4a have no physical meaning and are due to insignificant misalignments in the grating of our scanning double monochromator when measuring sequentially the up- and downwelling global radiation.

the wavelength in the near infrared was expected since it has been reported by all prior spectral measurements of the albedo over snow-covered surfaces [9,18–23].

Note that the albedo measurements shown in figure 4 are slightly lower than the values measured at Neumeyer station [9]. The difference maybe linked with characteristic of the snow, which is ultimately determined by the local meteorology. While at Neumeyer station the albedo of fresh snow was often measured, our measurements were carried out over a glacier covered by 2 m of old snow with grain radio of about 1 mm; old coarse-grained has lower albedo than fresh snow [20].

Measurements through the day have detected that the albedo decreased (up to 10%) as the SZA decreased through the day [9,21]. These diminishments range from 0.77 to 0.67 in the Arctic, and from 0.96 to 0.86 in the Antarctica [25]. The change in the albedo through the day has been attributed to changes in the snow characteristics [11]. Moreover, snow melting during daytime and refreezing during night may lead to albedo asymmetries (i.e. differences in the albedo measured at the same SZA, at different times of day) [25]. However, we did not detect significant chances in the albedo through the day at Union Glacier Camp. We attributed the apparent albedo stability to the relatively small variations through the day (from −12°C to −9°C) in the air temperature logged by the ALE's AWS during our albedo measurements.

Note that our measurements were carried out about 3 km away from the closest peak (1450 m altitude). Therefore, the slopes of the surrounding mountains were in the field of view of the input optics. However, the complex topography surrounding Union Glacier Camp did not lead to any measurable effect on our ground-based measurements. Indeed, we checked the possibility of asymmetries (i.e. differences between spectral measurements carried out at the same SZA, at different times of day). We found that the differences were always within the uncertainty bounds of our measurements [32].

The surrounding topography did affect satellite data. Figure 5a depicts the surface reflectance map of Antarctica from OMI Lambertian equivalent reflectivity (LER) at 354 nm. LER is the required reflectance of an isotropic surface needed to match the observed top of the atmosphere (TOA) reflectance in a pure Rayleigh scattering atmosphere [73]. Therefore, it can be used to roughly estimate the actual surface albedo. Figure 5b shows the average of the surface reflectance data at 354 nm computed for December by using the data over the period 2005–2009 [73]. This is the albedo used by OMI algorithm to compute the UV surface irradiance (see Table 1). As expected the general pattern shows a very high surface albedo over almost all the Antarctic continent (with values roughly equal to or greater than 0.95). Lower reflectivities can be observed over the Transantarctic Mountains (close to the Ross Ice Shelf) and over the Ellsworth Mountains. Due to the katabatic winds the Antarctic mountain slopes are often snow-free and showing their underlying dark bedrock. Considering the relatively large OMI pixel (maps in figures 5a and 5b have a 0.5×0.5 degree resolution), the difference between ground-based measurement of albedo (about 0.95 in the UV and visible) and the lower reflectivity recorded by satellite (about 0.9 at 354 nm, see Table 1) may be partly explained by these small dark areas inside satellite bright pixels and by the Lambertian approximation.

Figure 5b shows the annual average of the reflectance data at 354 nm of snow-covered scenes (with relaxed cloud-screening criteria [74]). In contrast to figure 5a, clear-sky OMI scenes have been selected by using cloud and aerosol data from the MODIS/Aqua satellite instrument that flies 12 min ahead of OMI. The result is a reflectivity dataset that does not rely on statistical methods to eliminate cloud effect [74]. Despite some minor differences in absolute values (likely related to of the different periods of time considered in the datasets), the patterns are nearly the same.

3. Albedo Effect on UV

In order to weight up the effect of the albedo, we computed the UVI by using the UVSPEC model under the conditions observed at Union Glacier Camp at 11:40 LT on 06.12.2012, 2012 (SZA = 60°, AOD = 0.02; SSA = 0.99; altitude = 700 m; cloudless conditions) but assuming different ozone values as well as different albedo levels. As pointed out above, as radiative transfer solver was used the DISORT solver [69]; the extraterrestrial spectrum was quoted from Gueymard [68]. Figure 6 shows that despite of the ozone, a surface reflectivity greater than 0.9 in the UV (as that measured at Union Glacier Camp) led to enhancements in the UVI under cloudless conditions of about 50% in comparison with snow free surfaces (albedo = 0.1).

The effect of the albedo can be also weighted up by comparing our ground-based measurements at Union Glacier Camp with other spectra measured at snow free locations (albedo = 0.1). We used some of our measurements carried out in the austral summer at Paranal Observatory (2,635 m altitude, 24°37′S, 70°24′W) and in Santiago de Chile (500 m altitude, 33°27′ S, 70°41′ W).

Figure 7a shows the UV index computed from spectral measurements carried out at Union Glacier Camp on 05.12.2012, at Paranal Observatory on 09.01.2013, and in Santiago de Chile on 10.12.2011. Mostly due to the latitude (and in turn to the SZA), peak UVI values at Union Glacier Camp are significantly lower than those in Santiago and at Paranal Observatory. The figure changes when further comparisons are carried at common SZA.

Figure 7b shows the ratio between UV spectra at SZA = 60° measured at Union Glacier Camp (on 05.12.2012) and at Paranal Observatory (on 09.01.2013). Figure 7c shows the ratio between

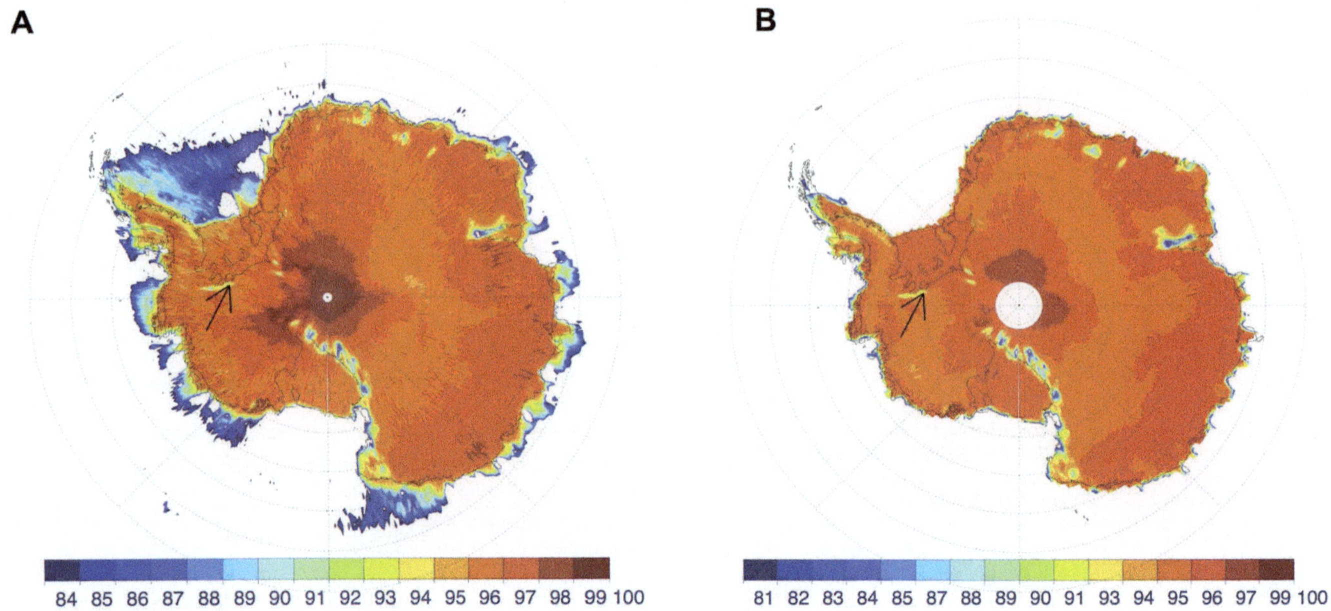

Figure 5. Satellite Albedo. a) Climatology albedo from OMI surface reflectance data (LER) at 354 nm for December. Arrow indicates the location of the Union Glacier Camp. b) Climatology albedo from OMI reflectance data (LER) at 354 nm for snow-covered scenes. Arrow indicates the location of the Union Glacier Camp. See the text for details.

UV spectra at SZA = 60° measured at Union Glacier Camp (on 05.12.2012) and at USACH station in Santiago de Chile (on 10.12.2011). In figures 7b and 7c we also show the ratio between modeled UV spectra computed by using the UVSPEC model under the conditions observed at the moment of the ground-based measurements (see details in the captions).

It can be seen in figures 7b that the UVA irradiance at Union Glacier Camp is significantly greater than at Paranal Observatory (the difference is 60% at 330 nm). Most of this difference is due to the albedo, and it would be even greater but the altitude of the observatory attenuates the difference. In figures 7c, it can be observed that the UVA irradiance at Union Glacier Camp is also significantly greater than in Santiago (the difference is 100% at 330 nm). Although again most of the difference is due to the albedo, the aerosol load in the case of Santiago boosts the difference. As shown in figures 7b and 7c, the differences in the UVB range are counterbalanced by the ozone (about 30% lower in Santiago and at Paranal than at Union Glacier Camp, see captions for details). However, despite the relatively high ozone, the UVI at Union Glacier Camp on 05.12.2012 at SZA = 60° (330 DU), was greater than in Santiago on 10.12.2011 at SZA = 60° (265 DU) and than at Paranal observatory on 09.01.2013 at SZA = 60° (260 DU).

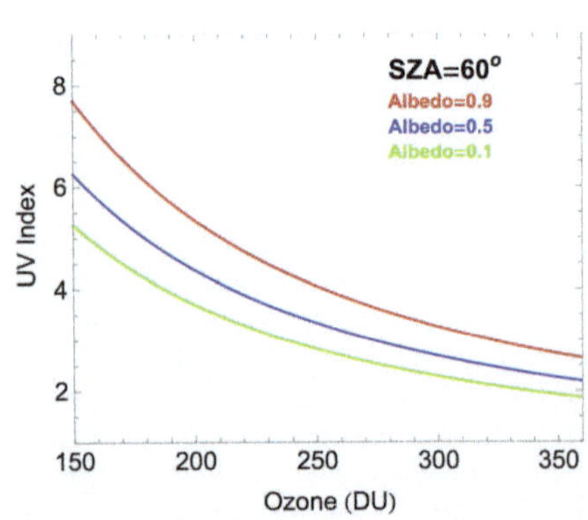

Figure 6. Change in the UVI with the albedo. The UVI was computed by using the UVSPEC model under the conditions observed at Union Glacier Camp at 11:40 LT on 06.12.2012 (SZA = 60°, AOD = 0.02; SSA = 0.99; altitude = 700 m) but assuming different ozone column values as well as different albedo levels (Red line: 0.9; Blue line: 0.5; Green line: 0.1).

Summary and Conclusions

We report on the first quality-controlled spectral measurements of the albedo and the UV irradiance at Union Glacier Camp, located in the southern Ellsworth Mountains on the broad expanse of Union Glacier. The measurements were carried out in December 2012 by using a double monochromator-based spectroradiometer during a campaign meant to weight up the effect of local albedo on the UV irradiance.

The UVI, the TOC, the AOD and the SSA, were computed from our spectra. We found that our surface measurements roughly agreed with OMI estimates of UVI and TOC; the detected differences ranged from 0.1% to 3% (i.e. up to about 10 DU) in the case of the TOC, and from 2% to 10% in the case of the UVI. Moreover, the AOD and the SSA retrieved from our surface measurements allowed us to conclude that the aerosol loading at Union Glacier Camp is extremely low (i.e. the aerosol influence in the area is negligible).

We also found that the spectral albedo measured at noon was about 0.95 in the UV and the visible part of the spectrum. We did not detect significant asymmetries (i.e. differences in the albedo measured at the same SZA, at different times of day). The differences between albedo values sampled at different times of day were always within the uncertainty bounds of our measurements.

In order to weight up the effect of the albedo, we compared our ground-based measurements at Union Glacier Camp with other spectra measured at snow-free locations. We used some of our measurements carried out in the austral summer at Paranal

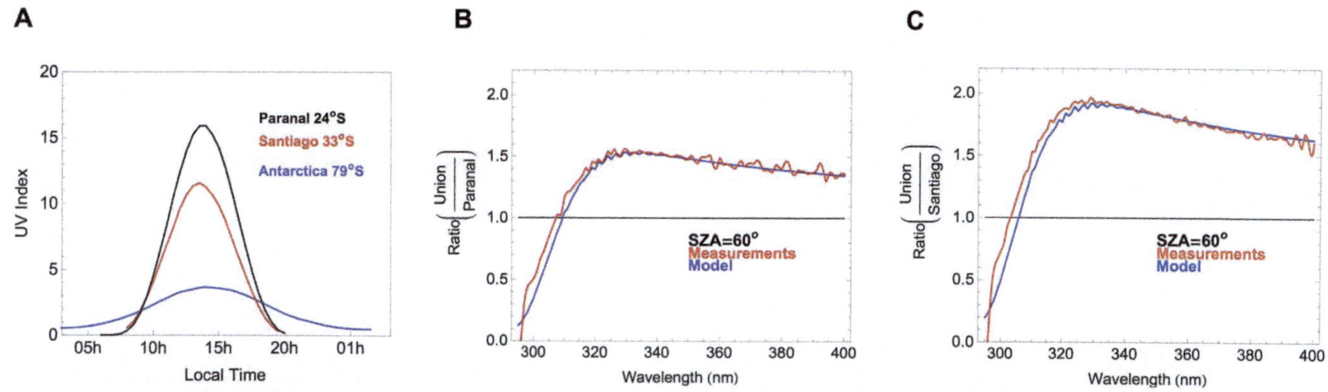

Figure 7. UV index at different locations. a) UV index computed from spectral measurements carried out by using the USACH spectroradiometer at Paranal Observatory on 09.01.2013 (Black line; minimum SZA = 2.6°), in Santiago de Chile on 10.12.2011 (Red line; minimum SZA = 11°), and at Union Glacier Camp on 05.12.2012 (Blue line; minimum SZA = 56.7°). b) Red line: ratio between UV spectra at SZA = 60° measured at Union Glacier Camp (on 05.12.2012) and at Paranal Observatory on (09.01.2013). Blue line: ratio between modeled UV spectra at Union Glacier Camp (albedo = 0.95; AOD = 0.02; SSA = 0.99; altitude = 700 m; ozone = 330 DU) and at Paranal Observatory (albedo = 0.1; AOD = 0.15; SSA = 0.7; altitude = 2,635 m; ozone = 260 DU). c) Red line: ratio between UV spectra at SZA = 60° measured at Union Glacier Camp (on 05.12.2012) and at USACH station in Santiago de Chile (on 10.12.2011). Blue line: ratio between modeled UV spectra at Union Glacier Camp (albedo = 0.95; AOD = 0.02; SSA = 0.99; altitude = 700 m; ozone = 330 DU) and in Santiago de Chile (albedo = 0.1; AOD = 0.3; SSA = 0.7; altitude = 500 m; ozone = 265 DU).

Observatory (2,635 m altitude, 24°37′S–70°24′W) and in Santiago de Chile (500 m altitude, 33°27′ S–70°41′ W). Further modeling allowed us to confirm that the surface reflectivity similar to that at Union Glacier Camp around noon (albedo = 0.95) led to enhancements in the UVI under cloudless conditions of about 50% in comparison with snow free surfaces (albedo = 0.1).

Author Contributions

Conceived and designed the experiments: RRC AD FL. Performed the experiments: JF JJ MT. Analyzed the data: RRC AD JC DL. Wrote the paper: RRC AD FL.

References

1. Weller R, Minikin A, Petzold A, Wagenbach D, König-Langlo G (2013) Characterization of long-term and seasonal variations of black carbon (BC) concentrations at Neumayer, Antarctica. Atmos Chem Phys 13: 1579–1590.
2. Bernhard G, Booth CR, Ehramjian JC (2005) UV climatology at Palmer Station, Antarctica. In: Ultraviolet Ground- and Space-based Measurements, Models, and Effects V, Bernhard G, Slusser JR, Herman JR, Gao W editors, Proceedings of SPIE, 5886, 51–62.
3. Cordero RR, Damiani A, Seckmeyer G, Riechelmann S, Laroze D, et al. (2013) Satellite-derived UV Climatology at Escudero Station (Antarctic Peninsula). Antarctic Science 25(6): 791–803.
4. Bernhard G, Booth CR, Ehramjian JC (2004) Version 2 data of the National Science Foundation's Ultraviolet Radiation Monitoring Network: South Pole. J Geophys Res doi: 10.1029/2004JD004937.
5. Bernhard G, Booth CR, Ehramjian JC, Nichol SE (2006) UV climatology at McMurdo Station, Antarctica, based on version 2 data of the National Science Foundation's Ultraviolet Radiation Monitoring Network. J Geophys Res doi: 10.1029/2005JD005857.
6. Bernhard G, Booth CR, Ehramjian JC, Stone R, Dutton EG (2007) Ultraviolet and visible radiation at Barrow, Alaska: Climatology and influencing factors on the basis of version 2 National Science Foundation network data. J Geophys Res doi:10.1029/2006JD007865.
7. Flemming J, Inness A, Jones L, Eskes HJ, Huijnen V, et al. (2011) Forecasts and assimilation experiments of the Antarctic ozone hole 2008. Atmos Chem Phys 11: 1961–1977.
8. Bernhard G, Booth CR, Ehramjian JC (2010) Climatology of Ultraviolet Radiation at High Latitudes Derived from Measurements of the National Science Foundation's Ultraviolet Spectral Irradiance Monitoring Network. In: UV Radiation in Global Climate Change: Measurements, Modeling and Effects on Ecosystems, Gao W, Schmoldt DL, and Slusser JR editors, Springer-Verlag and Tsinghua University Press, ISBN: 978-3-642-03312-4.
9. Wuttke S, Seckmeyer G, Koenig-Langlo G (2006) Measurements of spectral snow albedo at Neumayer, Antarctica. Annales Geophysicae 24: 7–21.
10. Lenoble J (1998) Modelling of the influence of snow reflectance on ultraviolet irradiance for cloudless sky. Appl Opt 37: 2441–2447.
11. Pirazzini R (2004) Surface albedo measurements over Antarctic sites in summer. J Geophys Res doi: 10.1029/2004JD004617.
12. Hall A (2004) The Role of Surface Albedo Feedback in Climate. J Climate 17: 1550–1568.
13. Winton M. (2006) Surface Albedo Feedback Estimates for the AR4 Climate Models. J Climate 19: 359–365.
14. Vaughan DG, Marshall GJ, Connolley WM, King JC, Mulvaney R (2001) Devil in the Detail. Science 293: 1777–1779.
15. Aoki T, Kuchiki K, Niwano M, Kodama Y, Hosaka M, et al. (2011) Physically based snow albedo model for calculating broadband albedos and the solar heating profile in snowpack for general circulation models. Journal of Geophysical Research doi: 10.1029/2010JD015507.
16. Pedersen CA, Roeckner E, Lüthje M, Winther JG (2009) A new sea ice albedo scheme including melt ponds for ECHAM5 general circulation model. Journal of Geophysical Research doi: 10.1029/2008JD010440.
17. Collins WD, Bitz CM, Blackmon ML, Bonan GB, Bretherton CS, et al. (2006) The community climate system model version 3 (CCSM3). Journal of Climate 19 (11): 2122–2143.
18. Grenfell TC, Warren SG, Mullen PC (1994) Reflection of solar radiation by the Antarctic snow surface at ultraviolet, visible, and near-infrared wavelengths. Journal of Geophysical Research doi: 10.1029/94JD01484.
19. Grenfell TC, Perovich DK (2004) Seasonal and spatial evolution of albedo in a snow-ice-land-ocean environment. Journal of Geophysical Research Oceans doi: 10.1029/2003JC001866.
20. Brandt RE, Warren SG, Worby AP, Grenfell TC (2005) Surface albedo of the Antarctic sea ice zone. Journal of Climate 18(17): 3606–3622.
21. Meinander O, Kontu A, Lakkala K, Heikkilä A, Ylianttila L, et al. (2008) Diurnal variations in the UV albedo of arctic snow. Atmos Chem Phys 8: 6551–6563.
22. Meinander O, Kazadzis S, Arola A, Riihelä A, Räisänen P, et al. (2013) Spectral albedo of seasonal snow during intensive melt period at Sodankylä, beyond the Arctic Circle. Atmos Chem Phys 13: 3793–3810.
23. Cordero RR, Damiani A, Ferrer J, Rayas J, Jorquera J, et al. (2013) Downwelling and Upwelling Radiance Distributions sampled under Cloudless Conditions in Antarctica, Applied Optics 52(25): 6287–94.
24. Li S, Zhou X (2003) Assessment of the Accuracy of Snow Surface Direct Beam Spectral Albedo under a Variety of Overcast Skies Derived by a Reciprocal Approach through Radiative Transfer Simulation. Appl Opt 42: 5427–5441.
25. Meinander O, Wuttke S, Seckmeyer G, Kazadzis S, Lindfors A, et al. (2009) Solar Zenith Angle Asymmetry Cases in Polar Snow UV Albedo. Geophysica 45(1–2): 183–198.
26. Seckmeyer G, Bais A, Bernhard G, Blumthaler M, Booth CR, et al. (2001) Part 1: Spectral instruments Instruments to Measure Solar Ultraviolet Radiation WMO-GAW No. 125 (Geneva, Switzerland: World Meteorological Organization).

27. Wuttke S, Seckmeyer G, Bernhard G, Ehramjian J, McKenzie R, et al. (2006) New spectroradiometers complying with the NDSC standards. J Atmos Ocean Technol 23(2): 241–251.
28. Cede A, Herman J, Richter A, Krotkov N, Burrows J (2006) Measurements of nitrogen dioxide total column amounts using a Brewer double spectrophotometer in direct sun mode. J Geophys Res doi: 10.1029/2005JD006585.
29. Gröbner J, Albold A, Blumthaler M, Cabot T, de la Casinière A, et al. (2000) The variability of spectral solar ultraviolet irradiance in an Alpine environment. J Geophys Res doi: 10.1029/2000JD900395.
30. Gröbner J, Blumthaler M, Kazadzis S, Bais A, Webb A, et al. (2006) Quality assurance of spectral solar UV measurements: result from 25 UV monitoring sites in Europe, 2002 to 2004. Metrologia 43: S66–S71.
31. Bais AF, Gardiner BG, Slaper H, Blumthaler M, Bernhard G, et al. (2001) The SUSPEN intercomparison of ultraviolet spectroradiometers. J Geophys Res 106: 12509–12525.
32. Cordero RR, Seckmeyer G, Pissulla D, DaSilva L, Labbe F (2008) Uncertainty Evaluation of Spectral UV Irradiance Measurements. Meas. Sci. Technol. 19: 1–15.
33. Gröbner J, Hülsen G, Vuilleumier L, Blumthaler M, Vilaplana JM, et al. (2007) Report of the PMOD/WRC-COST calibration and intercomparison of erythemal radiometers.
34. Antón M, Serrano A, Cancillo ML, Vilaplana JM (2011) Quality assurance of broadband erythemal radiometers at the Extremadura UV Monitoring Network (Southwestern Spain). Atmospheric Research 100: 83–92.
35. McKinlay AF, Diffey BL (1987) A reference action spectrum for ultraviolet induced erythema in human skin. Commission Internationale de l'Eclairage Journal 6: 17–22.
36. Cordero RR, Seckmeyer G, Pissulla D, Labbe F (2008) Uncertainty of experimental integrals: application to the UV index calculation. Metrologia, 45: 1–10.
37. Balis D, Kroon M, Koukouli ME, Brinksma EJ, Labow G, et al. (2007) Validation of Ozone Monitoring Instrument total ozone column measurements using Brewer and Dobson spectrophotometer ground-based observations. Journal of Geophysical Research doi: 10.1029/2007JD008796.
38. McPeters R, Kroon M, Labow G, Brinksma E, Balis D, et al. (2008) Validation of the Aura Ozone Monitoring Instrument total column ozone product. Journal of Geophysical Research doi: 10.1029/2007JD008802.
39. Damiani A, De Simone S, Rafanelli C, Cordero RR, Laurenza M (2012) Three years of ground-based total ozone measurements in Arctic: comparison with OMI, GOME and SCIAMACHY satellite data. Remote Sensing of Environment 127: 162–180.
40. Antón M, Cachorro VE, Vilaplana JM, Toledano C, Krotkov NA, et al. (2010) Comparison of UV irradiances from Aura/Ozone Monitoring Instrument (OMI) with Brewer measurements at El Arenosillo (Spain) - Part 1: Analysis of parameter influence. Atmos Chem Phys 10: 5979–5989.
41. Damiani A, Cabrera S, Muñoz RC, Cordero RR, Labbe F (2013) Satellite-derived UV irradiance for a region with complex morphology and meteorology: comparison against ground measurements in Santiago de Chile. International Journal of Remote Sensing doi: 10.1080/01431161.2013.796101.
42. Douglass A, Fioletov V, Godin-Beekmann S, Müller R, Stolarski RS, et al. (2011) Stratospheric ozone and surface ultraviolet radiation 1–80; Global Ozone Research and Monitoring Project-Report No. 52, 516 pp., Geneva, Switzerland.
43. Tanskanen A, Manninen T (2007) Effective UV surface albedo of seasonally snow-covered lands. Atmos Chem Phys 7: 2759–2764.
44. Cabrera S, Ipiña A, Damiani A, Cordero RR, Piacentini RD (2012) UV index values and trends in Santiago, Chile (33.5°S) based on ground and satellite data. Journal of Photochemistry and Photobiology B: Biology 115: 73–84.
45. Kazadzis S, Bais A, Balis D, Kouremeti N, Zempila M, et al. (2009) Spatial and temporal UV irradiance and aerosol variability within the area of an OMI satellite pixel. Atmos Chem Phys 9: 4593–4601.
46. Kazadzis S, Bais A, Arola A, Krotkov N, Kouremeti N, et al. (2009) Ozone Monitoring Instrument spectral UV irradiance products: comparison with ground based measurements at an urban environment. Atmos Chem Phys 9: 585–594.
47. Ialongo I, Buchard V, Brogniez C, Casale GR, Siani AM (2009) Aerosol Single Scattering Albedo retrieval in the UV range: an application to OMI satellite validation. Atmos Chem Phys Discuss 9: 19009–19033.
48. Buchard V, Brogniez C, Auriol F, Bonnel B, Lenoble J, et al. (2008) Comparison of OMI ozone and UV irradiance data with ground-based measurements at two French sites. Atmos Chem Phys 8: 4517–4528.
49. Tanskanen A, Krotkov NA, Herman JR, Arola A (2006) Surface ultraviolet irradiance from OMI. IEEE transactions on Geoscience and Remote Sensing 44(5): 1267–1271.
50. Tanskanen A, Lindfors A, Määttä A, Krotkov N, Herman J, et al. (2007) Validation of daily erythemal doses from Ozone Monitoring Instrument with ground-based UV measurement data. J Geophys Res doi: 10.1029/2007JD008830.
51. Weihs P, Blumthaler M, Rieder HE, Kreuter A, Simic S, et al. (2008) Measurements of UV irradiance within the area of one satellite pixel. Atmos Chem Phys 8: 5615–5626.
52. Ialongo I, Casale GR, Siani AM (2008) Comparison of total ozone and erythemal UV data from OMI with ground-based measurements at Rome station. Atmos Chem Phys 8: 3283–3289.
53. Cordero RR, Seckmeyer G, Pissulla D, DaSilva L, Labbe F (2007) Uncertainty evaluation of the spectral UV irradiance evaluated by using the UVSPEC Radiative Transfer Model. Optics Communications 276: 44–53.
54. Cordero RR, Seckmeyer G, Damiani A, Labbe F, Laroze D (2013) Monte Carlo-based Uncertainties of Surface UV Estimates from Models and from Spectroradiometers. Metrologia 50: L1–L5.
55. Cordero RR, Damiani A, Laroze D, Dasilva L, Labbe F (2013) Spectral UV radiance measured at a coastal site: a case study. Photochemical & Photobiological Sciences 12: 1193.
56. Blumthaler M, Kreuter A, Webb A, Bais A, Kift R, et al. (2009) Albedo Effect on UV Irradiance, MOCA Joint Assembly, Montreal, Quebec, July 2009, Available: http://www.moca-09.org.Accessed 1 October 2013.
57. Kylling A, Persen T, Mayer B, Svenøe T (2000) Determination of an effective spectral surface albedo from ground-based global and direct UV irradiance measurements. J Geophys Res 105(D4): 4949–4959.
58. Smolskaia I, Masserot D, Lenoble J, Brogniez C, de la Casinière A (2003) Retrieval of the Ultraviolet Effective Snow Albedo During 1998 Winter Campaign in the French Alps. Appl Opt 42(9): 1583.
59. Badosa J, McKenzie RL, Kotkamp M, Calbó J, González JA, et al. (2007) Towards closure between measured and modelled UV under clear skies at four diverse sites. Atmospheric Chemistry and Physics 7: 2817–2837.
60. Satheesh SK, Srinivasan J, Vinoj V, Chandra S (2006) New Directions: How representative are aerosol radiative impact assessments? Atmospheric Environment 40(16): 3008–3010.
61. Mayer B, Kylling A (2005) Technical note: The libRadtran software package for radiative transfer calculations - description and examples of use. Atmos Chem Phys 5: 1855–1877.
62. Dahlback A, Stamnes K (1991) A new spherical model for computing the radiation field available for photolysis and heating at twilight. Planet Space Sci 39: 671–683.
63. Gueymard CA (2004) The sun's total and spectral irradiance for solar energy applications and solar radiation models. Solar Energy 76: 423–453.
64. van den Broeke MR, Reijmer CH, van de Wal RSW (2004) A study of the surface mass balance in Dronning Maud Land, Antarctica, using automatic weather stations. Journal of Glaciology 50(171): 565–582.
65. Parish TR, Bromwich DH (1987) The surface windfield over the Antarctic Ice Sheets. Nature 328: 51–54.
66. Bromwich DH, Nicolas JP, Hines KM, Kay JE, Key E, et al. (2012) Tropospheric Clouds in Antarctica. Rev Geophys doi: 10.1029/2011RG000363.
67. Grooß JU, Brautzsch K, Pommrich R, Solomon S, Müller R (2011) Stratospheric ozone chemistry in the Antarctic: what determines the lowest ozone values reached and their recovery? Atmospheric Chemistry and Physics 11: 12217–12226.
68. Stamnes K, Slusser J, Bowen M (1991) Derivation of Total Ozone Abundance and Cloud Effects from Spectral Irradiance Measurements. Applied Optics 30: 4418–4426.
69. Bernhard G, Booth CR, McPeters RD (2003) Calculation of total column ozone from global UV spectra at high latitudes. J Geophys Res doi: 10.1029/2003JD003450.
70. Cordero RR, Seckmeyer G, Pissulla D, Labbe F (2009) Exploitation of Spectral Direct UV Irradiance Measurements. Metrologia 46: 19–25.
71. Bais AF, Kazantzidis A, Kazadzis S, Balis DS, Zerefos CS, et al. (2005) Deriving an effective aerosol single scattering albedo from spectral surface UV irradiance measurements. Atmos Environ 39(6): 1093–1102.
72. Buchard V, Brogniez C, Auriol F, Bonnel B (2011) Aerosol single scattering albedo retrieved from ground-based measurements in the UV and visible region. Atmospheric Measurement Techniques 4: 1–7.
73. Kleipool QL, Dobber MR, de Haan JF, Levelt PF (2008) Earth surface reflectance climatology from 3 years of OMI data. J Geophys Res doi: 10.1029/2008JD010290.
74. O'Byrne G, Martin RV, van Donkelaar A, Joiner J, Celarier EA (2010) Surface reflectivity from the Ozone Monitoring Instrument using the Moderate Resolution Imaging Spectroradiometer to eliminate clouds: Effects of snow on ultraviolet and visible trace gas retrievals. J Geophys Res doi: 10.1029/2009JD013079.

10

Where Are All the Fish: Potential of Biogeographical Maps to Project Current and Future Distribution Patterns of Freshwater Species

Danijela Markovic*, Jörg Freyhof, Christian Wolter

Leibniz-Institute of Freshwater Ecology and Inland Fisheries, Berlin, Germany

Abstract

The dendritic structure of river networks is commonly argued against use of species atlas data for modeling freshwater species distributions, but little has been done to test the potential of grid-based data in predictive species mapping. Using four different niche-based models and three different climate change projections for the middle of the 21st century merged pairwise as well as within a consensus modeling framework, we studied the variability in current and future distribution patterns of 38 freshwater fish species across Germany. We used grid-based (11×11 km) fish distribution maps and numerous climatic, topographic, hydromorphologic, and anthropogenic factors derived from environmental maps at a finer scale resolution (250 m–1 km). Apart from the explicit predictor selection, our modeling framework included uncertainty estimation for all phases of the modeling process. We found that the predictive performance of some niche-based models is excellent independent of the predictor data set used, emphasizing the importance of a well-grounded predictor selection process. Though important, climate was not a primary key factor for any of the studied fish species groups, in contrast to substrate preferences, hierarchical river structure, and topography. Generally, distribution ranges of cold-water and warm-water species are expected to change significantly in the future; however, the extent of changes is highly uncertain. Finally, we show that the mismatch between the current and future ranges of climatic variables of more than 90% is the most limiting factor regarding reliability of our future estimates. Our study highlighted the underestimated potential of grid cell information in biogeographical modeling of freshwater species and provides a comprehensive modeling framework for predictive mapping of species distributions and evaluation of the associated uncertainties.

Editor: Juan A. Añel, University of Oxford, United Kingdom

Funding: Current research is funded by the European Commission BIOFRESH project - Biodiversity of Freshwater Ecosystems: Status, Trends, Pressures, and Conservation Priorities (7th FWP ref. 226874), http://www.freshwaterbiodiversity.eu/. The funders had no role in study design, data collection and analysis, decision to publish, or preparation of the manuscript.

Competing Interests: The authors have declared that no competing interests exist.

* E-mail: markovic@igb-berlin.de

Introduction

Predicting future distribution patters of species following climate change projections becomes increasingly important in environmental management, species conservation and restoration planning at global, regional and local scales. Studies dealing with future species distribution patterns generally relay on niche-based species distributions models (SDMs) which relate the current conditions to the current species distributions and project these using climate change models (e.g., [1]–[4]).

Recent studies on methodological aspects of SDMs have shown that application of consensus methods reduce model based uncertainty and increase reliability of the projections [1], [5]–[7]. In particular, Marmion et al. [5] have shown that consensus methods based on averaging of all methods provide robust projections and significantly increase the accuracy of species distribution forecasts. Additional important aspects of the predictive modeling process include a well substantiated predictor selection methodology and the investigation of the general predictability of future changes using the information on current conditions. Despite the overall high relevance of future projections, there is a general lack of studies comprising all these uncertainty aspects of the predictive modeling process.

Species distribution patterns are affected by a combination of environmental factors acting at different spatial and temporal scales [4], [8]–[10]. Climate is widely acknowledged as primary factor at the continental scale whilst topography, the land use and habitats become important at regional to local scales [11]. Because dendritic river network structures constrain species dispersal ability, describing spatial distribution patterns of freshwater species is inseparable from describing hierarchy, heterogeneity and lateral connectivity of river systems [12]. The dendritic river network structure is commonly argued against grid cell related data in biogeographical modeling of freshwater species such as fish (e.g. [2], [9]). However, most studies on fish distribution patterns are void of an explicit predictor selection process and combine detailed river and fish data at the site scale with the global environmental data (20 km grid cells) resulting in serious scale mismatches and biased data samples.

Freshwater biodiversity is particularly vulnerable to climate change not only because temperature is climate dependent, but also because other pressures on freshwater biodiversity such as human consumptions of ecological assets, nitrogen deposition and species invasions show increasing trends over recent decades [13]–[][][16]. Since the combined effect of climate and human stressors is likely to be amplified in the future compared to their individual effects, it is expected that there will be a considerable change in species composition and diversity loss [17].

Here, we faced the challenge of modeling distribution patterns of 38 freshwater fish species across Germany at the scale grain used for the national fish records (11 km×11 km grid cell). Grid cell related species distribution maps are common in presenting species occurrence information. Such data have already proven their potential to describe current and future distribution patterns of plant and terrestrial vertebrate species (e.g. [18], [19]). To our knowledge, the potential of fish occurrence maps in predictive biogeographical modeling at a comparable resolution has not been exploited. As the initial step we identified factors affecting species distributions. At the landscape to regional scale previous studies indicated the position along the upstream–downstream gradient, mean temperature and site elevation as major predictors of species distributions [2], [9], [20]. Across river basis at continental to global scale fish species distributions were commonly described by discharge, climate, topography, net primary productivity, dam characteristics, land use properties and population density [3], [8], [16]. In a fish species traits study based on literature data Goldstein and Meador [21] showed that fish distribution patterns vary according to stream size and substrate type. Consequently, we collected information on the substrate and stream size properties of each cell as well as information on climatic, hydromorphologic, topographic and anthropogenic properties.

The main objective of our study was the investigation of the potential of grid-cell data (atlas data) in biogeographical modeling of current and prediction of future distribution patterns of freshwater fish species. Our comprehensive modeling framework involved grid cell related fish and environmental data at a regional scale resolution, explicit predictor selection, different SDMs and different climate change projections. The potential of climate as a single factor describing current species distributions has been studied, as well as the importance of climate as a factor acting together with other properties of the studied cells. To ensure prediction reliability we restricted our analyses within the predictive modeling stage to only those SDMs considered in the validation process as good to excellent. We investigated variability in the estimated effects of climate change for the middle of the 21st century emerging from different stages of the modeling process for the consensus model and for all combinations of the different climate change models and the selected SDMs. Analysis of the variability in the predicted effects of climate change revealed important sources of uncertainty, some of which limit the possibility of providing reliable estimates of future changes in freshwater fish species distribution patterns.

Table 1. Fish species included in the analyses and their occurrence frequency.

Species code	Species name	Occurrence frequency
Abrarama	*Abramis brama*	0.385
Albuatus	*Alburnoides bipunctatus*	0.076
Alburnus	*Alburnus alburnus*	0.278
Anguilla	*Anguilla anguilla*	0.658
Aspipius	*Aspius aspius*	0.120
Barbrbus	*Barbus barbus*	0.279
Barbtula	*Barbatula barbatula*	0.482
Blicrkna	*Blicca bjoerkna*	0.283
Caraelio	*Carassius gibelio*	0.189
Carasius	*Carassius carassius*	0.274
Chonasus	*Chondrostoma nasus*	0.098
Cobienia	*Cobitis taenia*	0.183
Cottobio	*Cottus gobio*	0.420
Cyprrpio	*Cyprinus carpio*	0.406
Esoxcius	*Esox lucius*	0.523
Gastatus	*Gasterosteus aculeatus*	0.523
Gobiobio	*Gobio gobio*	0.616
Gymnrnus	*Gymnocephalus cernuua*	0.244
Lampilis	*Lampetra fluviatilis*	0.085
Lampneri	*Lampetra planeri*	0.337
Leucatus	*Leucaspius delineatus*	0.206
Leucidus	*Leuciscus idus*	0.181
Leucscus	*Leuciscus leuciscus*	0.384
Lotalota	*Lota lota*	0.165
Misgilis	*Misgurnus fossilis*	0.145
Percilis	*Perca fluviatilis*	0.576
Phoxinus	*Phoxinus phoxinus*	0.248
Pungtius	*Pungitius pungitius*	0.253
Rhodarus	*Rhodeus amarus*	0.159
Rutiilus	*Rutilus rutilus*	0.648
Salmutta	*Salmo trutta*	0.597
Salvalis	*Salvelinus fontinalis*	0.104
Sanderca	*Sander lucioperca*	0.248
Scarlmus	*Scardinius erythrophthalmus*	0.334
Siluanis	*Silurus glanis*	0.124
Squaalus	*Squalius cephalus*	0.447
Thymllus	*Thymallus thymallus*	0.319
Tincinca	*Tinca tinca*	0.481

Methods

Species distribution data

The German society for ichthyology (GFI, www.fishartenartlas.de) and the Federal Agency for Nature Conservation (BfN, www.bfn.de) have kindly provided actual fish species occurrence data in form of actual presence indicators for grid cells of the topographic map of Germany at scale of 1:25 000 (TK25). Each TK25 cell covers an area of approximately 11×11 km size, forming a mesh of 2945 cells. The analyses have been reduced to 2935 cells due to missing environmental information for nine coastal cells. The grid cell based fish occurrence data are a result of merging presence/absence information from expert-verified surveys and catch reports of recreational and commercial fisheries across each of the German Federal States over the last two to three decades. We modeled only those fish species with sufficient reliable positive occurrences (more than 50), resulting in a set of 38 fish species (see Table 1).

Environmental data

Baseline data. The baseline data consist of approximately 50 climatic, hydromorphologic, topographic and anthropogenic predictors considered potentially relevant for general physiological and behavioral properties of studied species. To minimize the effect of spatial redundancy in the environmental information, we used only those data sets with scale grains ≤1.1 km, i.e., at least ten times smaller than the scale grain of our study. For calculation of the TK25 cell specific values we used ArcView, version 3.2 (ESRI).

Climate related predictors were based on the 30 arc-seconds resolution (approx. 1 km) of the WorldClim (www.worldclim.org) data set. The WorldClim data set consist of 19 bioclimatic predictors, monthly mean, minimum and maximum temperature and monthly total precipitation derived from the station time series data for the second half of the 20th century [22]. The bioclimatic predictors summarize the effects of annual trends, seasonality and potentially limiting environmental factors on the species distributions. Actual evapotranspiration is estimated using the Turc's formula [23] while the cell runoff resulted from the general water balance equation in which the groundwater storage is neglected due to the fact that the calculation period is several decades. Temperature averages for March–May and May–July were added to the climate related predictor data set to account for pronounced temperature requirements of fish during reproduction and early development stages.

The initial set of hydromorphological predictor variables considered the following TK25 cell indicators: maximum cumulative length of the upstream flow network, maximum and minimum basin area, dominant river type category and the maximum Strahler order. All listed indicators except the river type category were based on the CCM River and Catchment Database, version 2.1 (CCM2, [24]). Predictors related to basin area and upstream flow network length indirectly reflect basin heterogeneity and thus potential species diversity. The river type was derived from the German national typology classifying all flowing waters according to their morphological, geological, hydrological and biological properties into 24 river types [25]. For the purposes of our study, the categories were merged with focus on the major substrate type into following classes: (1) alpine streams; (2) boulder and gravel dominated large streams in the alpine foothills; (3) fine to coarse substrate dominated siliceous highland streams; (4) fine to coarse substrate dominated calcareous highland streams; (5) gravel dominated streams; (6) sand dominated streams; (7) organic substrate dominated streams; (8) loess and loam dominated streams; (9) marshland streams of the coastal plains; (10) small streams in riverine floodplains without dominant substrate type.

Germany has a pronounced North-South altitudinal gradient, with Alps in the South and the North- and the Baltic Sea in the North and North-East, respectively, wherefore the topographical predictor set included the average TK25 cell altitude as well as the mean and maximum slope as surrogates for the water velocity at a resolution of 250 m.

The anthropogenic predictor set included the population density data and the Corine Land Cover data (CLC) of the European Environment Agency (www.eea.europa.eu) at a resolution of 250 m. The initial 44 land use types were summarized into following ecologically meaningful land use categories: urban/commercial, agriculture, forest, grassland, freshwater, marine and other. For each TK25 cell dominant land use type as well as TK25 cell surface percentages covered by each category were estimated.

Future climate predictions. For evaluation of the potential future species distributions we used the downscaled climate projections for the middle of the 21st century (2040–2069, referred further as 2050s) from the CIAT data portal (www.ccafs-climate.org) at 30 arc-seconds spatial resolution. The CIAT data set is based on the general circulation models (GCMs) from the IPCC Fourth Assessment Report. Since the statistical downscaling framework is fixed to the baseline climate as defined by the WorldClim data set, consistency between our "baseline" and "future" climate datasets is assured. The most commonly used in the ecological studies are the projections of the HadCM3 model [26] for the IPCC's A2 and B1 storylines developed at the Hadley Centre for Climate Prediction (e.g. [1], [2], [4]. A2 storyline describes a very heterogeneous world with continuously increasing global population and regionally oriented economic growth [27]. B1 storyline describes a world characterized with global population that peaks in mid-century and declines thereafter, and introduction of clean and resource-efficient technologies [27]. As energy requirements are highest for the scenarios from the A1 family, we selected the A1b scenario of a world with continuously increasing global population, very rapid economic growth and maximum energy requirements that are balanced across all energy sources. In addition to HadCM3 we used projections of the ECHAM5 [26] model (Max Planck Institute for Meteorology, Germany) and the IPSL-CM4 [26] model (Institute Pierre Simon Laplace, France).

Analysis framework

Statistical analyses and model building were carried out using various own MATLAB (version 6.0.0.88) and R (version 2.14) codes combined with the BIOMOD library [28] (version 1.1–5) and the ME package [29] (version 3.3.3a).

Predictor variables selection. In order to avoid multicollinearity, for each of the predictor categories (climate, hydromorphology, topography and anthropogenic influences) using the principal component analysis (PCA) and the univariate analysis we selected a representative predictor set which satisfied the condition of pairwise correlations below 0.75. Further reduction of the predictor number was an iterative process supported with the univariate analysis of the relevance of each individual predictor and analysis of the predictor relevance within the multivariate model setup. Thereby, we used three selection criteria: (1) average univariate area under the receiver operating characteristic ≥0.65; (2) average permutation importance ≥5%; (3) at least for half of the considered species the respective parameter should be selected by the SDM's. If at least two out of these three criteria were fulfilled, the predictor was considered as a relevant factor affecting fish species distribution across the studied area at the studied scale.

Modeling current fish species distributions. Species distribution patterns across Germany were modeled using Generalised Linear Models (GLM), Generalised Additive Models (GAM), Random Forest (RF) and the Maximum Entropy Approach (ME). GLM is one of the best established techniques [10] and is an extension of the linear models that deal with the non-normal error distributions. GAM is a non-parametric extension of GLM widely used for analyses and description of the distribution patters of fish species (e.g. [7], [30]). Here, cubic spline smoothers with four degrees of freedom were used for fitting the GAM models. RF is a classification method which creates a certain number of trees (here 500) constructed using a different bootstrap sample from the original data [31]. This technique has shown to have a good ability to predict observed fish species distributions. In the ensemble modeling experiment by Grenouillet et al. [7] among eight tested statistical methods, RF came closest to the average model. Unlike the GLM, GAM and RF that use species presence and absence data, ME uses presence data only. Each site is assigned a probability value whereas the probability distribution itself is constrained by the properties of the original data [29]. Among all probability distributions which satisfy the given constraints, ME chooses the one with the maximum entropy.

For all SDMs except the ME, searching for a parsimonious model involved analyses of the model improvement based on the Akaike Information Criterion (AIC) through stepwise adding of new variables and leaving those out which do not significantly improve the fit. Estimates of the variable contributions in the individual models were based on the normalized correlations between the predictions using the original data and the predictions using the data in which the variable of interest is randomly permutated. The modeling process involved a random splitting of the data samples into the calibration (70%) and the validation data set (30%) repeated 10 times for each SDM. Species occurrence probabilities were transformed to the presence/absence information using the SDM specific thresholds which maximize both, sensitivity (the true positive rate) and the specificity (the true negative rate). The modeled presence/absence results were merged to a single model within a consensus modeling framework involving averaging across all SDMs including their replications (see [5]). Within the averaging procedure, a particular cell is assigned the presence indicator "1" only if the same indicator was estimated for more than 50% of the considered models. Agreement between the observed and modeled species distribution patterns was quantified by sensitivity, specificity and the coincidence rate. The later describes the ratio of the number of "correctly" predicted cells (presences and absences) to the total number of cells. General performance of the calibrated models was classified using the area under the receiver operating characteristic curve (ROC score) [e.g. 30]: 0.5–0.6 ("fail"), 0.6–0.7 ("poor"), 0.7–0.8 ("fair"), 0.8–0.9 ("good"), 0.9–1 ("excellent").

Patterns in the species' environmental responses. Identification of species groups according to their responses to different environmental predictors was performed using Fuzzy c-mean clustering technique (see [32]). Environmental responses were calculated as a ratio of the group based sum of predictor importance estimates and the sum of all importance estimates. In order to compare the variation of the environmental responses across different SDMs the clustering was applied to the results of each considered technique separately.

Future species distributions and associated uncertainties. For the species identified in the previous step as climate sensitive, potential future distribution maps for the period 2040–2069 under A1b scenario have been created according to different climate models and the SDMs identified as suitable for modeling fish species distribution at the particular scale resolution. A joint future consensus pattern was estimated through averaging of all pairwise combinations of SDMs (including repetitions) and climate models. Within the averaging procedure, species are considered to occur in a given cell if more than 50% of the models predicted species occurrence for the particular cell. The scope of our study involves not only derivation of future fish species distribution patterns, but also estimation of uncertainties and projections variability associated with the predictor variable selection, SDM, selected future climate projection and a relationship between the predictor ranges for the baseline and the future situation. We studied differences in the spatial distribution patterns of selected fish species in terms of distribution coincidence rates, differences in the estimated "habitat" losses and gains as well as percentages of study area where the future projection is unreliable due to mismatches between the predictor calibration and future range. Thereby, the "loss" ("gain") is defined as the ratio of grid cell number where the species is currently present (absent) but predicted to be absent (present) in the future to the total number of currently occupied cells. The coincidence rate between e.g. projections of two different climate models correspond to the ratio of coinciding presence and absence cells for these two models and the total number of TK25 cells.

Results

Predictor variables selection

From the baseline data set consisting of approximately 50 environmental predictors, through synthesis of the PCA, univariate and the correlation analysis we selected the major factors representing the climatic, hydromorphologic, topographic and anthropogenic properties of the studied grid cells (see Table 2 and Figure S1). Exemplarily, the first two PCA's of the precipitation related data set describing 74% and 24% of the joint variability were correlated most with annual mean precipitation (AnnPMean) and the precipitation seasonality (PSeason). Consequently, only these two precipitation related predictors were considered for further analysis. Also, due to high correlations with the mean annual temperature (AnnTMean) and AnnPMean, both evapotraspiration and runoff had to be excluded from the analysis scope. Within the subsequent analysis step we quantified the overall performance of the selected factors in describing spatial distribution patterns of the selected fish species and performed the final predictor selection. The results summarized in Table 2 indicate that mean temperature of wettest (TWetQuart) and driest quarter (TDryQuart) have lowest predictive power compared to other climatic factors and, also, that none of the considered anthropogenic factors hold significant predictive power. The univariate regression has additionally enabled us to identify the key factors of species's distributions. In particular, as shown in Figure 1, the mean altitude (AltMean) describes the spatial patterns of Lampetra fluviatilis (ROC score of 0.85), AnnTMean is the dominant factor in describing the spatial pattern for Silurus glanis (ROC score of 0.73) and habitat selection of Barbus barbus can be described using the Strahler stream order (ROC score of 0.8).

Finally, the "basic set of predictors" affecting spatial distribution of studied fish species entails the following characteristics of the studied grid cells (see Table 2 for abbreviation explanations): climatic (AnnTMean, Isotherm, TSeason and AnnPMean), topographic (AltMean) and hydromorphologic (CumLenkm, Strahler and RtypMost).

Modeling current fish species distributions

The model calibration process was initiated using the above described "basic set of predictors". Due to the stepwise predictor elimination in a search for the most parsimonious model, the final number of model predictors was commonly between four and six. Summary of the average statistical performance of the calibrated models per individual fish species is given in Table 3. The calibration ROC score for the RF based SDMs indicated "excellent" models (ROC score>0.9) for all 38 species, but the average validation score of 0.84 emphasized that there is a considerable performance drop when it comes to the validation sample. For GAM and GLM based SDMs, both average calibration and validation scores were around 0.8. ME had lowest average calibration and validation score (0.76 and 0.72, respectively) and, additionally, ME's performance has shown to be dependent on species's occurrence frequency. Namely, deviation of the ME based ROC validation score from the average validation ROC score (calculated as an average per species for all methods and all repetitions) was positively correlated with the occurrence frequency (Figure 2). For all SDMs applied, standard deviations of the ROC score estimates were ≤0.02 whereas sensitivities and specificities were >90 for RF and between 70 and 80 for all other SDMs (Table 3). Within the applied modeling

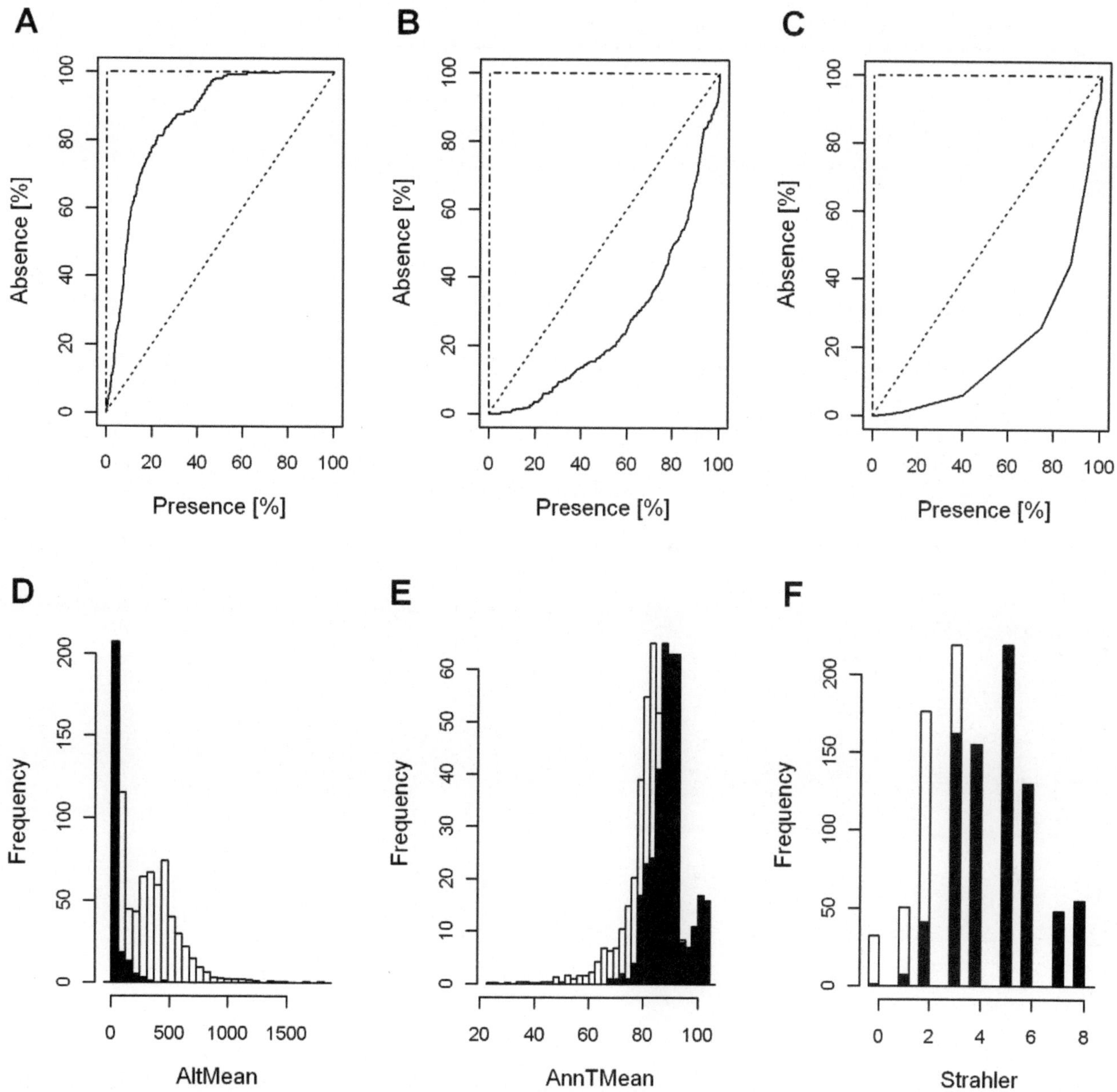

Figure 1. Lorenz Curve and the empirical probability distributions. (A, D) *L. fluviatilis*, (B, E) *S. glanis* (b, e) and (C, F) *B. barbus*. The ROC score of 0.85 for AltMean (*L. fluviatilis*), 0.73 for AnnTMean (*S. glanis*) and 0.8 for Strahler (*B. barbus*) indicate high discriminatory power of the individual predictors in describing species presence (shaded bars) and absence patterns.

approach both, percentage of presence and absence correctly predicted were simultaneously maximized, resulting in equal specificity and sensitivity estimates. Because our intention was to use the model results to investigate the potential influence of future changes, to align with assumption that species niches are stable over time, the model representations of those niches have to be robust and stable. Due to above described performance pitfalls, ME was considered inappropriate for the predictive modeling of the fish species distributions at the particular scale resolution and was not included in the consensus model building. The average performance of the consensus models (based on GAM, GLM and RF and their repetitions) is 0.83 ("good", see Table 3). To ensure robustness and reliability of our forecast, only those species with average GAM, GLM and RF based validation ROC score ≥0.8 (Table 3) were analyzed within the subsequent steps.

In order to identify which fish species out of those selected in the previous step are particularly sensitive to climate, we additionally ran all models using only climatic predictors. Again, RF based average ROC score was highest for both, calibration and the validation data set (0.97 and 0.80 respectively) whereas GAM and GLM had a similar performance with ROC scores around 0.75. Interestingly the average validation score based on GAM, GLM and RF for Aspius aspius, Chondrostoma nasus, Cobitis taenia, Cottus gobio, L. fluviatilis, Misgurnus fossilis, Pungitius pungitius and S. glanis was ≥0.8 suggesting that climate is an important factor that affects spatial distributions of these species. However, a comparison between the model performances using all considered factors (Table 3) and using climatologic factors only (Table 4) reveals a significant performance drop for the latter, except for C. gobio and S. glanis, which appeared as particularly climate sensitive.

Table 2. Environmental predictors and their discriminatory power.

Predictor	Description	ROC score	PI	No
AnnTMean	Annual mean temperature (°C)	0.66	7.6	26
DiuTRange	Mean diurnal temperature range (°C)	0.58	8.1	23
Isotherm	Isothermality (-)	0.58	7.7	21
TSeason	Mean temperature seasonality (°C^2)	0.58	9.9	23
Tmax	Maximum temperature of warmest month (°C)	0.61	3.0	18
TWetQuar	Mean temperature of wettest quarter (°C)	0.66	1.8	18
TDryQuar	Mean temperature of driest quarter (°C)	0.55	2.6	20
AnnPMean	Annual mean precipitation (mm)	0.65	6.1	29
PSeason	Mean precipitation seasonality (-)	0.56	3.6	19
AltMean	Mean altitude (m)	0.66	17.8	27
CumLenkm	Maximum cumulative length of the upstream flow network (km)	0.65	3.9	24
Strahler	Maximum Strahler order	0.64	12.9	36
RtypMost	Dominant river type	0.61	11.3	32
Lusemax	Dominant land use type	0.55	1.7	26
Popmean	Mean population density (Inhabitants/km^2)	0.57	2.0	24

"ROC score" is the mean ROC score based on the univariate analysis; "PI" is the mean permutation importance of each individual predictor for GLM, GAM, RF and ME based multivariate SDMs; "No" is the number of species for which the respective predictor variable was identified as statistically significant (the total number of considered species is 38).

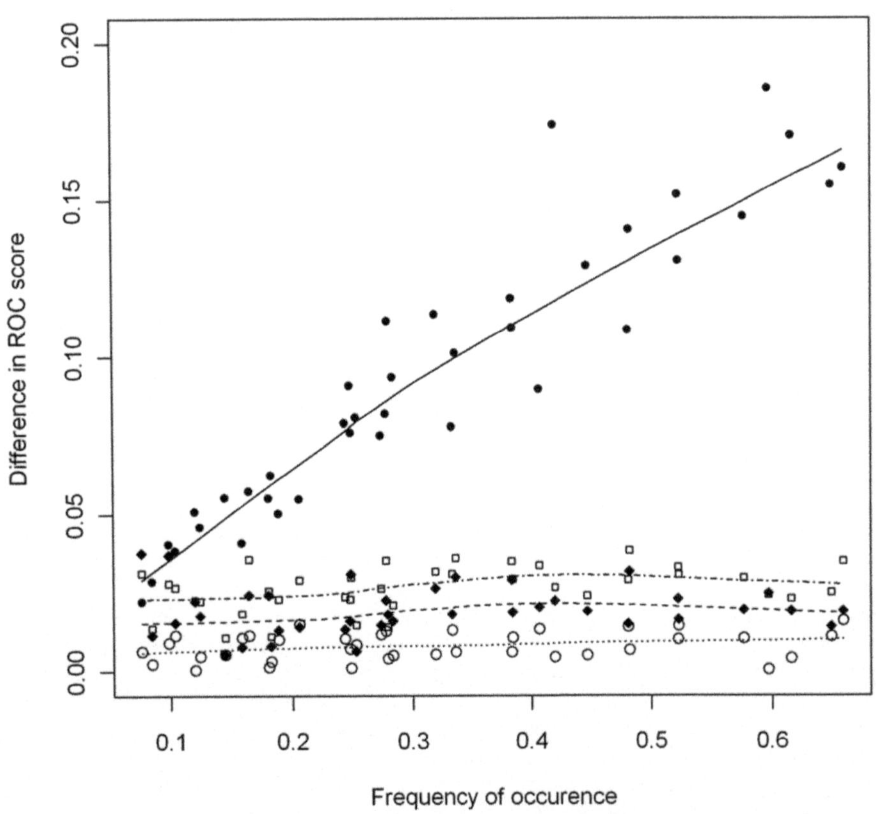

Figure 2. Relationship between species occurrence frequency and the ROC score. The ROC score difference is calculated as the absolute deviation of the average validation ROC score for the particular method from the total average validation ROC score for all applied SDMs (GLM, GAM and RF). The trend line is based on locally weighted scatterplot smoothing: ME (solid line), GAM (dotted line), GLM (dashed line), RF (dot-dash line).

Table 3. Summary of the model performance for the current conditions.

	Calibration ROC score				Validation ROC score				Sensitivity & Specificity				ROC
Species code	**GAM**	**GLM**	**RF**	**ME**	**GAM**	**GLM**	**RF**	**ME**	**GAM**	**GLM**	**RF**	**ME**	**CM**
Abrarama	0.79	0.78	0.97	0.72	0.78	0.77	0.82	0.68	71	70	92	69	0.82
Albuatus	0.90	0.85	0.98	0.90	0.88	0.83	0.90	0.85	81	76	93	86	0.85
Alburnus	0.83	0.81	0.98	0.78	0.81	0.80	0.86	0.74	75	74	93	75	0.84
Anguilla	0.74	0.73	0.97	0.62	0.73	0.73	0.78	0.58	67	67	91	61	0.80
Aspipius	0.92	0.90	0.99	0.89	0.91	0.88	0.93	0.86	84	82	95	86	0.89
Barbrbus	0.90	0.88	0.98	0.80	0.89	0.87	0.90	0.77	83	80	95	76	0.88
Barbtula	0.78	0.75	0.97	0.69	0.77	0.75	0.82	0.64	70	68	92	67	0.80
Blicrkna	0.81	0.80	0.97	0.76	0.81	0.80	0.84	0.72	73	71	92	73	0.83
Caraelio	0.78	0.77	0.97	0.76	0.77	0.76	0.80	0.72	72	71	90	73	0.83
Carasius	0.76	0.75	0.97	0.72	0.75	0.75	0.79	0.69	68	68	91	69	0.81
Chonasus	0.92	0.87	0.99	0.89	0.90	0.85	0.92	0.85	84	78	94	88	0.88
Cobienia	0.91	0.90	0.98	0.85	0.89	0.88	0.90	0.83	83	82	94	81	0.89
Cottobio	0.88	0.86	0.98	0.74	0.88	0.86	0.91	0.71	80	78	95	70	0.86
Cyprrpio	0.74	0.73	0.96	0.68	0.72	0.71	0.77	0.64	68	67	91	66	0.79
Esoxcius	0.77	0.75	0.97	0.66	0.74	0.74	0.79	0.62	70	69	91	64	0.81
Gastatus	0.79	0.77	0.97	0.68	0.78	0.77	0.83	0.64	72	71	92	66	0.82
Gobiobio	0.78	0.76	0.97	0.64	0.76	0.75	0.79	0.60	70	69	92	63	0.80
Gymnrnus	0.84	0.84	0.97	0.80	0.83	0.83	0.87	0.76	76	76	93	76	0.85
Lampilis	0.92	0.90	0.98	0.91	0.90	0.89	0.92	0.88	85	82	94	88	0.89
Lampneri	0.79	0.76	0.97	0.72	0.77	0.75	0.82	0.68	72	70	92	69	0.82
Leucatus	0.76	0.75	0.96	0.77	0.74	0.74	0.78	0.70	70	69	91	75	0.81
Leucidus	0.85	0.81	0.97	0.83	0.83	0.81	0.86	0.78	76	75	92	79	0.84
Leucscus	0.79	0.76	0.97	0.70	0.78	0.75	0.82	0.66	72	68	92	68	0.81
Lotalota	0.83	0.81	0.97	0.82	0.82	0.80	0.86	0.77	75	74	92	80	0.84
Misgilis	0.88	0.87	0.98	0.84	0.86	0.86	0.87	0.81	80	79	93	80	0.87
Percilis	0.76	0.75	0.97	0.65	0.75	0.74	0.79	0.61	70	68	92	63	0.81
Phoxinus	0.83	0.79	0.98	0.78	0.81	0.78	0.84	0.74	74	71	92	75	0.83
Pungtius	0.89	0.89	0.98	0.83	0.88	0.88	0.90	0.81	82	81	95	79	0.89
Rhodarus	0.81	0.80	0.97	0.78	0.78	0.78	0.81	0.75	74	73	91	73	0.84
Rutiilus	0.75	0.74	0.96	0.63	0.73	0.73	0.77	0.59	68	67	91	62	0.79
Salmutta	0.84	0.81	0.97	0.67	0.83	0.80	0.85	0.64	77	73	93	63	0.84
Salvalis	0.79	0.77	0.97	0.80	0.76	0.76	0.80	0.74	71	69	90	79	0.80
Sanderca	0.82	0.81	0.97	0.78	0.82	0.81	0.85	0.74	74	73	93	76	0.83
Scarlmus	0.75	0.74	0.96	0.71	0.73	0.72	0.77	0.66	67	66	91	69	0.80
Siluanis	0.88	0.86	0.98	0.86	0.86	0.85	0.89	0.82	79	77	93	84	0.86
Squaalus	0.81	0.78	0.97	0.70	0.79	0.77	0.82	0.66	73	70	92	67	0.82
Thymllus	0.84	0.81	0.98	0.76	0.83	0.81	0.86	0.72	76	73	93	72	0.83
Tincinca	0.73	0.72	0.96	0.66	0.71	0.71	0.76	0.62	66	65	91	64	0.79
Mean	**0.82**	**0.80**	**0.97**	**0.76**	**0.80**	**0.79**	**0.84**	**0.72**	**74**	**73**	**92**	**73**	**0.83**

The table values indicate averages over all model repetitions. Standard deviation of the ROC score estimates for the calibration and the validation data sets is less than 0.01, 0.02, respectively, for all methods. "ROC CM" is the ROC score of the consensus models.

Figure 3 shows the current species distribution patterns and their GAM, GLM and RF model representations for Alburnus alburnus, C. gobio, Lota lota, Sander lucioperca and S. glanis. Sensitivies and specificities were highest for the RF based models (Table 3 and Figure 3) implying that for this particular SDM, the modeled species distribution patterns are closest to the observed ones. The average difference between the observed and the estimated species' occurrence frequency was highest for GLM (+0.12), slightly lower for GAM (+0.11) and lowest for RF (+0.02), but all methods agreed that the currently occupied habitat is probably smaller that the potentially suitable habitat. Pairwise comparison of the model outputs revealed that GAM and GLM have a high agreement in terms of the average coincidence rate (93±1%) whereas between GAM and RF and between GLM and

Table 4. Summary of GAM, GLM and RF models based on AnnTMean, Isotherm, TSeason and AnnPMean.

	Validation ROC score			ROC score change			Sensitivity & Specificity		
Species code	**GAM**	**GLM**	**RF**	**GAM**	**GLM**	**RF**	**GAM**	**GLM**	**RF**
Albuatus	0.78	0.74	0.86	−0.09	−0.08	0.07	71	68	91
Alburnus	0.76	0.76	0.81	−0.04	−0.04	0.11	69	69	91
Aspipius	0.81	0.79	0.87	−0.1	−0.09	0.05	73	71	92
Barbrbus	0.75	0.74	0.81	−0.13	−0.12	0.07	71	68	92
Blicrkna	0.76	0.74	0.80	−0.04	−0.05	0.13	69	67	91
Chonasus	0.82	0.78	0.86	−0.07	−0.07	0.05	75	72	91
Cobienia	0.81	0.74	0.88	−0.08	−0.15	0.08	74	66	93
Cottobio	0.83	0.78	0.88	−0.05	−0.08	0.07	74	70	94
Gymnrnus	0.79	0.77	0.82	−0.03	−0.06	0.1	71	69	92
Lampilis	0.82	0.76	0.85	−0.07	−0.13	0.05	75	67	91
Leucidus	0.73	0.68	0.80	−0.09	−0.13	0.11	68	61	91
Lotalota	0.72	0.69	0.80	−0.09	−0.11	0.1	66	65	90
Misgilis	0.80	0.76	0.84	−0.05	−0.09	0.1	73	70	91
Phoxinus	0.72	0.65	0.80	−0.08	−0.13	0.13	67	60	91
Pungtius	0.81	0.78	0.86	−0.06	−0.09	0.08	75	71	93
Salmutta	0.78	0.71	0.82	−0.05	−0.09	0.12	71	66	93
Sanderca	0.77	0.75	0.81	−0.04	−0.05	0.12	70	70	91
Siluanis	0.83	0.81	0.85	−0.03	−0.04	0.08	75	73	92
Thymllus	0.76	0.72	0.82	−0.07	−0.09	0.11	70	66	92
Mean	**0.75**	**0.72**	**0.80**	**−0.07**	**−0.09**	**0.09**	**69**	**67**	**91**

ROC score, sensitivities and specificities indicate averages over all model repetitions. Standard deviation of the ROC score estimates for the validation data sets ranges from 0.01 to 0.03 for GLM and GAM based models and from 0.01 to 0.02 for RF. The "ROC score change" is the difference between the mean ROC for the models based on climatic factors and the mean ROC for models considering climatic, topographic and hydromorphologic predictors (see Table 3).

RF, the agreement was considerably lower (79±3% and 77±2%, respectively). Consequently, for the species distribution patterns shown in Figure 3 the average sensitivity and the specificity of the consensus model was lower (80±2%) than that of the RF model (98±1%) but higher than that of the GLM and GAM (76±2%).

Patterns in the species' environmental responses

Four distinct fish species groups have been identified according to their responses to climatic, hydromorphologic and topographic factors using the Fuzzy c-means clustering (see Table 5). Group 1 is mainly determined by altitude (e.g. P. pungtius, C. taenia, L. fluviatilis), group 2 by the hydromorphologic factors (e.g. Salmo Trutta, Phoxinus phoxinus, Thymallus thymallus), group 3 by a combination of hydromorphologic and climatic factors (e.g. S. lucioperca, S. glanis, A. alburnus, C. gobio), and group 4 by a rather equal contribution of all factors (e.g. C. nasus, A. aspipius, Gymnocephalus cernuua). Surprisingly, there was no group of species primarily determined by climatic factors.

Future species distributions and associated uncertainties

The average annual temperature across the studied TK25 grid cells for 2050s under A1b is according to the ECHAM5 model slightly lower (10.8°C) than the projections for the HadCM3 and IPSL-CM4 models (11.1°C, see Table 6) which is however, 2.4°C higher than AnnTMean of the second half of the 20th century (8.4°C). Regarding Isotherm, all climate models predict similar values and the variation is in almost the same range as that for the 20th century climate. The ECHAM5 model, unlike the other two models, predicts lower TSeason than that of the 20th century. AnnPMean was predicted to decrease by a similar amount for all climate models. Although PSeason and Tmax were shown to be not significant overall, they might have an effect for some of the selected species. Therefore, we added these to the "basic set of predictors" within this part of our analyses. Tmax was predicted to increase by 2.8°C (ECHAM5) to 4.1°C (HadCM3), while PSeason has lower (ECHAM5 and HadCM3) as well higher (IPSL-CM4) values then the average 20th century value for the study region.

Estimation of the future distribution patterns was performed for species from group (3), because for these species climate was more important than for all other species groups. From group (4), whose species are affected in general by all three studied factors (climatic, hydromorphologic and topographic), future distribution of the stenothermal L. lota was also assessed. Figure 4 shows potential future habitat suitability maps for C. gobio for different statistical approaches and different climate models. The apparent decrease in the habitat suitability is congruent with predicted temperature increase for all climate models (see Figure 3 (e–h) and Figure 4). However, Figure 4 and Figure 5 indicate significant variability for different statistical approaches as well as for the different climate models. In terms of the occurrence frequency, future projections for C. gobio vary from 0.10 (GAM & HadCM3) over 0.24 (consensus model merging all projections) to 0.47 (GLM & ECHAM5).The previous imply changes from a tremendous drop to a slight increase compared to the current occurrence frequency (0.42), and consequently underline the high uncertainty of the future projections. In terms of the coincidence rates, the situation is much more positive. Average coincidence rates and the corresponding standard deviations between the GAM and GLM,

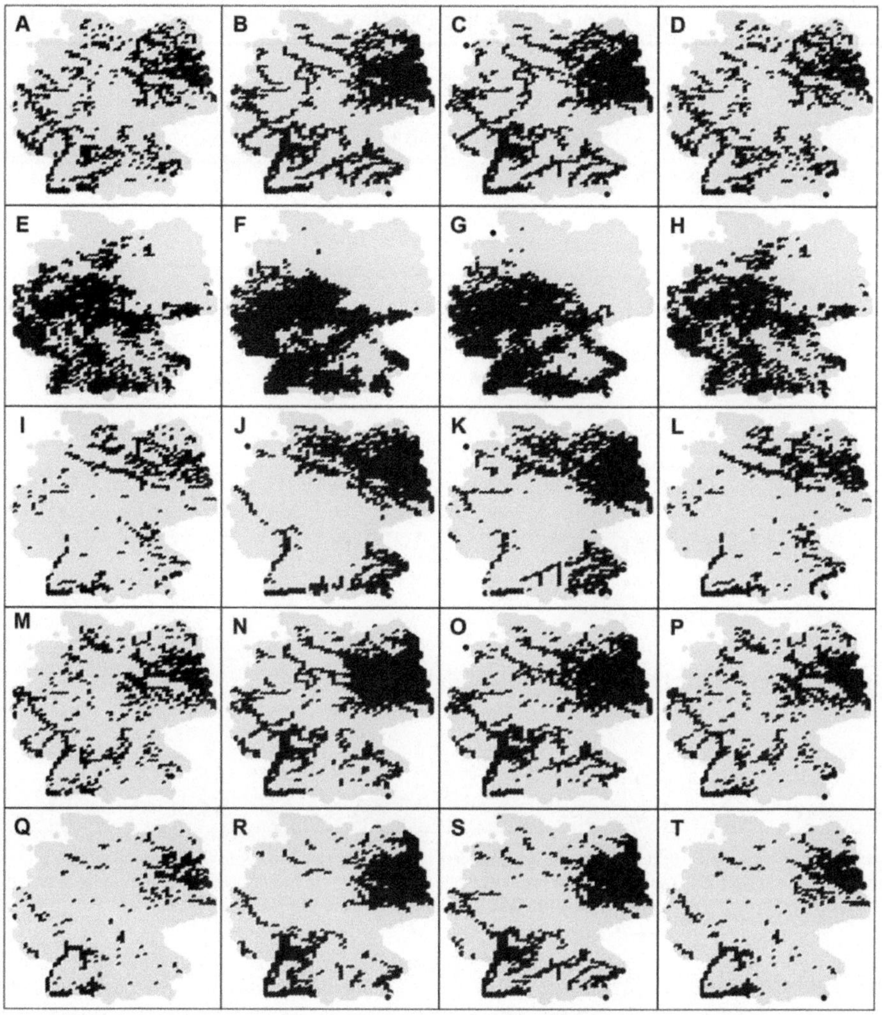

Figure 3. Current species distributions and their model representations. Each row shows species specific results: (A–D) *A. alburnus*, (E–H) *C. gobio*, (I–L) *L. lota*, (M–P) *S. lucioperca* and (Q–T) *S. glanis*. The first column indicates current species distribution and is followed by the model representations using GAM, GLM and RF, respectively. Gray dots indicate centroids of the grid cells across the whole study area (Germany).

GAM and RF and GLM and RF projections for species from group (3) were 92±7%, 88±6% and 87±6%, respectively. Differences in the projections of HadCM3 and IPSL-CM4 were rather negligible (average coincidence rate of 97±2%) whereas ECHAM5 projects slightly different patterns then the previous two models with lowest coincidence rate with the HadCM3 model (89±6%). Overall, differences among climate models were much less than differences among statistical methods.

Comparisons between the consensus models for the current and future situation calibrated using species specific predictor variables selected from the "basic set of predictors" indicated that the coldwater adapted species L. lota and C. gobio are expected to lose 45% and 71%, respectively of currently suitable cells (Table 7). The warm water adapted species A. alburnus, S. lucioperca and S. glanis are expected to significantly expand their distribution ranges. Including two additional factors in the model building process (PSeason and Tmax) for the coldwater adapted C. gobio resulted in a significant change of predicted distribution pattern from losing about 45% of the currently suitable cells to gaining about 29%. Thereby, RF estimates were considerably unproportional to that based on GAM and GLM.

Analysis of the current ranges of the model predictors and those predicted for 2050s revealed that AnnTMean is outside the calibration range for 82% of the study area (Table 8). For all other factors, the mismatch between the calibration and the future predictor range was lower than 2%. However, due to the multivariate model setup, on average, for more than 90% of the study area future estimates were based on at least one extrapolated species environmental responses.

Discussion

Methodological aspects

Overall, our study showed a high potential for the use of grid cell related information in biogeographical modeling of fish distribution patterns.

ME has shown to be excellent for modeling distribution patterns of species with low occurrence frequency, however, due to performance dependence on occurrence frequency it is considered inappropriate for biogeographical modeling at the particular scale resolution. Although RF was clearly outperforming GAM and GLM in the model calibration phase, all three methods had similar performance in the validation phase indicating that for

Table 5. Mean permutation importance of the predictor groups, cluster id's and the corresponding cluster membership grades.

	GAM					GLM					RF				
Species code	**C**	**H**	**HY**	**Fc**	**MG**	**C**	**H**	**HY**	**Fc**	**MG**	**C**	**H**	**HY**	**Fc**	**MG**
Albuatus	0.5	0.4	0.1	2	0.5	0.3	0.3	0.4	3	1.0	0.4	0.3	0.2	3	0.8
Alburnus	0.4	0.1	0.5	3	1.0	0.4	0.0	0.6	1	0.6	0.5	0.1	0.4	4	0.4
Aspipius	0.3	0.3	0.4	1	0.9	0.3	0.3	0.4	3	1.0	0.3	0.2	0.4	2	0.9
Barbrbus	0.2	0.3	0.5	1	0.8	0.3	0.1	0.6	1	0.9	0.2	0.2	0.6	2	0.5
Blicrkna	0.4	0.2	0.4	3	0.4	0.3	0.1	0.6	1	0.7	0.5	0.2	0.3	2	0.4
Chonasus	0.4	0.3	0.3	1	0.6	0.4	0.2	0.4	3	1.0	0.4	0.2	0.4	2	0.8
Cobienia	0.2	0.6	0.2	2	0.9	0.2	0.6	0.2	4	1.0	0.2	0.6	0.2	1	0.9
Cottobio	0.5	0.1	0.4	3	0.8	0.3	0.0	0.7	1	0.9	0.5	0.2	0.2	4	0.6
Gymnrnus	0.2	0.4	0.4	1	0.8	0.1	0.3	0.5	3	0.5	0.4	0.3	0.3	3	0.7
Lampilis	0.3	0.5	0.2	2	0.9	0.1	0.5	0.3	4	0.7	0.2	0.6	0.2	1	1.0
Leucidus	0.2	0.4	0.4	1	0.6	0.2	0.5	0.3	4	0.7	0.3	0.4	0.3	1	0.4
Lotalota	0.3	0.3	0.4	1	0.9	0.2	0.1	0.7	1	0.6	0.4	0.3	0.3	3	1.0
Misgilis	0.3	0.5	0.2	2	0.9	0.3	0.5	0.2	4	0.9	0.3	0.3	0.3	3	0.8
Phoxinus	0.2	0.2	0.6	4	0.8	0.1	0.0	0.9	2	1.0	0.4	0.3	0.3	3	1.0
Pungtius	0.1	0.8	0.1	2	0.7	0.1	0.8	0.1	4	0.9	0.3	0.6	0.1	1	0.9
Salmutta	0.3	0.1	0.6	4	0.9	0.1	0.0	0.9	2	1.0	0.5	0.2	0.3	4	0.4
Sanderca	0.4	0.0	0.6	3	0.8	0.4	0.0	0.6	1	0.9	0.6	0.1	0.4	4	0.9
Siluanis	0.5	0.0	0.5	3	0.9	0.5	0.1	0.5	1	0.6	0.7	0.1	0.2	4	0.8
Thymllus	0.3	0.1	0.7	4	0.9	0.0	0.0	1.0	2	0.9	0.4	0.2	0.4	2	1.0
Mean	0.3	0.3	0.4	-	-	0.2	0.2	0.5	-	-	0.4	0.3	0.3	-	-

Table values indicate average permutation importance in percent of each predictor group: climate (C), altitude (H) and hydromorphology (HY). Members of the C group are AnnTMean, Isotherm, TSeason and AnnPMean, in the HY group are Strahler, CumLenkm and RtypMost while H is based on the permutation importance of AltMean. Degree to which each species belong to a Fuzzy cluster Fc is indicated by the membership grade MG.

predictive purposes, they are all equally suitable. However, since GAM and GLM are both additive species distribution models, when used for conservation and restoration planning, the projections require a thorough revision. Due to the additive consideration of the contributing factors, one could calculate a high probability of species occurrence in a region where all factors except one are within the species environmental tolerance range. However, if this one factor points to the water temperature above species ultimate lethal limit, then it is obvious that the model will fail to give a reasonable estimate of the habitat suitability. A simple solution to this problem is adding an additional algorithm in the probability estimation, which reduces the species occurrence probability or the habitat suitability to zero for all locations where the species existential tolerance limits are exceeded for at least one factor.

The importance of using consensus methods among different SDMs towards assessment of the model inherent uncertainty is already addressed in recent studies (e.g. [1], [6], [7]). However, consensus methodology neither mitigates projection uncertainty due to a weak predictor selection nor projection uncertainty due to poor performing SDMs. To assure reliability of our future projections, explicit parameter selection was supplemented with application of the consensus methodology for only those species with "good" to "excellent" average validation ROC score. The

Table 6. Major characteristics of the bioclimatic predictor variables for the 20th century climate and for the future climate projections (2050s).

Predictor	Unit	Mean and range 20th century Worldclim	Mean and range for 2050s ECHAM5	Mean and range for 2050s HADCM3	Mean and range for 2050s IPSL-CM4
AnnTMean	°C	8.4 (2.5–10.4)	10.8 (4.8–12.9)	11.1 (5.1–13.2)	11.1 (5.0–13.1)
Isotherm	°C	3.1 (2.3–3.5)	3.2 (2.4–3.5)	3.1 (2.3–3.5)	3.1 (2.2–3.5)
TSeason	°C^2	63.7 (54.5–76.8)	62.7 (52.7–73.6)	66.8 (58.2–79.2)	65.3 (57.4–77.8)
AnnPMean	mm	732.1 (482.4–1414.1)	716.3 (473.3–1323.1)	715.3 (467.0–1363.6)	713.4 (473.6–1374.4)
Tmax	°C	*22.7 (19.4–25.5)*	*25.5 (21.3–28.7)*	*26.8 (22.4–30.1)*	*25.9 (22.2–28.6)*
PSeason	-	*20.7 (9.5–37.1)*	*15.0 (9.9–27.4)*	*17.2 (11.0–26.9)*	*22.5 (12.3–36.6)*

See Tableô 2 for abbreviation explanations.

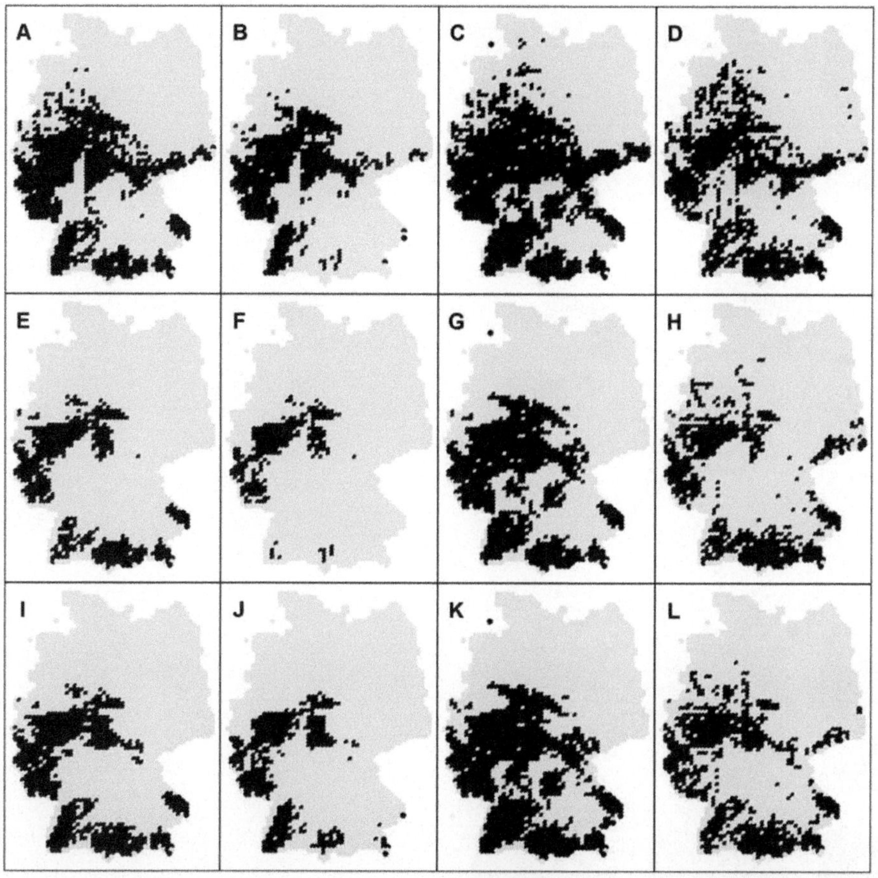

Figure 4. Potential distribution patterns for 2050s under A1b scenario for *C. gobio*. Each row shows the results of the particular climate model: (A–D) ECHAM5, (E–H) HADCM3 and (I–L) IPSL-CM4. The first column shows the result of the mean consensus model based on all repetitions of GAM, GLM and RF, followed by the mean future projections using all repetitions of the individual models: GAM, GLM and RF, respectively. Gray dots indicate centroids of the grid cells across the whole study area (Germany).

need for a well-grounded predictor selection process and detailed model validation was confirmed by the fact that RF provided "excellent" models for all calibration samples independent of the predictor data set used as well as with a significant change in the estimated future "losses" and "gains" following a change in the predictor setup. Namely, with including two more parameters in the modeling, due to low to moderate predictive power of the initially considered predictors, predictor hierarchical importance was significantly rearranged followed with changes in their contribution to the probability of occurrence. Consequently, estimates of climate change effects were significantly different from those of the initial model.

Uncertainties associated with the estimates of the climate change effects on fish distribution patterns across the TK25 grid cells are high and have multiple sources. Consistent with the results of Buisson et al. [6] we have shown that SDMs contribute to higher variability in future projections than the climate change models. Since most ecological studies on climate change effects were based on the combination of GAM with the HadCM3 model we outline that our results show that this particular combination led to the most pessimistic estimates of future habitat suitability, whereas the most optimistic one was the combination of GLM and ECHAM5 model.

As the most concerning uncertainty source we identified a mismatch between the current and future ranges of factors affecting species distributions. On average, for more than 90% of the studied area at least one factor is expected to be beyond the range considered in the model calibration, implying that this effect has to be necessarily included in the process of evaluation of the projection uncertainty, independently of the study grain (site, catchment or grid cells). We outline that in order to obtain reliable estimates of future changes, not only parameter-, SDM- and climate model selection should be performed with care but also selection of the study area. Extending the study area to regions with high overlap of current environmental conditions with the future expected conditions in the area of interest is a potential solution for avoiding the above described pitfall.

Species' environmental responses

Despite the uniqueness of our scale resolution, our results with respect to the type of climatic and topographic factors affecting fish species distributions were in accordance with recently observed general trends (see [2], [3], [9]). For instance, using GAM models calibrated on site data for 30 fish species from major French rivers Buisson et al. [2] have shown that the upstream-downstream gradient accounts for the major portion of the variability in species niche separation. The previous comply with our results as well as the conclusion that the mean annual temperature should be considered for both cold and warm water species. Our analyses indicate that neither human population density nor land use have considerable predictive power with respect to the distribution patterns of the analyzed fish species. The fact that Germany has

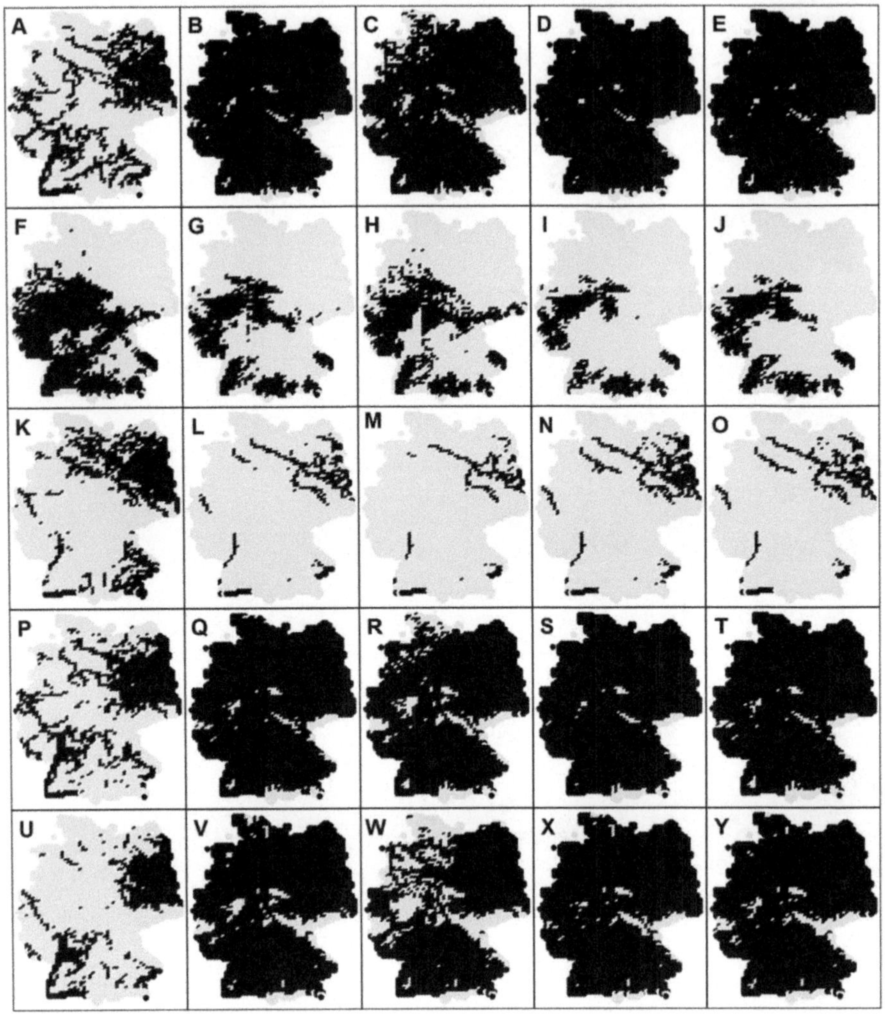

Figure 5. Current and the potential future distribution patterns of selected species for 2050s under A1b scenario. Each row shows species specific results: (A–E) *A. alburnus*, (F–J) *C. gobio*, (K–O) *L. lota*, (P–T) *S. lucioperca* and (U–Y) *S. glanis*. The first column is the consensus model of current species distribution based on all repetitions of GAM, GLM and RF, the second shows the result of the ensemble forecasting based on the combination of all SDM's (including repetitions) and all climate projections, followed by the maps showing the mean consensus model based on all SDM's (including repetitions) and ECHAM5, HADCM3 and IPSL-CM4 climate projections, respectively.

low population density variations as compared with other European countries might be the reason for the low predictive power of population density. Regarding land use, we assume that the size of our analysis unit was on one hand too small to catch the accumulated effects of the land use in the basin (e.g. agricultural land use and consequent nutrient loads) and on the other hand, too coarse to catch the local effects (e.g. shadow effects caused by trees). The necessity of describing basin heterogeneity when describing fish species distribution patterns is confirmed by the predictive power and the importance of the cumulated upstream river length and the Strahler order. The cumulated upstream river length describes the general biodiversity potential whereas the Strahler order describes the complexity level of the stream, and, consequently, the habitat heterogeneity level regarding the flow regime. Strahler order provided a suitable proxy for the upstream-downstream gradient and river type by integrating over all hydro-physical characters along the river network. Our result that the Strahler order alone describes a large portion of spatial distribution pattern of *B. barbus* is not surprising considering that this species generally occupy middle stream reaches. Also, significant univariate ROC score and high predictive importance for the factor describing dominant substrate type underlined the importance of this factor in biogeographical modeling of fish species and confirmed substrate preference as an important fish trait [21].

Through analysis of the species' environmental responses we have shown that *P. pungitus*, *C. taenia* and *L. fluviatilis* occupy similar environmental niche well described by the TK25 cell altitude. Actually, all of these species tend to occupy stream reaches close to the river mouth [33]. Lassale et al. [4] have shown that the spatial distribution pattern of *L. fluviatilis* across Europe is mainly dependent on the longitude at the mouth. Considering that Germany has a pronounced North-South altitudinal gradient, this is comparable to our finding that the average altitude of the studied region is the most important environmental predictor. *Salmo trutta*, *P. phoxinus* and *T. thymallus* were found strongly related with the hydromorphologic factors. They all require clean gravel substrates [34] and tend to occupy the upstream basin parts characterized by fisheries as "trout zone". *S. lucioperca*, *S. glanis*, *A. alburnus* and *C. gobio* showed to be driven by climatic and

Table 7. Summary of the mean expected loss and gain.

	GAM		GLM		RF		ECHAM5		HADCM3		IPSL-CM4		Mean	
Code	**loss**	**gain**	**loss**	**gain**	**loss**	**gain**	**loss**	**gain**	**loss**	**gain**	**loss**	**gain**	**loss**	**gain**
Alburnus	0	160	0	159	9	125	0	142	0	161	0	160	0	158
*Alburnus**	0	140	0	161	13	53	0	122	0	141	0	157	0	145
Cottobio	67	0	25	2	46	4	32	3	61	0	50	0	45	0
*Cottobio**	13	6	5	15	0	109	1	45	3	22	6	8	0	29
Lotalota	60	0	77	0	65	9	80	0	60	0	66	0	71	0
*Lotalota**	16	23	67	0	4	132	42	5	15	18	14	20	21	8
Sanderca	0	166	0	157	0	169	0	147	0	170	0	166	0	166
*Sanderca**	0	137	0	159	0	165	0	138	7	168	0	163	0	162
Siluanis	0	261	0	269	2	257	0	208	0	283	0	275	0	269
*Siluanis**	0	227	0	277	2	220	0	178	11	276	0	267	0	262

Table values indicate mean expected loss and gain (percentage of suitable cells) as compared to the consensus model for the 20th century climate. Asterisk in the species code (*) indicate that the predictor data set was extended by Tmax and PSeason. Model specific values (GAM, GLM and RF) are based on averaging across all future projections of the three climate models. Climate model specific values (ECHAM5, HADCM3 and IPSL-CM4) are based on averaging across all corresponding projections of the three SDMs. "Mean" indicates loss and gain resulting from comparison between the consensus future projection merging all SDMs and all climate change models and the consensus model for the 20th century climate.

hydromorphological factors occurring in stream middle reaches, but the first three tend to occupy deeper, low flowing waters in large streams, while *C. gobio* is restricted to fast-flowing water of small stream to medium-sized rivers [33]. Within the analyses of the potential of climate as a single factor describing fish species distribution both, *C. gobio* and *S. glanis* have been identified as particularly climate sensitive wherefore the finding on the importance of hydromorphology for these two species adds significant information with regard to the potential conservation pathways.

Expected changes in fish distribution patterns with climate change

Climate change is expected to cause alterations in both temperature and precipitation in Germany. We have shown that the climate alterations are likely to affect fish distribution ranges, manifested as narrowing of ranges of cold water fish (*C. gobio* and *L. lota*) and expansions for warm water species; however, the extent of these changes is highly uncertain.

Analogous to marine fish populations (e.g. [35]) we expect that as an implication of the climate change effects, freshwater organisms will face complex changes at organismal-, individual-, population and ecosystem levels of biological organisation, which will vary in magnitude and area among the different regions. Individual-level changes will involve movement into more suitable areas and consequently change species composition along the upstream-midstream-downstream gradient. Such a tendency has already been observed for the Upper Rhône River [36] where the increase of average water temperature in the last quarter of the 20th century by about 1.5° was followed by the replacement of cold water, northern species by thermophilic, southern species. Population level changes might be caused by local climatic changes. For example, Borgstrøm et al. [37] have shown that little or no snow, accumulated snow depth and summer temperatures have considerable effects on the population recruitment of brown trout in high mountain areas of Norway. Further, physiological and ecologic changes in populations of Atlantic salmon and brown trout involving time of spawning, egg survival rate, longevity and age and size at smelting are strongly affected by temperature [38], [39]. At the ecosystem level it is expected that the strength of trophic interactions between consumers and resources may become weakened or broken [15]. Also, increased temperatures have shown to lead to rejuvenation and increase of small fish abundance [40], [41], implying that higher temperatures might affect processes related to nutrient cycling. The importance of fish in nutrient cycling may be most pronounced during dry periods when external inputs are reduced [42], and drier periods are predicted to be more frequent, or of longer duration in many areas (ref to climate model). Woodward et al. [15] also noted that summer drought may lead to general habitat degradation through increase of pollutant concentrations. In addition, ecosystem

Table 8. Uncertainty estimation of the future species distributions.

scenario	AnnTMean	Isotherm	TSeason	AnnPMean	Tmax	PSeason	total
ECHAM5	74.0	0.1	3.6	0.3	50.0	0.0	84.0
HADCM3	85.6	0.1	0.4	0.6	83.1	0.0	96.3
IPSL	85.1	0.1	0.1	0.4	63.4	0.0	90.5
mean	81.6	0.1	1.4	0.4	65.5	0.0	90.3

Table values indicate percentage of cells with projected values outside the species calibration range (either lower than observed minimum or higher than the observed maximum). The column "total" is the total number of cells where for at least one model parameter the projected values are outside the species calibration range.

changes will vary among ecosystem types. For example, small streams, due to greater temperature and flow sensitivity, are expected to be more vulnerable than large streams to the future climate changes.

Secondary effects of the individual level changes, manifested as species establishments outside their current native spatial ranges and the subsequent invasions, might be even higher than direct temperature effects. For example, for common carp, currently non-native persistent but not established in England and Wales, it has been shown that the combination of temperature increase and carp invasion will result in habitat destruction, macrophyte loss and increased water turbidity [43]. Expansion of invasive species will modify biotic interactions by altering competitive dominance, increased predation rates, as well as enhancement of disease virulence [44].

Study limitations and recommendations

Despite high predictive performance of the SDMs based on grid cell related species information, we underline that the grid cell data should be seen only as an alternative in cases where stream network related data are not available. Especially in regions where the major factors affecting species distributions are dispersal limitations and local environmental peculiarities acting at spatial scales much smaller than the grid cell size, the stream network related data (site or river reach) have the primacy over the grid cell data. One of the major limitations of SDMs in general are the assumptions that species are in equilibrium with their environment and that there is no limitation to dispersal (see [45]). Fish species generally require different habitat properties for different life stages implying that a critical criterion for completing their life cycles is accessibility of functionally required habitats along the river systems [12]. For example, dams can block access to the historical spawning places and also induce habitat changes such as retention of nutrients and sediments, and can even lead to the genetic fragmentation of populations [46]. Also, possible adaptive shifts in environmental tolerances, especially when the species ability to disperse or migrate is limited, are ignored by the common SDMs [10]. However, even if all of the above listed SDM limitations have been carefully taken into account, the spatial study extent that covers only a fraction of species' realized niches [47] or a fraction of possible environmental characteristics of freshwater systems would seriously affect the prediction reliability of the effects of future ecosystem changes.

The previous considerations imply that, in order to account for the whole spectrum of climate change effects on fish populations, a multi-faceted approach is needed, including appropriate long term field data at a sufficiently broad spatial scale, information from experimental studies designed to understand species' environmental responses, establishment of a robust analysis framework and thorough evaluation of the uncertainties associated with estimates. Only with these major steps, will we achieve confidence in the ability of the methodology to model the current and predict the future effect of environmental changes.

Supporting Information

Figure S1 Predictor maps: annual mean temperature (AnnTMean), mean diurnal temperature range (DiuTRange), isothermality (Isotherm), mean temperature seasonality (TSeason), maximum temperature of warmest month (Tmax), mean temperature of wettest quarter (TWetQuar), mean temperature of driest quarter (TDryQuar), annual mean precipitation (AnnPMean), mean precipitation seasonality (PSeason), mean altitude (AltMean), maximum cumulative length of the upstream flow network (CumLenkm), maximum Strahler order (Strahler), dominant river type (RtypMost), dominant land use type (Lusemax) and mean population density (Popmean).

Acknowledgments

We thank three anonymous reviewers for helpful comments. We would like to thank the German society for ichthyology and the Federal Agency for Nature Conservation for kindly providing the fish occurrence data.

Author Contributions

Conceived and designed the experiments: DM. Performed the experiments: DM. Analyzed the data: DM. Contributed reagents/materials/analysis tools: DM. Wrote the paper: DM JF CW.

References

1. Thuiller W (2004) Patterns and uncertainties of species range shifts under climate change. Global Change Biology 10: 2020–2027.
2. Buisson L, Thuiller W, Lek S, Lim P (2008) Climate change hastens the turnover of stream fish assemblages. Global Change Biology 14: 2232–2248.
3. Lassale G, Crouzet P, Rochard E (2009) Modelling the current distribution of Europen diadromous fishes: an approach integrating regional antropogenic pressures. Freshwater Biology 54: 587–606.
4. Lassalle G, Beguer M, Beaulaton L, Rochard E (2008) Diadromous fish conservation plans need to consider global warming issues: An approach using biogeographical models. Biological Conservation 141: 1105–1118.
5. Marmion M, Parviainen M, Luoto M, Heikkinen RK, Thuiller W (2009) Evaluation of consensus methods in predictive species distribution modelling. Diversity and Distributions 15: 59–69.
6. Buisson L, Thuiller W, Casajus N, Lek S, Grenouillet G (2010) Uncertainty in ensemble forecasting of species distribution. Global Change Biology 16: 1145–1157.
7. Grenouillet G, Buisson L, Casajus N, Lek S (2011) Ensemble modelling of species distribution: the effects of geographical and environmental ranges. Ecography 34: 9–17.
8. Guegan JF, Lek S, Oberdorff T (1998) Energy availability and habitat heterogeneity predict global riverine fish diversity. Nature 391: 382–384.
9. Buisson L, Blanc L, Grenouillet G (2008) Modelling stream fish species distribution in a river network: the relative effects of temperature versus physical factors. Ecology of Freshwater Fish 17: 244–257.
10. Franklin J (2009) Mapping species distributions: spatial inference and prediction. New York: Cambridge University Press. 320 p.
11. Pearson RG, Dawson TP (2003) Predicting the impacts of climate change on the distribution of species: are bioclimate envelope models useful? Global Ecology & Biogeography 12: 361–371.
12. Wang L, Seelbach PW, Hughes RM (2006) Introduction to Landscape Influences on Stream Habitats and Biological Assemblages. American Fisheries Society Symposium 48: 1–23.
13. Heino J, Virkkala R, Toivonen H (2009) Climate change and freshwater biodiversity: detected patterns, future trends and adaptations in northern regions. Biological Reviews 84: 39–54.
14. Butchart SHM, Walpole M, Collen B, van Strien A, Scharlemann JPW, et al. (2010) Global Biodiversity: Indicators of Recent Declines. Science 328: 1164–1168.
15. Woodward G, Perkins DM, Brown LE (2010) Climate change and freshwater ecosystems: impacts across multiple levels of organization. Philosophical Transactions of the Royal Society B 365: 2093–2106.
16. Vörösmarty CJ, McIntyre PB, Gessner MO, Dudgeon D, Prusevich A, et al. (2010) Global threats to human water security and river biodiversity. Nature 467: 555–561.
17. Nelson KC, Palmer M, Pizzuto J, Moglen GE, Angermeier PL, et al. (2009) Forecasting the combined effects of urbanization and climate change on stream ecosystems: from impacts to management options. Journal of Applied Ecology 46: 154–163.
18. Thuiller W, Brotons L, Araujo MB, Lavorel S (2004) Effects of restricting environmental range of data to project current and future species distributions. Ecography 27: 165–172.
19. Araújo MB, Thuiller W, Williams PH, Reginster I (2005) Downscaling European species atlas distributions to a finer resolution: implications for conservation. Global Ecology & Biogeography 14: 17–30.
20. Buisson L, Grenouillet G (2009) Contrasted impacts of climate change on stream fish assembladges along an environmental gradient. Diversity and Distributions 15: 613–626.

21. Goldstein RM, Meador MR (2004) Comparisons of Fish Species Traits from Small Streams to Large Rivers. Transactions of the American Fisheries Society 133: 971–983.
22. Hijmans RJ, Cameron SE, Parra JL, Jones PG, Jarvis A (2005) Very high resolution interpolated climate surfaces for global land areas. International Journal of Climatology 25: 1965–1978.
23. Turc L (1961) Estimation of irrigation water requirements, potential evapotranspiration: A simple climatic formula evolved up to date. Annals of Agronomy 12: 13–49.
24. de Jager AL, Vogt JV (2010) Development and demonstration of a structured hydrological feature coding system for Europe. Hydrological Sciences Journal 55: 661–675.
25. Pottgiesser T, Sommerhäuser M (2004) Flließgewässertypologie Deutschlands: Die Gewässertypen und ihre Steckbriefe als Beitrag zur Umsetzung der EU-Wasserrahmenrichtlinie. In: Steinberg C, Calmano W, Wilken RD, Klapper H, editors. Handbuch der Limnologie 19. Landsberg: Ecomed Verlagsgesellschaft. pp. 7/04. VIII-2.1: 1–16.
26. IPCC (2007) Climate Change 2007: The Physical Science Basis. Contribution of Working Group I to the Fourth Assessment Report of the Intergovernmental Panel on Climate Change. In: Solomon S, Qin D, Manning M, Chen Z, Marquis M, et al., editors. Cambridge, UK: Cambridge University Press. 996 p.
27. Nakicenovic N, Swart R (2000) Emissions Scenarios: A Special Report of Working Group III of the Intergovernmental Panel on Climate Change. Cambridge, UK: Cambridge University Press.
28. Thuiller W (2003) BIOMOD – optimizing predictions of species distributions and projecting potential future shifts under global change. Global Change Biology 9: 1353–1362.
29. Phillips SJ, Dudik M (2008) Modeling of species distributions with Maxent: new extensions and a comprehensive evaluation. Ecography 31: 161–175.
30. Leatwick JR, Elith J, Hastie T (2006) Comparative performance of generalized additive models and multivariate adaptive regression splines for statistical modelling of species distributions. Ecological Modelling 199: 188–196.
31. Breiman L (2001) Random Forest. Machine Learning 45: 5–32.
32. Bezdek JC (1981) Pattern Recognition with Fuzzy Objective Function Algoritms. New York: Plenum Press. 272 p.
33. Kottelat M, Freyhof J (2007) Handbook of European Freshwater Fishes. Kottelat, Cornol, Switzerland, and Freyhof, Berlin, Germany. 646 p.
34. Oberdorff T, Pont D, Hugueny B, Porcher JP (2002) Development and validation of a fish-based index (FBI) for the assessment of 'river health' in France. Freshwater Biology 47: 1720–1734.
35. Rijnsdorp AD, Peck MA, Engelhard GH, Möllmann C, Pinnegar JK (2009) Resolving the effect of climate change on fish populations. ICES Journal of Marine Science 66: 1570–1583.
36. Daufresne M, Roger MC, Capra H, Lamouroux N (2003) Long-term changes within the invertebrate and fish communities of the Upper Rhône River: effects of climatic factors. Global Change Biology 10: 124–140.
37. Borgstrøm R, Museth J (2005) Accumulated snow and summer temperature – critical factors for recruitment to high mountain populations of brown trout (Salmo trutta L.). Ecology of Freshwater Fish 14: 375–384.
38. Pankhurst NW, King HR (2010) Temperature and salmonid reproduction: implications for aquaculture. Journal of Fish Biology 76: 69–85.
39. Johnson B, Johnson N (2009) A review of the likely effects of climate change on anadromous Atlantic salmon Salmo salar and brown trout Salmo trutta, with particular reference to water temperature and flow. Journal of Fish Biology 75: 2381–2447.
40. Jeppesen E, Meerhoff M, Holmgren K, González-Bergonzoni I, Teixeira-de Mello F, et al. (2010) Impacts of climate warming on lake fish community structure and potential effects on ecosystem function. Hydrobiologia 646: 73–90.
41. Daufresne M, Boët P (2007) Climate change impacts on structure and diversity of fish communities in rivers. Global Change Biology 13: 2467–2478.
42. Vanni M (2002) Nutrient cycling by animals in freshwater ecosystems. Annual Review of Ecology and Systematics 33: 341–70.
43. Britton JR, Cucherousset J, Davies GD, Godard MJ, Copp GH (2010) Non-native fishes and climate change: predicting species responses to warming temperatures in a temperate region. Freshwater Biology 55: 1130–1141.
44. Rahel F, Bierwagen B, Taniguchi Y (2008) Managing Aquatic Species of Conservation Concern in the Face of Climate Change and Invasive Species. Conservation Biology 22: 551–561.
45. Olden JD, Kennard MK, Leprieur F, Tedesco PA, Winemiller KO, et al. (2010) Conservation biogeography of freshwater fishes: recent progress and future challenges. Diversity and Distributions 16: 496–513.
46. Limburg KE, Waldman JR (2009) Dramatic declines in North Atlantic Diadromous Fishes. BioScience 59: 955–965.
47. Logez M, Bady P, Pont D (2012) Modelling the habitat requirement of riverine fish species at the European scale: sensitivity to temperature and precipitation and associated uncertainty. Ecology of Freshwater Fish 21: 266–282.

The Spatial Heterogeneity between Japanese Encephalitis Incidence Distribution and Environmental Variables in Nepal

Daniel E. Impoinvil[1,2]*, Tom Solomon[2], W. William Schluter[3], Ajit Rayamajhi[2,4], Ram Padarath Bichha[5], Geeta Shakya[6], Cyril Caminade[7], Matthew Baylis[1]

1 LUCINDA Group, Department of Epidemiology and Population Health, Institute of Infection and Global Health, University of Liverpool, Leahurst Campus, Neston, Cheshire, United Kingdom, **2** Brain Infections Group, Department of Clinical Infection, Microbiology and Immunology, Institute of Infection and Global Health, University of Liverpool, Liverpool, Merseyside, United Kingdom, **3** Programme for Immunization Preventable Diseases, World Health Organization Country Office for Nepal, Kathmandu, Nepal, **4** Department of Pediatrics, National Academy of Medical Sciences, Kanti Children's Hospital, Maharajgunj, Kathmandu, Nepal, **5** Child Health Division, Department of Health Services, Ministry of Health and Population, Government of Nepal, Kathmandu, Nepal, **6** National Public Health Laboratory, Department of Health Services, Ministry of Health and Population, Government of Nepal, Kathmandu, Nepal, **7** Department of Earth's Changing Environment, School of Environmental Science, University of Liverpool, Liverpool, Merseyside, United Kingdom

Abstract

Background: To identify potential environmental drivers of Japanese Encephalitis virus (JE) transmission in Nepal, we conducted an ecological study to determine the spatial association between 2005 Nepal JE incidence, and climate, agricultural, and land-cover variables at district level.

Methods: District-level data on JE cases were examined using Local Indicators of Spatial Association (LISA) analysis to identify spatial clusters from 2004 to 2008 and 2005 data was used to fit a spatial lag regression model with climate, agriculture and land-cover variables.

Results: Prior to 2006, there was a single large cluster of JE cases located in the Far-West and Mid-West terai regions of Nepal. After 2005, the distribution of JE cases in Nepal shifted with clusters found in the central hill areas. JE incidence during the 2005 epidemic had a stronger association with May mean monthly temperature and April mean monthly total precipitation compared to mean annual temperature and precipitation. A parsimonious spatial lag regression model revealed, 1) a significant negative relationship between JE incidence and April precipitation, 2) a significant positive relationship between JE incidence and percentage of irrigated land 3) a non-significant negative relationship between JE incidence and percentage of grassland cover, and 4) a unimodal non-significant relationship between JE Incidence and pig-to-human ratio.

Conclusion: JE cases clustered in the terai prior to 2006 where it seemed to shift to the Kathmandu region in subsequent years. The spatial pattern of JE cases during the 2005 epidemic in Nepal was significantly associated with low precipitation and the percentage of irrigated land. Despite the availability of an effective vaccine, it is still important to understand environmental drivers of JEV transmission since the enzootic cycle of JEV transmission is not likely to be totally interrupted. Understanding the spatial dynamics of JE risk factors may be useful in providing important information to the Nepal immunization program.

Editor: Benjamin J. Cowling, University of Hong Kong, Hong Kong

Funding: This study was supported by the Leverhulme Trust Research Leadership Award: "Predicting the effects of climate change on infectious diseases of animals", granted to Matthew Baylis. The funders had no role in study design, data collection and analysis, decision to publish, or preparation of the manuscript.

* E-mail: d.impoinvil@liverpool.ac.uk

Introduction

Japanese Encephalitis (JE) is a viral zoonotic mosquito-borne disease which causes 30,000 to 50,000 human cases per year, primarily in Asia, making it the most common cause of viral encephalitis [1]. Infection results in case fatalities of 0.3% to 60% [2,3]. JE is caused by the JE virus (JEV), a member of the Flaviviridae family, which also includes dengue virus, West Nile virus, and Yellow fever virus [4]. The virus is transmitted through a wide range of vectors, but the primary vector of JEV across Asia is the mosquito *Culex tritaeniorhynchus*. JEV infects a wide range of vertebrates; however its primary reservoir host is water birds such as egrets and herons [5]. Rice-fields are the preferred development sites for *C. tritaeniorhynchus* as well as the main foraging site for water birds; thus rice-fields provide an important transmission site for infectious and susceptible birds and mosquitoes to meet. Pigs are considered a secondary reservoir host because they produce high viraemias when infected, and readily infect pig-biting mosquitoes. Humans are considered tangential hosts because low viremia is produced when they are infected, which means they cannot infect mosquitoes when bitten; hence, the virus transmission cycle reaches an end when humans are infected. When mosquito populations become adequately high, JEV transmission spills over from the mosquito-bird-pig cycle into human populations. Two major risk factors of JEV transmission to humans are close proximity to rice fields and pigsties.

Major environmental drivers capable of amplifying JEV transmission are temperature and precipitation, which act primarily through their influence on mosquito vector life history such as development time of immature mosquito stages (i.e. larvae and pupae) and population abundance [6,7,8,9,10] and the extrinsic incubation period of the virus in the mosquito once it is acquired by bite from an infected vertebrate host [11]. However, with the concern of climate change threatening increased transmission of several arboviruses [12], advancing the knowledge base of JE-climate interactions is needed. Early studies of JE epidemiology demonstrated that JEV transmission was associated with high temperature and low rainfall [13]. Other studies in China [14] and Taiwan [15] have found significant temporal associations between climate and JEV transmission with different time lags. Despite these studies, a dearth of spatial studies exists that examine the associations between JEV transmission and climate variables. Other potential drivers of JEV transmission include land-use and land-cover [16,17,18].

While the influence of environmental drivers of JEV transmission has been described in other countries across Asia, few have investigated these drivers in Nepal, despite the heavy disease burden of JEV infection among human populations. JE was first recognized in Nepal in the terai region in 1978. The terai region is comprised of a belt of marshy grasslands, savannas, and forests at the base of the Himalayan Mountains. The altitude ranges from approximately 55 to 500 m. The transmission season in Nepal starts from June to October with a peak in August [19]. High case-fatality (15%–35%) due to JEV transmission has been reported in the terai since its discovery, making it the most common cause of encephalitis and a growing public health concern in the Nepal terai region [5]. Additionally, since 1997 there have been reports of JEV transmission expanding from the terai region to the hill region (altitude of 1,300 m) of Nepal with a significant number of cases appearing in the Kathmandu Valley [5,20,21]. It is not clear what is driving this expansion. Recent studies in animals in Nepal have found significant exposure to JEV in pigs, horses and ducks [22,23].

From 2004 to present, Nepal has been actively detecting JE cases through their national vaccine preventable diseases surveillance network for acute encephalitis syndrome (AES) conducted by the government of Nepal, with technical and financial support from the World Health Organization (WHO) [20]. Through the national surveillance program it has been confirmed that significant transmission of JEV is indeed occurring in the Kathmandu valley [21] and other locations in the hill and mountain districts [20] of Nepal and not only the terai region of Nepal. Given these findings, it has been suggested that the national JE prevention and control program should not only focus on the terai region, but should also give attention to other regions of Nepal [20]. In addition, under the Nepal Ministry of Health and Population, mass JE vaccination campaigns have been conducted annually in a phase-wise manner targeting 1 to 8 endemic districts each year beginning in 2006 with reported coverage >80% of the targeted population. Approximately half of the 23 districts have targeted only children 1 to 15 years of age, whereas the other districts have targeted all persons greater than 12 months of age. With the current surveillance program in place, it is now possible to examine, country-wide, the spatial patterns of JEV transmission in Nepal.

The expansion of JEV transmission into the hill and mountain regions raises questions on the ecology of transmission among the various regions of Nepal and the spatial patterns therein. In this study, we look at the changing trend of clustering of JE cases in the different districts of Nepal. In addition, in this study, we focus our environmental analysis on the 2005 JEV epidemic. This is due to the large number of cases that occurred in that year (669 laboratory-confirmed cases). Furthermore, we wanted to determine the impact of climate and land-use land-cover variables on JE cases without the influence of mass immunization. We conduct the first ecological spatial analysis in Nepal to determine the spatial association between Nepal JE incidence and two climate variables (mean monthly temperature and mean total precipitation) as well as two agricultural variables (percentage irrigated land, pig-to-human ratio) and three land cover variables (deciduous tree cover, grassland cover and cropland cover) at the district level using national laboratory-confirmed data. We tested the following hypotheses: 1) the environmental drivers of JEV transmission were the same before and after the 2005 epidemic, 2) lagged monthly temperature and precipitation variables predict JE incidence better than annual climate variables and 3) in Nepal, JEV transmission is related to agricultural, land-cover and land-use patterns (i.e. pig-to-human ratio, percentage irrigated land and deciduous tree cover, grassland cover and cropland cover) that are associated with transmission cycles involving the life history of the mosquito vector and availability of the reservoir host.

Methods

Japanese Encephalitis case detection

JE case detection in Nepal has been described previously [24]. Briefly, the Nepal model for JE case surveillance is based on experience implementing clinical surveillance of encephalitis. Because the symptoms of JE are similar to many other encephalitic diseases, patients are initially investigated as Acute Encephalitis Syndrome (AES) patients with the etiology of the encephalitis determined later. AES cases are defined as any patient presenting with acute onset of fever and a deterioration in mental status (e.g., confusion, disorientation, coma, or inability to talk) and/or new onset of seizures excluding simple febrile seizures. Using a structured reporting form, information on age, gender, district of residence, and whether the patient had been immunized against JEV are recorded. Clinical data including date of disease onset and symptoms (e.g., fever, neck rigidity, convulsions) as well as outcome at discharge (i.e., cured, referred, death, unknown) are recorded. Five mL of serum or 2 mL of cerebrospinal fluid (CSF) are obtained from each patient. The serum or CSF samples are labeled with the patient identification number and stored at 2°C to 8°C until transported in cold boxes with ice packs to the National Public Health Laboratory in Kathmandu. Laboratory confirmation is made from a single serum or CSF sample by detection of anti-JE immunoglobulin M (IgM) antibody titers using an IgM antibody capture enzyme- linked immunoassay.

Geographical and Meteorological data

District-level geographical information systems (GIS) shapefiles of Nepal were acquired from the International Centre for Integrated Mountain Development (geoportal.icimod.org/Downloads/) (See Figure S1. for a general map of Nepal districts). Nepal temperature and precipitation data were acquired from WorldClim Global dataset www.worldclim.org/). WorldClim is a set of global climate layers (climate grids) with a spatial resolution of a square kilometer. The WorldClim dataset, which spans from 1950 to 2000, is meant to serve as a high resolution source of climate data, which is interpolated from various climate sources [25]. The WorldClim station network over India – Nepal is relatively dense to have confidence in the final high resolution gridded product. Long-term climate normals were acquired for annual mean temperature and precipitation, as well as mean temperature and precipitation for the individual months of April and May, as well as the three month period from June to August. The WorldClim dataset is gridded and zonal estimation was used to summarize the values of the gridded climate datasets within Nepal district shapefiles using the Spatial Analyst module in ArcView GIS 9.3 ® (ESRI, Redlands, CA). In subsequent analysis we made the assumption that using the WorldClim dataset that spans from 1950 to 2000 would be able to predict JE incidence in the 2005 outbreak in Nepal.

Agricultural and Land-use Land-cover data

Nepal agricultural data (i.e. irrigated land area and pig number) was acquired from the Nepal Ministry of Agriculture and Cooperatives (www.moac.gov.np/home/statistics.php). All data were reported as "per district" and is based on 2008/2009 estimate from surveys. We assumed no major changes in pig number and irrigated land area in each of the districts from 2004 to 2008. We calculated the percent of irrigated land for each district by taking the total irrigated land area and dividing it by the total district areas size and multiplying by one hundred. We also calculated the ratio of pig-to-human population per district to account for differences in exposure in areas with high and low human and pig populations. Land-cover data was downloaded from Boston University's MODIS Land Cover and Land Cover Dynamics project (www-modis.bu.edu/landcover/). We used the dataset consisting of the International Geosphere Biosphere Programme (IGBP) classification scheme which is comprised of 17 land cover classes. After conducting a Pearson's coefficient correlation analysis with all land cover variables and 2005 district level JE incidence values, we selected the percentage of grassland cover ($r = -0.690$, $P<0.001$) and the cropland cover ($r = 0.609$, $P<0.001$) for further analysis because they had the highest negative and positive correlation with JE incidence respectively. We also included percentage of deciduous tree cover ($r = 0.442$, $P<0.001$), since other studies suggested this may be an important factor in mosquito niche selection [26].

Cluster analysis

Univariate Local Indicators of Spatial Association (LISA) were used to describe the spatial pattern of JE incidence clusters at the district level for the years of 2004 to 2008. LISA maps are particularly useful to assess the hypothesis of spatial randomness and to identify local hot and cold spots of a phenomenon [27,28]. The univariate LISA gives an indication of the degree of linear association (positive or negative) between the values for one value at a given location and the average of neighboring values in the surrounding location. Univariate LISA suggest two classes of positive spatial correlation, or spatial clusters (High-High and Low-Low), and two classes of negative spatial correlation, or spatial outliers (High-Low and Low-High). The LISA cluster maps indicate only the center of the cluster. The actual extent of the cluster includes the center and the surrounding neighbors as defined by a weights matrix. In GeoDa alpha release 0.9.8.14 (geodacenter.asu.edu/software/downloads), a randomization approach is used to generate a spatially random reference distribution to assess statistical significance. In order to do spatial analysis such as LISA analysis or other spatial analyses, a contiguity-based spatial weight matrix must be created to identify spatial relationships between geographic features, in our case Nepal district polygons. In this analysis, we used a first order Rook Contiguity which uses a common boundaries of polygons to define Nepal district polygon neighbors [27]. For 2005 and 2008 data, we also calculated the relative risk of JE incidence in high JE-incident clustered areas versus non-clustered areas using SatScan™. SatScan™ uses a variable-sized window to test for possible clusters noting the number of observed and expected observations inside the window at each location [29]. To do this purely spatial analysis, we used a circular spatial window and scanned for clusters with high rates using the discrete Poisson model.

Statistical model

In our analysis, we used a strategy similar to Wimberly et al., 2008 [30]. We used a three-step process to develop a model that identifies ecologically-important variables that are likely drivers of JEV transmission in Nepal using 2005 data only. The rationale of using this three-step process was first, to characterize how temperature and precipitation are associated with the distribution of JE cases, since these variables can have differential impacts on mosquito populations and subsequent virus transmission. For example, laboratory studies have found that there is an 'optimum range' for virus transmission in mosquitoes [11]. This can be modeled with quadratic polynomial terms; second, to expand the climate model to include agricultural and land-cover variables that have previously been identified as important risk factors of JEV transmission; third, to develop a parsimonious model that emphasizes the variables which are the most important for describing JEV distribution. In all models, a spatial lag regression model was done to account for spatial autocorrelation. The spatially lagged regression models determine whether the dependent variable covaries with its geographic neighbors [31]. The importance of accounting for spatial autocorrelation is to avoid spurious associations between dependent and independent variables when they share the same spatial clustering or are uniform in space [31].

Our first step was to compare competing models based on the climate variables mean temperature and mean precipitation. The models allowed for a) an interaction term between temperature and precipitation to determine if there were different responses between JE incidence and temperature for different levels of precipitation, b) a curvelinear or unimodal relationship between temperature and precipitation and JE incidence using a second-order trend surface of mean monthly temperature and mean monthly precipitation to determine if these climate parameters would provide a better fit to JE incidence and, c) a spatial lag term to account for spatial autocorrelation among the Nepal districts (i.e. account for JE clustering). Six models were compared based on the following:

1. Annual mean temperature and precipitation from 1950 to 2000
2. April mean monthly temperature and precipitation (2-month lag from transmission season)
3. May mean monthly temperature and precipitation (1-month lag from transmission season)
4. April temperature (2-month lag) and May precipitation (1-month lag)
5. April precipitation (2-month lag) and May temperature (1-month lag)

6. June-July-August mean temperature and precipitation (these 3 months represent the climate at peak JEV transmission)

The model used is below:

$$y_1 = \beta_0 + \beta_1 t_i + \beta_2 t_i^2 + \beta_3 p_i + \beta_4 p_i^2 + \beta_5 t_i p_i + \rho\omega y_i$$

where i represents each Nepal district, y_i is the JE incidence, t_i is the temperature variable, p_i is the precipitation variable, β is the parameter estimate and $\rho\omega_i y_i$ is the spatial lag term [31].

The months of April and May were chosen to encompass 2-month and 1-month climate lags respectively from the JEV transmission season which starts in June. The months of June to August were chosen since this is the time at which JEV transmission increases and peaks.

Using GeoDa, a spatial lag model approach was used to model JE incidence as a function of the competing sets of climate variables. Within each district, the incidence of JE was calculated by taking the number of JE cases in 2005 and dividing it by the 2005 population estimate for each district. District population estimates-were calculated by regressing a line through the projected Nepal population point estimates for 2001, 2006, 2011, 2016, and 2021; (www.cbs.gov.np/Gis_Maps/Population%20Projection_-Trends/Population%20Projection%202001%20-%202021.gif) from the regression line the population estimates were derived. Since these projections use 2001 Nepal census data as a baseline and extrapolate their projections based on population growth estimates, we found this approach to be satisfactory in providing estimates of the district population size. Due to the sizable variation of area and population density across these administrative units, the crude 2005 incidence rates for each district polygon was first estimated by using spatial empirical Bayes smoothing method available in GeoDa. This smoothing procedure was suggested to adjust for the potential bias resulting from estimating from a small population at risk [32]. The first order Rook contiguity-based spatial weight matrix method, mentioned in the cluster analysis section, was used for the both the spatial empirical Bayes smoothing and the spatial lag model. In order to avoid the violation of the constant variance assumption in linear regression, we used a negative-reciprocal transformation on the Spatial Empirical Bayes smoothed JE incidence. This drastically reduced heteroscedasticity (i.e. irregular variance) more than a log transformation. The dependent variable (i.e. smoothed JE incidence) was transformed using a negative reciprocal ($-1/(y_i+1)$) transformation. To further increase interpretability we added 1 to the newly transformed variable to make the values positive. To improve model convergence and interpretation, and reduce problems with collinearity (i.e. high correlation between independent variables) between second order and interaction terms, independent variables were centered by subtracting their mean [33]. The R^2, adjusted R^2 and Akaike's information criterion (AIC) were used as metrics to compare the candidate models. The model with the highest R^2 and lowest AIC was selected as having the best fit. To assist in visualizing the relationship between JE incidence and environmental variables, we made contour plots based on the spatial lag model in Microsoft ® Excel 2007 (Microsoft, Redmond, WA).

In the second step of our analysis, a full model using the 2005 data was created by adding the agricultural, land-use and land-cover variables (i.e. pig-to-human ratio, percentage irrigated land, percentage deciduous tree cover, percentage grassland cover and percentage cropland cover) to the best fitting temperature/precipitation model. Early exploratory analysis suggested that the addition of squared term of pig-to-human population ratio may improve subsequent linear regression models. To avoid problems with collinearity, we centered pig-to-human population ratio subtracting the mean from the variable. The new centered variable and the quadratic term of the new centered variable was entered into the model. We also found that taking the natural log of percentage irrigated land, deciduous tree cover, grassland cover and cropland cover also stabilized the variance between these variables and JE incidence.

In the third and final step of our analysis, a parsimonious model was created in IBM® SPSS® Statistics 17 using a backward selection method. The subsequent model variables found to be important in IBM® SPSS® Statistics 17 (IBM SPSS, Chicago, IL) were used to create the spatial lag model in GeoDa. This was done because at present there is no backward selection method in GeoDa and it would have been computationally tedious to execute this method using partial F tests with GeoDa.

To test whether environmental driver trends of 2005 still remain constant, despite the changes in JE incidence cluster patterns due to mass vaccination programs, we used the variables in the 2005 parsimonious model to create subsequent models using 2004, 2006, 2007 and 2008 JE incidence data. We evaluated the respective models by comparing the R^2 values and parameter estimates of 2005 with the other years.

Results

Seasonal transmission of JE

Figure 1. shows a time series of laboratory-confirmed JE cases from June 2004 to April 2009 and the annual seasonal transmission pattern of laboratory-confirmed JE cases. Figure 1A. shows that in 2006 there was a drop in the number of JE cases. However, a substantial number of JE cases still occurred (over 100 laboratory confirmed cases across all years). Figure 1B. confirms the general trend of JEV transmission is an initial increase of transmission from June followed by a peak in August and a decrease in transmission thereafter.

Cluster analysis

Figure 2. shows the distribution of JE incidence and the most likely clusters in Nepal from 2004 to 2008 using SatScan. The general trend shows JEV transmission concentrated in the Mid-West and Far-West terai regions with high values of JE incidence from 2004 to 2005 but gradually decreasing in the terai region. LISA analysis of JE epidemics that occurred from 2004 to 2008 identified different foci of JE transmission (Figure 3). The shift can be seen when looking at the JE incidence distribution of the 2005 outbreak year and the later year of 2008 (Figure 3), where high values in 2008 cluster around the Kathmandu valley. Using SatScan, we found that in 2005 the relative risk for populations inside the JE cluster (which included the following districts: Achham, Banke, Bardiya, Dailekh, Dang, Doti, Jajarkot, Kailali, Kalikot, Kanchanpur, Salyan, and Surkhet; see Figure 2.) compared to those outside the cluster was 9.82 (95% Confidence interval: 8.83–11.51). While in 2008, we found that the relative risk for populations inside the JE cluster (which included the following districts: Bhaktapur and Kathmandu) compared to those outside the cluster was 3.93 (95% Confidence Interval: 3.00–5.13 see Figure 2.). This suggests higher risk of JE in these clustered areas versus outside of these clusters.

Statistical modeling

The model based on May temperature and April precipitation had the best fit as indicated by the high R^2 value and low AIC (Table 1.). Temperature and the spatial lag term were the only variables found to be highly significant with a positive association with JE incidence. Quadratic and the interaction terms did not

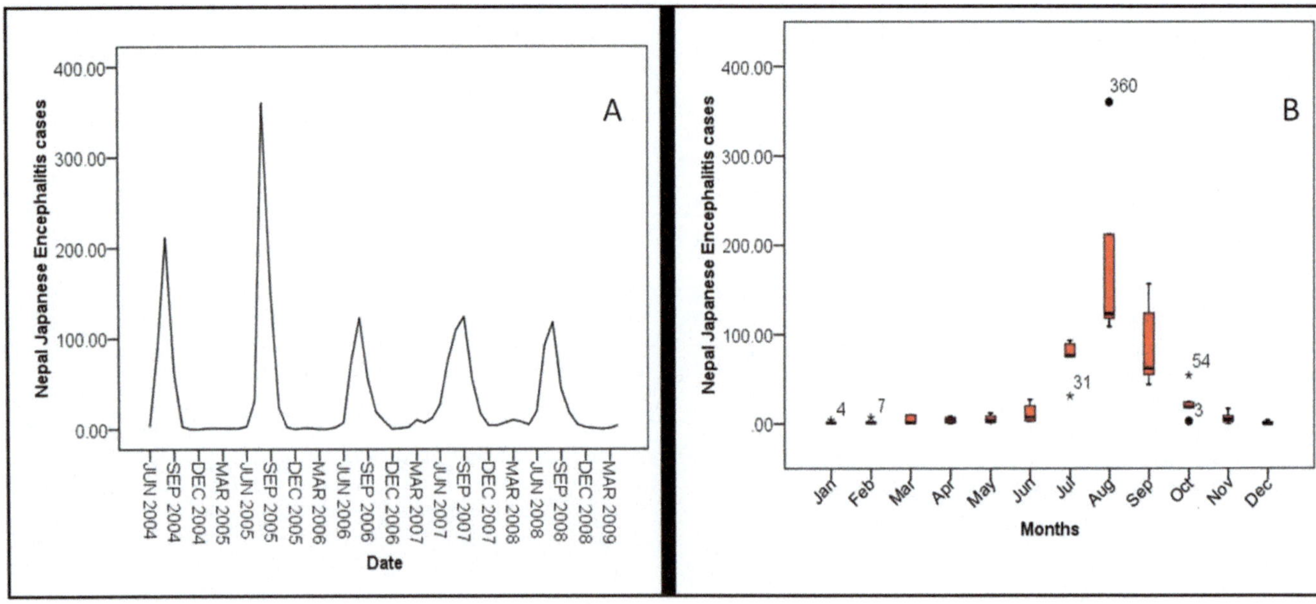

Figure 1. Time Series graph (A) of laboratory confirmed Japanese Encephalitis (JE) cases and a box plot (B) of seasonal transmission of confirmed JE cases from June 2004 to April of 2009. The dark line in the middle of the boxes is the median value; the bottom and top of the boxes indicates the 25th and 75th percentile respectively; whiskers represents 1.5 times the height of the box; and dots with numbers represent value of outlier cases and asterisks with number represent extreme values outlier cases (i.e. more than three times the height of the boxes).

have high statistical significance. The other models had weaker fits, though the model with April temperature and precipitation was comparable in R^2 value and AIC. Though precipitation was not statistically significant, the 2-months and 1-month lag (April and May) models did show a negative relationship with JE incidence, with high number of JE cases occurring when there is low precipitation in the preceding months (Figure 4C–F.). In the Annual and June–July–August models, precipitation had a positive, albeit non-statistically significant, association with JE incidence (Table 1. and Figure 4A and B.).

Compared to the crude climate model (i.e. May temperature/April precipitation alone), higher R^2 and higher AIC were obtained after incorporating the variables pig-to-human ratio and its quadratic term, the natural log-percentage of irrigated land and deciduous tree cover, grassland cover and cropland cover into the model (see Figure S2.). for the distribution of environmental variable values in Nepal used in full and parsimonious models). However, the only variable which was statistically significant in the full model was the interaction term and the spatial lag term (Table 2.). There was high collinearity in the full model, specifically with temperature (Variance Inflation Factor >16; see Table S1 for the Pearson's correlation coefficient matrix of the variables used in the full model for 2005 JE incidence). Therefore, we developed a parsimonious model using backward selection method in which precipitation, the percentage of irrigated land, the percentage of grassland cover and the pig-to-human ratio remained in the model, but not temperature, its quadratic term or the quadratic term of precipitation (Table 3.). The parsimonious model was found to be comparable to the full model in the R^2 value, but the parsimonious model had a lower AIC than the full model indicating a better explanatory model. Precipitation was negatively associated with JE incidence, while percentage of irrigated land was positively associated with JE incidence. Neither the percentage of grassland cover, pig-to-human ratio nor its quadratic term was significantly associated with JE incidence.

When we compared the observed JE incidence with the predicted JE incidence we found that the model heavily underestimated the JE incidence in Bardiya, Banke, Dang and Kailali (observed: 25.50, 21.69, 17.25 and 15.17 vs. predicted: 2.80, 2.92, 2.25 and 1.99 respectively). However a similar pattern in JE incidence distribution was still seen between the observed and predicted maps (see Figure S3. for the predicted smoothed JE incidence in each of the districts and the prediction error using the parsimonious model).

When we applied our parsimonious model to the other years, we found that 2004, 2006 and 2007, R^2 values, AIC, parameter coefficients were similar. Significance level among the environmental variables showed variability from year to years (Table 4.). For example, April precipitation was statistically significant during 2004 and 2005, percentage irrigated land was significant in 2005 and 2007 and percentage grassland cover was significant in 2004 and 2006. In 2007, a statistically significant unimodal relationship between JE incidence and pig-to-human ratio emerged. In 2008, April precipitation took on a positive trend, albeit not significant. The only variable consistently significant throughout the years was the spatial lag term.

All regression β coefficients in Table 1–4 represent a non-linear increase (or decrease when the coefficient value is negative) in JE incidence when there is a 1-unit increase in each respective predictor variable.

Discussion

In this paper we conducted one of the first spatial analyses of JEV transmission in Nepal. Using laboratory confirmed JE cases from the national AES surveillance program in Nepal, we identified spatial clusters of JE incidence in 2004 to 2008 and developed a spatial statistical model with climate and land-cover variables. We identified key variables that describe the spatial distribution of JE cases in Nepal.

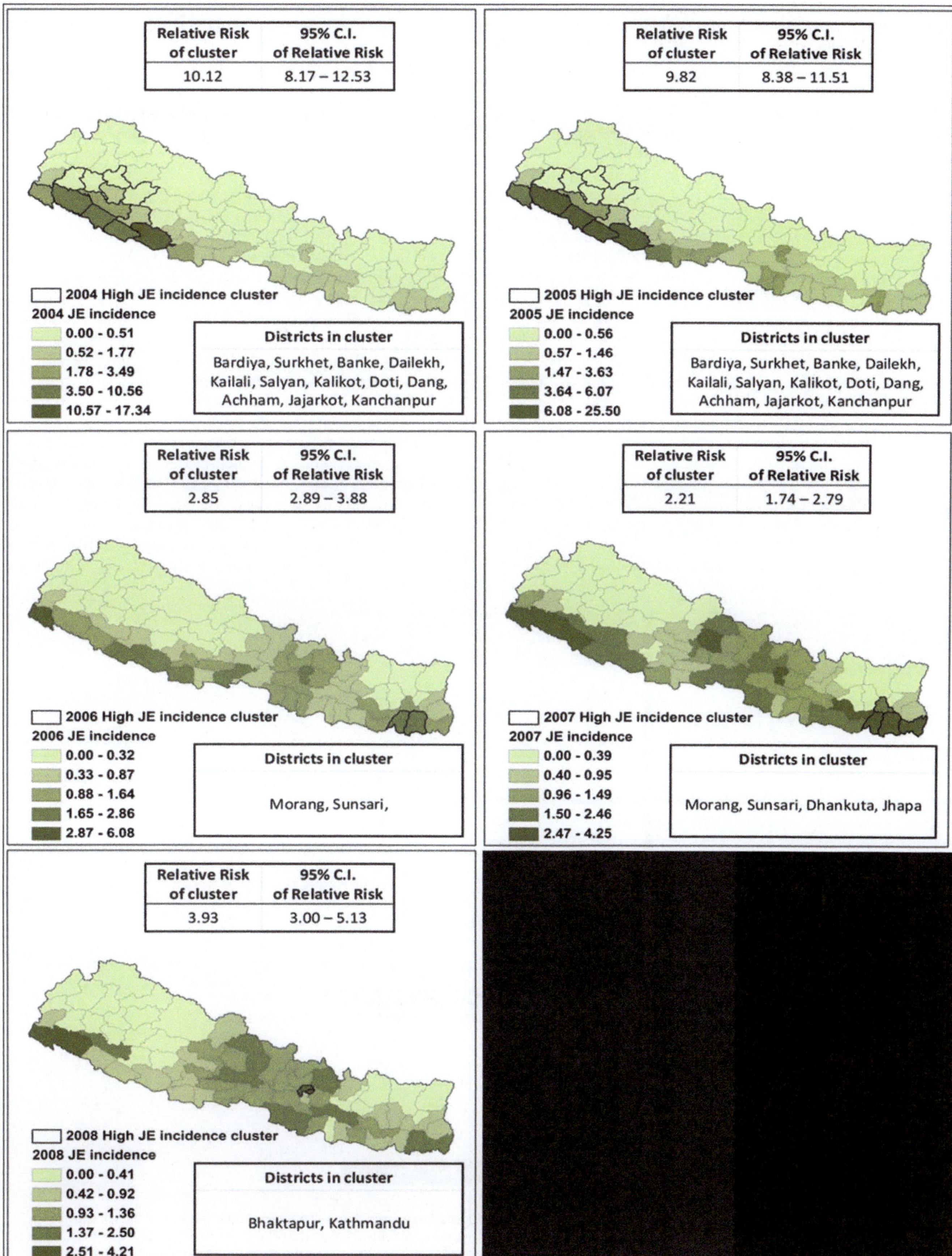
Relative Risk of cluster | 95% C.I. of Relative Risk
10.12 | 8.17 – 12.53
2004 High JE incidence cluster
2004 JE incidence
0.00 - 0.51
0.52 - 1.77
1.78 - 3.49
3.50 - 10.56
10.57 - 17.34
Districts in cluster
Bardiya, Surkhet, Banke, Dailekh, Kailali, Salyan, Kalikot, Doti, Dang, Achham, Jajarkot, Kanchanpur
Relative Risk of cluster | 95% C.I. of Relative Risk
9.82 | 8.38 – 11.51
2005 High JE incidence cluster
2005 JE incidence
0.00 - 0.56
0.57 - 1.46
1.47 - 3.63
3.64 - 6.07
6.08 - 25.50
Districts in cluster
Bardiya, Surkhet, Banke, Dailekh, Kailali, Salyan, Kalikot, Doti, Dang, Achham, Jajarkot, Kanchanpur
Relative Risk of cluster | 95% C.I. of Relative Risk
2.85 | 2.89 – 3.88
2006 High JE incidence cluster
2006 JE incidence
0.00 - 0.32
0.33 - 0.87
0.88 - 1.64
1.65 - 2.86
2.87 - 6.08
Districts in cluster
Morang, Sunsari,
Relative Risk of cluster | 95% C.I. of Relative Risk
2.21 | 1.74 – 2.79
2007 High JE incidence cluster
2007 JE incidence
0.00 - 0.39
0.40 - 0.95
0.96 - 1.49
1.50 - 2.46
2.47 - 4.25
Districts in cluster
Morang, Sunsari, Dhankuta, Jhapa
Relative Risk of cluster | 95% C.I. of Relative Risk
3.93 | 3.00 – 5.13
2008 High JE incidence cluster
2008 JE incidence
0.00 - 0.41
0.42 - 0.92
0.93 - 1.36
1.37 - 2.50
2.51 - 4.21
Districts in cluster
Bhaktapur, Kathmandu

Figure 2. Japanese Encephalitis incidence (2004–2008) in Nepal. Spatial Empirical Bayes smoothed JE incidence expressed as the number of cases per 100,000. Maps of 2004 to 2008 show the dominate district JE cluster, the relative risk of inside the cluster to outside the cluster and 95% confidence using SatScan.

JE clustered mostly in the Far-West and Mid-West terai Regions of Nepal in 2005. In 2008, a large JE cluster was found to exist in the central hill area, primarily the Kathmandu, Lalitpur, Bhaktapur districts (i.e. Kathmandu Valley). It has been suggested that the shifts in JEV transmission distribution may be due to 1) the mass immunization campaigns conducted in Nepal, which have been introduced since 2006 and are being conducted in a phase-wise manner starting with highest priority districts assigned by Nepal MOH or 2) conferred widespread immunity resulting in fewer susceptible people in the terai because of the large JE epidemic in 2005 [21].

However, the discovery that higher elevation locations in the hill areas, in particular the Kathmandu Valley, was found to have moderate levels of JEV transmission is a considerable finding, given that this region was not previously thought to have a significant JE problem [21]. Under conventional wisdom, the altitude (1,500 m) and subsequent lower temperature of the valley would preclude the possibility of transmission due to low likelihood the mosquito vectors could survive in this location and the temperature threshold for transmission of JEV in the mosquito could be achieved. However, studies have found potential mosquito vectors in the hill and mountain areas of Nepal [34,35,36]. Also, JEV transmission has been shown to still occur at temperatures as low as 20°C [11], which is a temperature that can be experienced in the Kathmandu Valley. Though JEV transmission in the Kathmandu Valley has been reported previously [37], detection was limited to only one hospital over a limited timeframe [21]. Further investigation in this area is required to understand the dynamics of transmission in this area to rule any effects of other confounding circumstances such as human movement of infected individuals into the region.

With regards to climate variables, we used the WorldClim dataset which represent climate normals over a 50-year period (1950–2000) rather than weather data at the time of the 2005 outbreak. Wimberly et al found that climate normals provided a better model fit to West Nile incidence than annual weather and weather anomalies [28], suggesting that long-term trends in weather pattern, better predicts subsequent pathogen transmission. For that reason we make the assumption that the predictors of JE incidence used in our analysis are fairly robust and climate shapes the distribution of JE incidence in Nepal. In our study, lagged monthly climate variables better predicted JE incidence rather than annual or concurrent (June–July–Aug) monthly climate variables. Districts with high temperature in the preceding months of the JEV transmission season had higher JE incidence. Similar patterns have been reported from China, Taiwan and

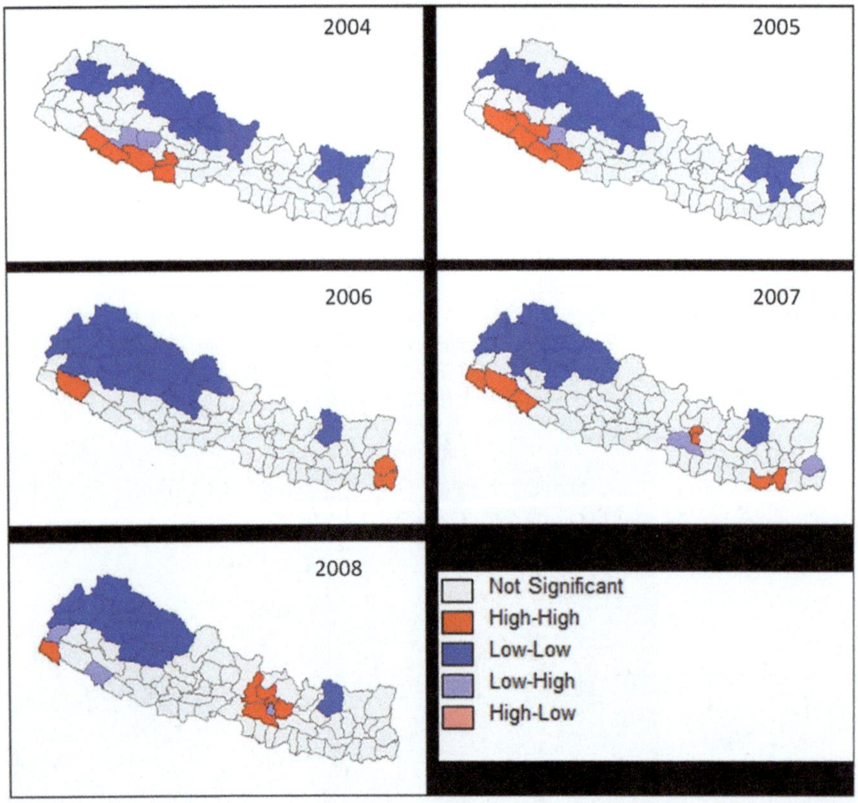

Figure 3. Univariate Local Indicators of Spatial Association (LISA) cluster maps from 2004 to 2008 (A–E) for JE incidence using GeoDa. The cluster map only shows the center of the cluster in color. High-High represents clustering of high JE incidence values, Low-Low represents clustering of low JE incidence values, Low-High represent low values of JE incidence clustered around high values of JE incidence, and High-Low represents high values of JE incidence clustered around low values of JE incidence.

Table 1. Trend surface models of 2005 district-level JE incidence as a function of temperature and precipitation.

	Model β coefficients†					
Variable	**Annual**	**Apr**	**May**	**Jun–Jul–Aug**	**Apr/May**	**May/Apr**
Intercept	0.0528*	0.0489	0.0607*	0.0651**	0.0590*	0.0497
t	0.0128***	0.0101***	0.0131***	0.0132***	0.0128***	0.0109***
t^2	0.0006	0.0005*	0.0008**	0.0002	0.0007**	0.0006*
p	0.0085	−0.0095	−0.0049	0.0036	−0.0062	−0.0066
p^2	−0.0036*	0.0028	0.0004	−0.0004**	0.0005	0.0020
pt	0.0021	−0.0029*	−0.0012*	0.0012*	−0.0012*	−0.0029*
ρ	0.7942***	0.7531***	0.7307***	0.7737***	0.7366***	0.7464***
R²	0.8072	0.8207	0.8170	0.8113	0.8168	0.8213
Adjusted R²	0.7932	0.8077	0.8037	0.7976	0.8035	0.8084
AIC	−83.3025	−90.7539	−90.2047	−85.9352	−89.8923	−91.3228

Apr/May = April temperature/May precipitation, May/Apr = May temperature/April precipitation, t = mean temperature, p = mean precipitation, *Rho* (ρ) = spatial lag parameter.
*statistically significant at α-value 0.1,
**statistically significant at α-value 0.05,
***statistically significant at α-value 0.01.
†All regression β coefficients represent a non-linear increase (or decrease when the coefficient value is negative) in JE incidence when there is a 1-unit increase in each respective predictor variable.

Japan [13,14,15]. The rationale was that high temperature decreased development time for larval and pupal mosquito stages. However, in our study, districts with low precipitation in the preceding months of the JEV transmission season also had higher JE incidence. It has been suggested by Mogi 1983 [13] that low precipitation prevented the loss of mosquito immature stages from high water currents. Similarly, it has been suggested in the US that drought periods induce large outbreaks of West Nile virus transmission [38,39,40,41]. It is thought that drought increases the association between mosquito vectors and birds by reducing the number of water sites available, hence concentrating birds and mosquitoes in one area. Alternatively, during these drought periods, predators of mosquitoes are significantly reduced; hence natural regulation of mosquito populations is significantly impeded. While this is true for WNV it is not clear if the same exists for JEV vectors. In contrast, a study in India found that rice-fields, primary sites for JEV transmission, are relatively stable ecosystems and sudden increases in mosquito populations as a function of climate are uncommon [42]. In our study, we lacked entomological data to complement the JE incidence data and given the somewhat contradictory results from US-Japan and India studies, we are unable to conclude with confidence that the associations with climate identified in Nepal occur via effects on mosquito populations. Still, the results of this study give insights into the relationship between JEV transmission and climate in Nepal.

Percentage irrigated land was a significant variable up until the 2008 parsimonious model and this is possibly because irrigated land provides a habitat for mosquito development and water bird foraging in Nepal. Further studies in Nepal should look more closely at the role played by irrigation as it may permit more targeted distribution of JEV vaccine or allow alternative JEV control mechanisms to be considered, such as reducing human proximity to irrigated land. Surprisingly, pig-to-human ratio did not significantly impact JEV transmission despite other studies showing this to be a crucial component to JEV transmission to humans [43,44]. Furthermore, JE incidence seemed to have a unimodal relationship with pig-to-human ratio. Further study in Nepal would be useful to explain the relationship.

Our model generally did well in predicting JE incidence in Nepal though it underestimated the magnitude of JE incidence in 4 districts. It is still not clear why transmission is particularly high in these areas given we used several explanatory variables in our model. Additional factors, other than those used in this study, may be involved in driving the high level of JE in these districts. When applying our model based on 2005 to the other years (2004, 2006, 2007 and 2008) we see variation in the model fit (i.e. R^2, adjusted R^2 and AIC) and the coefficients. This is to be expected as the association between JE transmission and environmental drivers are likely to have statistical noise due year to year variability in JE incidence. In addition, the impact of the immunization campaign may confound the associations.

Our study had 3 major limitations related to spatial demography, ecological fallacy and spatial-temporal analysis of JE transmission.

- Our study utilized the best available data on reported JE incidence at a national scale. However, it is important to recognize that the dataset is unlikely to have captured all JE occurrences in Nepal [20]. There is likely to be some under representation due to limited access to health care, death before seeking health care, health care–seeking behavior in general or reporting to health facilities not included in the Nepal AES surveillance network. Furthermore we did not factor into our analysis the effects of human movement on JE incidence in each district. However, we have no evidence of any systematic biases in these aspects of JE reporting that are likely to have affected the results of our spatial analysis.
- In this study, we did not take scale into consideration (i.e. district size) which may lead to a modifiable areal unit problem (MAUP) which is a sub-class of ecological fallacy [45,46]. It is possible that if we used a finer areal unit scale in Nepal such as the village development committee (VDCs), we may have obtained different results. For example, the association between JE cases and pig-to-human ratio may have been

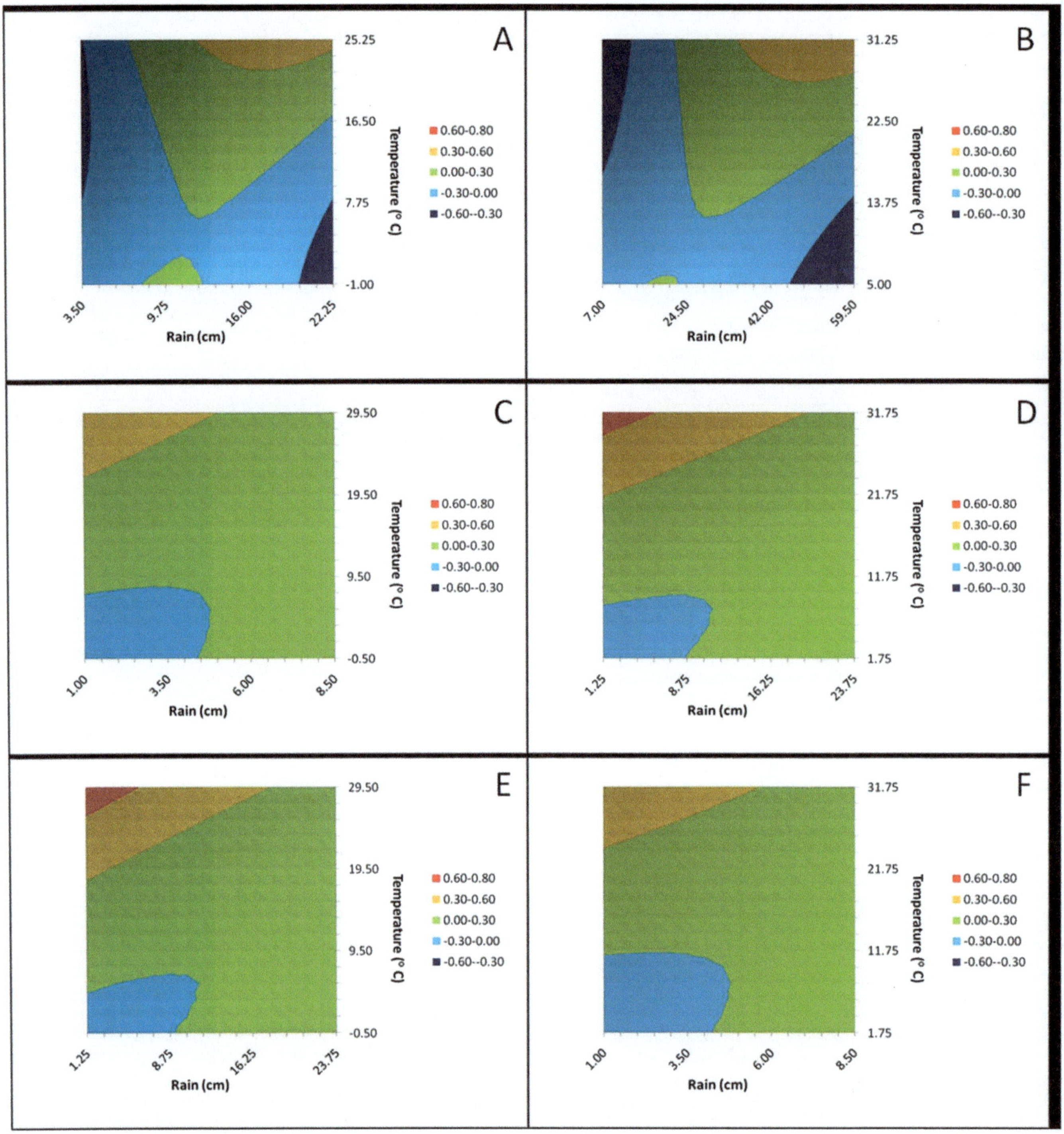

Figure 4. JE incidence as a function of climate variables. The response is the negative reciprocal of JE incidence. A) Annual climate model, B) June–July–August model, C) April climate model, D) May climate model, E) May temperature/April precipitation climate model and, F) April temperature/May precipitation climate model.

found to be significant or the model may have better predicted the occurrence of JE in Nepal, particularly in Bardiya, Banke, Dang and Kailali. Still, our results seem to agree with the findings of other studies [13,38,40,41] suggesting similar relationships exist in Nepal. Nonetheless, future analysis will involve using more robust techniques such as conditional autoregressive models to tease out these associations.

- In our regression analysis, we only focused on 2005 data. This was done since this year was an outbreak period with a high number of JE cases (669 cases) and mass vaccination had not been implemented yet, hence not serving as a confounder. While spatial-temporal analysis would have been more informative, we used this study as an initial assessment into the association between JE incidence and climate and land-

Table 2. Full model of 2005 district-level JE incidence as a function of climate, agriculture and land-use.

Variable	β Coefficient†	S.E.	95% Confidence Interval (β)		P-value
Intercept	0.124	0.068	−0.009	0.258	0.068
May – *t*	0.004	0.007	−0.008	0.017	0.501
May – t^2	0.001	0.001	0.000	0.001	0.137
April – *p*	−0.008	0.015	−0.037	0.021	0.581
April – p^2	0.001	0.004	−0.007	0.010	0.738
pt	−0.004	0.002	−0.007	0.000	0.031
Irrigated	0.013	0.017	−0.020	0.047	0.431
Pig-human	0.586	0.529	−0.450	1.622	0.267
Pig-human²	−7.180	3.962	−14.945	0.585	0.070
Deciduous	0.000	0.036	−0.070	0.070	0.999
Grassland	−0.031	0.033	−0.095	0.033	0.347
Cropland	0.000	0.025	−0.049	0.049	0.996
ρ	0.682	0.088	0.509	0.855	<0.001
R^2	0.830				
Adjusted R^2	0.800				
AIC	−85.762				

t = mean temperature, *p* = mean precipitation, ***Irrigated*** = natural log of percentage of irrigated land per district, ***Pig-human*** = pig-to-human ratio, ***Deciduous*** = natural log of percentage deciduous tree cover per district, ***Grassland*** = natural log of percentage grassland cover per district, ***Cropland*** = natural log of percentage cropland cover per district, ***Rho*** (ρ) = spatial lag parameter.
S.E. = Standard error of â coefficient.
†All regression β coefficients represent a non-linear increase (or decrease when the coefficient value is negative) in JE incidence when there is a 1-unit increase in each respective predictor variable.

Table 3. Parsimonious model of 2005 district-level JE incidence as a function of climate, agriculture and land-use.

Variable	β Coefficient†	S.E.	95% Confidence Interval (β)		Probability
Intercept	0.292	0.084	0.128	0.457	<0.001
April-*p*	−0.032	0.009	−0.051	−0.014	<0.001
Pig-human	0.637	0.516	−0.374	1.648	0.217
Pig-human²	−6.053	3.856	−13.610	1.505	0.117
Irrigated	0.033	0.012	0.010	0.056	0.005
Grassland	−0.033	0.022	−0.077	0.010	0.134
ρ	0.689	0.086	0.520	0.859	<0.001
R^2	0.8122				
Adjusted R^2	0.7986				
AIC	−89.933				

p = mean precipitation, ***Irrigated*** = natural log of percentage of irrigated land per district, ***Pig-human*** = pig-to-human ratio, ***Grassland*** = natural log of percentage grassland cover per district, **Rho** (ρ) = spatial lag parameter.
S.E. = Standard error of â coefficient.
†All regression β coefficients represent a non-linear increase (or decrease when the coefficient value is negative) in JE incidence when there is a 1-unit increase in each respective predictor variable.

Table 4. Comparison of parsimonious 2004 to 2008 models of district-level JE incidence as a function of climate, agriculture and land-use.

	Model β coefficient†				
Variable	**2004**	**2005**	**2006**	**2007**	**2008**
Intercept	0.2943**	0.2924**	0.2026**	0.1533*	0.1037
April – *p*	−0.0361**	−0.0323**	−0.0132	−0.0070	0.0012
Irrigated	0.0223*	0.0331**	0.0302**	0.0281**	0.0181
Grassland	−0.0497*	−0.0332	−0.0341**	−0.0221	−0.0229
Pig-human	0.3403	0.6372	0.5790	0.8329	0.0781
Pig-human²	−2.1978	−6.0526	−5.6163	−8.5296*	−2.0550
ρ	0.7330**	0.6892**	0.6952**	0.7801**	0.7725**
R^2	0.8476	0.8122	0.8415	0.8305	0.7535
Adjusted R^2	0.8366	0.7986	0.8300	0.8182	0.7356
AIC	−105.0710	−89.9330	−123.1960	−103.8150	−97.3233

p = mean precipitation, *Irrigated* = natural log of percentage of irrigated land per district, *Pig-human* = pig-to-human ratio, *Grassland* = natural log of percentage grassland cover per district, ρ = spatial lag parameter.
*statistically significant at α-value 0.05,
**statistically significant at α-value 0.01.
†All regression β coefficients represent a non-linear increase (or decrease when the coefficient value is negative) in JE incidence when there is a 1-unit increase in each respective predictor variable.

use/land-cover. Future study will involve spatial-temporal analysis of JE transmission in relation to climate, environment, vaccine coverage and demography.

In conclusion, April precipitation and percentage irrigated land provided the best explanation for JEV transmission spatial pattern in Nepal in 2005. However, with the introduction of the sequential phased mass immunization campaigns that prioritized highest risk districts, it becomes difficult to tease out the associations separate from the impact of the vaccination campaigns. Nonetheless, percentage of irrigated land remains a statistically significant variable influencing JEV transmission despite vaccination efforts. Therefore, increasing our understanding of the interactions between irrigated land and JEV transmission would be valuable in refining our knowledgebase of underlying factors that drive transmission process, such as bird foraging and predator-prey interactions.

Spatial analysis could be an important tool for decision makers to re-prioritize areas which require attention as disease clusters change due to immunization efforts or other changes in agricultural practices and climate change. Our ecological study provides initial but fundamental information that may be useful to Nepal decision makers at the district level in how they address JEV transmission through vaccine distribution or other alternative methods. Further study is needed to refine the findings of this study.

Supporting Information

Figure S1 Map of Nepal district by development zone and region.

Figure S2 Environmental variables used in full and parsimonious models. A) Mean May temperature, B) Mean April precipitation, C) percentage grassland cover, D) percent area of irrigated land and E) pig-to-human ratio.

Figure S3 Predicted smoothed JE incidence (A) and prediction error (B) for the 2005 dataset using the parsimonious model. The prediction error was calculated as the observed 2005 JE incidence (see Figure 2 in main text) minus the 2005 predicted smoothed JE incidence. The negative values in the prediction error represent overestimation of the model while the positive values represent underestimation of the model. The values were back-transformed from the model using the equation $(-1/(y_i-1))-1$ to retrieve the predicted JE incidence.

Table S1 Pearson's correlation coefficient matrix of variables used in the full model for 2005 JE incidence.

Acknowledgments

We would like to thank the Nepal Ministry of Health and World Health Organization for giving us access to this data and colleagues in the Department of Epidemiology and Population Health, University of Liverpool for comments and suggestions. We also would like to thank 3 anonymous reviewers for their insightful comments. This study was conducted and written under the approval of the Government of Nepal.

Author Contributions

Conceived and designed the experiments: DEI CC MB. Performed the experiments: DEI. Analyzed the data: DEI. Wrote the paper: DEI MB TS WWS CC AR RPB GS. Conducted the laboratory testing and compilation of the data: GS RPB AR. Interpreted the results: DEI MB TS WWS.

References

1. Solomon T (2006) Control of Japanese encephalitis–within our grasp? N Engl J Med 355: 869–871.
2. Umenai T, Krzysko R, Bektimirov TA, Assaad FA (1985) Japanese encephalitis: current worldwide status. Bull World Health Organ 63: 625–631.
3. Center for Disease Control and Prevention (Accessed 2011) Questions and Answers About Japanese Encephalitis. http://www.cdc.gov/ncidod/dvbid/jencephalitis/qa.htm.
4. Lindenbach BD, Rice CM (2001) Flaviviridae: The viruses and their replication. In: Knipe DMHP, ed. Fields Virology. 4th ed. Philadelphia: Lippincott Williams & Wilkins. pp 991–1042.
5. Endy TP, Nisalak A (2002) Japanese Encephalitis Virus: Ecology and Epidemiology. In: Mackenzie JS, Barrett ADT, Deubel V, eds. Japanese Encephalitis and West Nile Viruses. Berlin Heidelberg: Springer-Verlag. pp 11–48.
6. Gingrich JB, Nisalak A, Latendresse JR, Sattabongkot J, Hoke CH, et al. (1992) Japanese encephalitis virus in Bangkok: factors influencing vector infections in three suburban communities. J Med Entomol 29: 436–444.
7. Olson JG, Atmosoedjono S, Lee VH, Ksiazek TG (1983) Correlation between population indices of *Culex tritaeniorhynchus* and *Cx. gelidus* (Diptera: Culicidae) and rainfall in Kapuk, Indonesia. J Med Entomol 20: 108–109.
8. Vythilingam I, Oda K, Mahadevan S, Abdullah G, Thim CS, et al. (1997) Abundance, parity, and Japanese encephalitis virus infection of mosquitoes (Diptera:Culicidae) in Sepang District, Malaysia. J Med Entomol 34: 257–262.
9. Reisen WK, Aslamkhan M, Basio RG (1976) The effects of climatic patterns and agricultural practices on the population dynamics of *Culex tritaeniorhynchus* in Asia. Southeast Asian J Trop Med Public Health. pp 61–71.
10. Murty US, Rao MS, Arunachalam N (2010) The effects of climatic factors on the distribution and abundance of Japanese encephalitis vectors in Kurnool district of Andhra Pradesh, India. J Vector Borne Dis 47: 26–32.
11. Takahashi M (1976) The effects of environmental and physiological conditions of *Culex tritaeniorhynchus* on the pattern of transmission of Japanese encephalitis virus. J Med Entomol 13: 275–284.
12. Gould EA, Higgs S (2009) Impact of climate change and other factors on emerging arbovirus diseases. Trans R Soc Trop Med Hyg 103: 109–121.
13. Mogi M (1983) Relationship between number of human Japanese encephalitis cases and summer meteorological conditions in Nagasaki, Japan. Am J Trop Med Hyg 32: 170–174.
14. Bi P, Zhang Y, Parton KA (2007) Weather variables and Japanese encephalitis in the metropolitan area of Jinan city, China. J Infect 55: 551–556.
15. Hsu SM, Yen AM, Chen TH (2008) The impact of climate on Japanese encephalitis. Epidemiol Infect 136: 980–987.
16. Richards EE, Masuoka P, Brett-Major D, Smith M, Klein TA, et al. (2010) The relationship between mosquito abundance and rice field density in the Republic of Korea. Int J Health Geogr 9: 32.
17. Henrich TJ, Hutchaleelaha S, Jiwariyavej V, Barbazan P, Nitatpattana N, et al. (2003) Geographic dynamics of viral encephalitis in Thailand. Microbes Infect 5: 603–611.
18. Masuoka P, Klein TA, Kim HC, Claborn DM, Achee N, et al. (2010) Modeling the distribution of *Culex tritaeniorhynchus* to predict Japanese encephalitis distribution in the Republic of Korea. Geospat Health 5: 45–57.
19. Pant SD (2009) Epidemiology of Japanese Encephalitis in Nepal. J Nepal Paediatr Soc 29: 35–37.
20. Bhattachan A, Amatya S, Sedai TR, Upreti SR, Partridge J (2009) Japanese encephalitis in hill and mountain districts, Nepal. Emerg Infect Dis 15: 1691–1692.
21. Partridge J, Ghimire P, Sedai T, Bista MB, Banerjee M (2007) Endemic Japanese encephalitis in the Kathmandu valley, Nepal. Am J Trop Med Hyg 77: 1146–1149.
22. Pant GR (2006) A serological survey of pigs, horses, and ducks in Nepal for evidence of infection with Japanese encephalitis virus. Ann N Y Acad Sci 1081: 124–129.
23. Pant GR, Lunt RA, Rootes CL, Daniels PW (2006) Serological evidence for Japanese encephalitis and West Nile viruses in domestic animals of Nepal. Comp Immunol Microbiol Infect Dis 29: 166–175.
24. Wierzba TF, Ghimire P, Malla S, Banerjee MK, Shrestha S, et al. (2008) Laboratory-based Japanese encephalitis surveillance in Nepal and the implications for a national immunization strategy. Am J Trop Med Hyg 78: 1002–1006.
25. Hijmans RJ, Cameron SE, Parra JL, Jones PG, Jarvis A (2005) Very high resolution interpolated climate surfaces for global land areas. Int J Climatol 25: 1967–1978.
26. Ellis AM (2008) Incorporating density dependence into the oviposition preference-offspring performance hypothesis. J Anim Ecol 77: 247–256.
27. Wong DWS, Lee J (2005) Statistical Analysis of Geographic Information with ArcView GIS and ArcGIS ArcView GIS and ArcGIS Hoboken. New Jersey: John Wiley and Sons, Inc.
28. Anselin L Exploring Spatial Data with GeoDa™: A Workbook.
29. Kulldorff M, Huang L, Pickle L, Duczmal L (2006) An elliptic spatial scan statistic. Stat Med 25: 3929–3943.
30. Wimberly MC, Hildreth MB, Boyte SP, Lindquist E, Kightlinger L (2008) Ecological niche of the 2003 West Nile virus epidemic in the northern great plains of the United States. PLoS One 3: e3744.
31. Ward MD, Gleditsch KS (2008) Spatial Regression Model. Thousand OaksCalifornia: Sage Publications, Inc.
32. Marshall RJ (1991) Mapping disease and mortality rates using empirical Bayes estimators. J R Stat Soc Ser C Appl Stat 40: 283–294.
33. Kleinbaum DG, Kupper LL, Muller KE, Nizam A (1998) Applied Regression Analysis and Multivariable Methods Duxbury Press.
34. Darsie RF, Jr. (1997) Description of the pupa of *Aedes pseudotaeniatus* (Giles) (Diptera: Culicidae). J Am Mosq Control Assoc 13: 286–288.
35. Darsie RF, Jr., Courtney GW, Pradhan SP (1996) Notes on the mosquitoes of Nepal. IV. Results of the 1994 collecting in the Midwestern Region, including new country records and voucher confirmation (Diptera, Culicidae). J Am Mosq Control Assoc 12: 130–134.
36. Darsie RF, Jr. (2009) Description of the pupa of the Himalayan mosquito, *Ochlerotatus pulchriventer*. J Am Mosq Control Assoc 25: 248–250.
37. Zimmerman MD, Scott RM, Vaughn DW, Rajbhandari S, Nisalak A, et al. (1997) Short report: an outbreak of Japanese encephalitis in Kathmandu, Nepal. Am J Trop Med Hyg 57: 283–284.
38. Landesman WJ, Allan BF, Langerhans RB, Knight TM, Chase JM (2007) Inter-annual associations between precipitation and human incidence of West Nile virus in the United States. Vector Borne Zoonotic Dis 7: 337–343.
39. Shaman J, Day JF, Komar N (2010) Hydrologic conditions describe West Nile virus risk in Colorado. Int J Environ Res Public Health 7: 494–508.
40. Shaman J, Day JF, Stieglitz M (2005) Drought-induced amplification and epidemic transmission of West Nile virus in southern Florida. J Med Entomol 42: 134–141.
41. Wang G, Minnis RB, Belant JL, Wax CL (2010) Dry weather induces outbreaks of human West Nile virus infections. BMC Infect Dis 10: 38.
42. Das PK, Sivagnaname N, Amalraj DD (2006) Population interactions between *Culex vishnui* mosquitoes and their natural enemies in Pondicherry, India. J Vector Ecol 31: 84–88.
43. Hanna JN, Ritchie SA, Phillips DA, Shield J, Bailey MC, et al. (1996) An outbreak of Japanese encephalitis in the Torres Strait, Australia, 1995. Med J Aust 165: 256–260.
44. Liu W, Gibbons RV, Kari K, Clemens JD, Nisalak A, et al. (2010) Risk factors for Japanese encephalitis: a case-control study. Epidemiol Infect 138: 1292–1297.
45. Openshaw S (1984) The modifiable areal unit problem. Concepts and Techniques in Modern Geography 38: 41.
46. Gregory D, Johnston R, Pratt G, Watts MJ, Whatmore S (2009) The Dictionary of Human Geography. 5th ed. HobokenNJ, , USA: Wiley-Blackwell.

12

A New Tool for Exploring Climate Change Induced Range Shifts of Conifer Species in China

Xiaojun Kou[1,2], Qin Li[3], Carl Beierkuhnlein[4], Yiheng Zhao[2], Shirong Liu[5]*

1 State Key Laboratory of Earth Surface Processes and Resource Ecology, Beijing Normal University, Beijing, PR China, **2** College of Life Sciences, Beijing Normal University, Beijing, PR China, **3** Department of Botany, University of British Columbia, Vancouver, B.C., Canada, **4** Department of Biogeography, University of Bayreuth, Bayreuth, Germany, **5** Institute of Forest Environment and Ecology, Chinese Academy of Forestry, Beijing, PR China

Abstract

It is inevitable that tree species will undergo considerable range shifts in response to anthropogenic induced climate change, even in the near future. Species Distribution Models (SDMs) are valuable tools in exploring general temporal trends and spatial patterns of potential range shifts. Understanding projections to future climate for tree species will facilitate policy making in forestry. Comparative studies for a large number of tree species require the availability of suitable and standardized indices. A crucial limitation when deriving such indices is the threshold problem in defining ranges, which has made interspecies comparison problematic until now. Here we propose a set of threshold-free indices, which measure range explosion (I), overlapping (O), and range center movement in three dimensions (Dx, Dy, Dz), based on fuzzy set theory (Fuzzy Set based Potential Range Shift Index, F-PRS Index). A graphical tool (PRS_Chart) was developed to visualize these indices. This technique was then applied to 46 Pinaceae species that are widely distributed and partly common in China. The spatial patterns of the modeling results were then statistically tested for significance. Results showed that range overlap was generally low; no trends in range size changes and longitudinal movements could be found, but northward and poleward movement trends were highly significant. Although range shifts seemed to exhibit huge interspecies variation, they were very consistent for certain climate change scenarios. Comparing the IPCC scenarios, we found that scenario A1B would lead to a larger extent of range shifts (less overlapping and more latitudinal movement) than the A2 and the B1 scenarios. It is expected that the newly developed standardized indices and the respective graphical tool will facilitate studies on PRS's for other tree species groups that are important in forestry as well, and thus support climate adaptive forest management.

Editor: Keping Ma, Institute of Botany, Chinese Academy of Sciences, China

Funding: This research was carried out with the support of the Special Research Program for Public-Welfare Forestry (Grant No. 201404201) and NSFC's Project (Grant No. 31370486). The funders had no role in study design, data collection and analysis, decision to publish, or preparation of the manuscript.

Competing Interests: The authors have declared that no competing interests exist.

* Email: Liusr@caf.ac.cn

Introduction

The need to predict potential changes in species distribution in response to climate change has stimulated the development of Species Distribution Models (SDMs), or more generally named Ecological Niche Models (ENMs), which combine current species occurrence data with recorded environmental data layers to detect the distribution of and fundamental niche for a given species [1,2,3]. SDMs are reported to have moderate to high accuracy in reproducing current distributions based on training data sets [4]. Although still controversial regarding their transferability to future climates [5,6,7,8], SDMs are the most effective and widely used approach to forecast future ranges of species [2].

In addition to the current progress in theoretical or methodological studies [9,10,11,12,13], SDMs have been increasingly applied in practice-oriented evaluations and assessments for climate adaptive management. This includes the conservation of endangered species [14,15,16], natural reserve networks [3,10,17], biodiversity changes [18,19,20], and climate change driven invasion processes [21]. In ecological or biodiversity studies, there is the trend that a large number of species are modeled to make inferences or test hypotheses. These species are chosen either from eminent taxonomical groups, or from major life forms of some regional species pools [20,22,23].

A systemic exploration of potential range shifts in response to future climates for large numbers of tree species is still not feasible (but see [4,24,25]). Studies on individual tree species range shifts are invaluable for climate adaptive forest management, but systematic studies on large groups of species or at least on a set of key species in a focal area are even more important. Most species are expected to experience range shifts to some extent; forest policy should understand the likelihood of changes by considering the sensitivity of species to a modified climate, the intensity of response, and the future availability of niche space. Identifying potential range changes, or ideally the patterns (spatially or taxonomically) of changes within a region, will guide forest policy in management [25].

Unlike in biodiversity studies, which are based on established indices for diversity and species turnover [23], species range studies lack standardized measures and quantitative indices. Approaches that are comparative, cumulative, or aggregated across species are less developed, however this is exactly what is needed in practice when a large set of species is concerned (as in forestry) and arbitrary preferences must be avoided.

SDM's can be realized for hundreds or thousands of species, but it is difficult to generalize results that subsume a large number of species projections. In addition, the projections are usually not illustrated in traditional binary map format but rather in continuous value format, which is suitable for representing potential habitat suitability or probability of occurrence [1]. Thresholds are adopted to convert these range maps to binary ones [23,26,27,28].

The currently used threshold method makes interspecies comparisons of range shifts problematic. Firstly, how to best choose a threshold is still an unsolved problem. Furthermore, even if some standard species-based threshold selection procedure were widely accepted, the resultant thresholds would still be species specific, making comparisons among different species unreasonable. The development of taxonomical group level threshold selection procedure could be more complicated and is not foreseeable. Thus, a set of standard indices, which are threshold-free, are essential for both the researchers' and policy makers' perspectives. Fuzzy set theory is providing a promising theoretical concept for this purpose [29].

The adaptation of management directives to novel or changing conditions depends on successful communication between researchers and policy makers. Graphical representation is one of the most effective ways of facilitating interdisciplinary communication [30]. In the case of range shift studies, maps are a common method of graphical representation. However, preparing hundreds to thousands of individual maps for policy makers is not practical. Critical characteristics of range shifts and generality cannot be identified and the construction of simplified figures is challenging as well. It is difficult to summarize key patterns of range change and still maintain the information on individual species. Combining range shift indices with intuitive and easy-to-understand diagrams, based on precise data, would support the implementation of knowledge in climate adaptive forest management.

Due to the longevity of trees, forest management and adaptation must be proactively implemented as soon as possible if climatic features are changing. Conifers constitute perhaps the most important taxonomic group of trees in the world as much for their spatial extent as for their role in economy. This is also true for China, where ambitious afforestation programs are aiming at the maintenance and improvement of ecosystem services such as erosion control, water management, and the generation of biotic resources.

Pinaceae are representing the largest extant conifer family, including more than 230 species and with an extensive distribution [31,32,33,34]. Most members of this family are trees that are usually dominant or co-dominant in their ecosystems [35]. The world's largest biome and ecosystem type, the boreal forests, is characterized by Pinaceae trees. Over a large scale, these conifers form the northern tree line and are thus of major importance when discussing temperature-dependent biome shifts. Pinaceae species also dominate mountain forests at lower latitudes in the Northern hemisphere, and are major components of temperate mixed forests [31,33,35]. Needle-leaved trees represent an important Plant Functional Type in global vegetation models [36,37]. As mostly evergreen species (with the exception of the genus *Larix*), conifers are closely linked to climatic conditions throughout the year. Changing climate can affect conifers throughout all seasons.

Forests that are dominated by conifers are of global relevance in terms of economic resources, habitats for many species, and for the global carbon pool [31,33,35]. Thus, Pinaceae constitute one of the most important taxonomical groups in global forest ecosystems and in world forestry. Case studies on the future potential distribution of these trees are urgently needed.

In this study, we first project future ranges of the most abundant Chinese Pinaceae species using SDM's. We then develop and apply a quantitative index and a novel graphical tool to illustrate species potential range shifts. In a next step, we explore generality in the response of the whole group of Pinaceae species to changing climatic features. Finally, we discuss potential implications of this new approach in climate adaptive forest management.

Materials and Methods

Ethics statement

We confirm that no specific permissions were required for these study sites, the location is not privately-owned or protected in any way, and the field studies did not involve endangered or protected species.

Diversity of the Pinaceae family in China

Eastern Asia is one of the most important origins for Pinaceae, and one of the major refuges during the Pleistocene glaciation periods for this family that is concentrated to the Holarctic realm. However, during their evolution, the Pinaceae family has experienced considerabe shifts in latitudinal and altitudinal distribution [38]. Today, China is a hotspot of species diversity for Pinaceae. Among the 11 genera of Pinaceae, the following genera are represented in China: *Abies*, *Keteleeria*, *Nothotsuga*, *Tsuga*, *Pseudolarix*, *Larix*, *Cathaya*, *Pseudotsuga*, *Picea*, and *Pinus*. Only the genus *Cedrus* is not represented. A great variety of Pinaceae species are distributed throughout the Chinese mainland, Taiwan and the Hainan islands, spanning tropical to subarctic climate zones [33,35]. More than half (about 130 of 230) of the world's Pinaceae species occur in China [35].

Forests between the Great Xing'an Mountains and the Himalayas, and between Taiwan and the Altaic mountains, are dominated or co-dominated by Pinaceae species. In particular, four genera (*Abies*, *Picea*, *Larix* and *Pinus*) are common and species-rich in China, and constitute the major coniferous forests. The species of these four genera exhibit broad ranges in both geographical and environmental space. Single species are limited in their distribution areas, however, the taxonomic group is represented in across a wide range of climatic conditions. A clear geographical substitution sequence exists along the temperature gradient. For example, in the genus *Pinus*, *P. latteri*, *P. massoniana*, *P. tabulaeformis*, *P. koraiesis*, and *P. sylvestris* var. *mongolica* range from tropical to subarctic zones and exhibit a geographical substitution pattern [35]. The genus *Larix* is adapted to extremely cold temperatures during winter, as the species of this genus are deciduous and summer-green, avoiding frost damage especially in continental regions. Precipitation gradients are especially important at the margins of distribution, both as limiting factors of tree growth and as drivers of competition with other tree species.

Species and environmental data

Species distribution data. We extracted distribution maps of the species from the digital version of the *Atlas of Chinese Vegetation 1:1,000,000* [39] using ArcInfo Workstation version 8.3 (ESRI, Redlands, CA, USA). Within the four genera under focus, species were chosen that dominated or co-dominated forests based on a minimum threshold of thirty locations of presence. Rare and less abundant species were excluded in this study that is aiming towards ecosystem functions and forest management. A location of presence was defined as a $5' \times 5'$ (geographical coordinates) cell that contained at least 1% forest area dominated or co-dominated by the target species. This treatment ensures that sample points are at least 5′ apart, without missing small patches (above 1% of the cell area) of species distribution within the cell. We modeled the distribution ranges where the species were dominant and disregarded the non-dominant distribution or scattered distribution areas of individual species. Whether a species dominated or co-dominated a forest was based on whether the vegetation type was named after the target species at the unit of 'formation', as defined in the *Vegetation of China* [35].

Based on these criteria, twelve *Abies* species, twelve *Picea* species, six *Larix* species and sixteen *Pinus* species (summing up to 46 tree species) were selected for modeling as quantitatively relevant and representative species out of the total of 130 Chinese Pinaceae species. General information on the prevalence and centroid locations for all species is shown in Figure 1 and listed in Table S1.

Environmental data. We extracted the predictive dataset from BioPlant (download from http://www.zoology.ubc.ca/~qinli/BioKou/), a world plant bioclimatic dataset with a 10′ (latitude/longitude) sample resolution [40] and downscaled it to 5′ using change factor downscaling techniques. The BioPlant dataset calculates 15 layers of variables derived from monthly temperature and precipitation data, which were downscaled from general circulation model (GCM) predictions. Change factor downscaling techniques were adopted to obtain a fine scale dataset [41], which fully considered the effects of elevation on climatic variables.

Climatic data for calculating bioclimatic variables representing the present were downloaded from the WorldClim dataset [42], which has a resolution and interpolating schedule consistent with those used for our future scenarios. The same procedure was applied to calculate the 15 plant biological variables. Aridity was defined as the natural logarithm of the quotient of total potential evapotranspiration to total precipitation in the growing season (monthly temperature greater than 0°C in this case). Because the original BioPlant dataset did not contain this layer, we calculated it using the GRID module in ArcInfo Workstation v8.3. The time period in this study was set to the end of this century, and the future environmental layers were the 20-year means (from 2081 to 2100).

Species Distribution Model (SDM) implementation

There are various methods of creating an ensemble of single models to reduce prediction uncertainties. In this paper, we emphasize General Circulation Model (GCM) induced and predictor selection-caused uncertainties. We used MaxEnt to build the ensembles without taking into account uncertainties related to modeling techniques.

Each ensemble was composed of 25 combinations from five GCM's and five predictor sets. The five GCM's were extracted from twenty-three GCM-derived climatic predictions from the BioPlant data set. The five GCMs were EH5, HAD, IM3, MER and PC1, of which MER and PC1 predicted the highest and lowest temperature rise, EH5 and IM3 predicted the most divergent local patterns of annual precipitation, and HAD predicted intermediate temperature and precipitation [40].

For predictor selection, we choose the most ecologically relevant bioclimatic variables based on general plant distribution theory and our understanding of the possible factors limiting Pinaceae distribution [34,43,44]. The five predictor sets were 1) GDD0, and Aridity; 2) T_mean, P_total, and Aridity; 3) T_cold, GDD0, and Aridity; 4) T_mean, P_total, P_season, TP_syn, and Aridity; and 5) T_cold, GDD0, P_season, TP_syn, and Aridity. In the predictor sets, GDD0, Aridity, T_mean, P_total, T_cold, P_season, TP_syn represent growing degree days above 0°C, aridity of growing season, annual temperature mean, annual total precipitation, mean temperature of the coldest month, precipitation seasonality, and synchronicity between temperature and precipitation, respectively [40].

There are considerations on the spans of predictors: the number of predictors (2, 3, and 5), whether the variables represent annual mean or seasonal extremes (Tmean vs T_cold), temperature or precipitation constraints, and whether seasonal patterns are important (P_season, TP_syn). It is evident that aridity is a major factor limiting Pinaceae distribution, thus this variable is always included in the predictor sets.

Ensembles were built for all species and scenarios (integrating 5 GCMs plus 5 predictor sets for each species). MaxEnt was used to build individual SDM's for the current distribution of each species. Then the predicted future occurrence values based on the different climate models was summarized by the median. This provided the consensus prediction of the ensemble on a cell-by-cell basis [10]. We adopted logistic output format, and the default settings for all model parameters in MaxEnt (version 3.31). Scripts were compiled to perform a batch run with MaxEnt [45,46].

Measurements of model performances

The model performances were measured by the area under the receiver operating characteristic curve (AUC), maximum Kappa value (*Max* κ), and the maximum true skill score (*Max TSS*) for each species. All κ and *TSS* values were calculated with a series of thresholds ranging from 0.05 to 0.95, and having an interval of 0.05. The *Max* κ and *Max TSS* were selected from those values for each species. Arc Macro language (AML) (in ArcInfo Workstation version 8.3) scripts were adopted to implement the calculations. Means and standard deviations of AUC, *Max* κ and *Max TSS* indices were aggregated to the genera level (*Abies, Picea, Larix*, and *Pinus*).

Definitions for the Fuzzy Set based Potential Range Shift (F-PRS) set of indices and graphical representation

Threshold-free range shift indices. To quantify trends in range changes, two indices were used to measure the fundamental aspects of range shifts. The range increment index (I) measures the change in potential range area under a future climate scenario compared with the current climate. A negative value indicates range contraction and a positive value indicates range expansion. The range overlap index (O) measures the proportion of range overlap under a future climate scenario with the range overlap under the current climate. A value of 0 indicates that no overlap exists and a value of 1 indicates complete congruence. The translocation of the species range center is measured by three indices: Dx, Dy, and Dz, with the first two representing horizontal movements along longitudinal and latitudinal directions, and Dz representing the vertical movement.

The definition of a specific species range can be ambiguous when predictions are made in terms of continuous values, such as suitability or probability of occurrence [1]. Arbitrary threshold selection or selecting the 'best threshold' may cause incongruence in what constitutes a range, resulting in deviating predictions [23,26,27,47].A threshold-free range definition based on fuzzy set theory avoids this problem. Fuzzy set theory does not perform a discrete classification among members of a data set. Actually, it numerically estimates how well a member fits into certain categories. Thus, a member can belong to several categories to some degree as measured by its memberships [48].

This study focuses on how many cells could be classified as having a specific species listed as "present." A cell is not defined as

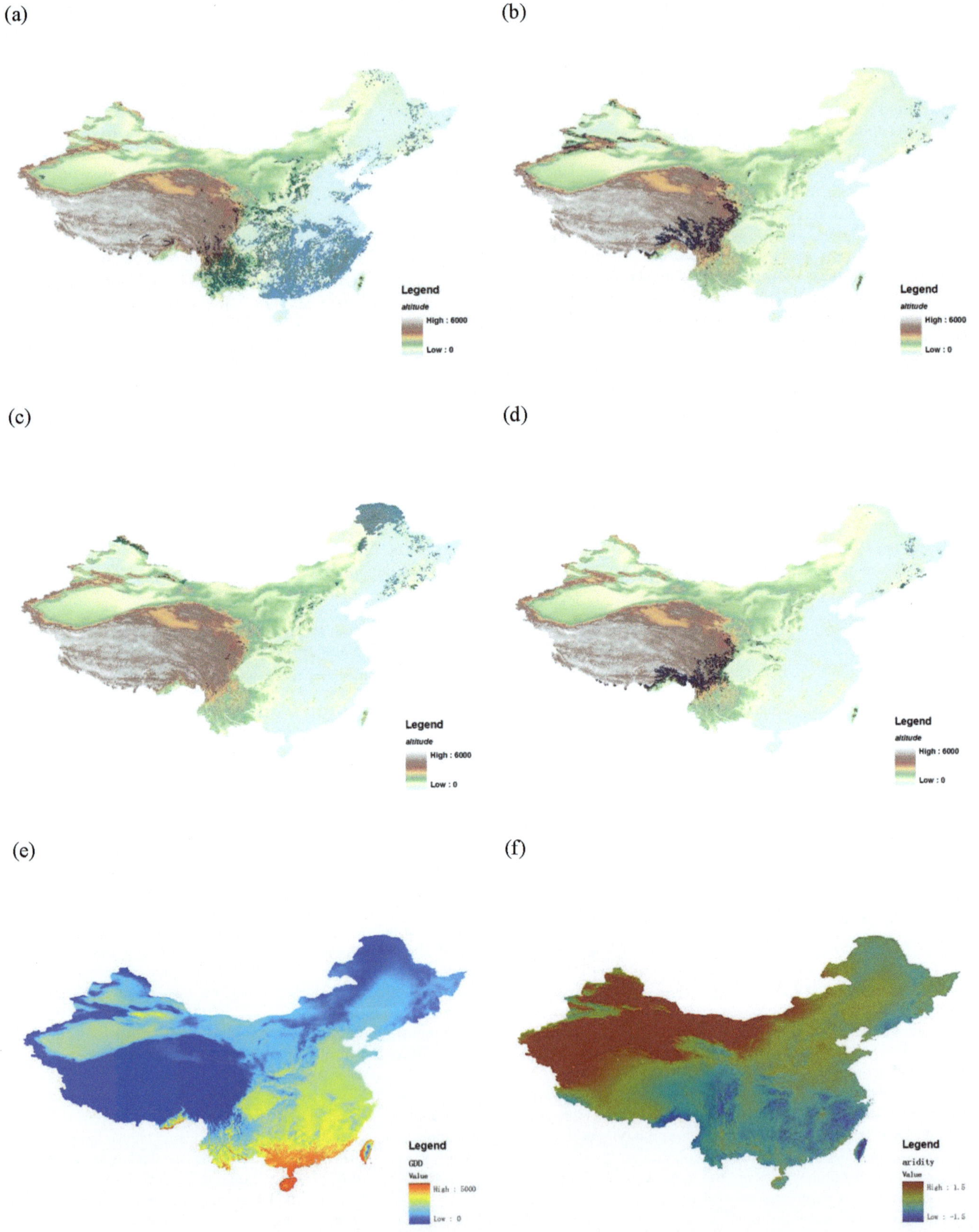

Figure 1. Species distributions and environmental layers. Figure (a–d) are distribution maps of species pooled to genera *Pinus*, *Abies*, *Larix* and *Picea*, respectively, superimposed over background digital elevation model (DEM) maps with 50% transparence. Figure (e) and (f) are maps of the environmental factors gross degree days (GDD) and aridity respectively.

'belonging' or 'not belonging' to the range but rather partially belongs to the range with a degree of membership [8,29]. Membership is a function of suitability or probability of occurrence, respectively. Here, I and O indices based on the membership function were applied [29]. The equations to calculate these indices are:

$$I = (c(F_f) - c(F_p))/c(F_p).$$

$$O = c(F_f \cap F_p)/c(F_f \cup F_p).$$

where, c represents fuzzy set cardinality, F_p represents the fuzzy set of the current distribution, and F_f is the fuzzy set of the future distribution. The indices were calculated using AML scripts in ArcInfo Workstation (Appendix File S1).

The calculated species range movements were based on the movement of the centers of each species range between time periods, where the range center was defined as the geometrical centroid of the entire potential range. To avoid the threshold selection problem, a fuzzy set style centroid was defined, instead of a traditional centroid, which did not depend on partitioning between "present" or "absent". All cells in a grid map that contributed to the centroid were weighted by the membership of those cells that pertain to a specific species. Cells with a membership of zero do not actually contribute to the centroid but can be dealt with in the same way as any other cell.

The three dimensions (x, y, z) of the centroid can be calculated by:

$$Cx = \sum(M_{ij} \times X_{ij}) / \sum X_{ij}$$

$$Cy = \sum(M_{ij} \times Y_{ij}) / \sum Y_{ij}.$$

$$Cz = \sum(M_{ij} \times Z_{ij}) / \sum Z_{ij}.$$

where Cx, Cy, and Cz are the x, y, z coordinates of the centroid. The i and j are the column number and row number of a cell in a grid map. X_{ij}, Y_{ij}, and Z_{ij} are the coordinates of the $cell_{ij}$, and M_{ij} is a binary variable indicating whether a focal species is present or absent in $cell_{ij}$. $\sum$ are the sums of all cells within a map.

Below, Dx, Dy, and Dz represent east/west, north/south, and higher/lower elevation movements, respectively. The range movement indices were defined as:

$$Dx = Cx_{end} - Cx_{start}$$

$$Dy = Cy_{end} - Cy_{start}$$

$$Dz = Cz_{end} - Cz_{start}$$

where the subscripts *end* and *start* were attached to the coordinates of centroids to indicate time periods.

In this study, membership is set equal to the logistic output of MaxEnt. Here, the units of Dx and Dy were angular degrees (longitude and latitude, respectively), and the unit of Dz was meters a.s.l.. All calculations were completed using ArcInfo Workstation version 8.3 (Appendix File S2).

Potential Range Shift chart (PRS_Chart) tool. A diagram tool named Potential Range Shift chart (PRS_Chart) (Figure 2) was developed to illustrate the five indices of Potential Range Shifts (I, O, Dx, Dy, and Dz). A unit circle was used to represent current distribution area for the focal species, and another circle with a radius proportional to the ratio of distribution areas between current and projected future distributions was used to indicate the future distribution area. Thus, I value was represented by $(Area_{black} - Area_{blue})/Area_{black}$. A sector of the unit circle with proportional area equal to the O value was used to demonstrate range overlap.

Dx and Dy combined were represented by a directional arrow line. The direction of the arrow represents the actual movement direction of distribution centers between current and future climates. The length of the arrow tail represents the distance of the movement. In order to depict a larger range of movement distance in the chart, logarithm transformation can be applied. In this study, a base 10 logarithm transformation was used, and the unit length of the arrow represents 1.0 degree (latitude and longitude) of movement. In this manner, a 10 degree movement was drawn 2 units long. Similarly, the distance and direction of vertical movement of the range center (Dz) was represented by a vertical arrow. It is also recommended to conduct a logarithm transformation to movement distance. A base 10 logarithm transformation was used with a unit length representing 100 meters of vertical movement.

An R script file was developed to draw the PRS_Chart (Appendix 2). Range shift charts of all combinations of species and climate scenarios were drawn using this script file (Appendix File S3).

Statistical analyses

The mean and standard deviation of I and O indices were calculated for each genus and for the three climate change scenarios under consideration. The mean and standard deviation were also calculated at the family level for each scenario. Grand means and standard deviations were finally calculated by pooling all species and scenarios.

Double-sided student distributions were used for all means of I, Dx, Dy, Dz values to test if these indices differed significantly from zero ($\alpha = 0.05$). Analysis of Variance (ANOVA) testing was conducted to determine whether the difference of the five indices among scenarios was significant ($\alpha = 0.05$). If any of the ANOVA analyses were significant, the least significant difference t-test (LSD-t) test was conducted for scenario to scenario comparisons ($\alpha = 0.10$). This test was applied to all four genera and to the whole Pinaceae family by pooling the four genera together. All calculations were conducted in Microsoft Visual Fortran (with IMSL Stat Library).

Results

Performances of the SDM modeling

The high AUC values illustrate very good discriminating power of the applied models. The lowest mean AUC value among the four genera, found in the genus *Pinus*, reached 0.975. Standard deviation was generally low (with highest standard deviation of 0.018, also in *Pinus*) (Table 1). The MaxTSS values were slightly lower than those of AUC (with lowest mean MaxTSS of 0.897 in *Pinus*) and were more variable (with highest standard deviation of 0.066 in *Pinus*) (Table 1). *Max* κ, however, indicated only fair performances (with the lowest mean of 0.371 in *Abies*, and highest

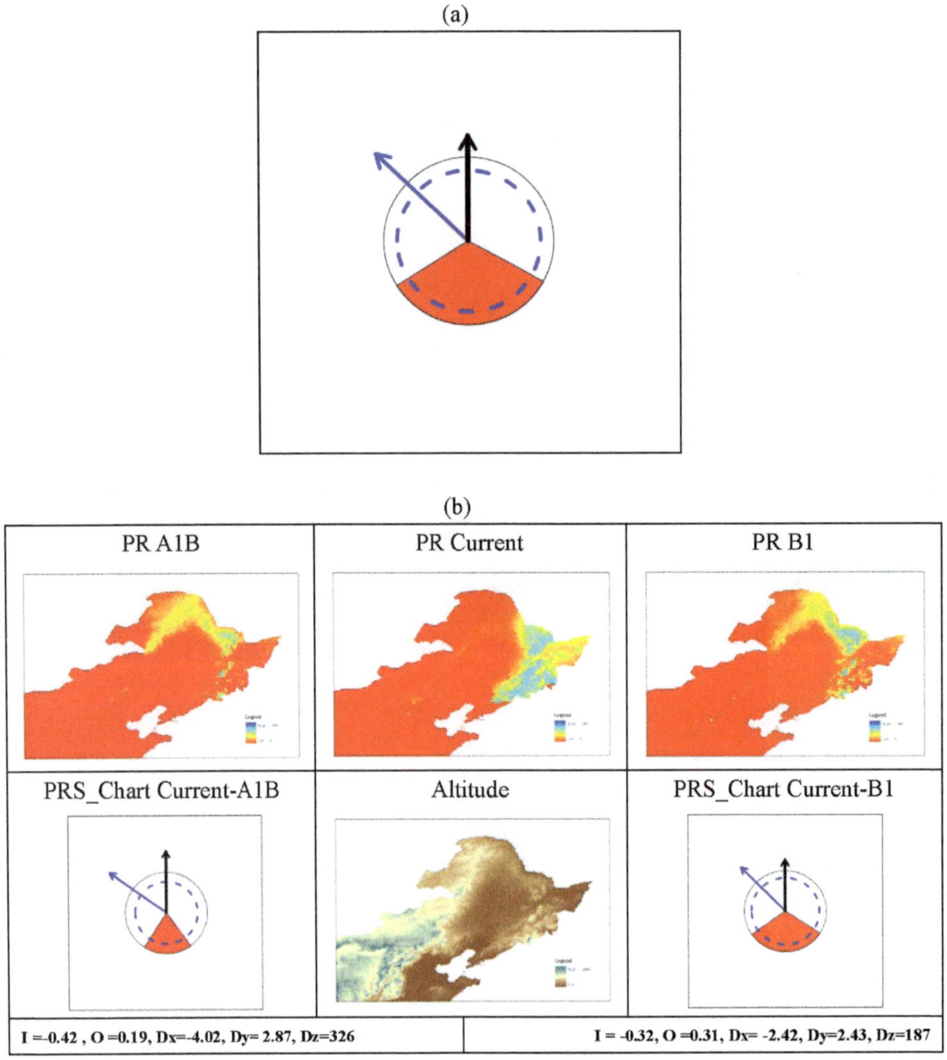

Figure 2. PRS_Chart and an example for *Pinus koraiensis*. Figure (a) is an example potential range shift chart (PRS_Chart) to visualize PRS indices. A unit circle (with thin black continuous line) was used to represent current distribution area for the focal species, and another circle (with broad blue broken line) with radium proportional to the ratio of distribution areas between current and projected future distributions was used to represent future distribution area. A sector of the unit circle (in red) with proportional area equal to the **O** value was used to represent range overlapping. **Dx** and **Dy** combined were represented by a directional arrow line (in blue). The distance and direction of vertical movement of range center (**Dz**) was represented by a vertical arrow (in black). Figure (b) are maps and PRS_Charts that served as an example for spices *Pinus koraiensis*. Maps in first row are potential range maps for climate scenario A1B (in the end of this century), current climate, and B1, respectively. The map in the middle of second row is digital elevation map, and charts in the left and right are PRS_Charts for A1B and A2 respectively. The values listed in bottom row are values of the PRS indices for A1B and B1.

standard deviation of 0.180 in *Pinus*) (Table 1), although it should be noted that the scales for AUC and *MaxTSS* are from 0.5 to 1.0, representing completely random prediction to perfect prediction, while the scale for *Max* κ is from 0.0 to 1.0.

General patterns of Potential Range Shifts (PRS's)

According to the PRS_Charts (Figure 3, 4, 5, 6), the following general traits of spatial patterns can be detected:

(1) Considerable interspecies variation exists in PRS's for each of the five indices;

(2) PRS indices were very consistent among the three climate change scenarios for all species.

(3) No clear inter-genera differences appeared in all five indices.

(4) No clear range expansion or contraction trends could be detected in the I index.

(5) The values of O index were generally low, mostly below 0.5 and peaking around 0.3 to 0.4. The A1B scenario was related to smaller or similar O values compared to those of A2 and B1.

(6) As to horizontal movements (Dx, Dy), there was a general trend for northward spreading, but no clear trend in eastward or northward direction. Thus, the prevalent moving directions were northeast, north, and northwest.

(7) As to vertical movement (Dz), almost all species moved upward with very few exceptions.

(8) Species that moved south were always accompanied by exceptionally strong upward shift (large Dz), indicating a

Table 1. Performance of SDM modeling measured by model evaluation indices.

	Abies	*Picea*	*Larix*	*Pinus*
AUC	0.984 (0.009)	0.983 (0.012)	0.982 (0.005)	0.975 (0.018)
Max TSS	0.935 (0.036)	0.913 (0.053)	0.920 (0.029)	0.897(0.066)
Max κ	0.371 (0.096)	0.375 (0.140)	0.372 (0.150)	0.380(0.180)

Notes: Table content is means and standard deviations of model evaluation indices of species distribution model (SDM) aggregated from all combinations species within the genra and climate scenarios (A1B, A2, and B1). **AUC** is area under the receiver operating characteristic curve. **Max TSS** is maximum true skill score. **Max κ** is maximum Kappa value.

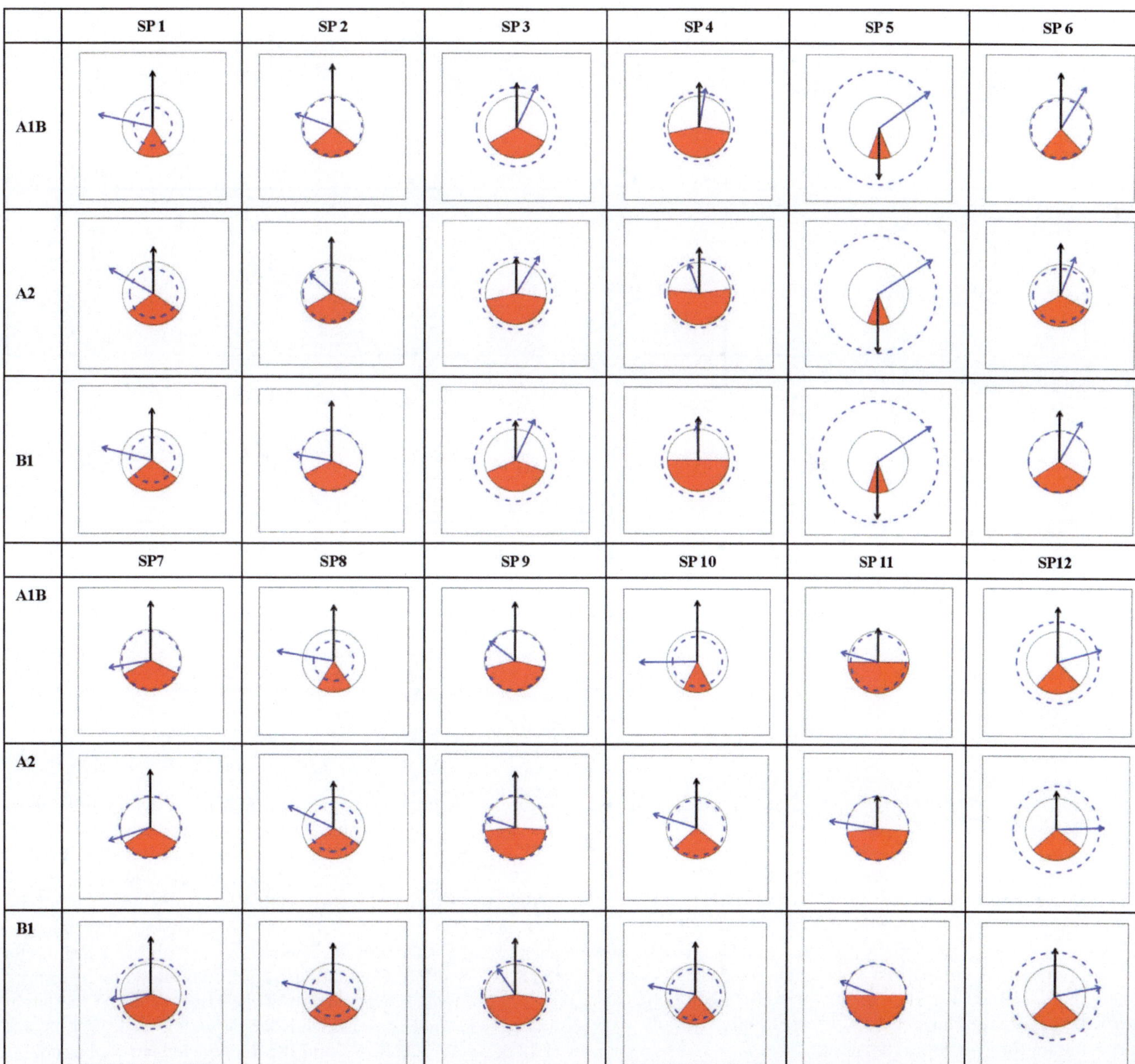

Figure 3. PRS_Chart for *Abies*. Each chart is a Potential range shift chart (PRS_Chart) that is explained in Figure 2. Charts in the tabular style figure representing combinations of species (SP1-12) in *Abies* and climate scenarios (A1B, A2, and B1). General species information can be found in Table S1.

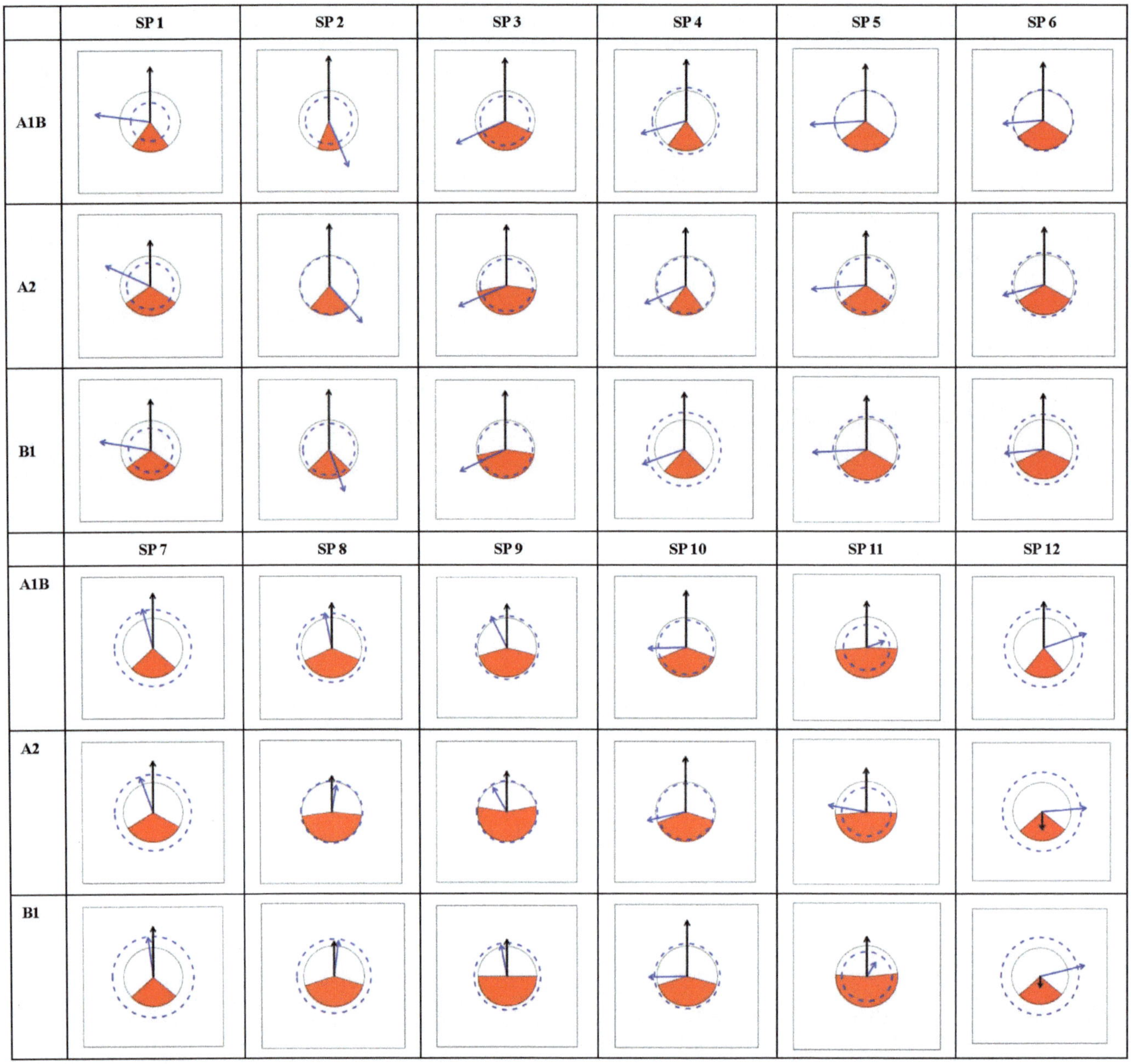

Figure 4. PRS_Chart for *Picea*. The same as legend of Figure 3 except the species number is for *Picea*.

compensatory effect between upward and northward movement trends.

Quantitative summarization of Potential Range Shifts

The mean range size changes (I index in table 2) for Pinaceae species were relatively small (0.09, 0.12, and 0.22 for A1B, A2, and B1 respectively) in comparison with large standard deviations (0.59, 0.59, and 0.63 for A1B, A2, and B1). No significant ($\alpha = 0.05$) range size changes were detected both at family level and genus level for all three climate scenarios.

The range overlapping (O index in Table 2) showed low values (with means 0.29, 0.36 and 0.35 for A1B, A2, and B1) and relatively small interspecific variations (with standard deviations 0.14, 0.13 and 0.13 for A1B, A2, and B1) at family level and a similar pattern at the genus level.

The longitudinal location of the range center (Dx index in Table 2) was found to move slightly westward (with mean −1.15, −0.60 and −0.70 degrees for A1B, A2, and B1) but this trend was not statistically significant and was connected with large interspecific variation (standard deviation 4.12, 3.52, and 3.72 degrees for A1B, A2, and B1). Comparable results were found for all genera.

The latitudinal displacement of the range center at the family level (Dy in Table 2) was slightly larger (with mean1.27, 1.03, and 1.13 degrees for A1B, A2, and B1) than the longitudinal trend, but had much smaller interspecific variations (with standard deviation 2.18, 1.86, and 1.84 degrees for A1B, A2, and B1). The northward movement was highly significant ($\alpha = 0.01$) at the family level. Comparable trends were found in *Pinus* and *Abies* but did not

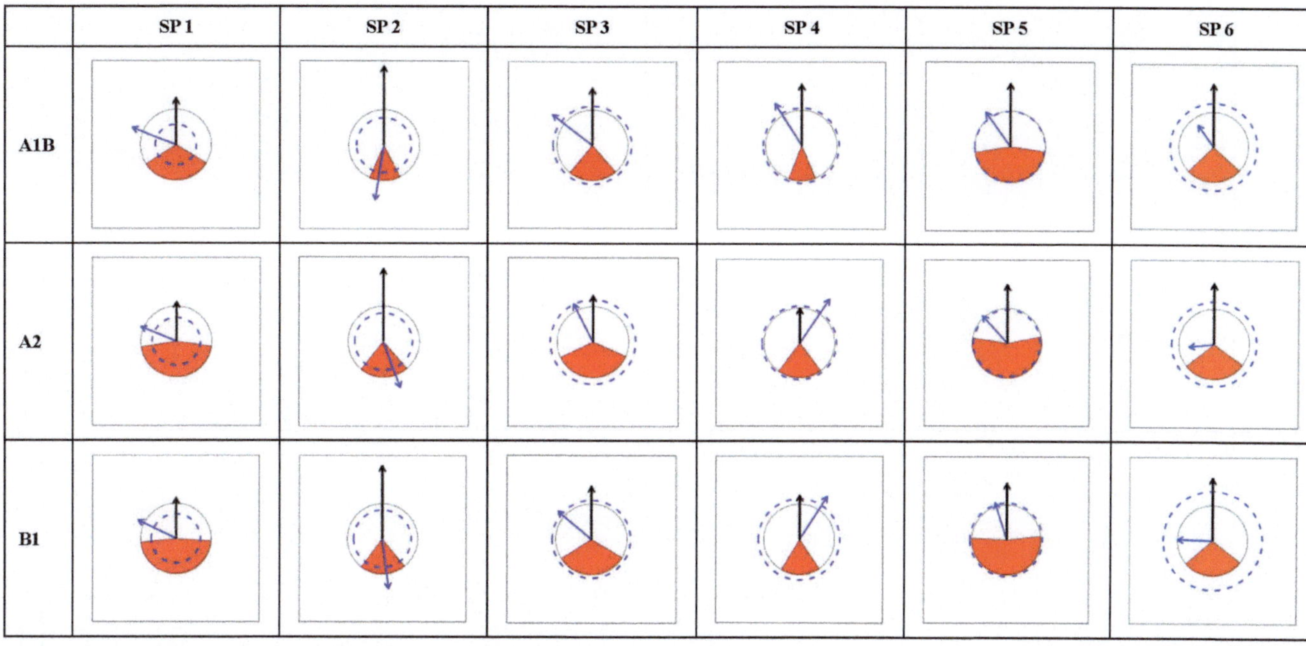

Figure 5. PRS_Chart for *Larix*. The same as legend of Figure 3 except the species number is for *Larix*.

emerge in *Picea* and *Larix*. This indicates that the northward trend at the family level can be attributed mainly to pine and fir species.

The vertical movement (Dz in Table 2) trends were highly significant at the family level and were also highly significant in *Pinus* and *Picea*, but the significance of *Abies* and *Larix* depends on climate scenarios. Mean movement distances of the family were 727, 339, and 364 meters, with standard deviations 392, 345, and 342 meters for A1B, A2, and B1, respectively.

Inter-scenario comparisons

The ANOVA (Table 3) indicated no significant difference ($\alpha = 0.05$) among scenarios in horizontal directions (Dx and Dy) both at the genus and at the family level. There was also no significance among scenario differences in range size changes (O index) at family and genus level with the exception of *Picea*, which responded to scenario B1 with a significantly ($\alpha = 0.10$ by LSD-t test) larger range expansion compared to A1B or A2.

Generally, there were highly significant differences among scenarios ($\alpha = 0.10$ by ANOVA in Table 3) in range overlap (O index) and vertical movement (Dz) at family level. O values in A1B scenario were found to be significantly smaller than those in A2 and B1, and Dz values in A1B were significantly larger than those in A2 and B1. A similar pattern was also found at the genus level but with a few cases of ANOVA results being significant instead of highly significant. An exception was the genus *Larix*, which did not differ in vertical movement among scenarios. Consistent lower O values and larger Dz values indicated that a realization of scenario A1B would lead to the strongest range shifts.

Discussion

Performance of the novel set of indices for tree species range shift

Modeling results on the future potential optimum distribution of tree species are urgently needed for forest ecosystem management and silviculture in order to avoid the loss of resources and ecosystem services. Climate change projections are still vague to a certain degree mainly due to uncertainties in global economic development, mitigation policy, and technological progress. However, general trends emerge especially when scenarios are compared.

Here, a new quantitative index is presented as a tool for forestry, ecology, and biogeography that circumvents restrictions and disadvantages of former species distribution modeling approaches according to their focus on selected species, scenarios and directions, and intensity of shift. This research contributes to the rapid development and progress of biogeographical modeling in response to the ongoing global changes [49,50,51,52,53,54]. In this study, models are based on data for grid cells where the species were common or even dominant in order to avoid bias from small species populations on special sites with extraordinary microclimatic conditions that were not representative for the climate at the landscape scale of the modeled grid cells. Distributional data of this type did not differ substantially from total species ranges inferred from other map sources [39,55].

As a novel and robust approach, the Fuzzy Set based Potential Range Shift Index (F-PRS Index) is introduced in this study. Here, the application is focused on dominant, widespread, and economically important conifer species of the Pinaceae family in China. However, the method can be implemented to other species groups and regions in the future.

The F-PRS Index: (1) helps detect patterns that hide behind large amounts of information, and thus deepens understanding of the studied phenomena; (2) delivers a powerful tool for the building of statistically testable hypotheses; and (3) supports the efficient implementation of meta-analyses. Here, the novel F-PRS Set of Indices performance is demonstrated according to the first two aspects.

One important piece of information on the upward movement trends becomes evident by the Dz index, which integrates three levels of spatial information (current distribution, future distribution, and altitude). It is expected that this will contribute to the

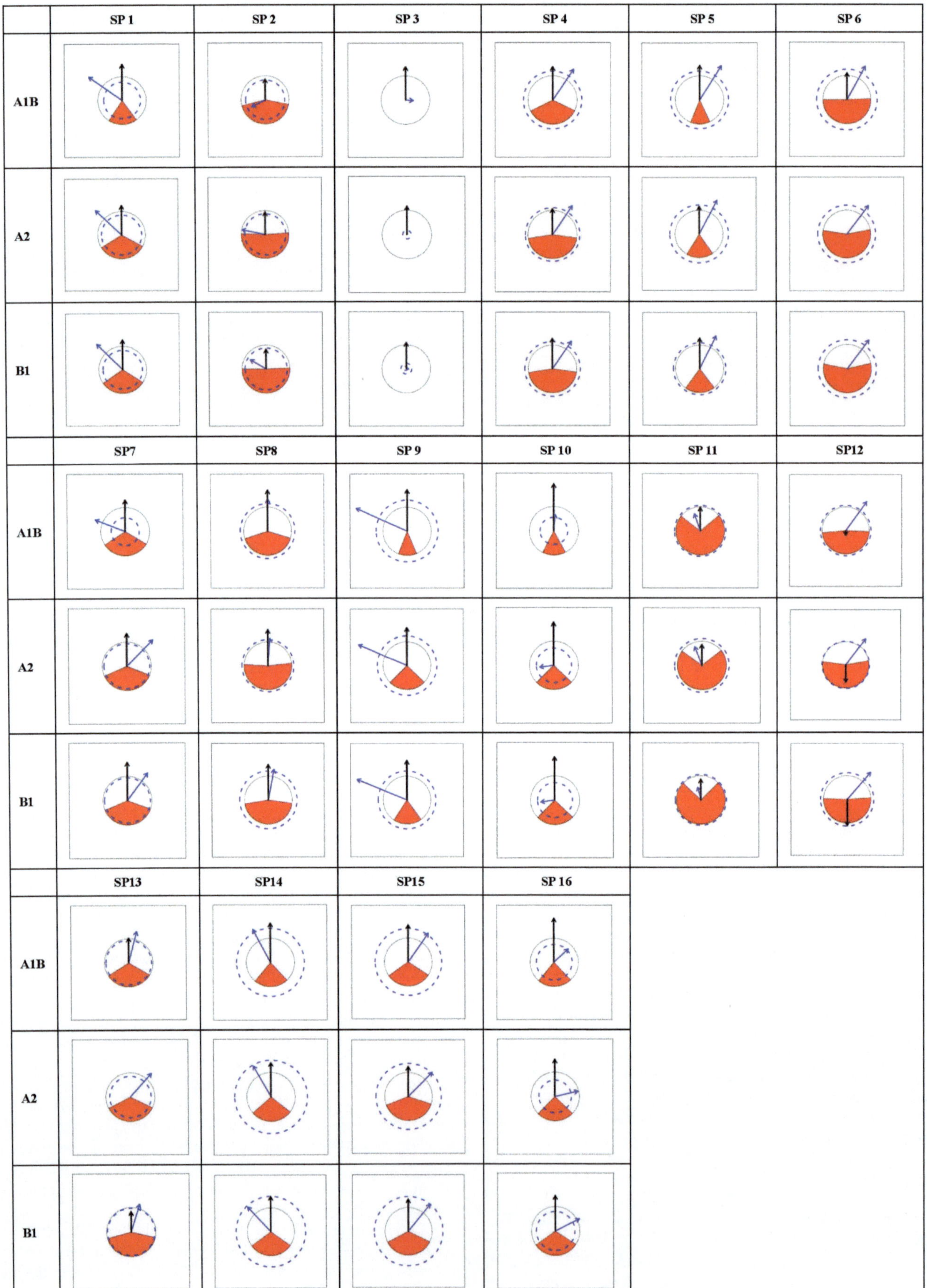
SP 1
SP 2
SP 3
SP 4
SP 5
SP 6
A1B
A2
B1
SP7
SP8
SP 9
SP 10
SP 11
SP12
A1B
A2
B1
SP13
SP14
SP15
SP 16
A1B
A2
B1

Figure 6. PRS_Chart for *Pinus*. The same as legend of Figure 3 except the species number is for *Pinus*.

third F-PRS Index aspect, and delivers an excellent basis for meta-analyses after wider applications.

General requirements for indices are: (1) The ability to extract key attributes that are essential to the studied phenomena out of complex interactions and processes; (2) The capacity to reach standardization and comparability among studies; (3) The stability against random perturbations; and (4) The independence of the members of the indices set.

The F-PRS indices set introduced in this study definitively fulfills the first criterion because the five aspects represented by the five member indices are all critical attributes for range studies. Species range size change is one of most concerning topics in climate change biology [56,57]. Range size conservation, and range size being identified as species level trait to test species level selection have instigated a large number studies in evolutionary ecology [58,59]. Range overlapping that measures the ratio of area shared by the current range and future range has a profound application in conservation biology [60]. Range position change is a hot topic in biogeography.

The drivers of northward and upward movements for species in response to climate warming are intensely studied both empirically and theoretically [61,62,63,64,65,66]. The temperature related "altitude for latitude" movement hypothesis is intensely debated in biogeography [67]. However, longitudinal movement, for instance in response to changes in precipitation regimes, is a little touched topic due to a lack of methodological concepts. Here, a tool that enables the detection of anisotropy in temporal trends in all directions is presented.

One preeminent advantage of the F-PRS indices set is that they are standardized and can be adopted to make comparisons among different studies. Although it requires that the SDM's predicted suitability maps are consistent, which is not granted in all cases, it can at least be used to make comparisons among studies using the same SDM. Ensemble models that integrate various species distribution modeling algorithms such as BIOMOD can serve as a standard pre-analysis in order to avoid deviations between specific SDM's [13].

The fuzzy set style indices presented here are more stable than former threshold defined measurements of range shifts [29]. Two sources of mechanisms may produce unstable results for threshold measurements. The first one is that small perturbations (such as random processes during SDM modeling and different strategies in selecting pseudo absent data) may cause large differences in selecting "best" threshold values, thus resulting in deviating range maps. The second source of uncertainty in traditional approaches comes from the fact that even small perturbations in threshold can cause large differences in range maps in case that the predicted suitability map contains a large area of very low values. The fuzzy set indices are certainly free from the first source of instability, because they are threshold free, and performed much better in dealing with the second source [29].

Lastly, it is evident that all the five indices in the F-PRS indices set have clearly differentiated meanings. Thus, we believe the new F-PRS indices set is good based on the above criterions.

Applicability in range change studies and forest managements

In practice, it is critical to achieve mutual understanding between researchers and policy makers via readily available communication techniques. Visualization is one of the most emphasized strategies to facilitate communication, the exchange of information, and its implementation [30]. In this case study, a graphical representation of range shifts (Figure 3, 4, 5, 6) is introduced, which will contribute to an efficient climate adapted management policy for coniferous forest in China. The effects that climate change is likely to bring about to Chinese forests can be instantly recognized. Consequences for climate adapted management becomes obvious.

Visualization is not only beneficial for forest management but also for research. Although it seems that all the information presented in the graphs has already been contained in the calculated indices, all the spatial information that is emerging in the patterns may not be revealed from numerical indices. Graphs are more appropriate than single values or numbers to demonstrate large-scale spatial patterns, which is exemplified by the eight general patterns identified in this study (section3.2). Based on these spatial findings, more specific hypotheses can be raised to be tested in the future.

Unlike theoretical studies, which usually emphasize summarizing general patterns and further reveal underlying mechanisms that form a large amount of case studies [68,69,70,71], management sectors such as forestry and conservation biology need to keep balance between generality and detail [72,73]. At one end, overall policy to maintain forest ecosystems is highly emphasized, at the other end, not a single species can be neglected in forestry or nature conservation. As the former has been fully emphasized, it is necessary to draw some attention on details, especially a few exceptional cases that can be clearly seen from the graphs (Figure 3, 4, 5, 6). For example, AbiesT05 (*Abies spectabilis*) had exceptionally large range expansion (several times the current range) and showed a strong downward movement trend in each scenario. PinusT03 (*Pinus pumila*) seemed to lose almost its complete potential range in each of the scenarios. PinusT11 (*Pinus massoniana*) had very large overlap between scenarios and little movement, indicating a stable range position. *Picea* speciesT11 (*Picea morrisonicola*) had very similar horizontal movement trends between scenarios A1B and B1, but was very different in A2, and PiceaT12 (*Picea spinulosa*) displayed large upward movement in A1B but moved slightly downward in A2 and B1. Apart from the fact that knowing the reasons that brought about those exceptions might have theoretic significance, knowing the facts themselves also has conservation and management importance.

Assisted migration of tree species in climate adaptive forest management

The results of this study show that for various species with considerable range displacement, assisted migration becomes more or less inevitable in order to maintain the functioning of ecosystems and to preserve the biological diversity of tree species. Intended translocation of endangered species has been proposed to avoid climate change caused extinction [74]. Immediate concerns about unintended side effects were expressed [75,76] and debates were aroused [77,78,79,80,81].

Here, extending the idea of assisted migration and endangered species translocation in biological conservation is proposed for the rearrangement of a larger amount of, as well rare, economically important tree species. However, the design of "novel ecosystems" is currently under debate and ecological and practical consequences are not solved, yet. Based on this case study on Pinaceae species in China, there are only around one third of overlapping

Table 2. Statistics of potential range shift indices.

Shift Indices	Emission Scenarios	*Abies*		*Picea*		*Larix*		*Pinus*		Family pooled	
		Mean	**Std**	**Mean**	**Std**	**Mean**	**Std**	**Mean**	**Std**	**Mean**	**Std**
I	**A1B**	0.17	0.82	0.02	0.42	−0.03	0.45	0.09	0.58	0.09	0.59
	A2	0.26	0.85	0.03	0.37	0.00	0.40	0.11	0.57	0.12	0.59
	B1	0.37	0.91	0.23	0.45	0.13	0.53	0.15	0.56	0.22	0.63
O	**A1B**	0.29	0.12	0.30	0.10	0.25	0.12	0.29	0.18	0.29	0.14
	A2	0.35	0.11	0.36	0.11	0.35	0.14	0.38	0.17	0.36	0.13
	B1	0.34	0.11	0.36	0.10	0.33	0.14	0.37	0.16	0.35	0.13
Dx (Degree)	**A1B**	−0.97	4.53	−1.87	3.14	−0.19	0.71	−0.72	5.28	−1.15	4.12
	A2	−0.17	4.14	−1.51	2.84	−0.08	1.14	−0.44	4.15	−0.60	3.52
	B1	−0.30	4.11	−1.41	2.80	−0.36	1.14	−0.61	4.72	−0.70	3.72
Dy (Degree)	**A1B**	1.56*	1.85	−0.13	1.82	0.42	2.09	2.42**	2.14	1.27**	2.18
	A2	1.45*	1.94	−0.25	1.52	0.52	1.64	1.87**	1.65	1.03**	1.86
	B1	1.51*	1.83	0.00	1.52	0.46	1.71	1.94**	1.72	1.13**	1.84
Dz (Meters)	**A1B**	526**	424	720**	380	669*	510	329**	238	527**	392
	A2	295*	408	542**	364	419*	379	188**	168	339**	345
	B1	357*	412	528**	351	472*	398	205**	173	364**	342

Notes: Table contents are mean(**mean**) and standard deviation (**Std**) of potential range shift (PRS) indices representing by combinations of **I**, **O**, **Dx**, **Dy**, **Dz** and CO_2 emission scenarios **A1B**, **A2**, and **B1** shown in first two columns. Superscripts ** after mean values indicating that the mean is very significant from zero ($\alpha = 0.01$), and * is significant different from zero ($\alpha = 0.05$) by using double sided student distribution. No superscript after mean value means it is not significant from zero, except the mean of O index which did not test for deviation from zero.

Table 3. Comparisons of Potential range shift (PRS) indices among climate scenarios.

Shift Indices	Tested Item	*Abies*	*Picea*	*Larix*	*Pinus*	Family pooled
I	ANOVA	X	P = 0.032	X	X	X
	A1B	/	0.02[a]	/	/	/
	A2	/	0.03[a]	/	/	/
	B1	/	0.23[b]	/	/	/
O	ANOVA	P = 0.034	P = 0.007	P = 0.018	P = 0.010	P = 0.000
	A1B	0.29[a]	0.30[a]	0.25[a]	0.29[a]	0.29[a]
	A2	0.35[b]	0.36[b]	0.35[b]	0.38[b]	0.36[b]
	B1	0.34[b]	0.36[b]	0.33[b]	0.37[b]	0.35[b]
Dx	ANOVA	X	X	X	X	X
	A1B	/	/	/	/	/
	A2	/	/	/	/	/
	B1	/	/	/	/	/
Dy	ANOVA	X	X	X	X	X
	A1B	/	/	/	/	/
	A2	/	/	/	/	/
	B1	/	/	/	/	/
Dz	ANOVA	P = 0.030	P = 0.026	X	P = 0.000	P = 0.000
	A1B	526[a]	720[a]	/	329[a]	527[a]
	A2	295[b]	542[b]	/	188[b]	339[b]
	B1	357[b]	528[b]	/	205[b]	364[b]

otes: The table contents after row **ANOVA** are results of analysis of variance for each genus comparing among climate scenarios, in which X representing not significant ($\alpha = 0.05$) or P-values is provided significant. Contents after row **A1B**, **A2**, and **B1** are results of multiple comparisons if the ANOVA result is significant by using least significant difference t test (LSD-t). the/means test is not carried out. Number and superscript represent mean value and multiple comparison result, with same superscript characters meaning not significant and different character meaning significant ($\alpha = 0.10$).

areas that are both suitable for the species at current climate and under climatic conditions in the end of this century, not accounting for species adaptive ability and other sources of uncertainty.

The huge overall changes in suitable ranges require that foresters track the potential ranges of most of the Pinaceae species, otherwise a tremendous decline of coniferous forest cannot be excluded. It is unfeasible and unrealistic to let natural processes (i.e. nature dispersal or local adaptation) fulfill this task for two reasons: firstly, current speed of anthropogenic caused climate change is much faster than natural climate variability in history [82], thus this will leave no chance for tree species to adapt in a short time period [83]. Secondly, the high fragmentation of natural habitats segregated by agricultural land uses, urban areas and road systems is suppressing dispersal and colonization of new habitats [84,85,86]. More broadly, forest management needs to identify and prepare in advance habitats for the establishment of larger numbers of the tree species close to but perhaps outside their previous "native" range, or even novel forest communities because they were formerly considered to be potentially natural communities, not representing future climate anymore [87].

Fundamental changes in forestry will become unavoidable. Although there are many risks connected to the introduction of species to new localities outside of their previous distribution, such as biological invasion or unwanted species interactions, keeping all species in their current locations will not be unfeasible either. In this study it isshown that about two-thirds of the coniferous forests in China may experience species loss and become restricted in their functioning.

We support the point made by Thomas [81] that the question is not on whether or not to make changes, but on how to manage forests in a way that promotes advantages and avoids disadvantages. This involves a broad and deep understanding of the processes and the functioning of ecosystems. Thus, the idea of climate adaptive management requires not only best practice and reliable projections for species range shifts, but also the integration of knowledge of all basic branches of biological science, such as ecology, biogeography, genetics and evolution.

Limits and perspectives

Species distribution models and climate envelope modeling features some intrinsic limitations, such the disregard of evolutionary processes and adaptation, or the ignorance of within-species genotypic and ecological variability. Experimental research must be designed to specifically address these issues that cannot be tackled by the actual nature of species distribution models. However, in the case of key tree species, rapid directives are needed in order to adapt species composition and structure of forests at an early stage to the environmental conditions that are expected in the near future during the life-time of the planted saplings. Evolutionary processes are irrelevant at this time scale. However, it is important to know genetic and ecological variability, especially for tree species with a large spatial distribution [88].

The introduced set of indices delivers a powerful and practical approach for adapting the Chinese forests to future developments that are forecasted in climate scenarios and global climate models. Uncertainty according to realized scenarios and climate model output must be considered in forest management [89]. Especially the role of climatic extremes that are expected to become more and more frequent and intense (IPCC 2012) is an issue that is important in forest practice but not satisfactory in its reflection in climate models. Climatic extremes are much more important for long-lived species such as trees than long-term trends in average values [90].

In consequence, flexible approaches are asked for which allow management to be adapted for the developing state of knowledge, to further develop societal and emission scenarios, and to improved climatic projections. The introduced set of indices, which can be applied for larger groups of tree taxa, offers these options and will allow the implementation of new findings, in order to identify their relevance in forestry.

In perspective, it is not only the application of such models but also the assignment and upscaling of species distribution models to global vegetation models is required. Key species for important biomes, such as the members of the Pinaceae family, are neglected in the current state-of-the-art approaches for global vegetation models that are the basis for the calculation of future carbon fluxes and budgets. The introduced approach in this study brings these modelling philosophies one step closer and may contribute to make vegetation models that integrate plant functional types such as needle-leaved trees more reliable.

Conclusions

As a consequence of the long-term life cycles of trees, there is an urgent need in forestry for reliable forecasts on future ranges of tree species. Species Distribution Models are providing such projections but at continental scales, many species must be taken into account. The Pinaceae family represents a high number of ecologically and economically important tree species in China. For this reason, this family was chosen to develop and implement a new set of indices that enable the detection of directions and intensity of range shifts at the species, genus and family level.

This new threshold free set of indices for tree range shifts features several advantages compared to traditional approaches. Besides its capacity to integrate across a large number of species and to identify general patterns of range shifts for large species groups, the PRS_Chart graphical tool offers an intuitive visualization tool for the science-policy interface which will support the translation of research into practice.

Particularly, the predicted low range overlap between current and future tree species ranges combined with a significant northward as well as upward movement trend for the whole group of species implies that assisted migration can become increasingly important in climate adaptive forest management.

Supporting Information

File S1 AML file for calculating PRS indices (I, O).An AML script file under ARCGIS 8.3 Workstation for calculating PRS indices I and O.

File S2 AML file for calculating PRS indices (Dx, Dy, Dz).An AML script file under ARCGIS 8.3 Workstation for calculating PRS indices Dx, Dy, and Dz.

File S3 R file for plotting PRS_Chart. A R script file under R version 2.15.1 for plotting PRS_Chart.

Table S1 General information of species studied.Notes: SP.No.: species number; Prev.: species prevalence indicated by ratios*10^{-4} of cells occupied by that species to total study area; LG and LT: longitude and latitude of the species centroid.

Acknowledgments

This research was carried out with the support of the Special Research Program for Public-Welfare Forestry (Grant No. 200804001) and NSFC's Project (Grant No.31370486).

Author Contributions

Conceived and designed the experiments: XK SL. Performed the experiments: XK QL YZ. Analyzed the data: XK QL YZ. Contributed reagents/materials/analysis tools: XK QL. Wrote the paper: XK CB SL.

References

1. Elith J, Graham CH, Anderson RP, Dudik M, Ferrier S, et al. (2006) Novel methods improve prediction of species' distributions from occurrence data. Ecography 29: 129–151.
2. Peterson AT, Soberon J, Pearson RG, Anderson RP, Martinez-Meyer E, et al. (2011) Ecological Niches and Geogrpahical Distributions Princeton, Oxford: Princeton University Press.
3. Araujo MB, Cabeza M, Thuiller W, Hannah L, Williams PH (2004) Would climate change drive species out of reserves? An assessment of existing reserve-selection methods. Global Change Biology 10: 1618–1626.
4. McKenney DW, Pedlar JH, Rood RB, Price D (2011) Revisiting projected shifts in the climate envelopes of North American trees using updated general circulation models. Global Change Biology 17: 2720–2730.
5. Guisan A, Thuiller W (2005) Predicting species distribution: offering more than simple habitat models. Ecology Letters 8: 993–1009.
6. Hampe A (2004) Bioclimate envelope models: what they detect and what they hide. Global Ecology and Biogeography 13: 469–471.
7. Ibanez I, Clark JS, Dietze MC, Feeley K, Hersh M, et al. (2006) Predicting biodiversity change: Outside the climate envelope, beyond the species-area curve. Ecology 87: 1896–1906.
8. Real R, Marquez AL, Olivero J, Estrada A (2010) Species distribution models in climate change scenarios are still not useful for informing policy planning: an uncertainty assessment using fuzzy logic. Ecography 33: 304–314.
9. Araujo MB, Guisan A (2006) Five (or so) challenges for species distribution modelling. Journal of Biogeography 33: 1677–1688.
10. Araújo MB, New M (2007) Ensemble forecasting of species distributions. Trends in Ecology and Evolution 22: 42–47.
11. Austin MP, Van Niel KP (2011) Improving species distribution models for climate change studies: variable selection and scale. Journal of Biogeography 38: 1–8.
12. Elith J, Graham CH (2009) Do they? How do they? WHY do they differ? On finding reasons for differing performances of species distribution models. Ecography 32: 66–77.
13. Thuiller W, Lafourcade B, Engler R, Araujo MB (2009) BIOMOD - a platform for ensemble forecasting of species distributions. Ecography 32: 369–373.
14. Engler R, Guisan A, Rechsteiner L (2004) An improved approach for predicting the distribution of rare and endangered species from occurrence and pseudo-absence data. Journal of Applied Ecology 41: 263–274.
15. Jaeschke A, Bittner T, Reineking B, Beierkuhnlein C (2013) Can they keep up with climate change? - Integrating specific dispersal abilities of protected Odonata in species distribution modelling. Insect Conservation and Diversity 6: 93–103.
16. Wilson CD, Roberts D (2011) Modelling distributional trends to inform conservation strategies for an endangered species. Diversity and Distributions 17: 182–189.
17. Bittner T, Jaeschke A, Reineking B, Beierkuhnlein C (2011) Comparing modelling approaches at two levels of biological organisation - Climate change impacts on selected Natura 2000 habitats. Journal of Vegetation Science 22: 699–710.
18. Araujo MB (2002) Biodiversity hotspots and zones of ecological transition. Conservation Biology 16: 1662–1663.
19. Botkin DB, Saxe H, Araujo MB, Betts R, Bradshaw RHW, et al. (2007) Forecasting the effects of global warming on biodiversity. Bioscience 57: 227–236.

20. Thuiller W, Lavorel S, Araujo MB, Sykes MT, Prentice IC (2005) Climate change threats to plant diversity in Europe. Proceedings of the National Academy of Sciences of the United States of America 102: 8245–8250.
21. Bradley BA, Oppenheimer M, Wilcove DS (2009) Climate change and plant invasions: restoration opportunities ahead? Global Change Biology 15: 1511–1521.
22. Hickling R, Roy DB, Hill JK, Fox R, Thomas CD (2006) The distributions of a wide range of taxonomic groups are expanding polewards. Global Change Biology 12: 450–455.
23. Thuiller W (2004) Patterns and uncertainties of species' range shifts under climate change. Global Change Biology 10: 2020–2027.
24. Chambers D, Perie C, Casajus N, de Blois S (2013) Challenges in modelling the abundance of 105 tree species in eastern North America using climate, edaphic, and topographic variables. Forest Ecology and Management 291: 20–29.
25. Iverson LR, Prasad AM, Matthews SN, Peters M (2008) Estimating potential habitat for 134 eastern US tree species under six climate scenarios. Forest Ecology and Management 254: 390–406.
26. Jimenez-Valverde A, Lobo JM (2007) Threshold criteria for conversion of probability of species presence to either-or presence-absence. Acta oecologica 31: 361–369.
27. Liu CR, Berry PM, Dawson TP, Pearson RG (2005) Selecting thresholds of occurrence in the prediction of species distributions. Ecography 28: 385–393.
28. Liu CR, White M, Newell G (2013) Selecting thresholds for the prediction of species ocurrence with presence-only data. Journal of Biogeography 40: 778–789.
29. Kou XJ, Li Q, Liu SR (2011) Quantifying species' range shifts in relation to climate change: A case study of Abies spp. in China. Plos ONE 6(8): e23115. doi:23110.21371/journal.pone.0023115.
30. Ficarra FVC, Nicol E, Ficarra VM (2011) Research and Development: Business into Transfer Information and Communication Technology. In: Ficarra FVC, Lozano CD, Jimenez MP, Kratky ENA, CipollaFicarra M, editors. Advances in New Technologies, Interactive Interfaces, and Communicability: First Iinternational Conference. pp.44–61.
31. Eckenwalder JE (2009) Conifers of the World. Portland: Timber Press.
32. Farjon A (2008) A Natural history of Conifers. Portland: Timber Press.
33. Farjon A (2010) A handbook of the World's Conifers. Boston: Brill.
34. Smith WK, Hinckley TM, editors (1995) Ecophysiology of coniferous forests. San Diego Academic Press.
35. Wu ZY (1980) Vegetation of China. Beijing: Science Press.
36. Kattge J, Diaz S, Lavorel S, Prentice C, Leadley P, et al. (2011) TRY - a global database of plant traits. Global Change Biology 17: 2905–2935.
37. Prentice IC, Cramer W, Harrison SP, Leemans R, Monserud RA, et al. (1992) A GLOBAL BIOME MODEL BASED ON PLANT PHYSIOLOGY AND DOMINANCE, SOIL PROPERTIES AND CLIMATE. Journal of Biogeography 19: 117–134.
38. LePage BA (2003) The evolution, biogeography and palaeoecology of the Pinaceae based on fossil and extant representatives. In: Mill RR, editor. Proceedings of the Fourth International Conifer Conference: Conifers for the Future? pp.29–52.
39. Hou XY (2001) The Vegetation Atlas of China(1:1000 000). Beijing: Science Press.
40. Kou XJ, Li Q, Liu SR (2011) High-resolution bioclimatic dataset derived from future climate projections for plant species distribution modeling. Ecological Informatics 6: 196–204.
41. Tabor K, Williams JW (2010) Globally downscaled climate projections for assessing the conservation impacts of climate change. Ecological Applications 20: 554–565.
42. Hijmans RJ, Cameron SE, Parra JL, Jones PG, Jarvis A (2005) Very high resolution interpolated climate surfaces for global land areas. International Journal of Climatology 25: 1965–1978.
43. Harrison SP, Prentice IC, Barboni D, Kohfeld KE, Ni J, et al. (2010) Ecophysiological and bioclimatic foundations for a global plant functional classification. Journal of Vegetation Science 21: 300–317.
44. Woodward FI (1987) Climate and plant distribution. Cambridge UK: Cambridge University Press.
45. Elith J, Phillips SJ, Hastie T, Dudik M, Chee YE, et al. (2011) A statistical explanation of MaxEnt for ecologists. Diversity and Distributions 17: 43–57.
46. Phillips SJ, Dudik M (2008) Modeling of species distributions with Maxent: new extensions and a comprehensive evaluation. Ecography 31: 161–175.
47. Allouche O, Tsoar A, Kadmon R (2006) Assessing the accuracy of species distribution models: prevalence, kappa and the true skill statistic (TSS). Journal of Applied Ecology 43: 1223–1232.
48. Zadeh LA (1965) Fuzzy sets. Information and Control 8: 338–353.
49. Ives AR, Helmus MR (2010) Phylogenetic Metrics of Community Similarity. American Naturalist 176: E128–E142.
50. Kupfer JA (2012) Landscape ecology and biogeography: Rethinking landscape metrics in a post-FRAGSTATS landscape. Progress in Physical Geography 36: 400–420.
51. Muenkemueller T, de Bello F, Meynard CN, Gravel D, Lavergne S, et al. (2012) From diversity indices to community assembly processes: a test with simulated data. Ecography 35: 468–480.
52. Swenson NG (2011) Phylogenetic Beta Diversity Metrics, Trait Evolution and Inferring the Functional Beta Diversity of Communities. Plos One 6.
53. Tuomisto H (2010) A diversity of beta diversities: straightening up a concept gone awry. Part 1. Defining beta diversity as a function of alpha and gamma diversity. Ecography 33: 2–22.
54. Uuemaa E, Mander U, Marja R (2013) Trends in the use of landscape spatial metrics as landscape indicators: A review. Ecological Indicators 28: 100–106.
55. Fang JY, Wang ZH, Tang ZY (2009) Atlas of Woody Plants in China. Beijing: Higher Education Press.
56. Thomas CD, Cameron A, Green RE, Bakkenes M, Beaumont LJ, et al. (2004) Extinction risk from climate change. Nature 427: 145–148.
57. Thomas CD, Franco AMA, Hill JK (2006) Range retractions and extinction in the face of climate warming. Trends in Ecology & Evolution 21: 415–416.
58. Jablonski D (2008) Species Selection: Theory and Data. Annual Review of Ecology Evolution and Systematics. pp. 501–524.
59. Webb TJ, Gaston KJ (2003) On the heritability of geographic range sizes. American Naturalist 161: 553–566.
60. Mouillot D, Gaston K (2009) Spatial overlap enhances geographic range size conservatism. Ecography 32: 671–675.
61. Colwell RK, Brehm G, Cardelus CL, Gilman AC, Longino JT (2008) Global warming, elevational range shifts, and lowland biotic attrition in the wet tropics. Science 322: 258–261.
62. Crimmins SM, Dobrowski SZ, Greenberg JA, Abatzoglou JT, Mynsberge AR (2011) Changes in Climatic Water Balance Drive Downhill Shifts in Plant Species' Optimum Elevations. Science 331: 324–327.
63. Dobrowski SZ, Crimmins SM, Greenberg JA, Abatzoglou JT, Mynsberge AR (2011) Response to Comments on "Changes in Climatic Water Balance Drive Downhill Shifts in Plant Species' Optimum Elevations". Science 334.
64. Feeley KJ, Silman MR, Bush MB, Farfan W, Cabrera KG, et al. (2011) Upslope migration of Andean trees. Journal of Biogeography 38: 783–791.
65. Frei E, Bodin J, Walther G-R (2010) Plant species' range shifts in mountainous areas-all uphill from here? Botanica Helvetica 120: 117–128.
66. Lenoir J, Gegout JC, Marquet PA, de Ruffray P, Brisse H (2008) A significant upward shift in plant species optimum elevation during the 20th century. Science 320: 1768–1771.
67. Jump AS, Matyas C, Penuelas J (2009) The altitude-for-latitude disparity in the range retractions of woody species. Trends in Ecology & Evolution 24: 694–701.
68. Engler R, Randin CF, Thuiller W, Dullinger S, Zimmermann NE, et al. (2011) 21st century climate change threatens mountain flora unequally across Europe. Global Change Biology 17: 2330–2341.
69. Reyer CPO, Leuzinger S, Rammig A, Wolf A, Bartholomeus RP, et al. (2013) A plant's perspective of extremes: terrestrial plant responses to changing climatic variability. Global Change Biology 19: 75–89.
70. Bellard C, Bertelsmeier C, Leadley P, Thuiller W, Courchamp F (2012) Impacts of climate change on the future of biodiversity. Ecology Letters 15: 365–377.
71. Thuiller W, Lavergne S, Roquet C, Boulangeat I, Lafourcade B, et al. (2011) Consequences of climate change on the tree of life in Europe. Nature 470: 531–534.
72. Curtis JMR, Vincent ACJ (2008) Use of Population Viability Analysis to Evaluate CITES Trade-Management Options for Threatened Marine Fishes. Conservation Biology 22: 1225–1232.
73. Kery M, Royle JA, Schmid H, Schaub M, Volet B, et al. (2010) Site-Occupancy Distribution Modeling to Correct Population-Trend Estimates Derived from Opportunistic Observations. Conservation Biology 24: 1388–1397.
74. Thomas CD (2011) Translocation of species, climate change, and the end of trying to recreate past ecological communities. Trends in Ecology & Evolution 26: 216–221.
75. Vila M, Hulme PE (2011) Jurassic Park? No thanks. Trends in Ecology & Evolution 26: 496–497.
76. Webber BL, Scott JK, Didham RK (2011) Translocation or bust! A new acclimatization agenda for the 21(st) century? Trends in Ecology & Evolution 26: 495–496.
77. Carrete M, Tella JL (2012) Is assisted colonization feasible? Lessons from past introductions. Frontiers in Ecology and the Environment 10: 12–13.
78. Corlett RT (2012) Climate change in the tropics: The end of the world as we know it? Biological Conservation 151: 22–25.
79. Kreyling J, Bittner T, Jaeschke A, Jentsch A, Steinbauer MJ, et al. (2011) Assisted Colonization: A Question of Focal Units and Recipient Localities. Restoration Ecology 19: 433–440.
80. Schwartz MW, Hellmann JJ, McLachlan JM, Sax DF, Borevitz JO, et al. (2012) Managed Relocation: Integrating the Scientific, Regulatory, and Ethical Challenges. Bioscience 62: 732–743.
81. Thomas CD (2011) Anthropocene Park? No alternative. Trends in Ecology & Evolution 26: 497–498.
82. Dobrowski SZ, Abatzoglou J, Swanson AK, Greenberg JA, Mynsberge AR, et al. (2013) The climate velocity of the contiguous United States during the 20th century. Global Change Biology 19: 241–251.
83. Etterson JR, Shaw RG (2001) Constraint to adaptive evolution in response to global warming. Science 294: 151–154.
84. Pearson RG, Dawson TP (2005) Long-distance plant dispersal and habitat fragmentation: identifying conservation targets for spatial landscape planning under climate change. Biological Conservation 123: 389–401.
85. Summers DM, Bryan BA, Crossman ND, Meyer WS (2012) Species vulnerability to climate change: impacts on spatial conservation priorities and species representation. Global Change Biology 18: 2335–2348.

86. Feeley KJ, Rehm EM (2012) Amazon's vulnerability to climate change heightened by deforestation and man-made dispersal barriers. Global Change Biology 18: 3606–3614.
87. Chiarucci A, Araujo MB, Decocq G, Beierkuhnlein C, Maria Fernandez-Palacios J (2010) The concept of potential natural vegetation: an epitaph? Journal of Vegetation Science 21: 1172–1178.
88. Oney B, Reineking B, O'Neill G, Kreyling J (2013) Intraspecific variation buffers projected climate change impacts on Pinus contorta. Ecology and Evolution 3: 437–449.
89. Millar CI, Stephenson NL, Stephens SL (2007) Climate change and forests of the future: Managing in the face of uncertainty. Ecological Applications 17: 2145–2151.
90. Jentsch A, Beierkuhnlein C (2008) Research frontiers in climate change: Effects of extreme meteorological events on ecosystems. Comptes Rendus Geoscience 340: 621–628.

13

Early Evidence of Acheulean Settlement in Northwestern Europe - La Noira Site, a 700 000 Year-Old Occupation in the Center of France

Marie-Hélène Moncel[1]*, Jackie Despriée[1], Pierre Voinchet[1], Hélène Tissoux[1,2], Davinia Moreno[1], Jean-Jacques Bahain[1], Gilles Courcimault[1], Christophe Falguères[1]

1 Department of Prehistory-UMR 7194 CNRS, National Museum of Natural History, Institut de Paléontologie Humaine, Paris, France, **2** Bureau de Recherches Géologiques et Minières, GEO/G2R, BP 36009, Orléans, France

Abstract

The human settlement of Europe during Pleistocene times was sporadic and several stages have been recognized, both from paleaoanthropological and archaeological records. If the first phase of hominin occupation (as early as 1.4 Ma) seems mainly restricted to the southern part of the continent, the second phase, characterized by specific lithic tools (handaxes), is linked to Acheulean settlements and to the emergence of *Homo heidelbergensis*, the ancestor of Neanderthals. This phase reached northwestern Europe and is documented in numerous sites in Germany, Great Britain and northern France, generally after 600 ka. At la Noira (Brinay, Central France), the Middle Pleistocene alluvial formation of the Cher River covers an archaeological level associated with a slope deposit (diamicton). The lithic assemblage from this level includes Large Cutting Tools (LCTs), flakes and cores, associated with numerous millstone slabs. The lithic series is classified as Acheulean on the basis of both technological and typological analyses. Cryoturbation features indicate that the slope deposits and associated archaeological level were strongly frozen and disturbed after hominin occupation and before fluvial deposition. Eight sediment samples were dated by the electron spin resonance (ESR) method and the weighted average age obtained for the fluvial sands overlying the slope deposits is 665±55 ka. This age is older than previous chronological data placing the first European Acheulean assemblages north of 45th parallel north at around 500 ka and modifies our current vision of the initial peopling of northern Europe. Acheulean settlements are older than previously assumed and the oldest evidences are not only located in southern Europe. La Noira is the oldest evidence of Acheulean presence in north-western Europe and attests to the possibility of pioneering phases of Acheulean settlement which would have taken place on a Mode 1-type substratum as early as 700 ka. The lithic assemblage from la Noira thus provides behavioral and technological data on early Acheulean occupation in Europe and contributes to our understanding of the diffusion of this tradition.

Editor: Fred H. Smith, Illinois State University, United States of America

Funding: This study has been financially supported by a French ANR project (Agence Nationale pour la Recherche) PremSeptAcheu (BLAN 2006 01) directed by M-H. Moncel, a Collective Research Project for study the first occupations of the Middle Loire Basin directed by J. Despriee, the Centre Region, the Department of Prehistory of the National Museum of National History and the Ministry of Culture. The authors also wish to thank the Conseil Régional d'Ile-de-France and the Ministère de la Culture, Sous-direction de l'Archéologie, Service Régional de l'Archéologie of the Region Centre (France) for their assistance in the acquisition of an ESR spectrometer and a portable gamma spectrometer, respectively. The English version has been corrected by an official translator Louise Byrne. The funders have no role in the study, data collections and analyses, decision to publish or preparation of the manuscript and illustrations.

Competing Interests: The authors have declared that no competing interests exist.

* E-mail: moncel@mnhn.fr

Introduction

Discoveries made over the past two decades attest to a large diversity of Mode 1-type assemblages as early as 1.4 Ma ago in Europe (e.g. Orce, Atapuerca, Pirro-Nord [1,2]). The scarcity of sites over so long a period of time suggests short-lived dispersal events and probably a source-sink dynamic from the south with phases of depopulation and recolonization. Assemblages without evidence of bifacial technology persist after 1 Ma and new discoveries in the center of France and in England enhance our vision of human colonization in northern parts of the continent [3,4,5,6,7]. The first assemblages with Large Cutting Tools (LCTs) (Mode 2, cf. Clark [8]) are described in East Africa from 1.8 Ma and in the Levant and India from 1.5 Ma [9,10,11,12,13,14], and relationships between African, Levantine and Indian sites are acknowledged [15]. In China, LCTs at 800 ka suggest a local or allochthonous origin [16,17].

The first assemblages with bifacial tools in Europe are much younger. The usual hypothesis is that the first Acheulean evidence appears at around 600 ka in southern Europe, followed by a rapid colonization of the northwest at 500 ka [18,19,20,21]. This first occurrence could be older [22], but this has not yet been firmly established (sites of Solana del Zamborino and Cueva Negra Quipar). The Acheulean tradition would thus have been the first to continuously occupy northern latitudes.

The few hominin fossils dating between 800 and 500 ka (Gran Dolina TD6, Mauer, Boxgrove) are attributed to either *Homo antecessor* or *Homo heidelbergensis* and diversity and ambiguities of anatomical features are noted [23,24,25,26,27,28,29,30,31]. Dental analysis points to longitudinal migrations of new hominin groups from Asia, and genetic data suggest speciation events in Africa or Eurasia prior to 600 ka [32,33,34].

The 800 to 500 ka timescale is a key period in Europe since series with (Notarchirico and Arago levels P-Q) and without bifacial tools (Pakefield, Happisburg I, Soleilhac, Isernia la Pineta, Atapuerca Gran Dolina TD6) co-occurred. New discoveries at la Noira site, located beyond the 45th parallel north in the center of France, enhance our vision of both the northern colonization of Western Europe and the earliest bifacial traditions [35,36,37]. Evidence from la Noira raises the question of the relationship between *Homo heidelbergensis* dispersals and the onset of the Acheulean and LCTs.

The study of the lithic assemblage of la Noira using technological and typological analytic methods (e.g. [38,39,40,41,42] to define the techno-economic processes, manual dexterity, technological sequences (*chaîne opératoire*) and technological skills involved in tool production contributes to our knowledge of hominin group cognition in Europe during the period when *Homo heidelbergensis* appeared.

Materials and Methods

1.The Middle Cher River Valley

The site of la Noira is located in the Middle Cher River Valley, a tributary of the Loire River in the Middle Loire Basin (Centre region, France) near Brinay village (Figures 1, 2). In this area, the fossil fluvial system of the Cher River is composed of nine stepped alluvial formations [37] deposited during successive fluvial incisions and aggradations as a response to climatic cyclicity during the Early and Middle Pleistocene [35]. This stepped disposition is also related to a slow uplift of the Paris Basin, as described by Antoine et al. [43] for river systems in the northern Paris Basin.

2.The site of la Noira

The site of la Noira was discovered at the beginning of the 1970s in a quarry. Since 2003, systematic geological and geochronological studies have been applied to the Pleistocene depositional sequence, supported by the French Ministry of Culture and an ANR project. Field work permission for la Noira site was granted to Jackie Despriée and Marie-Hélène Moncel by the French Ministry of Culture. The quarry owner granted permission to conduct fieldwork and this study. The site is located in the Les Fougères Formation, which is one of the stepped Pleistocene fluvial sheets deposited by the Cher River (sheet D) lying on the western slope (left bank) of the valley. It is a thick fluvial sandy formation, about 1.250 km long and 0.5 km wide. This terrace was deposited on an Oligocene lacustrine limestone block delimited by two west-east ground features. The fluvial formation is deposited between a relative altitude of +13 m and +21 m above the last level of incision under the present alluvial plain.

Above the substratum of Stampian lacustrine limestone weathered in green clay, the Les Fougeres Quaternary formation bears five successive strata: a coarse slope deposit or diamicton (stratum a) covered by two sequences of sandy alluvial layers (stratum b), then a rubble level (stratum c) and a silty soil (stratum d). The archaeological level studied in this article is located in stratum a (Figure 3).

The sedimentation history seems similar to that described by Antoine et al for the Somme valley [43]. In the la Noira quarry, the observed sedimentary formation was approximately six to eight meters thick (see Figure 3). Firstly, the basal layer (stratum a) was deposited during an early glacial stage after the down cutting of the valley. It is a coarse slope deposit which contains masses of local millstone slabs deposited on the clayey bedrock (Tertiary limestone), mixed with magmatic, metamorphic, crystalline and sedimentary rocks, and covered with coarse quartz sand in a brown rubefied clayey matrix with numerous iron pisoliths and weathered blocks or pebbles of endogenous rocks. Some of the siliceous slabs, which range in size from several centimeters to several decimeters, bear scars attesting hominin activities [6] (Figure 4). These materials (granite, gneiss, quartz, sandstones, lacustrine millstones…) were spread and accumulated during Tertiary epochs (Oligocene to Pliocene) by torrents or wadis on the peneplain areas at the southern edge of the Massif Central. These detrital formations were lateritized during Oligocene stages. During the Quaternary period, after each down cutting of the Cher River, at the beginning of a cold climatic stage, these coarse sediments came down slope by solifluction from the interfluve or the edges of the plateau, and were stocked on the floor of the new alluvial plain. The artifacts are associated with abundant millstone slabs, generally lying flat on the clayey floor.

After hominin activities, part of stratum a - enclosing the archaeological pieces - and the underlying limestone bedrock were cryoturbated before the sandy alluvial deposition. Despite this reworking, an undisturbed archaeological area from stratum a, containing many artefacts, is currently undergoing excavation. After this human occupation, the Cher River deposited gravels and sands as new slope coarse deposits (gelifuction) above the diamicton, during the same glacial stage and the next interglacial period, before the following incision of the valley.

About 300 Electronic Spin Resonance (ESR) samples were taken in the fossil sandy alluvial deposits of the three middle valleys of the Loir, Cher and Creuse Rivers in the Loire catchment basin. Dating was carried out at the National Museum of Natural History, Paris, France.

Results

1.Dating

The results enable us to place the stepped formations of each fluvial system (and the prehistoric sites covered by these formations) within Early and Middle Pleistocene chronological frameworks based on oxygen isotope variations (Figure 5). In the Cher Valley, ESR applied to optically bleached quartz was used to date more than 80 samples [35,6,37].

At la Noira, the very acidic quartzose deposits are not conducive to the preservation of organic remains, and the age of the site was therefore determined using the ESR method applied to optically bleached sedimentary quartz grains. Since 2003, ESR samples have been taken from stratum b at la Noira site. The ESR age obtained for stratum b, corresponding to a fluvial deposit, provided a mean age value of 655±55 ka, [35,37] (Table 1). The la Noira site is the only currently known prehistoric site discovered during the time period ranging between 700 and 600 ka in the Centre Region [6].

The hominin occupation occurred between the end of the river incision and the fluvial deposits and suggests that hominins were present during the early glacial stage, just before the pleniglacial phase and before interglacial fluvial deposition [6]. This glacial/interglacial cycle can be assigned to the MIS 16/MIS 15 cycle, according to the ESR results obtained for the fluvial layers.

2.Lithic assemblage of la Noira, stratum a

The lithic assemblage is located in the coarse stratum a beneath the sandy fluvial sequence.

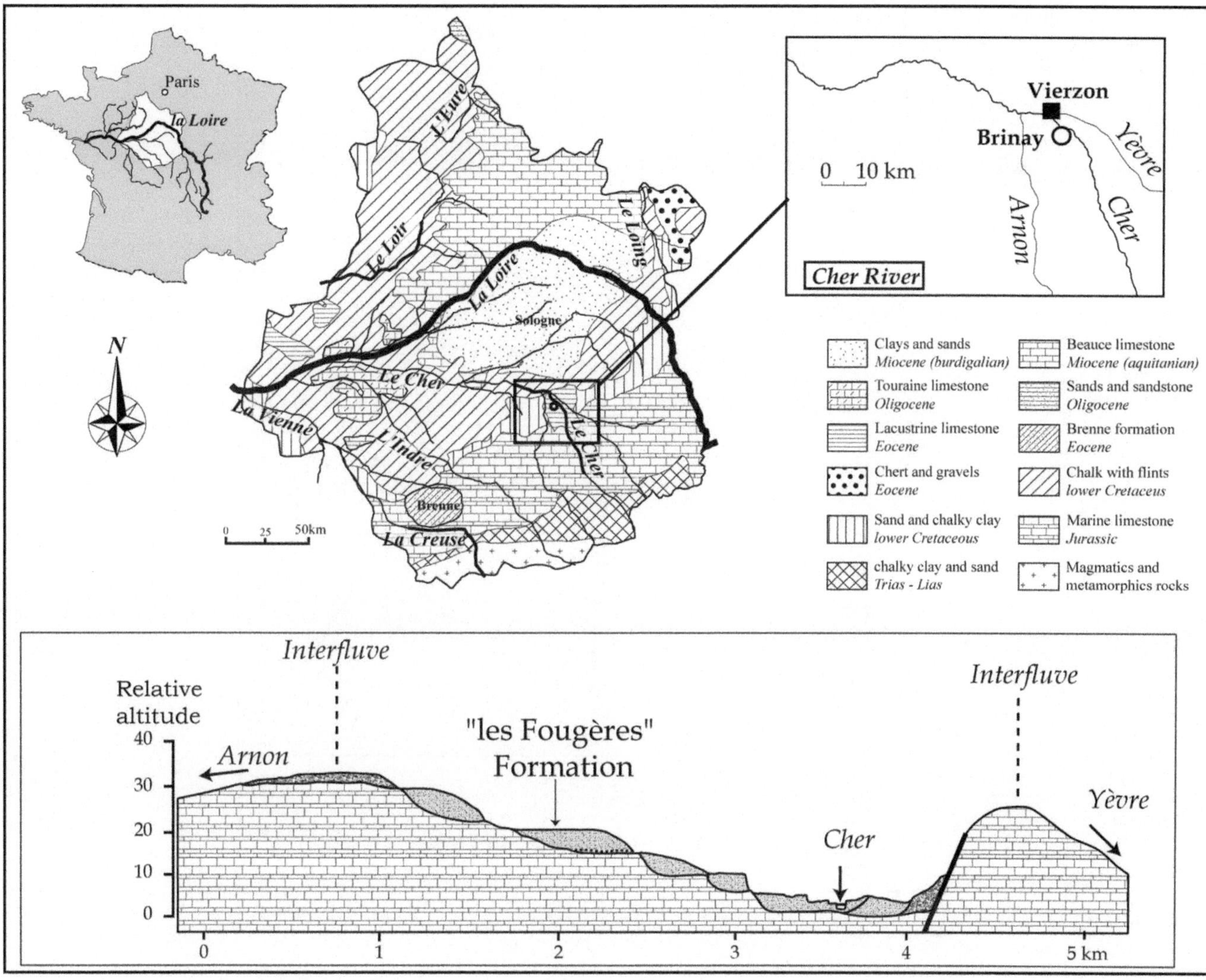

Figure 1. Geographic and geological maps showing the location of the la Noira Site at Brinay, (Cher, France) in the Middle Cher River valley and the location of the Les Fougères fluvial formation on the western slope of the valley.
The figure is for illustrative purposes only.

The stratum a level yielded a series of 340 items found *in situ* in two sectors of the site during fieldwork conducted between 2003 and 2012. Only pieces with evidence of manufacture have been analyzed and five rolled cores from an upper and older formation were discarded. The assemblage is composed of cores, flakes, crude core-tools and Large Cutting Tools (LCTs), which make up 17% of the series.

2.1.Raw materials. The main raw materials used are local millstone slabs (90%) (Table 2). The millstone slabs shaped or flaked by hominins are issued from diagenetic silicifications included in the Oligocene lacustrine limestone that forms the substratum of the Vierzon area, and, at Brinay, the plateau and the slopes of the Cher valley. These silicifications form parallelepiped slabs of up to several decimeters, generally from five to ten centimeters thick.

These silicified slabs are found lying in a horizontal position in the lacustrine limestone formation. The process of silicification fossilized original and various facies types deposited at the bottom of the lacustrine area in calm and shallow freshwater. Brecciated structures, algal and stromatolitic deposits, vermiculated cavities, burrows and tubing with casts of roots and plant debris are visible in the millstone and affect its thickness, but it can nonetheless be flaked or shaped. Some slabs with a very fine and homogeneous siliceous mass bear no irregularities and have very good flaking qualities.

Millstones moved downslope after being exposed during river incision in the limestone formation. They were carried over a very short distance to the slope bottom, and only rare marks shorter than one millimeter are visible on ridges. No surface crushing is observed. These slope formations thus provided raw material and were used as an outcrop. Available millstone slabs were exploited *in situ* by hominins and artefacts are associated with this deposit.

Rare Jurassic oolithic chert nodules were also used by hominins. These nodules were mixed with millstone slabs, and come from the oldest alluvial formations further up the slopes.

2.2.Crude cores or bifacial tools. Millstone slabs (80 to 240 mm long) at various stages of manufacture and one large flake with few scars were recovered. These pieces can be attributed to cores or unfinished bifacial tools (preforms): 1) incomplete management of convergent edges and a tip indicating shaping, 2) width/thickness ratio higher than 1.5 for LCT preforms, 3) thin and invasive series of removals on LCTs, deep and short on cores.

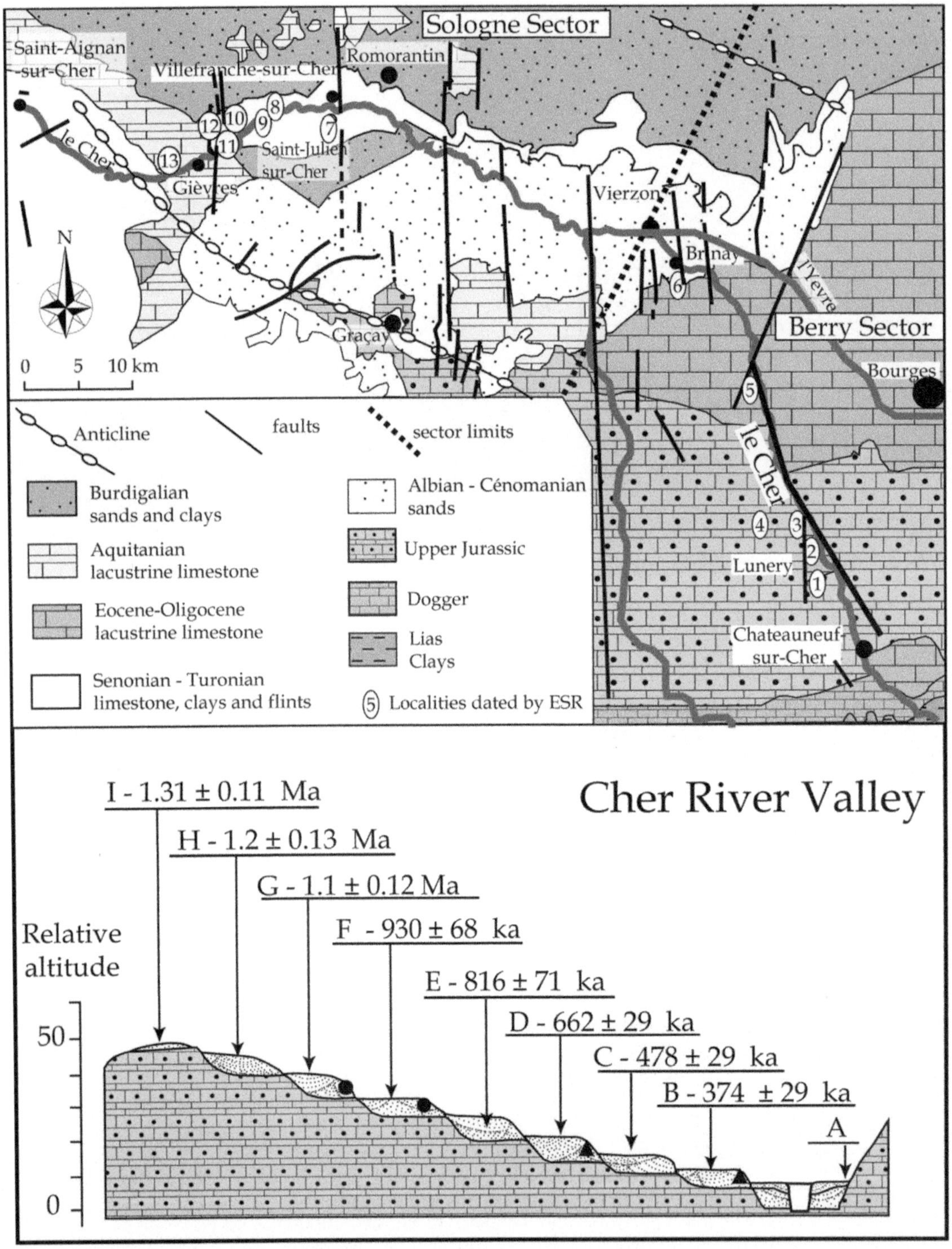

Figure 2. Geological map of the Vierzon area and situation of the Middle Cher Valley and theoretical transect of the Middle Cher valley, and ESR ages of the eight fossil fluvial formations. (rounds : Early Palaeolithic sites, triangles : Lower Palaeolithic sites). The figure is for illustrative purposes only.

They attest to the presence of workshops with expedient *débitage* and partial shaping (Table 2, Figure 6). Hominins may have been attracted to this zone by the large quantity of available raw materials on the river bank. This site might have also recorded evidence of domestic activities, as indicated by six broken bifacial tools and the frequent partial crushing of cutting edges not due to taphonomic processes. Some LCTs were not broken during shaping; they are largely worked and retouched. We observe either a lateral percussion impact or a flexion on the fracture (from the flatter face). Moreover, a fragment of a bifacial tool was reworked as a point. The others do not display evidence of reworking.

2.3.Core technology. The core technology of the assemblage indicates complex and organized techno-economic patterns and is focused on the *débitage* of medium-large flakes. Part of this *débitage* is devoted to the production of large flakes for both flaking (n = 2) and shaping (n = 2). Large flakes are

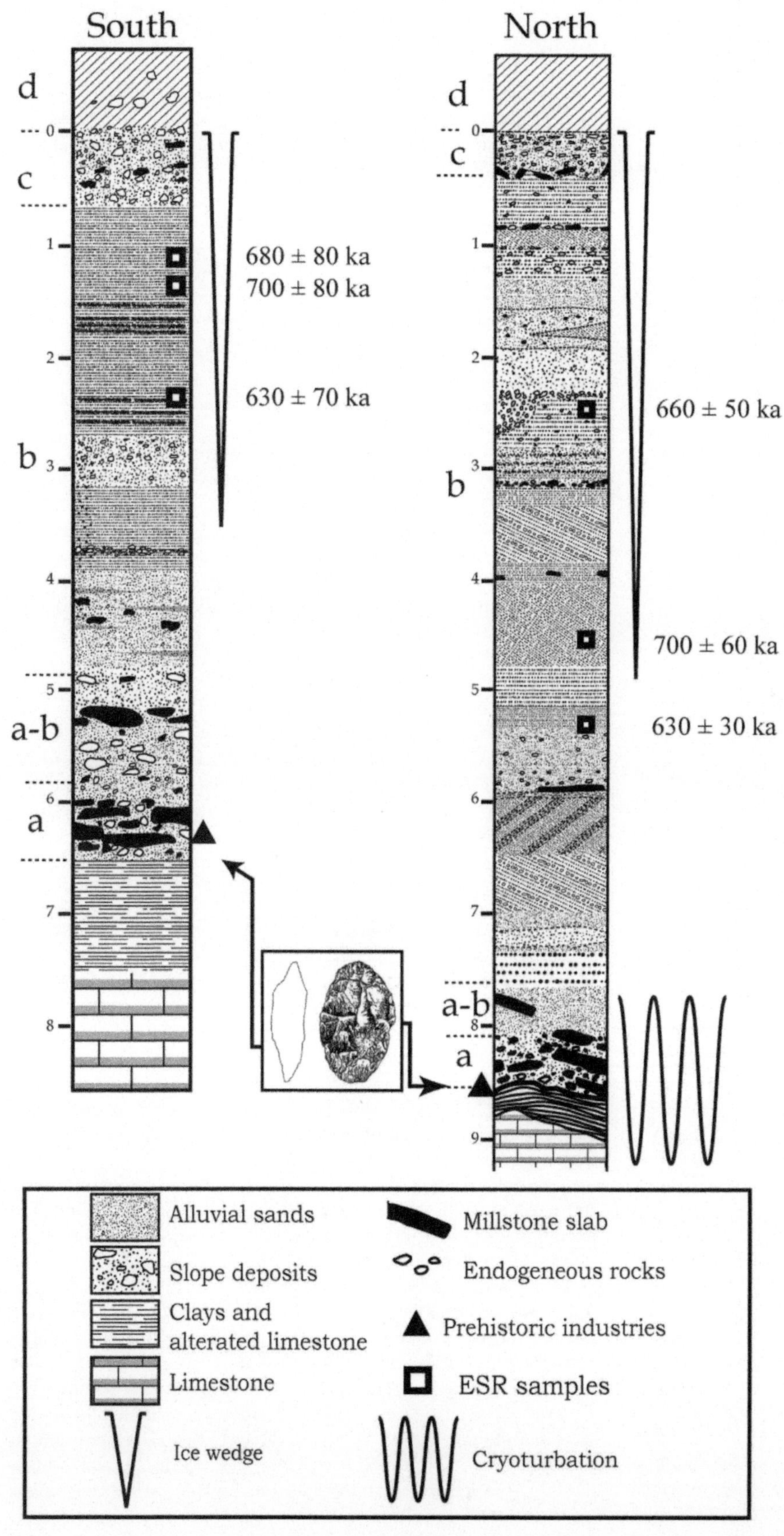

Figure 3. Positions of ESR samples (black squares) taken in 2003 (South Log) and in 2006 (Northern Log) and ages obtained. The black triangles indicate the location of the prehistoric levels observed in situ during quarrying, sampling or excavating. The figure is for illustrative purposes only.

1

2

Figure 4. Photos of the stratum a with a biface *in situ*. 1.View of stratum a, with cryoturbated masses of millstone slabs and artefacts on weathered limestone. 2. View of a biface *in situ* at the bottom of a cryoturbated feature (cf. Despriee et al., 2007).

distinguished from small *débitage* by their dimensions: length or width >10 cm [11,14,43].

The high number of cores enables us to appraise the quantitative and qualitative modalities of the knapping technology and volume management. Cores have been identified by the organization of removals. A panoply of opportunistic or structured methods using hard hammer direct percussion has been identified. Two cores bear strong crushing marks on their edges and some large quartz and metamorphic rock pebbles found *in situ* document percussion activities and have been considered to be a set of hard hammers. The cores can be divided into several groups of varied modalities with regard to the shape of the slab and the best opportunities offered. This indicates awareness of the reactions of the millstone slab properties and the relatively minor role of natural slab shape, except at the beginning of the reduction sequence.

The main method used is partial bifacial flaking on millstone slabs (44%). Each cortical surface of the slab is knapped by a single series of short deep removals using alternate or face by face flaking.

Two large groups of partial bifacial discoid cores can be distinguished (Table 3 and Figures 7, 8):

1) symmetrical cores with alternating flaking resulting in a sinusoidal cutting edge (angles between 75–85°), 2) unsymmetrical cores with one flat surface due to a face by face debitage. The flat surface bears more invasive and elongated (deep or not) centripetal or crossed removals opposite another surface with short scars due to the perpendicular knapping of the slab slice (striking platform?). The secant cutting edges are more regular than for the former type. Some final removals could be produced on the abrupt surface if favorable platforms existed. Cutting edge angles are close to 70–80°. No evidence of core preparation is apparent and flaking stopped abruptly, either due to the large quantity of available slabs on the site or to the fact that the flaking surface was no longer functional with suitable angles. The core-flakes are cortical and thick (40 mm) with abrupt removals on the upper face of the flake and some scars on the lower surface (bulb).

Three groups of total bifacial cores exist:

1) pyramidal cross-section with deep and elongated or short removals on one face (85°),
2) alternate flaking on bi-pyramidal surfaces (60–70°),
3) flat surface with thin and more or less invasive and centripetal removals opposite a pyramidal surface bearing the last removals (60°). The first and second categories can be described as bifacial flaking using the most suitable angles with each previous core scar resulting in an irregular peripheral cutting edge. For the third type, as one of the surfaces is convex with centripetal removals, a flaking hierarchy is suggested. A few pieces suggest a hierarchy of the two surfaces between the platform and the flaked surface. One isolated case displays a "*chapeau de gendarme*" platform (89°) and a flat flaked surface (pre-Levallois concept *e.g.* Tyron et al. [45]. Two abrupt removals were produced beforehand on the lateral part of the core before a final invasive removal. These two lateral removals could suggest previous preparation of the lateral edge of the core. However the poor quality of the stone does not enable us to identify any kind of predetermination in core management (although this piece bears some of the technical criteria developed by the Levallois methodology). In this case, we only have evidence of hierarchical management and we are unable to distinguish stages of full debitage from possible phases of convexity preparation (Figure 7b).

Trifacial (three orthogonal flaking surfaces or a mixture of one surface and a partial bifacial flaking) and unifacial (one flaking surface) cores display expedient debitage, parallel or perpendicular to cortical slab planes. Overexploited multifacial cores are rare.

Most of the cores measure between 50 and 180 mm, indicating the coexistence of both small and large debitage. The few cores measuring close to or more than 200 mm were flaked by limited removals. The single chert core is one of the smallest pieces and was exploited in the same way as the millstone cores. Since most of the cores bear preserved cortical patches on both surfaces, we can estimate the average thickness of the flaked slabs between 40 and 90 mm, which is higher than for the LCTs. Double patina on some bifacial cores indicates recurrent occupations and the re-use of abandoned cores, including the most complex ones.

Flakes are consistent with core management and are issued from all stages of the reduction sequence (Figures 8, 9). Millstone flakes are the largest products (up to 100 and 200 mm long). Five flakes present features related to bifacial shaping using a soft hammer (evidence of lipped platform, curved profile and thin section).

21% of the flakes are retouched (n = 39), most of them being the largest millstone flakes. Thin and direct retouch does not denote the pursuit of a specific morphology. Tools are mainly single (76%), and are largely comprised of scrapers (30%), Clactonian notches and denticulates (27%).

2.4.Large Cutting Tools (LCTs). The main feature of the series is the diversity of the LCTs, most of which are made on 20–40 mm thick millstone slabs. LCTs are distinguished from retouched flakes by their dimensions [11,45,14]. They generally

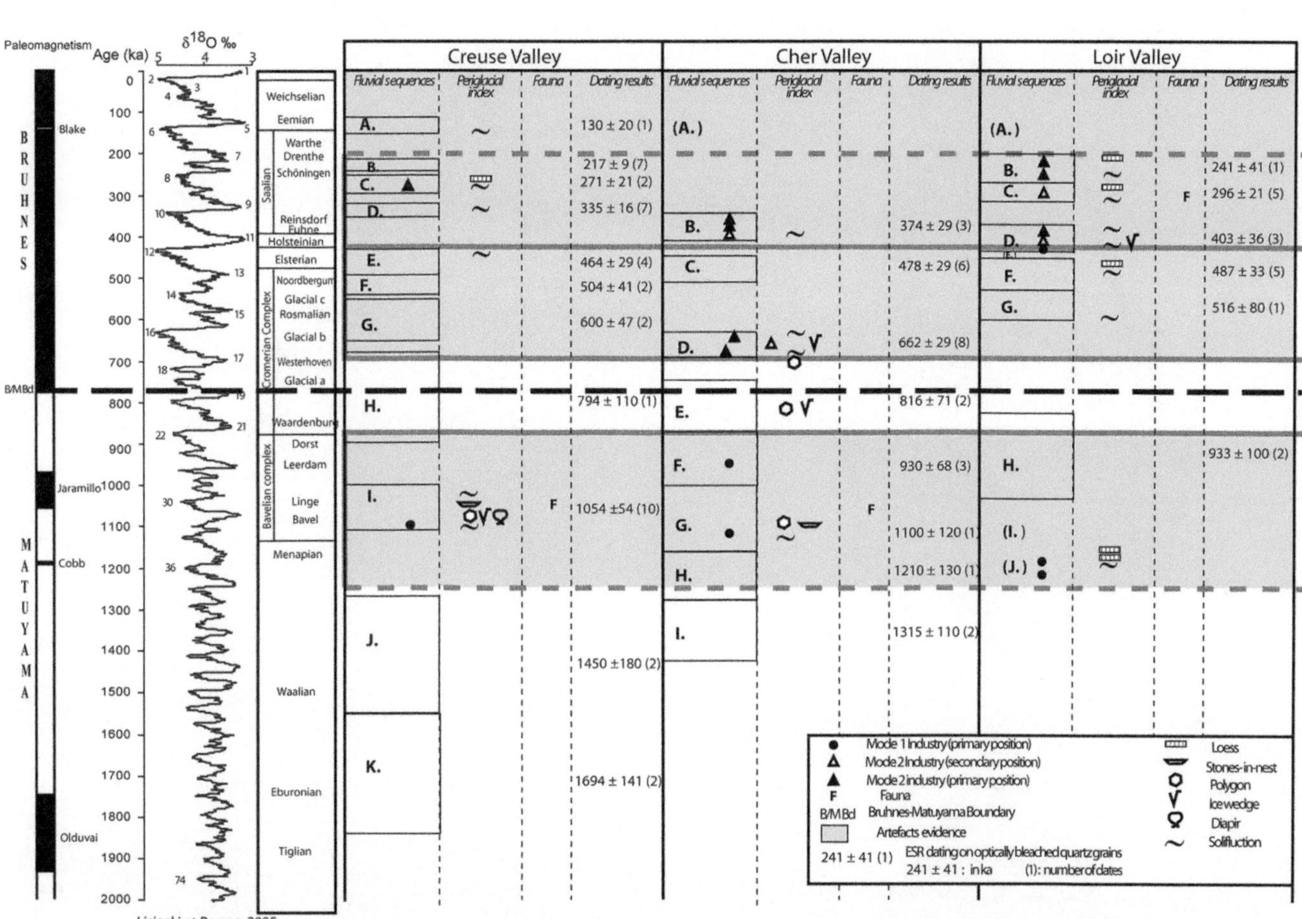

Creuse Valley : **A.** SAINT-MARCEL les Chambons, **B.** SAINT-MARCEL, LE PONT CHRETIEN, LE PECHEREAU, **C.** CIRON les Champs-de-Chaumes, **E.** CEAULMONT le Petit-Multon, SCOURY les-Grands-Champs, SAINT-MARCEL les Pommeurs B/les Mersants B, **F.** SAINT-MARCEL les Coutures, **G.** GARGILESSE Bourny, la Garenne, **H.** CIRON les-Brandes-de-Montflamme, **I.** EGUZON-CHANTÔME Fressignes, Pont-de-Lavaud, GARGILESSE-DAMPIERRE le Cerisier-Montcocu, **J.** EGUZON-CHANTÔME Rue Jean-Jaures, **K.** GARGILESSE-DAMPIERRE le Pin.
Cher Valley : **B.** GIEVRES la Morandière, **C.** SAINTE-THORETTE le Grand-Bretigny, **D.** BRINAY la Noira, GIEVRES la Genetière, la Plaine de la Morandière, **E.** LUNERY-ROSIERES la Terre-des-Sablons 1, la croix de Bagneux, **F.** LUNERY-ROSIERES la Terre-des-Sablons 2, SAINT-JULIEN-SUR-CHER La Marelle, G. LUNERY-ROSIERES la Terre-des-Sablons 3, H. MARMAGNE Bois-aux-Vaches , I. MARMAGNE Bois-Saint-Aubin.
Loir Valley : **B.** MOREE Villeprovert, **C.** THORE les Champronds, MONTOIRE la Maison-Blanche, **D.** PEZOU les Grouais-de-Chicheray, **E.** LIGNIERES Courcelles, **F.** SAINT-FIRMIN-DES-PRES Moncé, TRÔO Maison-Petit, **G.** PEZOU Chêne-Carré, **H.** PONCE RUILLE, **I.** LISLE Belle- Assise, **J.** SAINT-HILAIRE-LA-GRAVELLE le Pont-de-la-Hulauderie.

Figure 5. Chronological framework for the river systems of the Creuse, Cher and Loir valleys, ESR ages of river aggradation, location of prehistoric artefacts in primary or secondary position, and evidence of periglacial phenomena. La Noira Acheulean workshops are situated in MIS 16, as shown by evidence of periglacial phenomena, before covering by interglacial fluvial deposits during MIS 15.

measure between 60 to 160 mm long but some attain 170–235 mm. The two flake-supports are cortical, belong to the middle-sized group and are issued from centripetal or unidirectional *débitage*. The presence of reduced (47%) or large cortical surfaces (38%) facilitates the reading of the successive removals and suggests voluntary size variations for LCTs.

Table 1. ESR dates from the la Noira sequence.

Samples	Stratum	Depth	ESR ages
Cher 2004-12	Stratum b (top)	1.10 m–1.20 m	680±80 ka
Cher 2004-11	Stratum b (top)	1.20 m–1.45 m	700±80 ka
Cher 2004-10	Stratum b (top)	2,40 m–2.50 m	630±70 ka
Cher 2006-2	Stratum b (median)	2.50 m–2.60 m	660±50 ka
Cher 2006-1	Stratum b (median)	4.50 m–4.60 m	700±60 ka
Cher 2006-3	Stratum b (bottom)	5.20 m–5.30 m	630±30 ka

Several groups of tools have been distinguished according to the type of tip management and the bifacial equilibrium between the two faces (Table 4). These are: 1) knife-types, with or without worked tips and sometimes bearing partially thinned backs, 2) bifacial tools with peripheral additional sharpened areas and a "*déjeté*" bec with tip management included in the general shaping, 3) partial convergent tools made by few removals, 4) "classical bifaces" with preconceived shaping of a general bifacial volume with two convergent edges and a tip, 5) bifacial cleavers with a

Table 2. Main lithic types in stratum a at la Noira.

Artefacts	Millstone	Jurassic Oolithic chert
Crude pieces	20	-
Cores	57 (40 slabs, 2 flakes, 2 nodules)	1 (nodule)
Flakes	199	6
Large cutting tools	58 (including 2 flakes)	-

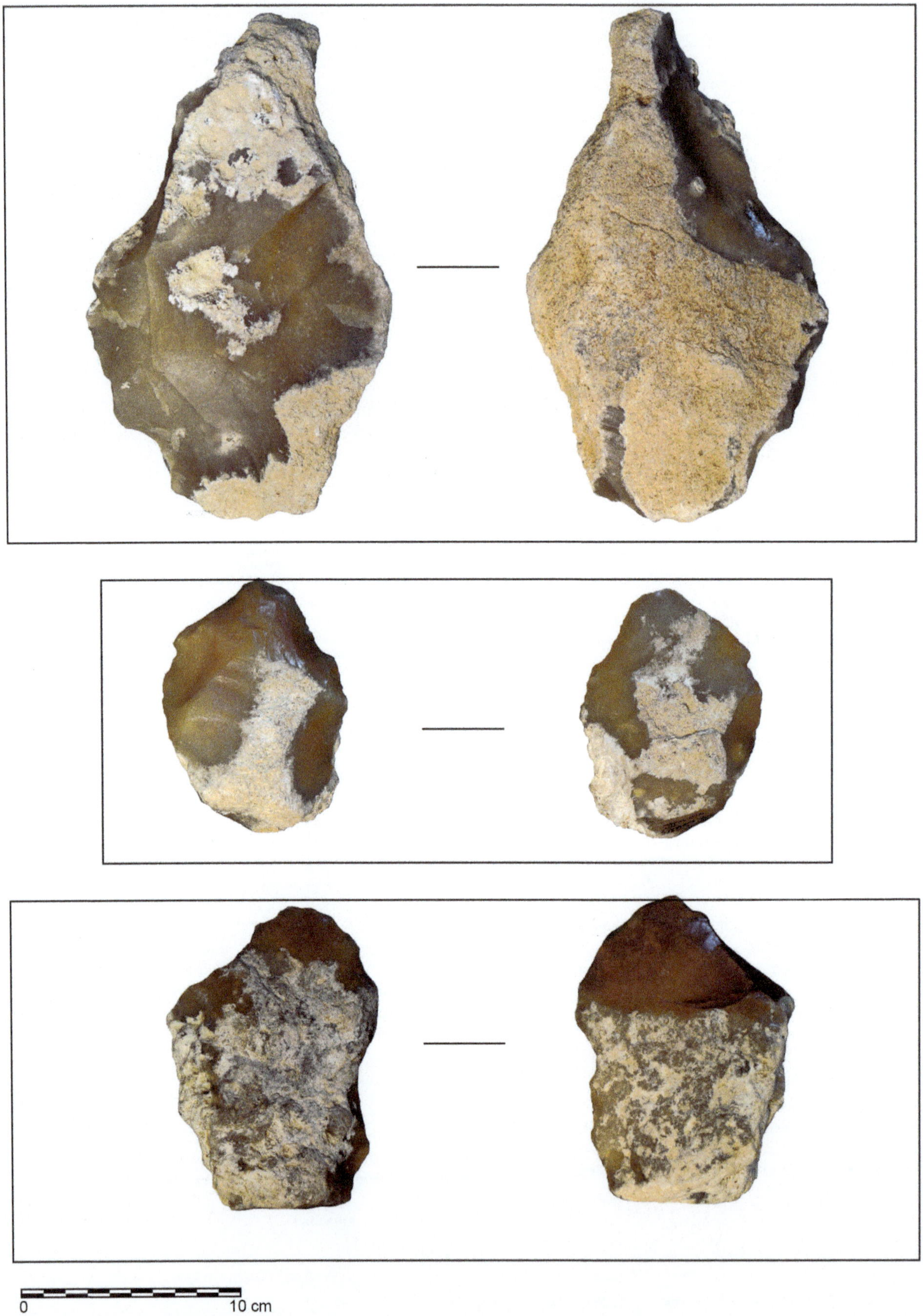

Figure 6. Examples of slabs with few removals. Crude cores or crude bifacial tools?

Table 3. Main core types from stratum a at la Noira.

	Millstone	Oolithic Jurassic chert	Total
Unifacial on one face or slab edge	6		6
Partial bifacial discoid cores	24	1	25
Total bifacial discoid cores	8		8
Bifacial discoid core (debitage of a surface)	1		1
Bifacial discoid or thick tool	1		1
Trifacial cores with few removals	4		4
Trifacial cores	3		3
Orthogonal simple or double	2		2
Multidirectional cores	5		5
Indeterminate cores (broken)	3		3
Total	57	1	58

transversal cutting edge created by a large lateral removal or several transversal removals included in the general bifacial shaping of the tool. In addition to these LCTs, the series also includes some massive "*rabot*" scrapers.

1)The backed biface (worked tip) and backed bifacial lateral tool are two variants (Figure 10). The first is a bifacial tool associating two convergent tools managed by bifacial removals and a partial natural lateral back sometimes partially reduced by deep and large removals. They are symmetrical while the section is either symmetrical or plano-convex. There are two successive series of removals, with the second one formatting the upper half of the two convergent edges. The pointed tip is rectified by final retouch. The edge opposite the back is also rectified by continuous retouch suggesting additional functional areas (knife-type) or prehensile parts.

The second variant displays alternate or face by face shaping by a short series of thin removals in order to manage a bifacial partial peripheral cutting edge. Cortical surfaces are largely preserved and tools are not symmetric in shape. The section is symmetrical or plano-convex, except for one twisted piece. The edge opposite the back (unworked or slightly worked) is reworked by a second series of unifacial retouch or short removals. Tool tip management is incorporated into the general shaping. In spite of secondary retouch, the cutting edge remains sinusoidal with angles of approximately 70–80°.

2)Two groups may be distinguished. Bifacial tools with distinct bifacial areas and a worked tip are characterized by peripheral and invasive bifacial shaping and convergent edges. The tip is carefully worked. Both the organization of the scars and the location and discontinuity of the final retouch suggest distinct sharpened areas. Two notches create a "*déjeté*" bec on the tool tip on two pieces (Figure 11).

Bifacial tools with distinct bifacial areas but without a specific tip are similar to the previous type (Figure 12). We observe: 1) Bifacial peripheral shaping by short and thin removals in one, two or three alternate or face by face series. Tip and base management are included in this general shaping. Micro-retouch is preferentially located on the two convergent edges and the base and three other areas may be distinguished. For one tool, a "nose scraper" is prepared on a lateral edge. Edges are sinusoidal and angle values are varied (from 55 to 85°). Tools are not always symmetric in the same way as the cross-section, 2) Bifacial and invasive shaping with large removals covering the two faces. A second series of short removals are sometimes located on one edge and the tip. Retouch is discontinuous on both edges.

3) Partial bifacial tools occupy the upper half or third of a slab (Figure 12, 13). Shaping is made by a small number of short removals except for one tool where a large unifacial removal has been detached in order to shape the tip. The tip morphology is oval and the broken tip of one tool suggests a functional use of the tool extremity. Additional retouch is rare and cutting edge angles vary between 50 and 80°. A second unifacial cutting edge is sometimes prepared on a lateral edge. Unworked single or double backs are preserved.

4) Bifaces *sensus stricto* are symmetrical tools demonstrating the pursuit of a symmetrical bifacial or bilateral equilibrium and a preconceived form (Figures 13, 14, 15). Removals cover both faces, except on flakes where the inferior face is less shaped than the upper surface (Figure 14). Three successive series of scars may be observed, made up of a first series of deep and invasive removals using face by face or alternate shaping mode, and a second series of shorter and thinner removals in order to finish formatting the bifacial volume. Then final retouch is made on parts of the sharp cutting edges and the tip which is worked by complementary short removals. Sections are plano-convex, symmetrical or twisted.

5) The bifacial cleavers are completely covered by numerous invasive bifacial removals on both faces (Figure 15). The transversal cutting edge is created by a large lateral predetermined removal from the initial shaping phase or several transversal removals included in the general bifacial shaping of the tool. The proximal part is oval or rectangular and worked by a series of removals. A second series of small removals rectifies some parts of the two lateral cutting edges and the transversal cutting edge (angle values are systematically close to 50°).

6) Finally, slab fragments (60 to 100 mm long and 11–40 mm thick) are made up of: 1) massive scrapers ("*rabot*" with abrupt or bifacial retouch including scalar retouch on one face). Opposite and inverse retouch on some massive scrapers may be interpreted either as the manufacture of a second tool or a prehensile part of the scraper, 2) partial or total oval bifacial tools, 3) bilateral unifacial tools, and 4) tools such as a "*bec*" made by abrupt retouch.

One to three series of removals may be identified and these rarely cover both faces completely. The first series is made up of deep hard hammer removals and the second is composed of thin and invasive removals in order to finish the shaping (Figure 16).

1

2

Figure 7. Examples of millstone cores of stratum a. 1. Partial bifacial core on millstone slab with deep removals and a partial peripheral edge. 2. Millstone bifacial core with a careful facetted configuration of the platform.

The shape of the second series of removals indicates the use of a soft hammer. The mode of shaping (alternate or face by face) is partially related to the slab cross-section for the first phases. When the blank is a flake, the inferior face is less covered by scars than the upper face and the tool is minimally shaped. Lastly, on most of the tools, unifacial retouch rectifies some areas of the edges and tip and does not systematically produce regular cutting edges (44% of the edges are sinusoidal). This ultimate retouch could represent resharpening in some cases. It attests to the control of tool edge management. Tips are often thinned by additional short removals. Butts are mainly worked by general shaping removals (38%) or else remain cortical (with a natural slab edge).

From a typological point of view, the LCT morphologies are diverse: cordiform, triangular, oval or amygdaloid-type regardless of the tool category. The tip is either oval or pointed, except for the three bifacial cleavers, whatever the shaping mode. The tip section is triangular, trapezoidal or oval. Neither size nor the elongation index (1–1.5) distinguishes the categories. The thickest part of the LCT is either the butt or the central part of the tool, depending on the aim and intensity of shaping (remains of large cortical surfaces of the slab).

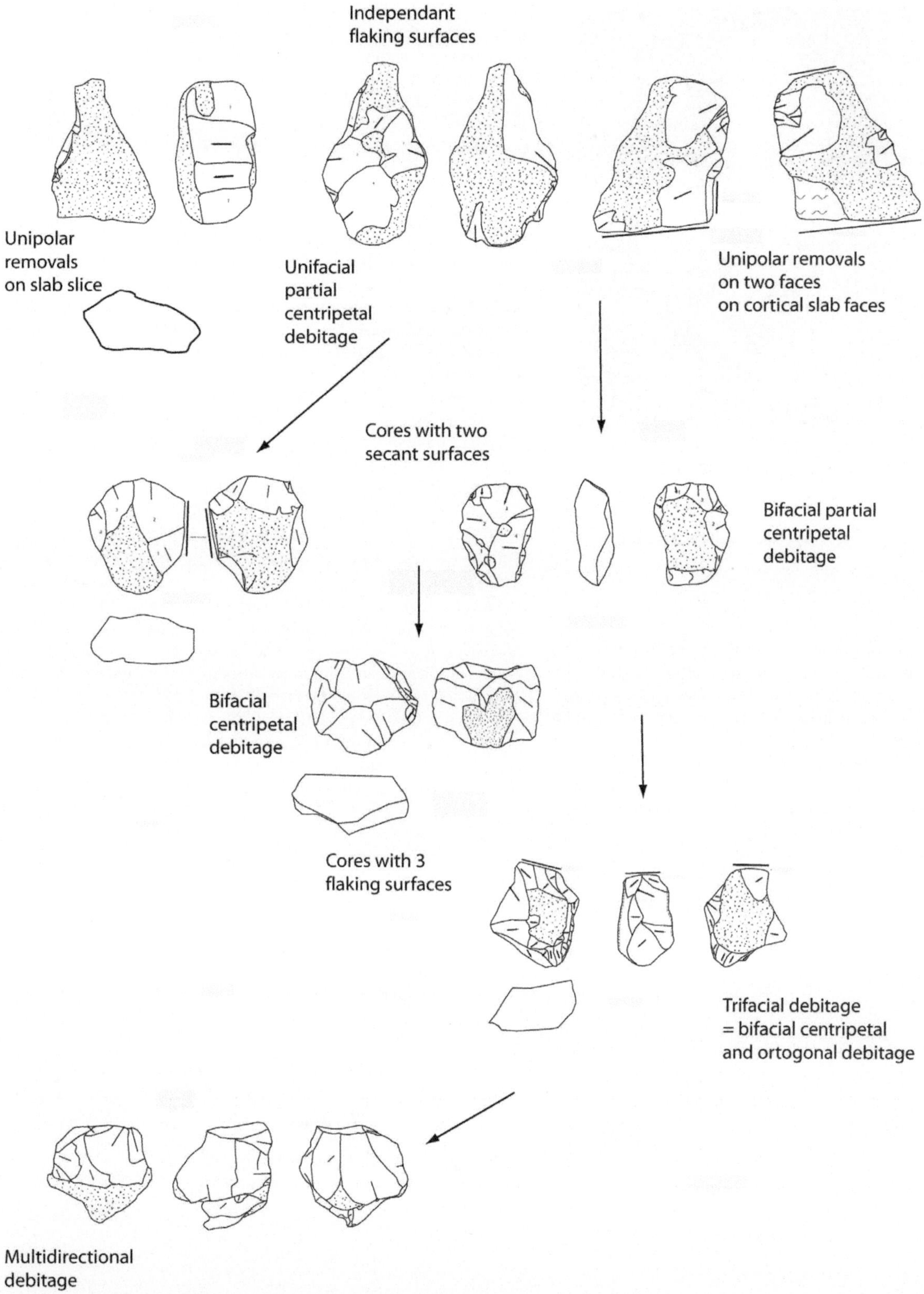

Figure 8. Debitage reduction sequence at stratum a at la Noira.

Specifically, 60% of the LCTs are asymmetrical in shape except for bifaces *s.s.* with a bifacial or bilateral equilibrium and a preconceived form. Cross sections are plano-convex or symmetric in relation to the shaping mode, whatever the categories and the final morphological results. They attest to the limited role of slab morphology even if the thinnest slabs were selected for LCTs.

Figure 9. Flakes issued from shaping and flaking.

Some are twisted. Cutting edge angles generally vary from 50 to 80°.

Discussion

In spite of the polymorphism of lithic series, whereby each site appears to present specific features, perhaps due to the limited number of available sites in Europe, the la Noira assemblage conforms well with the 600–500 ka European sites with LCTs and the Acheulean (Table 5).

1) As at la Noira, the ratio of LCTs is limited in Europe, and generally varies from 1 to 5%, except for some sites, such as Boxgrove [25], while it is as high as 15% in the Levant (except at the early site of Ubeidiya with rare LCTs) and East Africa.

2) LCT diversity at la Noira is similar to that observed at Arago in levels P-Q, in terms of morphology, diversity of tool types and shaping modes [20]. Some of the LCTs show tip management and a bifacial equilibrium between the two faces. However, flake-cleavers, extensively shaped tools and raw material diversity oppose Arago to la Noira, where no evidence of temporal and spatial fragmentation of the LCT reduction sequence was observed. Although large flakes could have been produced on slabs at la Noira, few were used for shaping and they remained unretouched on the site. LCT diversity and standardization are lower at Notarchirico (levels B, D, F), with both numerous convergent chopping-tools and some rare bifaces *s.s.* [48]. At Boxgrove, ovate tools are standardized with final shaping and possible resharpening using the "*coup de tranchet*" [25].

The same diversity of core technology on local stones is observed on 800–500 ka series without LCTs (Pakefield-700 ka and Happisburg III-900 ka in Great-Britain, Vallparadis and TD6 Gran Dolina-800 ka in Spain, Isernia-504/610 ka in Italy) and with LCTs (Notarchirico-640 ka in Italy, Arago levels P and Q-550 ka, Boxgrove-550 ka in Great-Britain) [46,3,4,20,47]. La Noira cores are mainly discoidal and they attest to the minor role of raw material shape, like for most of the LCTs of this site. Slabs were selected in accordance with shaping and flaking activities, with the thinnest slabs being reserved for LCTs.

Multiple examples in Europe and in Africa indicate that parameters such as activities [49,48,18], occupation duration, raw materials [50,51,52,53,54,55,64], environments or regional variants [38,56,57] affect the composition of assemblages, erase inter-site differences and may explain flexibility and some aspects of the diversity of the heavy-duty components (Clactonian *versus* Acheulean) and selective raw material provisioning. Stratum a at La Noira records a workshop context and the use of slabs explains common and specific features observed on the series, both on cores and LCTs. Itconcurs well with other northwestern European sites characterized primarily by the use of flint and the presence of bifacial cleavers, whereas southern European series indicate the use of complementary raw materials and the presence of cleavers on coarse-grained stone flakes.

Since the first identification of the Acheulean by Gabriel de Mortillet (1872) and the earliest use of the term biface by Vayson de Pradennes (1920) in the Somme Valley in France, many definitions and classifications have been proposed, leading to contradictory debates [58,38]. Some of these discussions are typological and focus on the biface and its morphology, whereby the biface is considered as a "*fossile directeur*". Others are linked to the handaxe ratio (for instance 40% for Kleindienst [59] or technological aspects such as the ability to produce large flakes which, associated with blocks or pebbles, are then used as blanks for a wide diversity of LCTs, unifacial and bifacial tools, which are either minimally shaped with distinct functional areas or extensively shaped in order to manage a bifacial volume and a

Table 4. Major types of LCTs at la Noira stratum a. Number and size.

	On slab			On flake	Indet.
	Number	Minimum length in mm	Maximum length in mm		
Biface *s.s.*	5	86	153	1 (149-91-35 mm)	2
Backed biface (worked tip)	3	108	153		
Backed bifacial tool	6	105	235	1 (118-82-44 mm)	
Convergent bifacial tool	2	66	120		
Distinct bifacial areas+worked tip	3	88	122		2
Double or multiple bifacial tools	3	90	170		4
Partial bifacial tool	4	57	160		
Bifacial cleaver	3	80	115		
Massive scrapers. Partial LCTs	9	51	115		4

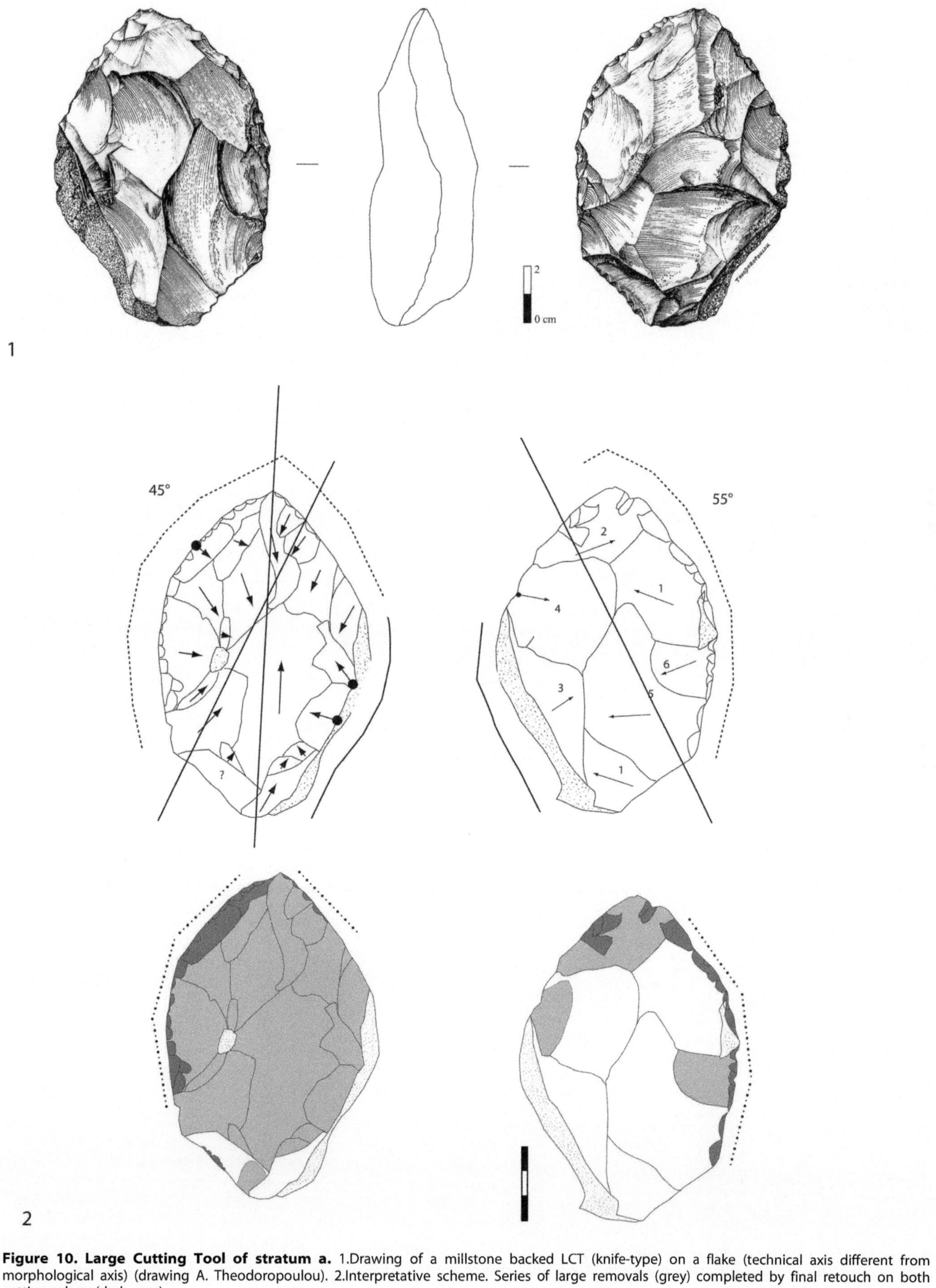

Figure 10. Large Cutting Tool of stratum a. 1.Drawing of a millstone backed LCT (knife-type) on a flake (technical axis different from morphological axis) (drawing A. Theodoropoulou). 2.Interpretative scheme. Series of large removals (grey) completed by final retouch on both cutting edges (dark grey).

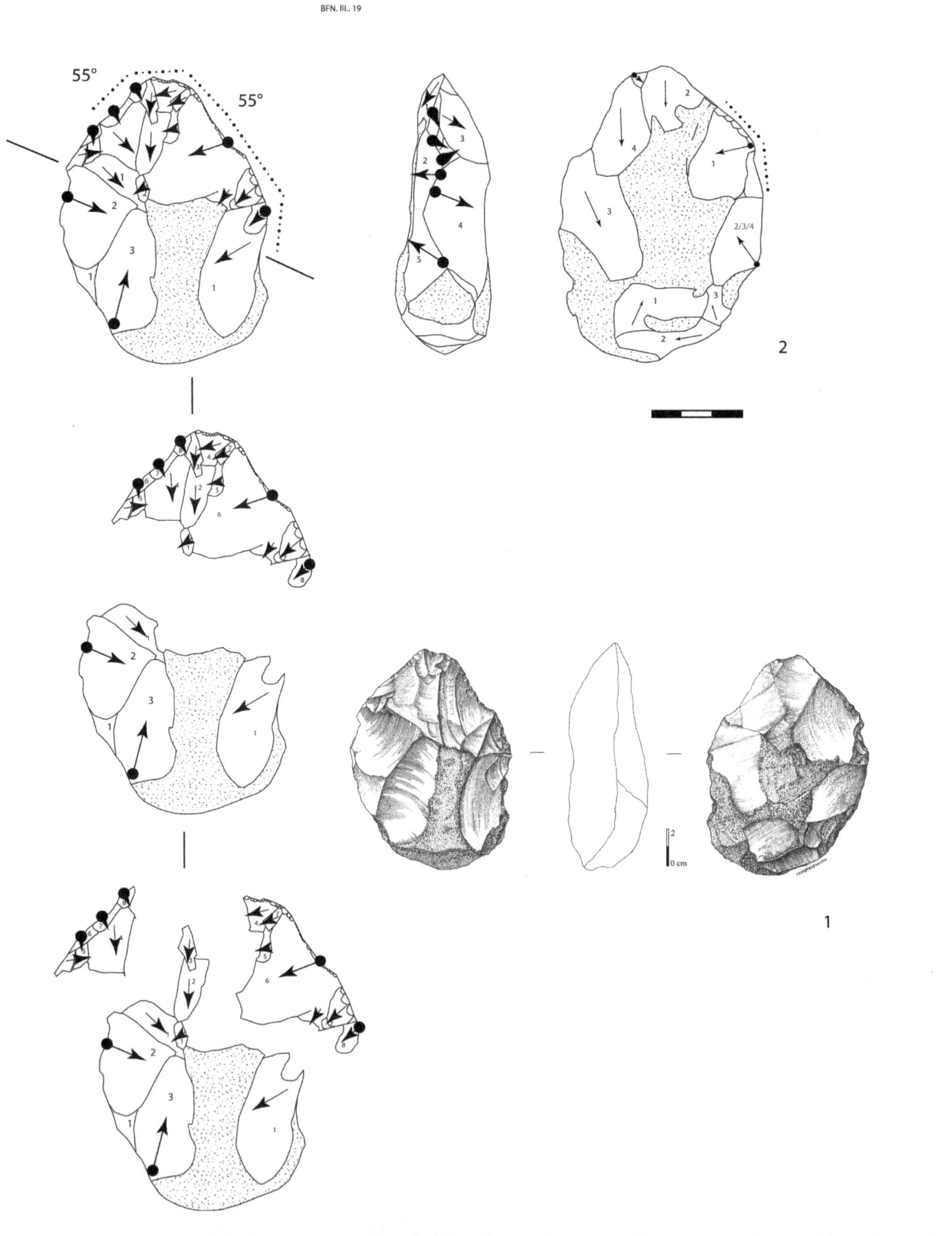

Figure 11. Large Cutting Tool of stratum a. 1.Drawing of a minimally shaped LCT on a millstone nodule with an oval shape (drawing A. Theodoropoulou). 2.Interpretative scheme. View of the different series of removals located on the two convergent edges and the tip.

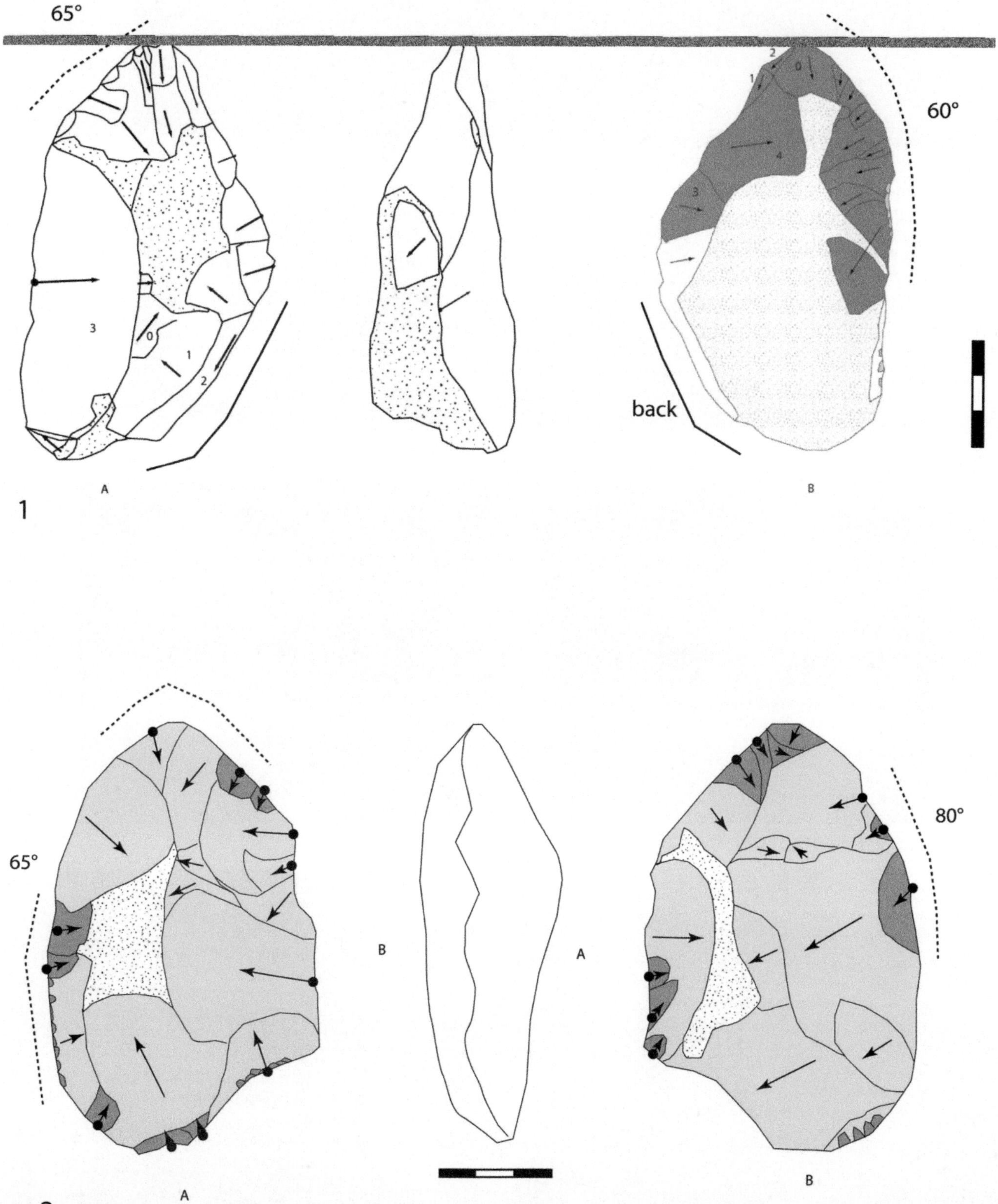

Figure 12. Large Cutting Tools of stratum a. 1. Minimally shaped pointed LCT in millstone. A series of large removals (dark grey) for managing the upper part of the tool. The distal convergent edges and the tip are carefully worked. 2. Extensively shaped pointed LCT in millstone with invasive removals (grey) and edge thinning (dark grey).

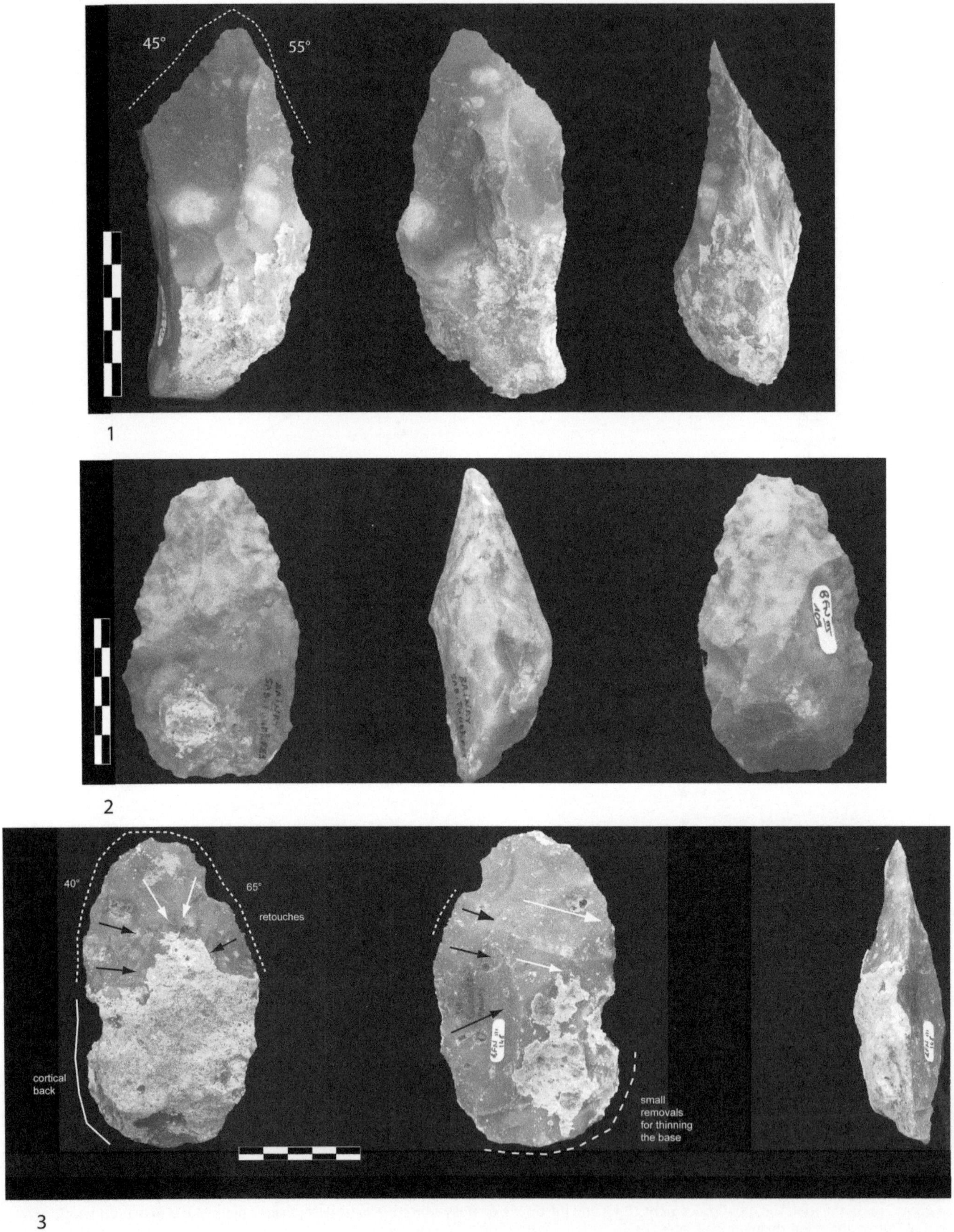

Figure 13. Large Cutting Tools of stratum a. 1. Pointed LCT with shaping of the tip by a series of more or less invasive removals. Diamond shaped cross-section. 2. Extensively shaped LCT with an oval tip. Asymmetrical cross-section. Large removals cover both faces by alternate shaping

completed by smaller thinning removals. 3. Minimally shaped symmetrical LCT on a thin millstone slab. Alternate shaping and invasive removals are completed by the thinning of the upper part of the tool. Partial cortical back.

tip. There was an application of preconceived morphological templates, which are flexible and adapted to locally available raw materials. This feature may or may not be associated with complex core technology. Since the dichotomy proposed by Leakey [57] between the Developed Oldowan and the Early Acheulean, a large diversity of assemblages and "savoir-faire" have been brought to light both in Africa and Eurasia and the authors of these lithics are commonly referred to as "Acheuleans" [39,8,60,56,61,62,41,63,64].

The Early Acheulean is not just defined by the manufacture of LCTs requiring specific and common skills (for instance, the ability to create bilateral equilibrium, maintenance of cross-section and plan-view symmetry and final shaping of the tip and cutting edge) or by the systematic production of large flakes (>10 cm long) using various methods. Many assemblages, where LCTs exist, also yield unsymmetrical tools and lack bifacial tools and minimally shaped LCTs on flakes or blocks. Other parameters are also required to designate the emergence of the Acheulean and new behavioral criteria occasionally materialize as early as 1.5 Ma [41,44]: higher mobility, more selective raw material gathering when necessary, site location, variety of ecological settings, mobile LCTs and curated pebble tools (multifunctional LCTs, e.g.[44].

Given the shaping technology of the la Noira assemblage, this is an Acheulean assemblage which can be considered as introduced. A local Mode 1 origin and the presence of *Homo antecessor* populations do not appear to be relevant here. It is important to underline the lack of evolutionary features in Mode 1 in Europe before the Acheulean. Very few European sites have yielded remains of pre-bifacial technology with large flakes and partial LCTs. Bogatyri (1.1 Ma) in Russia, Pradayrol (900 ka) and Soleilhac (700-500 ka) have recorded some bifacial endeavors, as observed in East Africa in Oldowan or Developed Oldowan assemblages [65,66]. The la Noira series does not exhibit Early African Acheulean features (Olduvai, Peninj and Gadeb) such as unifacial tools, picks and numerous heavy duty tools, such as bolas or spheroids [41,44,39]. The Acheulean technology at la Noira is advanced and very different from that of the African basal Acheulean. Shaping at la Noira differs from these African assemblages in that it aims to produce many tools with a general LCT bifacial shape by successive series of removals, even if there are also minimally shaped LCTs in the series. Moreover, secondary retouch for rectifying cutting edges and soft hammer use are attested here, whereas in Africa these are common from 1 Ma ago, leading to tool standardization at 700 ka [67,68,69]. Thus we cannot postulate that the emergence of new technical behavior at la Noira is issued from a local European substratum, as described in the Early African Acheulean.

Figure 14. Large Cutting Tools of stratum a. 1. LCT on a millstone flake with large removals on the upper part of the tool and thinning by shorter removals and final retouch of the distal cutting edges and tip. 2. Oval millstone LCT with an asymmetrical cross-section and alternate shaping.

Since the la Noira stratum a lithic assemblage shares some features with the 800-500 ka Levantine and African series, two scenarios for the arrival of bifacial technology at 700 ka in Europe are consequently relevant: 1) The arrival of new hominins from the Levant along the Mediterranean coast (*Homo heidelbergensis*?) or new traditions which remained long enough in this part of the world to acquire local features, either through a process of accumulation of behaviors by multiple arrivals or transit between Africa and the Levant inducing behavioral influences. 2) Direct arrivals from Africa through the Levant or North Africa. Direct dispersals from Asia can be excluded due to the specificity of the Indian series and the lack of LCTs in Central Asia and Central Europe. In the absence of clearly dated ancient sites in Spain [19] and even though periods of low sea level are confirmed at ≈700-650 ka [70], little evidence suggests transit from Africa to Gibraltar. The only basis for this is the presence of large flake-cleavers on both sides of the Mediterranean [52,71]. Flake-cleavers in Spain and southwestern France (including the site of Arago, levels P and Q) are considered as African features and constitute possible evidence of crossing through Gibraltar. In the North of Europe, cleavers are only made by bifacial removals on nodules and we are unable to discern whether they correspond to the same type of tool with a transversal cutting edge managed by a methodology adapted to other raw material shapes, a behavioral specificity of these areas or another kind of tool.

Between 1 Ma and 600 ka ago in Africa, diverse hominins, *Homo erectus* and *Homo rhodensiensis*, were capable of manufacturing bifacial tools [26] during a period considered to represent a technological gap ("classical" Acheulean, soft hammer at 1 Ma ago, occasional blade technology at 500 ka [72]). Assemblages including LCTs display inter-site diversity and tool standardization (e.g. Thomas Quarry I level L −1 Ma, Olorgesailie-900-600 ka, Garba XII level J-900 ka, Gombore II-800 ka, Tighenif-800 ka [56,73,74,75,76]. Twisted LCTs from la Noira resemble obsidian bifaces from Gombore [42] and shaping modes are similar to those employed at Isenya-700 ka [69]. From the Levantine corridor to the Arabian Peninsula, the dearth of sites prevents us from reconstructing Acheulean filiations. Even so, sites such as GBY-790 ka, El Meirah and Latamne-700-500 ka in Syria, Karakhach

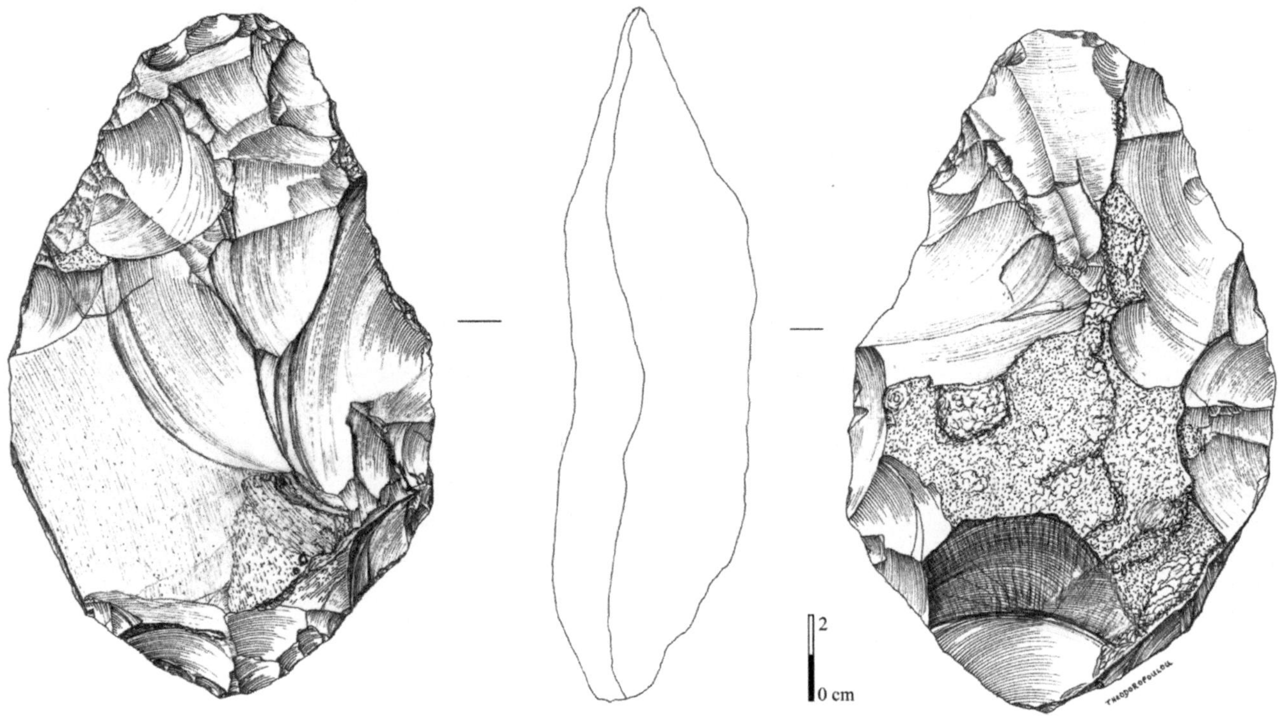
2
0 cm

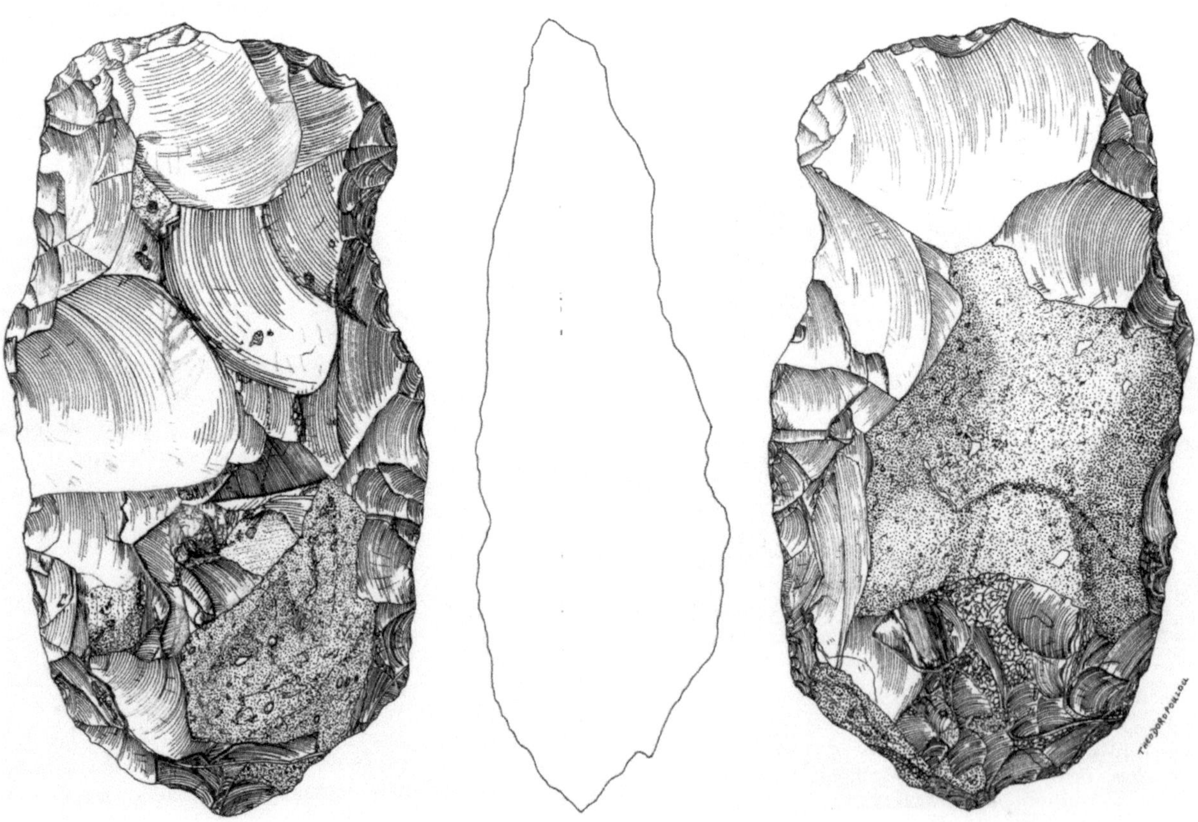

Figure 15. Large Cutting Tools of stratum a. 1.LCT on a millstone flake (drawing of LCT of figure 14). 2.Bifacial cleaver on a millstone slab (drawings A. Theodoropoulou).

I in Armenia, Joubb Jannine (900 ky) in Lebanon or El Kowm Basin localities suggest a mosaic of regional variants and layers of local traditions and traditions with African affinities [49,9,77,78,79,80].

Conclusion

The lithic assemblage of la Noira provides evidence of the Acheulean tradition at 700 ka in the center of France. The age of 700 ka establishes an early Acheulean presence in Europe and modifies our current vision of the initial peopling of northern Europe. The Acheulean is thus older than previously assumed and the oldest evidence is no longer located in southern Europe.

The hominins who made the LCTs found in stratum a were present at la Noira after the period of river incision, at the beginning of a glacial climatic period (MIS 16), before the pleniglacial phase, as demonstrated by the cryoturbation of the archaeological level. They settled at the bottom of the slope along the river and beside suitable local slabs. The archaeological remains were then overlaid by fluvial sand attributed to the next interglacial stage. Seven ESR ages obtained on these deposits point towards the attribution of the sequence to the MIS 16/15 climatic cycle.

The la Noira stratum a yielded wide LCT diversity and a structured core technology as early as 700 ka. The lithic assemblage suggests pioneering phases of bifacial technology before 500 ka in Europe. Low hominin densities, adaptation to Europe, distance from the core area and perhaps hominin diversity, like in East Africa, could account for this multiplicity [81,82]. LCT diversity, relative independence from the natural morphology of the raw material and different degrees of planning in the knapping process provide evidence of an external Acheulean tradition at la Noira, even though there is no temporal or spatial fragmentation of the LCT reduction sequence and the procurement system is geared almost exclusively towards local stone (workshop context).

These pioneering phases would have taken place on a Mode 1-type substratum, possibly encountering Mode 1 groups even if population densities were low, and adapting to different vegetal and animal landscapes than those in the Levant or Africa. Environmental and chronological data from northwestern sites suggest occupations during climatically favorable periods (temper-

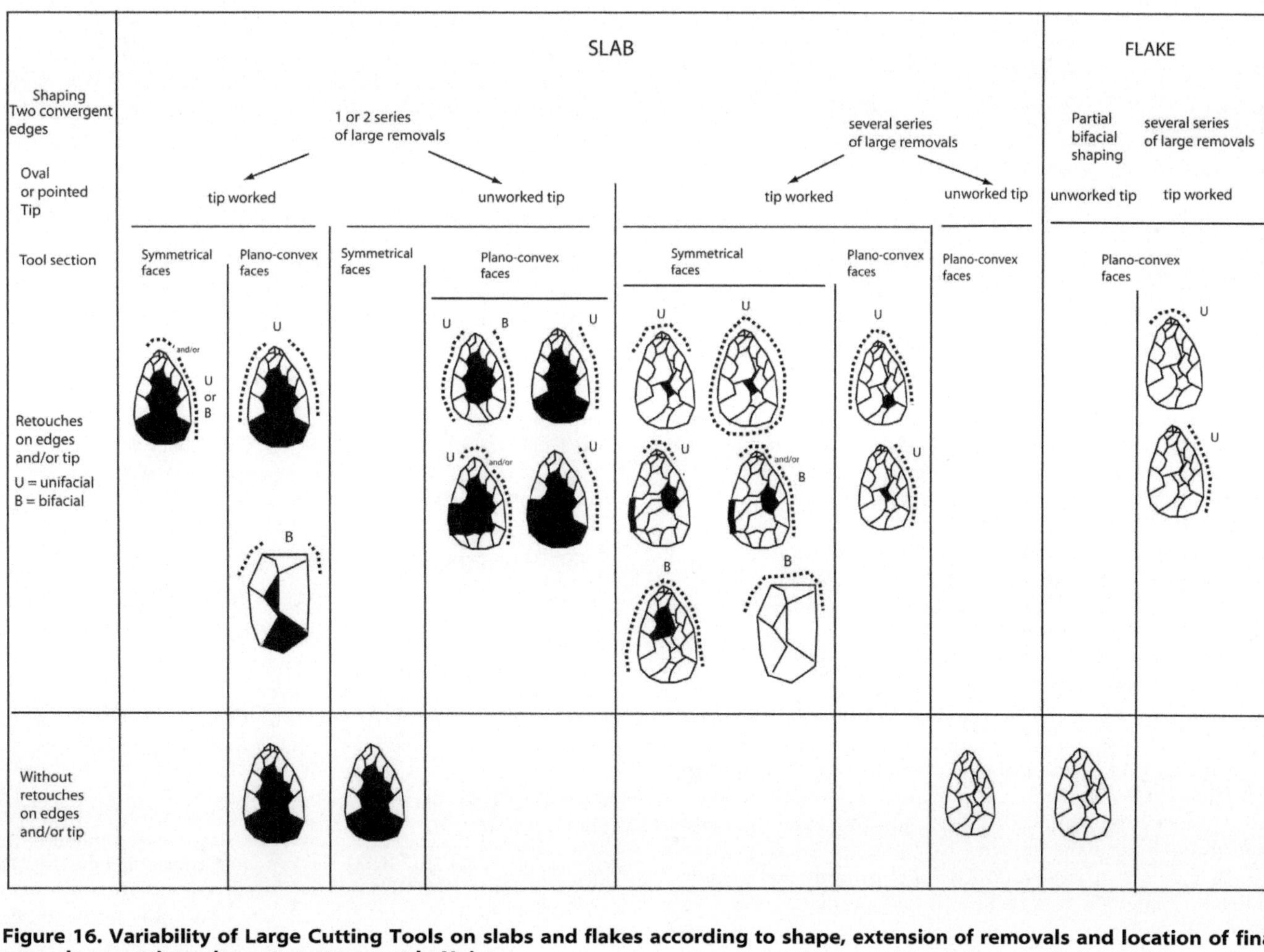

Figure 16. Variability of Large Cutting Tools on slabs and flakes according to shape, extension of removals and location of final retouch on cutting edges at stratum a at la Noira.

Table 5. 800-600 ka European sites [89–97].

Country	Sites	Ages	MIS	Methods	References
European sites without LCTs between 800 and 600 ka					
England	Happisburg 3	814–870 ka	21	Biostratigraphy	[4]
		970-936 ka	25	Paleomagnetism	
France	Pradayrol	900 ka	21, 22	Biostratigraphy	[65]
Spain	Gran Dolina TD6	731±63 ka	19,25	Biostratigraphy - TL - IRSL	[89,91,94]
		857-780 ka	21	Paléolagnetism - ESR/U-Th	
Spain	Vallparadis	790±230		Biochronology	[47,92,93]
		830±130 ka		ESR-ESR/US - Paleomagnetism	
England	Pakefield	700 ka		Paleomagnetism-Amino-acid-Biostratigraphy	[3]
France	Soleilhac	700-500 ka		Biostratigraphy – Paleomagnetism - Tephras	[46,90,97]
Italy	Isernia-la-Pineta	610 ka		ESR/US	[46]
		602±2 ka		Ar/Ar	
Sites with *Homo antecessor* remains					
Spain	Gran Dolina TD6	731±63 ka	19,25	Biostratigraphy-TL - IRSL	[89,91,94]
		857-780 ka	21	Paleomagnetism - ESR/U-Th	
European sites with LCTs between 700 and 500 ka					
Italy	Nortarchirico	700-600 ka		Tephras - Racemisation	[95]
		640±70 ka		TL	
France	Brinay, la Noira	665±55 ka	16	ESR	[6,36]
France	Morandière	610±60 ka	15	ESR	[6,35,37]
France	Villefranche/Cher	-	15	Stratigraphy	[6,35,37]
France	Tautavel-Arago	550 ka		Biostratigraphy – ESR/US - U series	[30]
England	Boxgrove	550 ka	13	Biostratigraphy	[96]
England	Happisburg 1	-	13	Amino-acid - Biostratigraphy	[4]
Sites with *Homo heidelbergensis* remains					
Germany	Mauer	609 ka		Biostratigraphy – ESR/US – IR/RF	[30]
France	Tautavel-Arago	550 ka		Biostratigraphy – ESR/US - U series	[20]

ate climate). The specific time period in question here is centered on the Middle Pleistocene Transition (MPT), after the Brunhes-Matuyama shift at 800 ka leading to an extension of the grassland habitat into higher latitudes, thereby opening or closing corridors [83,84,85,86,87]. Cyclical climate changes (glacial/interglacial) would have led to successive depopulation or extinction, and the subsequent recolonization before and between the MIS 16 and 12 cold events [27]. The second later transition (Mid-Brunhes Event –MBE), occurring between MIS 13 and 11, would have allowed for a more widespread diffusion of elaborate bifacial manufacture techniques. Vital information is missing for determining pathways and timing, and filiation appears to be impossible to resolve due to the influence of activities and environmental patterns on behavior. Relationships with faunal turnover across Eurasia are not always evident, due to a lack of clear mammalian dispersals from Africa to Eurasia and through the Saharan belts. Relationships between changes in carnivore associations (around 500 ka) and the demise of feline megahunters as early as 1 Ma ago, climatic changes and hominid dispersals advance contradictory arguments [88]. These environmentally linked changes could have promoted sporadic hominin expansion to Europe aided by new technics and new social organization [61,79]. It is not yet clear precisely which advantages *Homo heidelbergensis* held over *Homo antecessor*, although a larger brain; "brain expansion scenario", and a new technology may hold part of the answer.

Acknowledgments

The authors wish to thank the French ANR project (Agence Nationale pour la Recherche), the Department of Prehistory of the National Museum of National History, the Conseil Régional d'Ile-de-France and the Ministère de la Culture, Sous-direction de l'Archéologie, Service Régional de l'Archéologie of the Centre Region (France).

We would like also to thank warmly the editor Fred Smith and the two anonymous reviewers who helped us to improve this paper.

Author Contributions

Conceived and designed the experiments: JD MHM. Performed the experiments: JD PV JJB MHM HT DM GC CF. Analyzed the data: JD PV JJB MHM HT DM CF. Contributed reagents/materials/analysis tools: MHM JD GC. Wrote the paper: MHM JD PV JJB. Fieldwork and excavation direction: JD MHM. Lithic series analysis: MHM JD. Geological studies: JD PV JJB. Dating analysis: PV HT DM JJB CF. Mapping: GC. Illustrations: MHM JD PV JJB GC.

References

1. Carbonell E, Garcia-Anton MD, Mallol C, Mosquera M, Ollé A, et al. (1999) The TD6 lithic industry from Gran Dolina, Atapuerca (Burgos, Spain) production and use. J Hum Evol 37: 653–693.
2. Arzarello M, Marcollini F, Pavia G, Pavia M, Petronio C, et al. (2006) Evidence of earliest human occurrence in Europe: the site of Pirro Nord (Southern Italy). Naturwissenschaften 94: 107–112.
3. Parfitt SA, Barendregt RW, Breda M, Candy I, Collins MJ, et al. (2005) The earliest record of human activity in northern Europe. Nature 438(15): 1008–1012.
4. Parfitt SA, Ashton N, Lewis SG, Abel RL, Coope GR, et al CB (2010) Early Pleistocene human occupation at the edge of the boreal zone in northwest Europe. Nature 466: 229–233.
5. Carrión JS, Rose J, Stringer C (2011) Early Human Evolution in the Western Palaearctic: Ecological Scenarios. Quat Sci Rev 30(11–12): 1281–1295.
6. Despriée J, Voinchet P, Tissoux H, Bahain JJ, Falguères C, et al. (2011) Lower and Middle Pleistocene human settlements recorded in fluvial deposits of the middle Loire River Basin, Centre Region, France. Quat Sci Rev 30(11–12): 1474–1485.
7. MacDonald K, Martinez-Torres M, Dennell RW, Bermudez de Castro JM (2012) Discontinuity in the record for hominin occupation in south-western Europe: Implications for occupation of the middle latitudes of Europe. Quat Int 271(0): 84–97
8. Clark G (1969) World Prehistory: A New Outline (Cambridge University Press, London).
9. Bar-Yosef O, Goren-Inbar N (1993) The Lithic Assemblages of Ubeidiya. A Lower Palaeolithic site in the Jordan Valley (Quedem, Jerusalem). 266 p.
10. Doronichev VB (2008) The Lower Palaeolithic in Eastern Europe and the Caucasus: A Reappraisal of the Data and New Approaches, PaleoAnthropology: 107–157.
11. Sharon G (2010) Large flake Acheulian. Quat Int 223–224: 226–233
12. Lepre CL, Roche H, Kent DV, Harmand S, Quinn RL, et al. (2011) An earlier origin for the Acheulian. Nature 477: 82–85.
13. Pappu S, Gunnell Y, Akhilesh K, Braucher R, Taieb M, et al. (2011) Early Pleistocene Presence of Acheulian Hominins in South India. Science 331: 1596–1599.
14. Beyene Y, Katoh S, WoldeGabriel G, Hart WK, Sudo M, et al. (2013) The characteristics and chronology of the earliest Acheulean at Konso, Ethiopia. Proc Natl Acad Sci USA 110 (5): 1584–1591.
15. Goren-Inbar N, Feibel CS, Verosub KL, Melameb Y, Kislev ME, et al. (2000) Pleistocene Milestones on the Out-of-Africa Corridor at Gesher Benot Yakov, Israel. Science 289: 944–947.
16. Hou Y, Potts R, Baoyin Y, Zhentang G, Deino A, et al. (2000) Mid-Pleistocene Acheulean-like Stone Technology of the Bose Basin, South China. Science 287: 1622–1626.
17. Dennell RW (2009) Palaeolithic Settlement of Asia. Cambridge University Press, Cambridge. 548 p.
18. Tuffreau A, Lamotte A, Goval E (2008) Les industries acheuléennes de la France septentrionale. L'Anthropologie, Paris 112: 104–139.
19. Jiménez-Arenas JM, Santonja M, Botella M, Palmqvist P (2011) The oldest handaxes in Europe: fact or _artefacts? J Archaeol Sci 38(12): 3340–3349.
20. Barsky D, de Lumley H (2010) Early European Mode 2 and the stone industry from the Caune de l'Arago's archeostratigraphical levels "P". Quat Int 223–224: 71–86.
21. Herisson D, Airvaux J, Lenoble A, Richter D, Claud E, et al. (2012) Le gisement acheuléen de La Grande vallée à Colombiers (Vienne, France): stratigraphie, processus de formation, datations préliminaires et industries lithiques. Paleo 23: 1–18.
22. Scott GR, Gibert L (2009) The oldest hand-axes in Europe. Nature 461: 82–85.
23. Carbonell E, Bermùdez de Castro JM, Arsuaga JL, Allue E, Bastiret M, et al. (2005) An Early Pleistocene hominin mandible from Atapuerca-TD6, Spain. Proc Natl Acad Sci USA 102(16): 5674–5678.
24. Carbonell E, Bermudez de Castro J, Pares J, Perez-Gonzalez A, Cuenca-Bescos G, et al. (2008) The first hominin of Europe. Nature 452: 456–470.
25. Roberts M, Parfitt S (1999) Boxgrove: A Middle Pleistocene Hominid site at Eartham Quarry. English Heritage, West Sussex, England. 356 p.
26. Rightmire GP (2001) Patterns of hominid evolution and dispersal in the Middle Pleistocene. Quat Int 75: 77–84.
27. Manzi G (2004) Human Evolution at the Matuyama-Brunhes boundary. Evol Anthropol 13: 11–24.
28. Hublin JJ (2009) The origin of Neandertals. Proc Natl Acad Sci USA 106(38): 16022–16027.
29. Mounier A, Marchal F, Condemi S (2009) Is *Homo heidelbergensis* a distinct species? New insight on the Mauer mandibule. J Hum Evol 56: 219–246.
30. Wagner GA, Krbetschek M, Degering D, Bahain JJ, Shao Q et al. (2010) Radiometric dating of the type-site for *Homo heidelbergensis* at Mauer, Germany. Proc Natl Acad Sci USA 107(46): 19726–19730.
31. Stringer C (2012) The Status of *Homo heidelbergensis* (Schoetensack 1908). Evol Anthropol 21: 101–107.
32. Bischoff JL, Williams RW, Rosenbauer RJ, Aramburu A, Arsuaga JL, et al. (2007) High-resolution U-series dates from the Sima de los Huesos hominids yields 600±66 kyrs: implications for the evolution of the early Neanderthal lineage. J Archaeol Sci 34: 763–770.
33. Martinòn-Torres M, Bermùdez de Castro JM, Gòmez-Robles A, Arsuaga JL, Carbonell E, et al. (2007) Dental evidence on the hominin _dispersal during the Pleistocene. Proc Natl Acad Sci USA 104: 13279–13282.
34. Premo LS, Hublin JJ (2009) Culture, population structure, and low genetic diversity in Pleistocene hominins. Proc Natl Acad Sci USA 106(1): 33–37.
35. Despriée J, Voinchet P, Bahain JJ, Tissoux H, Falguères C, et al. (2007) Les nappes alluviales pléistocènes de la vallée moyenne du Cher (Région Centre, France): contexte morphosédimentaire, chronologie ESR et préhistoire. Premiers résultats. Quaternaire 18(4): 349–368.
36. Despriée J, Voinchet P, Tissoux H, Moncel MH, Arzarello M, et al. (2010) Lower And Middle Pleistocene Human Settlements In The Middle Loire River Basin, Centre Region, France. Quat Int 223–224: 345–359.
37. Voinchet P, Despriée J, Tissoux H, Falguères C, Bahain JJ, et al. (2010) ESR chronology of alluvial deposits and first human settlements of the Middle Loire Basin (Region Centre, France). Quat Geochronol 5(2–3): 381–384.
38. Bordes F (1961) Typologie du Paléolithique _inférieur et moyen (Delmas, Bordeaux). 460 p.
39. Leakey MD (1971) Olduvai Gorge: Excavations in Bed I and Bed II, 1960–1963 (Cambridge University Press, Cambridge).
40. Boëda E, Geneste JM, Meignen L (1990) Identification de chaînes opératoires lithiques du Paléolithique ancien et moyen. Paleo 2: 43–80.
41. Torre de la I, Mora R, Martinez-Moreno J (2008) The early Acheulean in Peninj (Lake Natron, Tanzania). J Anthropol Arch 27: 244–264.
42. Gallotti R, Collina C, Raynal JP, Kieffer G, Geraads D, et al. (2010) The Early Middle Pleistocene Site of Gombore II (Melka Kunture, Upper Awash, Ethiopia) and the Issue of Acheulean Bifacial Shaping Strategies. Afr Archaeol Rev 27: 291–322.
43. Antoine P, Limondin-Lozouet N, Chaussé C, Lautridou JP, Pastre JF, et al. (2007) Pleistocene fluvial terraces from northern France (Seine, Yonne, Somme): synthesis and new results. Quat Sci Rev 26: 2701–2723.
44. Tyron CA, Mc Breaty S, Texier PJ (2005) Levallois lithic technology from the Kapthurin Formation, Kenya: Acheulian origin and middle Stone Age diversity. Afr Archaeol Rev 22(4): 199–229
45. Torre de la I (2011) The Early Stone Age lithic assemblages of Gadeb (Ethiopia) and the Developed Oldowan/early Acheulean in East Africa. J Hum Evol 60(6): 768–812.
46. Coltorti M, Feraud G, Marzoli A, Peretto C, Ton-Thate T, et al. (2005) New 40Ar/39Ar, stratigraphic ans paleoclimatic data on the Isernia La Pineta Lower Palaeolithic site, Molise, Italy. Quat Int 131: 11–22.
47. Martinez K, Garcia J, Carbonell E, Agusti J, Bahain JJ, et al. (2010) A new Lower Pleistocene archaeological site in Europe (Vallparadis, Barcelona, Spain). Proc Natl Acad Sci USA 107(13): 5762–5767.
48. Piperno M, editor(1999) Notarchirico. Un sito del Pleistocene medio iniziale nel bacino di Venosa (Edizioni Osanna, Venosa, Italy). 621 p.
49. Clark JD (1967) The Middle Acheulian occupation of Latamne, Northern Syria. Quaternaria 9: 1–68.
50. Villa P (1981) Matières premières et provinces culturelles dans l'Acheuléen français. Quaternaria: 19–35
51. Ashton NM, White M (2003) Bifaces and raw materials: flexible flaking in the British Early Palaeolithic. In: Soressi M, Dibble H, editors. Multiple Approaches to the Study of Bifacial Technology. University Museum Monography, 115, University of Pennsylvania, Philadelphia. pp. 109–124.
52. Mourre V (2003) Implication culturelles de la technologie des hachereaux (Phd, Nanterre, Paris, Université de Paris X).
53. Sharon G (2008) The impact of raw material on Acheulian large flake production. J Archaeol Sci 35(5): 1329–1344.
54. Sharon G (2009) Acheulian Giant-Core Technology, Curr Anthropol 50(3): 335–367.
55. Sharon G, Alperson-Afil N, Goren-Inbar N (2011) Cultural conservatism and variability in the Acheulian sequence of Gesher Benot Yaaqov. J Hum Evol 60(4): 387–397.
56. Isaac G, Harris JWK, Kroll EM (1997) The Stone Artefact Assemblages/A comparative _study. In: Isaac G, editor. Koobi Fora Research Project. Oxford University Press. pp 262–362.
57. Leakey LSB (1951) Olduvai Gorge. A report on the evolution of the hand-axe culture in Beds I-IV (Cambridge University Press, Cambridge).
58. Breuil Abbé H, Kelley H (1954) Le Paléolithique ancien. Abbevillien. Clactonien. Acheuléen. Levalloisien. Bulletin de la Société Préhistorique Française LI(8): 1–18.
59. Kleindienst MR (1961) Variability within the late Acheulian assemblage in Eastern Africa. South African Archaeological Bulletin 16: 35–52.
60. Semaw S, Rogers M, Stout D (2009) The Oldowan-Acheulian Transition: Is there a "Developed Oldowan" Artifact Tradition? In: Camps M, Chauhan P, editors. Sourcebook of Paleolithic Transitions Springer. pp 173–194.
61. McPherron SP (2000) Handaxes as a mesure of the mental capabilities of early hominids. J Archaeol Sci 27(8): 655–663.
62. McPherron S (2006) What typology can tell us about Acheulian handaxe production. In: Goren – Inbar N, Sharon G, editors. Axe Age Acheulian Tool-making from Quarry to Discard. Equinox Publishing Ltd, London. pp 267–287

63. Winton V (2005) An Investigation of Knapping-skill Development in the Manufacture of Palaeolithic Handaxes. The necessary conditions for a uniquely hominin behavior. In: Roux V, Brill B, editors. Stone knapping. MacDonald Institute Monograph. pp 109–119.
64. Clark JD, Haynes CV (1969) An elephant butchery site at Mwanganda's village, Karonga, Malawi, and its relevance for Palaeolithic archaeology. World Archaeol 1: 390–411.
65. Turq A, Despriee J, Airvaux J, Texier JP, Maureille B (2012) A la conquête de l'ouest: il y a 1 million d'années en Europe (Ed. Maison de l'Histoire de France – Musée national de Préhistoire, Les Eyzies). 160 p.
66. Guadelli JL, Turq A, Seronie-Vivien R (2012) Le site de Pradayrol à Caniac-du-Causse (Lot). In: Turq A, Airvaux J, Despriée J, Texier PJ, editors. La conquête de l'Ouest Il y a un million d'années en Europe. Musée des Eyzies. Maison de l'Histoire de France. pp 136–137.
67. Roche H, Brugal JP, Lefevre D, Ploux S, Texier PJ (1988) Isenya: état des recherches sur un nouveau site acheuléen d'Afrique orientale. Afr Archaeol Rev 6: 27–55.
68. Roche H, Brugal JP, Delagnes A, Feibel C, Harmand S, et al. (2003) Les sites archéologiques plio-pléistocènes de la formation de Nachukui, Ouest-Turkana, Kenya: bilan synthétique 1997–2001. C R Palevol 2: 663–673.
69. Texier PJ, Roche H (1995) The impact of predetermination on the development of some acheulean chaînes opératoires. In: Bermudez de Castro J, Arsuaga JL, Carbonell E, editors. Evolucion humana en Europa y los yacimientos de la Sierra de Atapuerc. Junta de Castilla y Leon, Vallaloid. pp 403–420.
70. Santonja M, Villa P (2006) The Acheulian of Western Europe. In: Goren – Inbar N, Sharon G, editors. Axe Age Acheulian Tool-making from Quarry to Discard. Equinox Publishing Ltd. London. pp 429–479.
71. Sharon G (2011) Flakes Crossing the Straits? Entame Flakes and Northern Africa-Iberia Contact During the Acheulean. Afr Archaeol Rev 28: 125–140.
72. Wilkins J, Chazan M (2012) Blade production at 500 thousand years ago at Kathu Pan 1, South Africa: support for a multiple origins hypothesis for early Middle Pleistocene blade technologies. J Archaeol Sci 39(6): 1883–1900.
73. Raynal JP, Sbihi Alaoui FZ, Geraads D, Magoga L, Mohi A (2001) The earliest occupation of North Africa: the Maroccan perspective. Quat Int 75: 65–77.
74. Raynal JP, Sbihi-Alaoui FZ, Michib A, El Graoui M, Lefèvre D, et al. (2011) Contextes et âge des nouveaux restes dentaires humains du Pleistocène moyen de la carrière Thomas I à Casablanca (Maroc). Bulletin de la Société Préhistorique Française 108(4): 645–671
75. Texier PJ (2001) Du seuil de l'hominisation aux premières cultures: potentialité et apports de la taille expérimentale. Praehistoria 2: 51–61
76. Schick K, Clark JD (2003) Biface technological development and variability in the Acheulean industrial complex in the Middle Awash Region of the Afar Rift, Ethiopia. In: Soressi M, Dibble HL, editors. Multiples Approaches to the Study of Bifacial technologies. Museum of Archaeology and Anthropology. Philadelphia. University of _Pennsylvania. pp 1–30
77. Petraglia M (2003) The Lower Paleolithic of the Arabian Peninsula: occupations, adaptations, and dispersals. Journal of World Prehistory 17(2).
78. Yazbeck C (2004) Le Paléolithique du Liban: bilan critique. Paléorient 30(2): 111–126
79. Goren-Inbar N, Sharon G (2006) Axe Age: Acheulian Tool-making from Quarry to Discard (Equinox Publishing Ltd, London).
80. Jagher R, Le Tensorer JM (2011) El Kowm, a key area for the Palaeolithic of the Levant in Central Syria. In: Le Tensorer JM, Jagher R, Otte M, editors. The Lower and Middle Palaeolithic in the Middle East and Neighbouring. ERAUL, Université de Liège, University of Basel. pp. 197–209.
81. Lycett SJ, Gowlett JAJ (2008) On questions surrounding the Acheulean "tradition". World Archaeol 40(3): 295–315.
82. Lycett SJ, von Cramon-Taubadel N (2008) Acheulean variability and hominin dispersals: a model-bound approach. J Archaeol Sci 35: 553–562.
83. Guthrie RD (1984) Mosaics, allelochemics and nutrients: an ecological theory of Late Pleistocene megafaunal extinctions. In: Martin PS, Klein RG, editors. Quaternary extinctions: a prehistoric revolution. University of Arizona Press. pp. 259–298.
84. Almogi-Labin A (2011) The paleoclimate of the Eastern Mediterranean during the transition from early to mid Pleistocene (900 to 700 ka) based on marine and non-marine records: An integrated overview. J Hum Evol 60(4): 428–436.
85. Ashton N, Lewis SG (2012) The environmental contexts of early human occupation of northwest Europe: The British Lower Palaeolithic record. Quat Int 271: 50–64.
86. Messager E, Lebreton V, Marquer L, Russo-Ermolli E, Orain R, et al. (2011) Palaeoenvironments of early hominins in temperate and Mediterranean Eurasia: new palaeobotanical data from Palaeolithic key-sites and synchronous natural sequences. Quat Sci Rev 30(11–12): 1439–1447.
87. Abbate E, Sagri M (2012) Early to Middle Pleistocene Homo dispersals from Africa to Eurasia: Geological, climatic and environmental constraints. Quat Int 267(0): 3–19.
88. Belmaker M (2009) Hominin adaptability and patterns of faunal turnover in the Lower-Middle Pleistocene transition in the Levant. In: Camps M, Chauhan PR, editors. A sourcebook of Paleolithic transitions: methods, theories and Interpretations. Springer. pp. 211–227.
89. Berger GW, Perez-Gonzalez A, Carbonell E, Arsuaga JL, Bermudez de Castro JM, et al. (2008) Luminescence chronology of cave sediments at the Atapuerca paleoanthropological site, Spain. J Hum Evol 55: 300–311.
90. Bonifay MF, Brugal JP (1996) Biogéographie et biostratigraphie des grandes faunes du Pléistocène inférieur et moyen en Europe du sud: Apport des gisements français. Paléo 8: 19–29.
91. Cuenca Bencos G, Rofes J, Lopez-Garcia JM, Blain HA, De Marta RJ, et al. (2010) Biochronology of Spanish Quaternary small vertebrate faunas. Quat Int 212: 109–119.
92. Duval M, Grün R, Falguères C, Bahain JJ, Dolo JM (2009) ESR dating of Lower Pleistocene fossil teeth: Limits of the single saturating exponential (SSE) function for the equivalent dose determination. Radiat Meas 44(5–6): 477–48.
93. Duval M, Moreno D, Shao Q, Falgueres C, Bahain JJ, et al. (2011) Datación por ESR del yacimiento arqueológico del Pleistoceno inferior de Vallparadís (Terrassa, Cataluña, España) = ESR dating of the Early Pleistocene archaeological site at Vallparadís (Terrassa, Cataluña, Spain). Trabajo de Prehistoria 68(1): 7–24.
94. Falgueres C, Bahain JJ, Yokoyama Y, Arsuaga JL, Bermudez de Castro JM, et al. (1999) Earliest humans in Europe : the age of TD6 Gran Dolina, Atapuerca, Spain. J Hum Evol 37: 343–352.
95. Lefèvre D, Raynal JP, Vernet G, Kieffer G, Piperno M, (2010) Tephrostratigraphy and the age of ancient Southern Italian Acheulean settlements: The sites of Loreto and Notarchirico (Venosa, Basilicata, Italy). Quat Int 223–224: 360–368
96. Pope M, Roberts M, Maxted A, Jones P (2009) The Valdoe: Archaeology of a Locality within the Boxgrove. Palaeolandscape, West Sussex. Proceedings of the Prehistoric Society 75: 239–263.
97. Raynal JP, Paquereau MM, Daugas JP, Miallier D, Fain J, et al. (1985) Contribution à la datation du volcanisme quaternaire du massif central français par thermoluminescence de inclusions de quartz et comparaisons avec d'autres approches : implications chronostratigraphiques et paléoenvironnementales. Quaternaire 22(4): 183–207.

14

Pleistocene Climate, Phylogeny, and Climate Envelope Models: An Integrative Approach to Better Understand Species' Response to Climate Change

A. Michelle Lawing[1,2]*, **P. David Polly**[1]

1 Department of Geological Sciences, Indiana University, Bloomington, Indiana, United States of America, **2** Department of Biology, Indiana University, Bloomington, Indiana, United States of America

Abstract

Mean annual temperature reported by the Intergovernmental Panel on Climate Change increases at least 1.1°C to 6.4°C over the next 90 years. In context, a change in climate of 6°C is approximately the difference between the mean annual temperature of the Last Glacial Maximum (LGM) and our current warm interglacial. Species have been responding to changing climate throughout Earth's history and their previous biological responses can inform our expectations for future climate change. Here we synthesize geological evidence in the form of stable oxygen isotopes, general circulation paleoclimate models, species' evolutionary relatedness, and species' geographic distributions. We use the stable oxygen isotope record to develop a series of temporally high-resolution paleoclimate reconstructions spanning the Middle Pleistocene to Recent, which we use to map ancestral climatic envelope reconstructions for North American rattlesnakes. A simple linear interpolation between current climate and a general circulation paleoclimate model of the LGM using stable oxygen isotope ratios provides good estimates of paleoclimate at other time periods. We use geologically informed rates of change derived from these reconstructions to predict magnitudes and rates of change in species' suitable habitat over the next century. Our approach to modeling the past suitable habitat of species is general and can be adopted by others. We use multiple lines of evidence of past climate (isotopes and climate models), phylogenetic topology (to correct the models for long-term changes in the suitable habitat of a species), and the fossil record, however sparse, to cross check the models. Our models indicate the annual rate of displacement in a clade of rattlesnakes over the next century will be 2 to 3 orders of magnitude greater (430-2,420 m/yr) than it has been on average for the past 320 ky (2.3 m/yr).

Editor: David Nogues-Bravo, University of Copenhagen, Denmark

Funding: This work was supported by the National Science Foundation Research Grant EAR-0843935. The funders had no role in study design, data collection and analysis, decision to publish, or preparation of the manuscript.

Competing Interests: The authors have declared that no competing interests exist.

* E-mail: alawing@indiana.edu

Introduction

The Intergovernmental Panel on Climate Change reported that mean annual temperature will increase anywhere from 1.1 to 6.4°C by the end of the 21st century [1]. A shift in temperature as large as 6°C is comparable to the mean annual temperature difference between a glacial and an interglacial climate [2]. Many species, especially in temperate regions, shift their geographic distributions dramatically between glacial and interglacial cycles. These orbitally-forced geographic distribution dynamics influence speciation, range sizes, and latitudinal patterns [3]. Further, many species have significantly shifted their geographic distributions in the past few decades [4,5]. If a species is not able to track its suitable habitat or adapt to these changing climatic conditions, it will become extinct [6].

The response of species to climate change has often been studied in field settings by tracking a species' geographic movement through years or decades. In a meta-analysis of 1,700 species compiled from published field studies, habitat tracking poleward is on average 0.61 km yr^{-1} [7]. This approach provides a good estimate of the current rate of geographic displacement in response to climate change; however, it does not provide a framework for comparison with a background rate of change (an expected rate). An expectation can be established from estimating past rates of change and is an important indicator as to whether the current response is within the range of normality and if it is sustainable. Here we extend existing methods so that we can model suitable habitats for species in the geological past through a series of glacial-interglacial cycles. Our aim is to make use of the rich, continuous record of changes in global climate through the Quaternary that has emerged from stable isotope data and paleoclimate modeling. By using stable oxygen isotope data, we have derived a geographically explicit climate data set for the entire late Quaternary of North America by scaling between two general circulation climate models (GCMs) for glacial and interglacial periods. These data provide us with a set of North American paleoclimate maps for the last 300,000 years spaced ~4,000 years apart onto which we have projected phylogenetically scaled climate envelopes to estimate the geographic distributions of a clade of eleven rattlesnakes. The rates of historical waxing and waning of habits through several glacial-interglacial cycles provides a comprehensive non-anthropogenic context for evaluating current rates of geographic displacement due to anthropogenic climate change.

A climate envelope model generally characterizes a set of suitable habitats for a species derived from their present geographic location. The climate envelope models are constructed from the associations between the geographic position of a species' occurrence and its climate. Climate envelope models can be selected, trained, and tested under current climate conditions [8], but there is difficulty in testing these models under different climates. Insight can be gained from comparing projections of climate envelope models on paleoclimate maps with fossil occurrences [9], from projecting climate envelope models of invasive species onto continents that are being invaded [10,11], or from comparing projections of climate envelope models on paleoclimate maps with traditional molecular genetic phylogeographic predictions of potential distributional areas [12].

The climate envelope model has been critiqued for disregarding biotic interactions, dispersal, and evolutionary change [13,14,15]. In addition, climate envelope modeling takes into account species' occurrences, which only represent their realized niche. The realized niche is the current space occupied by a species resulting from biotic and abiotic interactions, where the fundamental niche is all possible space that supports viable populations within a species, but in which the species does not necessarily currently occupy [16]. Review of these criticisms suggests climate envelope modeling does provide a useful first approximation to study the dynamics of species' response to climate change when considered at the appropriate macro scale [17]. In addition, a study comparing climate envelope models and mechanistic models in 100 plant species suggest some climate envelope models can be used to predict changes in a species' geographic distribution under alternative climate scenarios [18].

An alternative approach to modeling suitable habitat for a species is mechanistic modeling [19,20,21]. A mechanistic model is based on the functional morphology, behavior and physiological requirements of a species and is independent of the species' geographic distribution and the climate in which the species lives [22]. These functional traits are then linked to climate and environmental variables, which can be mapped in geographic space. Functional trait data are required as input parameters in mechanistic models and specifically physiological data, such as temperature tolerances and energy, water, and mass balance, are not always readily available for species being studied. Identifying species' physiological requirements involves experimentation that is costly, time consuming, and impractical when dealing with many species.

Another aspect of climate envelope modeling concerns how to take phylogenetic effects into account. Phylogenetic comparative methods (PCM) have been used to account for phylogenetic relatedness and to elucidate past rates of change of climate tolerances by reconstructing ancestral climate envelopes [23,24,25]. The PCM approach models changes in the climate envelope of a species, without necessarily projecting those changes into geographic space by use of paleoclimate maps. By projecting the reconstructed envelope onto geographic space, the past location and size of the ancestor's suitable habitat can be estimated, but a contemporaneous paleoclimate map is needed, something that has heretofore been unavailable to PCM-based studies. We extend the PCM approach by modeling evolving envelopes along the branches of the phylogenetic tree, not just the ancestral nodes, and we project the evolving envelopes onto the series of paleoclimatic maps that we generate for every temporal step along the branch using the oxygen isotope data and GCM models.

Previous studies have estimated past geographic distributions at 6, 21, and 126 kya by projecting the climate envelopes of extant species' models onto GCMs for those times (reviewed in [26]). These snapshots show how suitable habitats of species are likely to have shifted geographically with climate change, assuming no evolution. However, several studies indicate that climatic niches of species can evolve over short evolutionary time scales [27,28,29], which suggests past climate envelope modeling that takes into account phylogeny might be more informative than those that do not. The further back in time a climate envelope model is extrapolated, the more critical it is to account for evolutionary adaptation in its construction [23,30].

GCMs are complex integrations of mathematical functions that describe atmospheric and oceanic circulation, sea ice, land surface properties, and atmospheric properties. GCMs use computationally intensive numerical models to simulate past, present, and future climates, including atmospheric and oceanic temperature and precipitation. Because of the computationally intensive nature of GCMs, only a limited set of past global climates have been modeled. Examples of such paleoclimate models for the Quaternary are for 0, 6, and 21 kya, produced by the Paleomodelling Intercomparison Project II [31], and for the last interglacial ~120–140 kya by Otto-Bliesner et al. [32].

In contrast, nearly continuous high-resolution estimates of global and local mean annual paleotemperatures are available for many parts of the world in the form of oxygen isotope values from sediment and ice cores for the entire Quaternary, and even the entire Cenozoic [33,34]. In order to estimate temperature and precipitation for past climates that may not have been modeled with GCMs, we use stable oxygen isotope values to interpolate between two time periods when climate is either known (as in the present) or has been modeled. We used this interpolation to construct paleoclimate maps at ~4 ky time intervals for the last 320 ky.

Here we developed phylogenetically-informed climate envelope models for 11 rattlesnake species (*Crotalus*). We project a phylogenetically reconstructed climate envelope onto a map of paleoclimate at many corresponding time intervals over the last 320 ky to determine the potential rate of geographic change in a species' suitable habitat. We refer to these models as paleophylogeographic models. By synthesizing both the evolutionary history of climatic tolerances and the climatic history in which these species evolved, we have arrived at a more detailed understanding of how they have responded to climate change and we predict how they might respond in the future.

Materials and Methods

Study System

Snakes are particularly useful for understanding the effects of climate change on terrestrial vertebrate species because their ectothermic physiology is highly dependent on the ambient temperature. We specifically chose rattlesnakes because the geographic distributions of some species extend north of former glacial margins, assuring that their geographic distributions have, in fact, changed over recent geological history [35]. The genus *Crotalus* originated in North America, most likely in the mid-Miocene, 13–15 mya [36], and its subsequent history on the continent is a complicated combination of dispersal, vicariance, and speciation in the context of changing climate and geography [37,38,39,40].

Species' Occurrences

Detailed geographic distributions of 11 rattlesnake species were obtained from published range maps [35]. The sampling scheme of Polly [41] was adopted, which is essentially a point based scheme (available at http://mypage.iu.edu/~pdpolly/Data.html). First, points spaced by 50 km were laid across North America and then points that overlap with a species' geographic distribution were selected to represent that species' geographic distribution. We call these range occurrences. We chose to base our models on an evenly spaced set of occurrence points sampled using equidistant 50 km points across each modern species' geographic distribution, instead of point occurrence data derived from museum collection events, such as the Global Biodiversity Information Facility, www.gbif.org, because our 11 species are poorly covered, are commonly misidentified in museum collection records or existing collection records have a strong geographic bias, especially in Mexico. Geographic distribution maps of these species [35], primarily stemming from sightings by experts, mark-recapture studies, and other longitudinal studies that do not produce museum voucher specimens, may provide a better representation of their full geographic and climatic ranges. The use of range polygons for habitat modeling has been criticized because some species are known to occur only in specific microhabitats whose climate differs from the surrounding landscapes encompassed in generalized range maps. We recognize the potential downfall of using range occurrences, but our goal is to estimate large-scale geographic changes over long temporal scales rather than to predict precisely where our species occur or do not occur on a small or medium scale landscape. Each of our points sampled at 50 km intervals from a modern geographic distribution has a coarseness that is appropriately similar to the geographic resolution of paleoclimate estimates and fossil assemblages [41,42]. Map building, intersection, extraction and manipulation were all performed in a combination of ArcGIS, Mathematica 7.0, and MySQL. All further analyses were performed in Mathematica 7.0.

Table 1. Nineteen bioclimatic variables derived from the WorldClim database.

Abbreviation	Variable Description
BIO1	Annual Mean Temp (C)
BIO2	Mean Diurnal Range (C)
BIO3	Isothermality (100 * BIO2 / BIO7)
BIO4	Temp Seasonality (100 * SD)
BIO5	Max Temp of Warmest Month (C)
BIO6	Min Temp of Coldest Month (C)
BIO7	Temp Annual Range (C) (BIO5–BIO6)
BIO8	Mean Temp of Wettest Quarter (C)
BIO9	Mean Temp of Driest Quarter (C)
BIO10	Mean Temp of Warmest Quarter (C)
BIO11	Mean Temp of Coldest Quarter (C)
BIO12	Annual Precip (mm)
BIO13	Precip of Wettest Month (mm)
BIO14	Precip of Driest Month (mm)
BIO15	Precip Seasonality (CV)
BIO16	Precip of Wettest Quarter (mm)
BIO17	Precip of Driest Quarter (mm)
BIO18	Precip of Warmest Quarter (mm)
BIO19	Precip of Coldest Quarter (mm)

Climate Envelope Modeling

We chose climate envelope models to describe potential suitable habitat for a species because a method for incorporating climate envelope models with phylogenetic comparative methods has been established [24,25]. We used 19 climatic variables derived from the 2.5 arc-minutes WorldClim database Version 1.4 to develop climate envelope models [43]. These 19 variables describe means and extremes of temperature and precipitation on a monthly, quarterly, and annual basis and are referred to as bioclimatic variables (Table 1). The bioclimatic variables were sampled at the 50 km point occurrences that represent each species' geographic distribution.

The specific bioclimatic variables important for each species differ in type and magnitude, so we included all 19 in the climate envelope models. One concern for including too many variables to model a species' suitable habitat is over-fitting the model to the data; however, this is not the case here because the climate data are not being used as independent variables in a statistical model with the aim of explaining variance in the independent data. Rather the variables are used to characterize a multivariate climate space inhabited by the species and the addition of correlated variables does not have the effect of inflating the climate envelope. The bioclimatic variables that are most closely associated with the species' geographic distribution will be the ones that emerge as influential in the suitable habitat models, whereas the ones that are not associated with the geographic distribution will have little or no effect on the suitable habitat models.

We evaluated two variants of a rectilinear climate envelope, BIOCLIM [44], and 5 variants of an ellipsoidal climate envelope, the generalized linear model, GLM [45], to determine which algorithm was the best predictor of the assumed known modern geographic distributions. For BIOCLIM, rectilinear climate envelopes were calculated for each species based on the maxima and minima of points in the 19-dimensional multivariate climate space. In the first variant we used the absolute maximum and minimum along each axis, in the second variant we used the 5th and 95th percentiles to minimize effects of climatic outliers. For the GLM method, a 19-dimensional ellipsoidal climate probability envelope was calculated, in which a parameter is estimated for each climate variable for location and affiliation with other climate variables. This is then transformed into a probability for each geographic point included in the climate envelope using a logit link function. The five variants of the GLM method included points in the distribution of suitable habitat when their p-values were greater than 0.1, 0.2, 0.3, 0.4, and 0.5, respectively. We did not use logistic models or genetic algorithms such as GARP [46] because these approaches are designed to find the best fit of occurrence data to a specific climate to determine where the existing climate is appropriate for the species. Because we are modeling across time, it is expected that correlations between climate variables will change from what they are in the modern climate; rectilinear climate envelopes do not rely on absence data nor do correlations between climate variables affect their boundaries, making the envelope model more appropriate for estimating changing suitable habitat distributions in past climates.

Quantitative descriptors of distribution models

We used three statistics to describe geographic distributions for the purposes of assessing the fit of suitable habitat model distributions to known distributions and for describing the rate and degree of change in the paleophylogeographic models. The three statistics are: 1) the geographic center, measured in longitude and latitude, and calculated as the mean of each coordinate for all points in the distribution; 2) the standard deviation of points in the distribution relative to the geographic center; and 3) the number of points included in the distribution.

Selection of the best suitable habitat distribution model

We tested the predictive power of the seven models by comparing their geographic centers, standard deviations, and numbers of points to known species' distributions. We used an independent two-sample t-test, where the geographic center was the point of comparison, the standard deviation served as the variance, and the numbers of points were the degrees of freedom. This method tests the null hypothesis that the modeled distribution is not significantly different from the known one. To test the robustness of the model distributions with respect to the known distributions and to account for non-normality in the spatially distributed dataset, we subsampled known species' distributions down to 25% of the original occurrences and rebuilt the modeled distributions 100 times. To further summarize the fit of each model distribution to the known species' distribution, we calculated an index of overlap (twice the number of shared points by the modeled and known distributions divided by the sum of number of points in both).

Phylogenetic regression of climate envelopes

To construct the evolving climate envelopes, we regressed the climate envelopes of 11 rattlesnake species onto a composite phylogeny [47] based on a mixed model Bayesian analysis [48] and maximum parsimony [49] (Figure 1). Branch lengths were calibrated by averaging divergence times suggested by fossil evidence [36] and genetic distances in molecular phylogeographic studies [37,38,39,40,50]. We used phylogenetic generalized linear model regression [51] to regress the 95% BIOCLIM envelopes onto the phylogeny using the maximum and minimum of the bioclimatic variables as traits and assuming a Brownian motion model of evolution [23]. This regression yields reconstructions of the most likely envelopes at each of the nodes of the tree. The most likely envelope at any point along a branch of the tree is simply a linear extrapolation between the envelopes at each end of the branch, scaled by the relative position of the point along the branch.

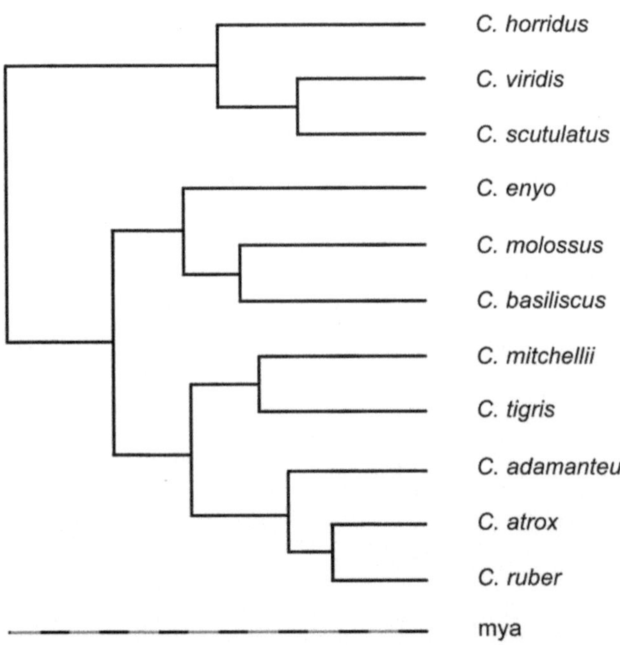

Figure 1. Composite phylogeny of 11 rattlesnakes in the genus *Crotalus*. Phylogenetic relationship of 11 rattlesnake species based on a composite phylogeny from a mixed model Bayesian analysis and maximum parsimony. The color offsets on the bar labeled millions of years ago (mya) are in million year increments.

We tested the conformity of the climatic variables with our assumption of a Brownian motion model. The expected absolute divergence of a trait scaled to time since divergence is $D = rt^a$, where D is the absolute divergence in the trait, r is a rate parameter, t is the time since divergence, and a is a coefficient related to the mode of evolution [52,53,54]. The parameter a ranges between 0 and 1, where 0 represents stabilizing selection, 0.5 represents perfect Brownian motion (randomly fluctuating selection or genetic drift), and 1 represents perfect diversifying selection. We used maximum likelihood to estimate r and a for each trait from the pairwise differences between the values at two tips relative to the interval of phylogenetic time connecting the two tips on the tree [53].

Interpolated paleoclimate maps

We used modern climate [43] and the GCM MICROC3.2 model of the LGM (~21 kya) from PMIP2 [31] to interpolate climates between and beyond these data. We tested our interpolated estimates of paleoclimate against independent GCMs for two times, 6 kya and 120 kya. We estimated mean annual temperature and annual precipitation by interpolating proportionally between our two end-member climate datasets, scaling the interpolation by composite benthic stable oxygen isotope ratios [33]. The stable oxygen isotope ratios are derived from multiple cores at a site in the North Atlantic Ocean on the western flank of the Mid-Atlantic Ridge. They are a proxy for northern hemisphere to global temperature and probably have less noise than terrestrial cores because they are deposited in a more stable environment.

We chose to test the climate interpolation at 6 kya because several GCMs have been used to model climate then (notably CCSM and MICROC3.2), allowing us to determine whether our interpolated model is as similar to the GCMs as they are to each other. We also compared the differences between a GCM from the last interglacial [32] with our climate interpolation model. Because our purpose is to facilitate the study of the continental-scale response of terrestrial animal habitats to climate change, we restricted our test to North America and used 50 km points as our level of resolution.

Construction of paleophylogeographic models and their relationship with change in temperature

Paleophylogeographic models were reconstructed in increments through the last 320 ky by projecting the phylogenetically-scaled climate envelopes onto concurrent paleoclimate maps (Figure 2). Increments were approximately 4 ky, but specifically determined by age estimates of oxygen isotope values [33]. We calculated the average change in geographic center (km) and areal extent (km^2) for all incremented time steps that included only phylogenetically scaled climate envelopes, only paleoclimate models, and both. *Crotalus enyo*, *C. adamanteus*, and *C. ruber* were excluded because of their intermittently undetectable suitable habitat at a 50 km scale. Because these species are modeled to have lost their entire suitable habitat at a 50 km scale during extreme climate, they are probably the most vulnerable to excessive climate change and excluding them from our further analyses provides us with a highly conservative estimate of geographic shifts due to expected

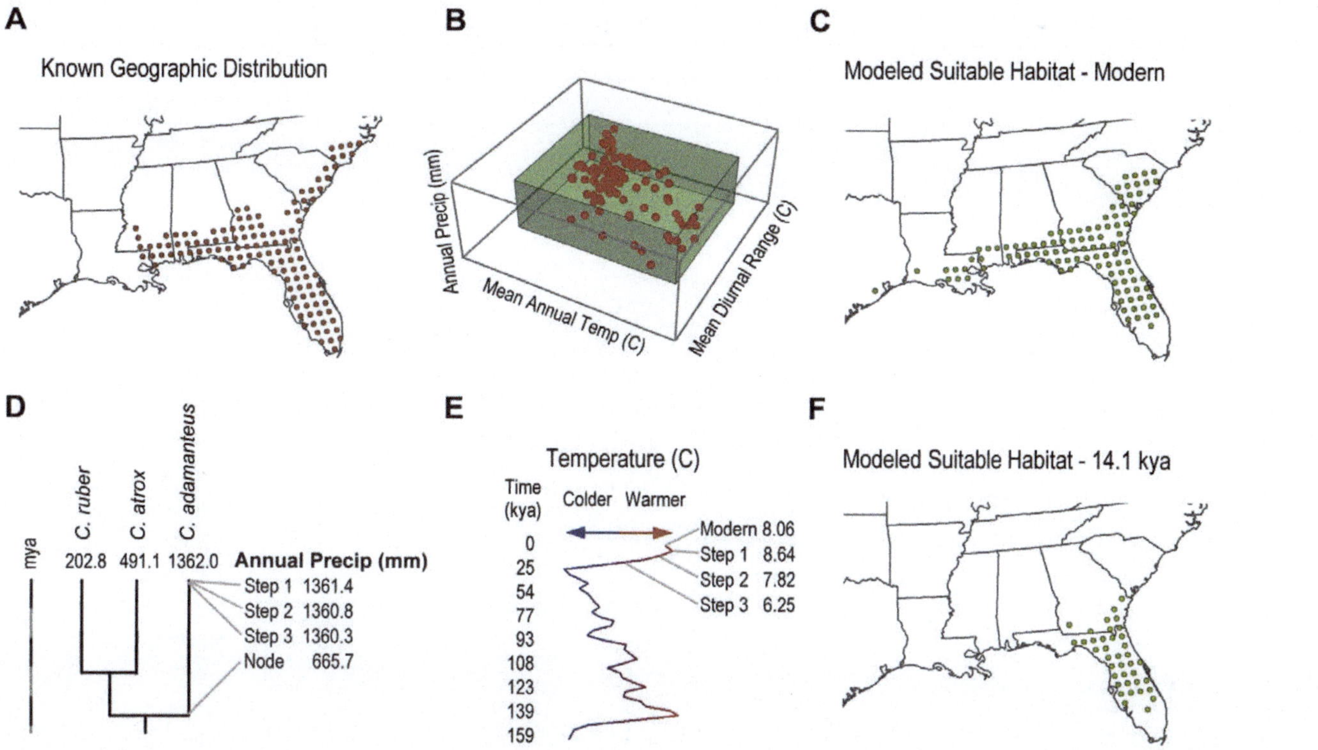

Figure 2. Illustration of a paleophylogeographic model. A, Modern geographic distribution for *Crotalus adamanteus*. B, Climate envelope for three bioclimatic variables. The red points represent the climate associated today with each red 50 km point from the modern geographic distribution in A. The green cube represents the climate envelope, the 5th and 95th percentile of each of the three bioclimatic variables. C, Suitable habitat modeled from the climate envelope on the modern climate. The green 50 km points on the map all fall within the green climate space in B and are considered suitable habitat for *C. adamanteus* today. D, Phylogeny and ancestral reconstructions. Annual Precipitation (mm) is regressed on the phylogeny. The first three steps of the reconstruction are shown at 4.7 kya, 9.4 kya and 14.1 kya. E, Temperature estimates for the North American continent derived from a composite oxygen isotope curve. The paleoclimate reconstruction for each step is scaled based on this curve. F, Phylogenetically scaled climate envelope projected onto isotopically scaled paleoclimate model at 14.1 kya. These 50 km points are considered suitable habitat for *C. adamanteus* 14.1 kya.

evolutionary change or climate change. All pairwise comparisons of average change in geographic center and areal extent were linearly regressed on corresponding change in mean annual temperature (°C) with a bootstrapped estimate of standard error and a randomization to account for non-normally distributed data.

Verification with the fossil record

To assess the validity of the paleophylogeographic models we compared the predicted past distribution of *Crotalus* to known occurrences in the Pleistocene fossil record by testing the overlap of past modeled distributions and fossil occurrences. Forty-one documented occurrences [36] were downloaded from the Paleobiology Database on 11 May, 2011, using the genus name '*Crotalus*'. Species level identification is not commonly reported for these rattlesnakes in the Paleobiology Database, so the comparison was generalized to the genus level. The clade of rattlesnakes that we modeled only represent about one third of the recognized species in this genus, so our sample does not represent the niche diversity of the genus. However, several of the species within this clade represent the northern extent of the current known distribution of the genus. Although the species' suitable habitat at the northern extent of the distribution of the genus does not necessarily represent the harshest climates (e.g., montane species in Mexico live in more extreme climates), they do represent the northern extent of the genus geographically. Much of the North American continent is covered by different species within this genus, so paleodistribution models of the northern extent of rattlesnake suitable habitat provides a testable hypothesis. If a fossil occurs north of the northernmost edge of the modeled suitable habitat, it is clear that the models are incorrect.

Projection of climate envelopes on future climate scenarios

Suitable habitats were modeled under two future climate scenarios for the year 2100 for 11 rattlesnake species. The two climate scenarios were developed for an increase in mean annual temperature by 1.1°C and by 6.4°C, the range of possible mean annual temperature increases reported by the IPCC [1]. Future climate maps were constructed by the climate interpolation method described above in 'Interpolated paleoclimate maps' using the change in stable oxygen isotope values as a proxy for change in temperature.

Results

Selection of the best climate envelope model

BIOCLIM envelope models fared better than GLM envelope models to quantitatively characterize the 11 rattlesnake species' known distributions, where all GLM models with varying parameters were statistically distinguishable from known distributions (Table S1 and S2). The 95% BIOCLIM model fared best (8

Table 2. Maximum likelihood estimate of evolutionary rate and mode.

Bioclimate Variable	*r*	P	*α*	P	Evo Mode
Annual Mean Temp	10.60	0.17	0.66	0.21	Random
Mean Diurnal Range	20.15	0.46	0	1	Stabilizing
Isothermality	3.69	0.58	0.39	0.70	Random
Temp Seasonality	805.38	0.83	0.41	0.49	Random
Max Temp of Warmest Month	7.57	0.38	0.51	0.71	Random
Min Temp of Coldest Month	20.13	0.76	0.55	0.08	Random
Temp Annual Range	26.69	0.95	0.33	0.46	Random
Mean Temp of Wettest Quarter	23.09	0.90	0.39	0.29	Random
Mean Temp of Driest Quarter	20.74	0.33	0.47	0.63	Random
Mean Temp of Warmest Quarter	6.79	0.20	0.65	0.24	Random
Mean Temp of Coldest Quarter	18.07	0.78	0.60	0.70	Random
Annual Precip	579.12	0.99	0	0.99	Stabilizing
Precip of Wettest Month	86.99	0.27	0	1	Stabilizing
Precip of Driest Month	30.57	0.47	0.05	1	Stabilizing
Precip Seasonality	17.27	0.29	0.29	0.99	Stabilizing
Precip of Wettest Quarter	240.34	0.15	0	1	Stabilizing
Precip of Driest Quarter	116.99	0.61	0.02	1	Stabilizing
Precip of Warmest Quarter	208.06	0.60	0	1	Stabilizing
Precip of Coldest Quarter	123.14	0.28	0	1	Stabilizing

Maximum likelihood estimate of evolutionary rate, *r*, a coefficient related to the mode of evolution, *α*, and the corresponding evolutionary mode (Evo Mode). Evolutionary mode is characterized as either stabilizing (stabilizing selection), random (randomly fluctuating selection or genetic drift), or diversifying (diversifying selection). P values are derived from 10,000 permutations of the maximum likelihood estimate.

of the 11 modeled habitat distributions were statistically indistinguishable from the documented modern species' geographic distribution) and we used it for the remaining analyses (Table S1 and S2).

Evolutionary mode of climate variables

All climate variables considered in our climate envelope models either evolved under a Brownian motion model of evolution or under stabilizing selection (Table 2). In either case, a Brownian

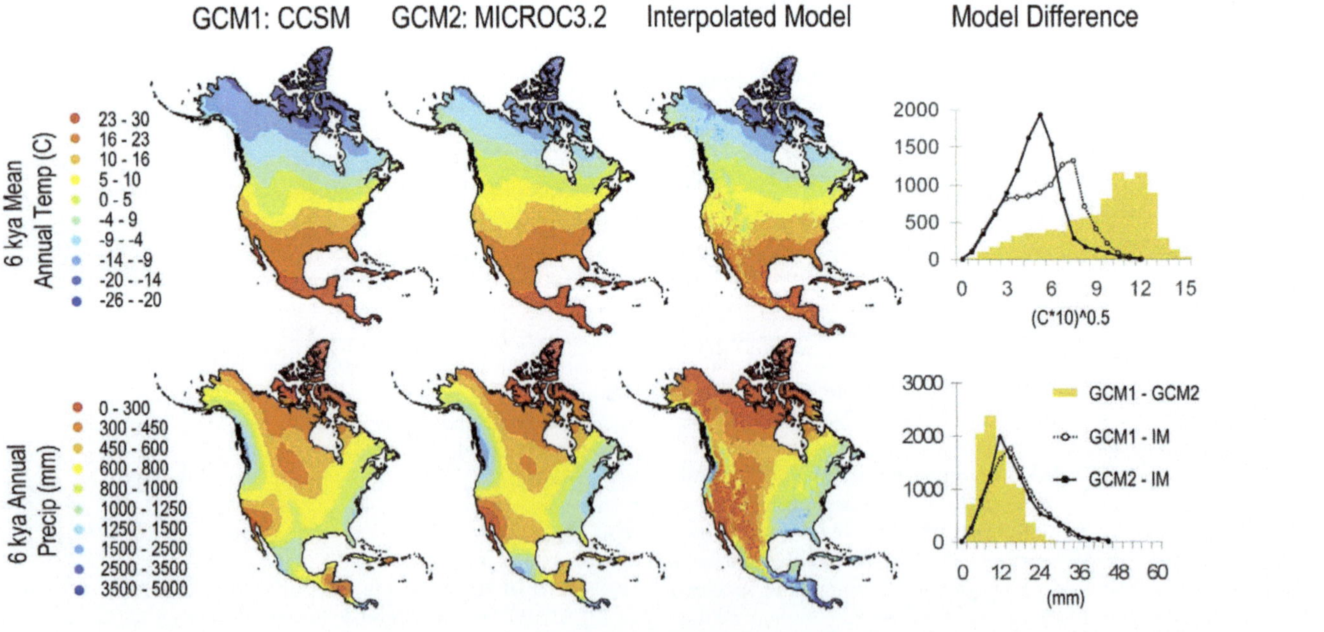

Figure 3. Mean annual temperature and annual precipitation modeled for 6 kya. Two general circulation models (GCM1 and 2) and our interpolation model (IM) are compared. Graphs at the right show histograms of the absolute differences between the two GCMs (yellow bars) and between our model and each of the GCMs (lines).

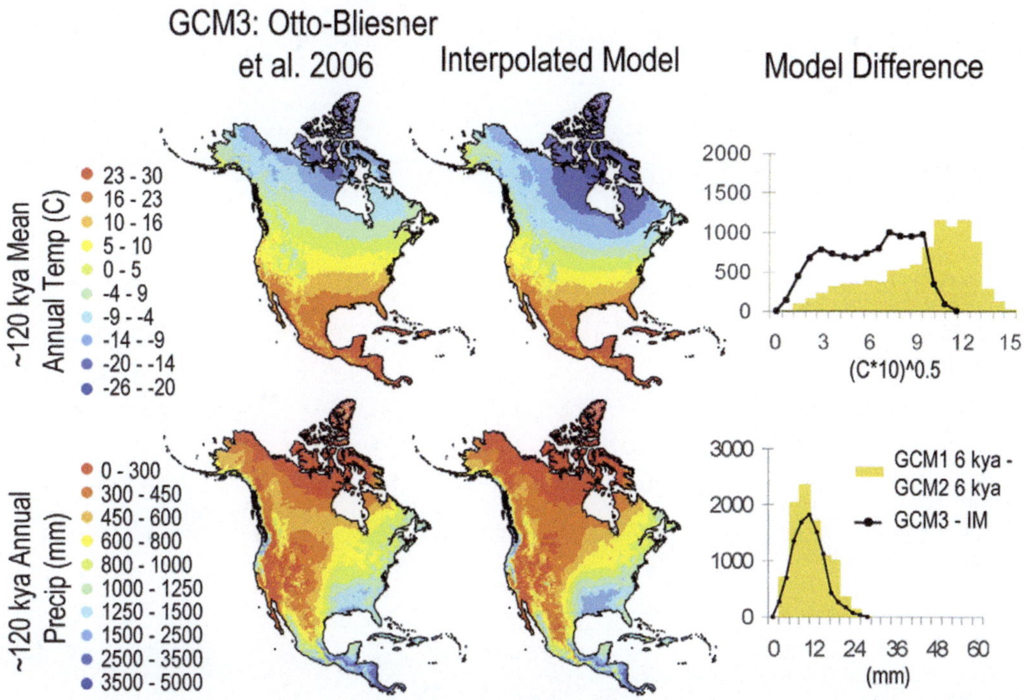

Figure 4. Mean annual temperature and annual precipitation modeled for ~120 kya. One GCM and an interpolation model (IM) are compared. Graphs at the right show differences between our model and the GCMs (line) with the differences between the two 6 kya GCMs for comparison (yellow bars).

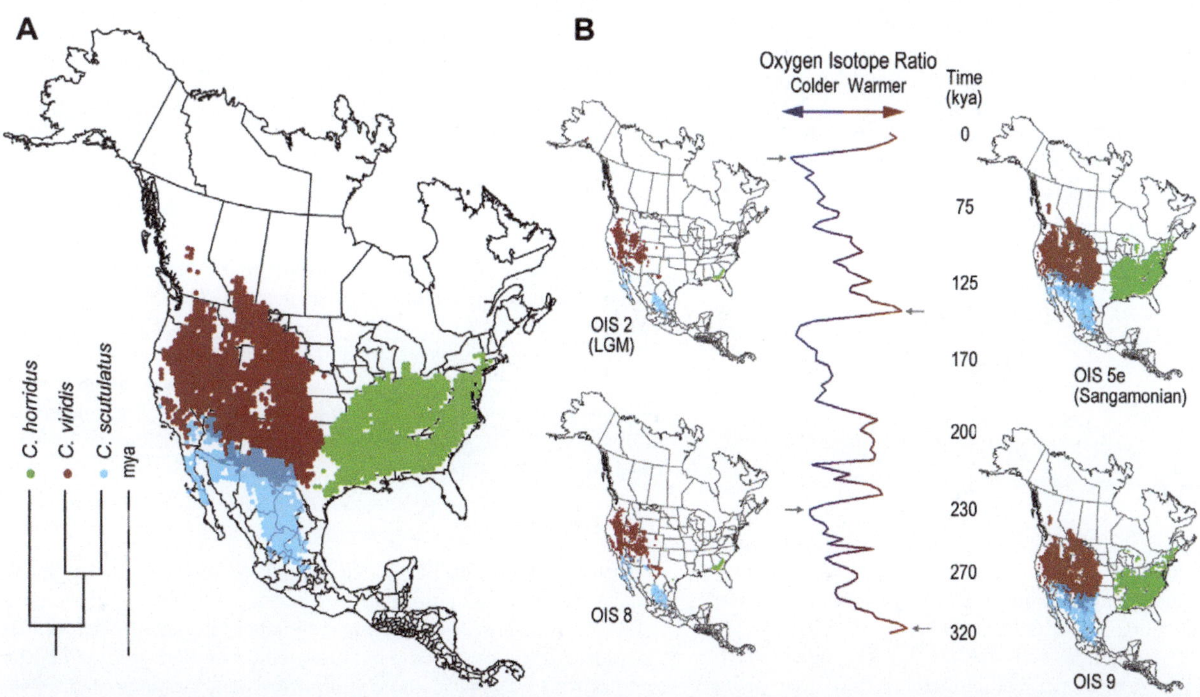

Figure 5. Paleophylogeographic distribution models for three species of rattlesnake (*Crotalus*). A, Phylogeny and modern geographic distribution models mapped onto modern climatic conditions. The dark gray curve represents the southern extent of glaciers during the LGM. B, Composite oxygen isotope curve for the last 320 ky inset with four paleophylogeographic reconstructions at four points, two glacial and two interglacial, to illustrate the effects of climate changes and phylogeny on the distribution of suitable habitats. Phylogenetically scaled climate envelopes were projected onto isotopically scaled paleoclimate models to generate these maps. Supplemental videos show animations of the paleophylogeographic distributions through the last 320 ky for these three species (Video S1) and for the remaining species (Video S2 and S3).

motion model of evolution is adequate to model the most likely ancestral nodes across a tree and was used to phylogenetically regress our climate envelope models.

Paleoclimate interpolation

Our interpolated model of mean annual temperature at 6 kya was consistent with the corresponding GCMs; in that, it was no more different from either GCM as the two GCMs were from each other (Figure 3). Note here that we are testing if the difference distribution between the interpolated model and either climate model is greater than the difference distribution between the two GCMs, not if the distributions match. That our interpolated model of mean annual temperature was accurate (the differences between the interpolated model and either GCM are less than or equal to the difference between the two GCMs) is no surprise since the stable isotope values used to make the interpolation are themselves a proxy of mean annual temperature, albeit one that does not contain information about geographic variation of temperature across the continent. More importantly, our interpolated model of annual precipitation, which is derived entirely from the spatial correlation between precipitation and temperature in our two end-member climate models and our temperature-based interpolation between them, also compared favorably with the two GCMs (Figure 3). Furthermore, our climate interpolations for the last interglacial, which lies outside our two end-member models, were as similar to the last interglacial GCM as our 6 kya interpolations were to their corresponding GCMs (Figure 4).

Paleophylogeographic models through time

Figure 5 shows how the paleophylogeographic distribution of suitable habitats of three rattlesnakes is expected to have changed through time. The potential for dramatic changes in the location and areal extent of suitable habitat are particularly apparent in the paleophylogeographic models for glacial times (Figure 5B). Among the eleven rattlesnakes species studied here, suitable habitats have expanded rapidly northward since the LGM in *Crotalus adamanteus*, *C. enyo*, *C. horridus*, and *C. tigris* (see Video S1, S2, and S3). At some times in the past, suitable habitat for a few species (*C. adamanteus*, *C. enyo*, and *C. basiliscus*) was so constricted that it was undetectable at the 50 km resolution of our models.

Fossil Record Occurrences

All described occurrences of *Crotalus* from the Quaternary fossil record are consistent with the predicted range of our paleophylogeographic models as far as temporal control on the fossil sites permits comparisons (Figure 6). However, we note that the fossil record of these snakes is poor and does not provide a powerful test of the details of our models, highlighting the need for further study of these paleoclimatically informative animals.

Paleophylogeographic models and their relationship to temperature change

Climate has contributed more to changes in modeled suitable habitat of these rattlesnakes than evolutionary change by two to three orders of magnitude over the last 320 ky (Figure 7). The historical change in mean annual temperature has been strongly correlated with both change in geographic center ($R^2=0.915\pm0.002$ s.e.; $p<0.0001$; $n=5{,}050$) and areal extent ($R^2=0.924\pm0.002$ s.e.; $p<0.0001$; $n=5{,}050$) of these species' suitable habitats (Figures 7B and 7D). Based on this correlation, we estimate that the centers of these habitats have on average been displaced by 34.93 km per°C and their areal extents have changed by 121,591 km^2 per°C. A pairwise comparison of average changes between time intervals ($n=5{,}050$) reveals a 0.0023 km/year average rate of displacement.

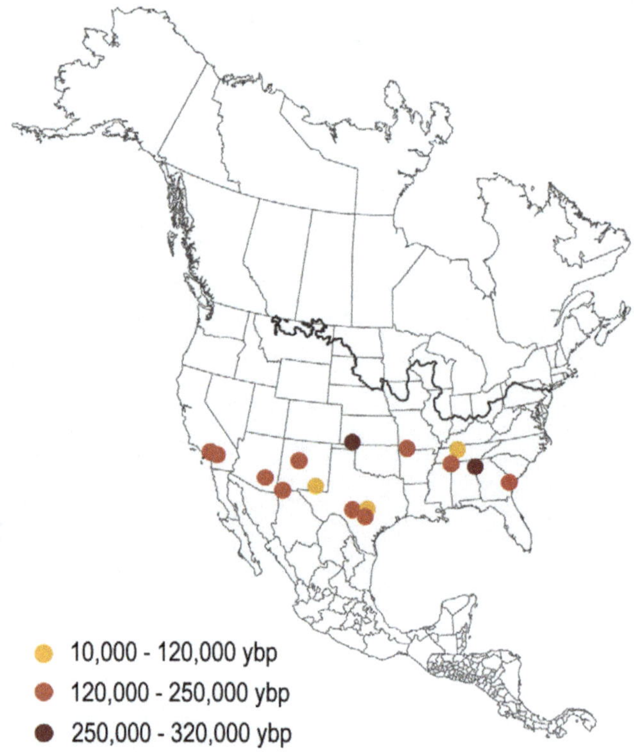

Figure 6. Fossil occurrences of *Crotalus* in North America for the last 320,000 years. Orange points show occurrence sites whose maximum ages are less than 120 kya, red points show sites with maximum ages between 120–250 kya, and brown points show sites with maximum ages between 250–320 kya. The data were downloaded from the Paleobiology Database (http://pbdb.org) on 22 May 2011, using the group name '*Crotalus*'.

Projection of climate envelopes on future climate scenarios

If we extrapolate from our estimates of Pleistocene rates of change per°C, the average displacement of the centers of species' suitable habitats in the next 90 years will be 38.31–217.49 km (0.43–2.42 km/yr) and their areal extents will change by 133,783–799,963 km^2 (Figure 8). The rates of current displacement are two to three orders of magnitude greater than the rate of change we measured through the dramatic climatic fluctuations of the last 320 ky, 0.0023 km/year.

Discussion

Reasonable multi-parameter spatial estimates of past Quaternary climates can be produced by interpolation from end-member GCMs using a proxy for a single climate parameter (mean annual temperature), provided that the end-members represent the extremes of the paleoclimates that are being estimated. Despite good agreement of our interpolated paleoclimate models with the GCM of the last interglacial, it is likely that our method will produce increasingly inaccurate models the further they are extrapolated from the end-member climate data. In addition, interpolated paleoclimate models would have an increased accuracy with the inclusion stable oxygen isotope geographic variation. This computationally straightforward tool does not

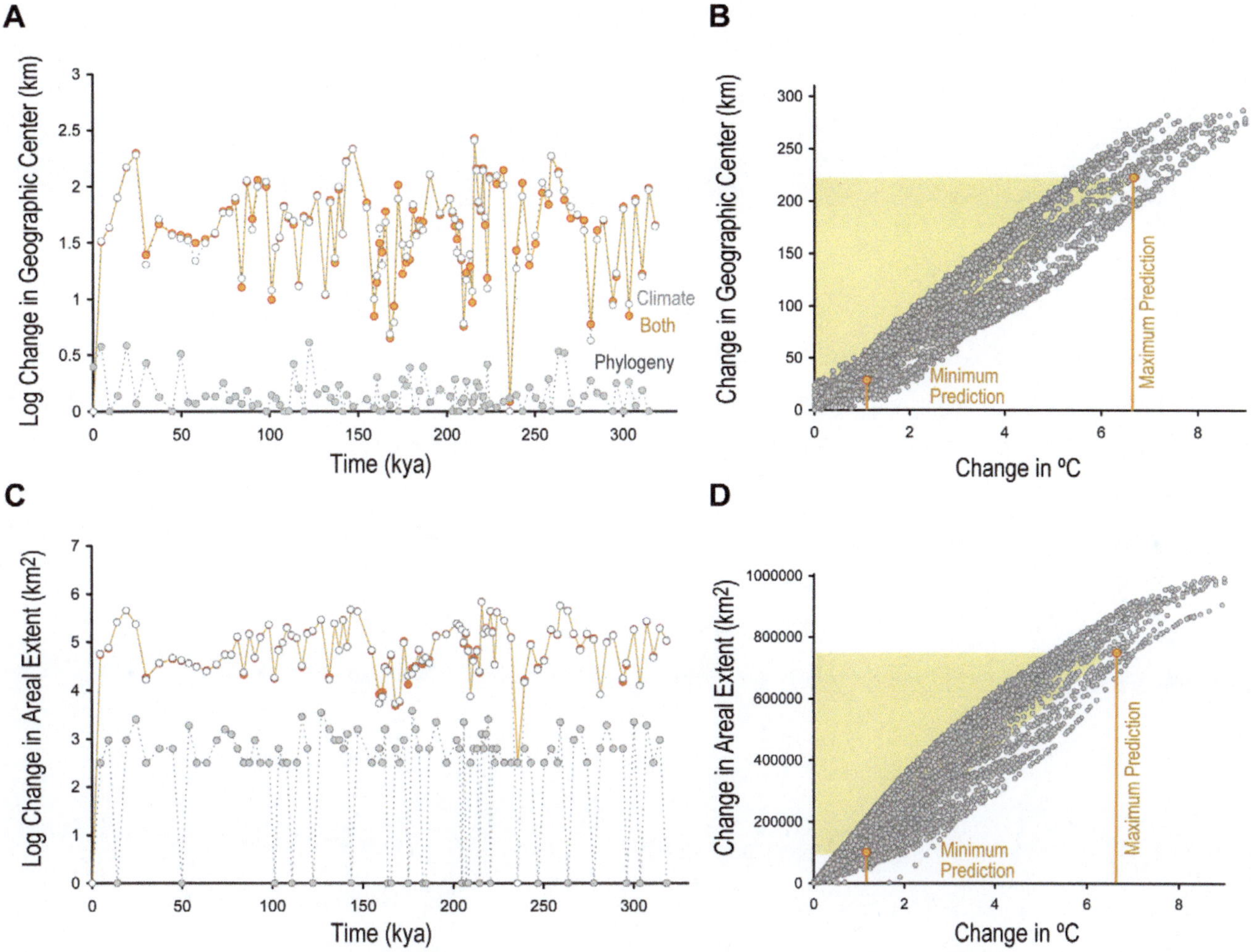

Figure 7. Average change in species' distributions of suitable habitat over the last 320 ky. A, Time series of change in geographic center (km). Change due to climate and phylogeny are modeled separately to identify the contributions of each in a model incorporating change due to both climate and phylogeny. B, Change in geographic center as it relates to temperature (°C). C, Time series of change in areal extent (km^2). D, Change in areal extent as it relates to temperature change (°C). The shaded area indicates increase in global temperature by the end of the 21st century predicted by IPCC 2007.

replace the need for GCMs, but it facilitates the modeling of dynamic changes in past climates at a continental scale, which is especially useful for studies of species' responses to past climate change.

During the last 320 ky, three major glacial cycles have come and gone, the global mean annual surface temperature has varied by 6 to 8°C, and some of the warmest and coldest global temperatures of the last million years have occurred, forcing species to repeatedly shift either their geographic distributions or their climatic tolerances. We suggest that physiological limits or climatic tolerances did not shift rapidly, because of the phylogenetic component of this analysis. The variance in the bioclimatic variables across the entire clade does not necessarily encompass the extreme variations of a fixed 50 km point through the glacial-interglacial cycles (Table 3). For example, the annual mean temperature in Bloomington, IN is 11.5°C. Several of the rattlesnakes encompass that temperature as part of the range of the annual mean temperatures that they currently inhabit. However, the annual mean temperature during the LGM in Bloomington, IN was −8.5°C and none of the species come close to including that temperature within their range of annual mean temperatures. The same is true for many of the bioclimatic variables and in Table 3 we show these ranges for annual mean temperature, minimum temperature of the coldest month, maximum temperature of the hottest month, and temperature seasonality.

Large fluctuations in climate from the glacial-interglacial cycles did not erase the phylogenetic signal present in bioclimatic variables describing species' climate envelopes. We tested the assumption of a Brownian motion-like model of evolution for the phylogenetic regression of bioclimatic variables (Table 2). The mean diurnal range and all variables related to precipitation have an evolutionary mode of stabilizing selection, suggestive of phylogenetic niche conservatism [55]. The remaining temperature variables have between species variation that is consistent with a Brownian motion-like model of evolution, suggestive of phylogenetic signal and not conservatism. In either case, diversifying or divergent selection was not identified for any bioclimatic variable describing the climate envelopes, and ancestral climate envelopes could be reconstructed.

The rate at which climate has forced change in suitable habitat in the rattlesnakes we studied has been much more rapid than changes in climate envelopes associated with macroevolution or

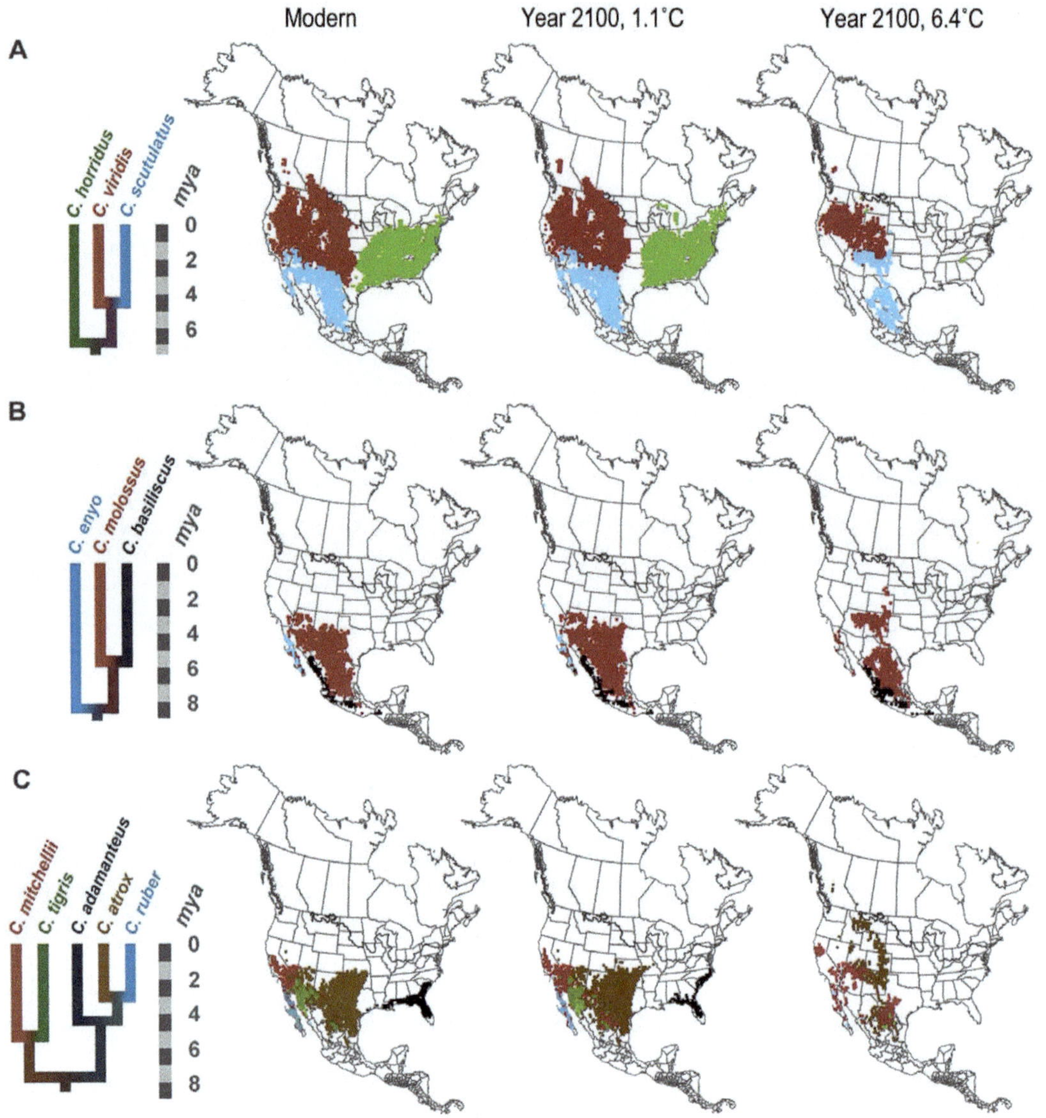

Figure 8. Current and future predictions of suitable habitat. Suitable habitat distributions were modeled under two future climate scenarios for the year 2100 for 11 rattlesnake species. The two climate scenarios are derived from an increase of mean annual temperature by 1.1°C and by 6.4°C. A, B, and C, Phylogeny and modeled suitable habitat distributions by clade.

speciation (Figure 7). However, our paleophylogeographic models do not incorporate anagenesis apart from a gradual linear change since speciation; therefore, the phyletic evolution modeled is dampened by modeling an average trait value for the climate envelopes through time. The phylogeny we use to model phyletic evolution has roughly estimated divergence dates attached to it and we do not incorporate a range of possible divergence time estimates in our phylogenetic regression. It is likely that our calculations of the change in available suitable habitat will minimally change if we incorporated ranges in divergence times. We estimate ancestor nodes for divergences that happened millions of years ago, so the extrapolation to the last 320 ky will only minutely change the effects of phylogeny from the estimated ancestor.

In this same regard, the results in Figure 7 are expected because phylogenetic change is modeled with a constant rate since the spilt of the last common ancestor millions of years ago. So the modeled change on 4 ky time increments eventually produces the larger shifts that we see between species' climatic envelopes. Although we would expect this outcome for this particular group of species, due to the deeper divergences, it is not necessarily known what to expect for other species groups. Our intention here is to develop an interdisciplinary framework that can be extended to deeper time contexts and for other species groups.

The amount of evolutionary change is probably underestimated in our paleophylogeographic models, because adaptive changes that accumulate within glacial or interglacial periods may effectively be erased by the cycling climate and not incorporated in macroevolutionary change [56]. Accounting for potential phyletic evolution increases model complexity and is beyond the scope of this study. Future investigations should incorporate a stochastic parameter to model within species variation per generation, because evolutionary dynamics can happen over even short time scales [57,58]. In addition, future study might investigate the evolution of dispersal ability, which potentially has a large effect on the geographic limits of a species' distribution

Table 3. Example of climate fluctuations for Bloomington, IN and species' ranges.

	Annual Mean Temp	Min Temp of Coldest Month	Max Temp of Warmest Month	Temp Seasonality
Bloomington, IN				
Present	11.5	−7.3	29.9	950
LGM	−8.5	−29.1	11.6	910
120 kya	9.7	−22.8	38.4	1408
Species' Ranges				
C. horridus	2.5 - 21.5	−17.9 - 6.9	21.0 - 35.9	562 - 1212
C. viridis	−3.8 - 22.1	−22.6 - 5.6	15.7 - 38.0	543 - 1245
C. scutulatus	5.9 - 24.2	−9.6 - 14.7	14.9 - 43.2	106 - 949
C. enyo	15.2 - 23.9	3.4 - 12.3	25.8 - 37.1	336 - 502
C. molossus	5.9 - 28.2	−11.0 - 16.5	14.9 - 41.4	106 - 864
C. basiliscus	15.0 - 28.8	3.1 - 17.7	24.6 - 39.9	141 - 512
C. mitchellii	7.5 - 24.3	−6.7 - 12.3	25.4 - 42.2	336 - 913
C. tigris	15.0 - 25.7	−1.7 - 11.3	31.7 - 41.2	512 - 827
C. adamanteus	15.9 - 23.8	−0.8 - 13.7	31.5 - 33.8	332 - 776
C. atrox	6.7 - 27.6	−10.6 - 19.6	23.3 - 42.2	145 - 974
C. ruber	10.1 - 23.9	−3.6 - 12.3	25.8 - 38.9	336 - 740

Example of the variance of 4 bioclimatic variables for one 50 km point in the Midwest, Global ID #139912, (Bloomington, IN) and the range of these variables in the modern climate envelopes for 11 rattlesnake species.

[59]. Regardless, the velocity with which climate is now changing will shift entire biomes rapidly and individual species will have to shift geographically at very high rates to keep up [60].

The future projected displacements in rattlesnake suitable habitat we calculated of 0.43–2.42 km/yr are similar to those observed over recent decades in other species (0.61 km/yr) [7]. Nevertheless, these projected rates are two to three orders of magnitude greater than the expected or background rate of displacement in rattlesnakes (0.0023 km/yr) that we modeled through the dramatic climate fluctuations of the last 320 ky. Those species whose suitable habitat changed the most markedly through the Pleistocene climate cycles may be ones that are particularly sensitive to current climatic fluctuations and habitat alterations. Our paleophylogeographic models suggest that *C. adamanteus*, the Eastern diamondback rattlesnake, is one of these particularly sensitive species. Conservation studies have already documented a decline in this species' abundance in association with habitat destruction and fragmentation [61,62].

Based on the rate in the fastest changing species (*C. horridus*), future displacement of the center of a species' suitable habitat could be as high as 144.91–711.32 km (1.61–7.90 km/yr) and a change in areal extent as much as 257,243–1,526,050 km^2. Geographic displacements of this magnitude are many times larger than the size of current natural preserves; for example, the largest US National Park, Wrangell-St. Elias Preserve in Alaska, is only 53,418 km^2. Although suitable habitat may increase (or decrease) in areal extent as global temperature increases, the location of that habitat may be displaced faster or farther than a species is able to track, and may result in a population bottleneck that could lead to extinction.

Every species has its own particular relation to climate and environment, of course, but our results are generalizable beyond rattlesnakes. Any species that has a similar sized geographic distribution today at similar latitudes will have experienced the same climatic changes in the past and will experience the same ones in the near future. We predict that the rapidity and magnitude of changes in other terrestrial vertebrates will be similar, on average, to what we found in rattlesnakes, a prediction that is substantiated by the similar rates of geographic expansion that have been observed in other species in recent decades (e.g., [7], but see [63]).

The modeling of species' distributions through time illustrates how dramatically they can contract during glacial periods, expand in interglacial periods, and fragment and coalesce in complicated ways in between. The ability of these 11 rattlesnake species to persist through the last 320 ky, when suitable habitat was repeatedly fragmented and reorganized, suggests that conservation strategies that create habitat corridors [64] or rely on managed relocations may be successful [60,65]. The ways in which species responded to past climatic shifts provides a framework for how we expect species to respond in the future. Knowledge about the past allows for more informed conservation decisions in the present, which are vital for the preservation of species' biodiversity in our rapidly changing global climate.

Supporting Information

Table S1 Quantitative descriptors of known geographic distributions and climate envelope models for 11 rattlesnake species. Standard errors were estimated by recalculating each statistic from 100 random subsamples of 90% of the points in each known distribution. Geographic center (GC) is the centroid of the geographic distributions reported in latitude and longitude, WGS 1984. Areal extent is the number of 50 km points that the geographic distribution covers.

Table S2 Tests for the fit and stability of climate envelope models. Independent two sample t-tests for departure of modeled distributions from the known species' distribution (Model Fit) and for departure of 100 randomly subsampled models, each at 25% of the original data, from the known species' distribution (Model Stability). For model fit, the P values indicate the probability that the model is identical to the known distribution apart from chance. An asterisk indicates the model is significantly different from the known distribution. For model stability, the P value indicates the mean probability that the subsampled models are identical to the climate envelope model except by chance. Overlapping proportion (OP) is the proportion of points in the model and the known distribution.

Video S1 Paleophylogeographic model for three rattlesnake species (*Crotalus horridus*, *C. viridis*, and *C. scutulatus*). A, Phylogeny and suitable habitat models mapped onto climatic conditions from corresponding time intervals to illustrate the effects of climate and phylogenetic changes on the distribution of suitable habitats. Phylogenetically scaled climate envelopes were projected onto isotopically scaled paleoclimate models to generate these maps. The dark gray curve represents the southern extent of glaciers during the LGM. B, Composite oxygen isotope curve for the last 320 ky inset with a yellow circle to indicate the oxygen isotope ratio and the time interval used in the model on the adjacent map.

Video S2 Paleophylogeographic model for three rattlesnake species (*Crotalus enyo*, *C. molossus*, and *C. basiliscus*). A, Phylogeny and suitable habitat models mapped onto climatic conditions from corresponding time intervals to illustrate the effects of climate and phylogenetic changes on the distribution of suitable habitats. Phylogenetically scaled climate envelopes were projected onto isotopically scaled paleoclimate models to generate these maps. The dark gray curve represents the southern extent of glaciers during the LGM. B, Composite oxygen isotope curve for the last 320 ky inset with a yellow circle to indicate the oxygen isotope ratio and the time interval used in the model on the adjacent map.

Video S3 Paleophylogeographic model for five rattlesnake species (*Crotalus mitchellii*, *C. tigris*, *C. adamanteus*, *C. atrox*, and *C. ruber*). A, Phylogeny and suitable habitat models mapped onto climatic conditions from corresponding time intervals to illustrate the effects of climate and phylogenetic changes on the distribution of suitable habitats. Phylogenetically scaled climate envelopes were projected onto isotopically scaled paleoclimate models to generate these maps. The dark gray curve represents the southern extent of glaciers during the LGM. B, Composite oxygen isotope curve for the last 320 ky inset with a yellow circle to indicate the oxygen isotope ratio and the time interval used in the model on the adjacent map.

Acknowledgments

We thank S.C. Brassell, A.K. Bormet, C. Casola, E.S. Gercke, R.M. Green, C.C. Johnson, E.P. Martins, J.M. Meik, L.C. Moyle, K.D. Nold, J.-Y. Peterschmitt, and L.M. Pratt for assistance, advice and thoughtful discussion. J.M. Meik kindly provided an insightful review of this manuscript, as well as two anonymous reviewers. We acknowledge the Paleobiology Database (paleodb.org) and the Laboratoire des Sciences du Climat et de l'Environnement (pmip2.lsce.ipsl.fr/) for collecting and archiving data. The PMIP2 Data Archive is supported by CEA, CNRS, MOTIF, and the Programme National d'Etude de la Dynamique du Climat.

Author Contributions

Conceived and designed the experiments: AML PDP. Performed the experiments: AML. Analyzed the data: AML. Contributed reagents/materials/analysis tools: PDP. Wrote the paper: AML. Contributed to manuscript revision: AML PDP.

References

1. Solomon S, Qin D, Manning M, Chen Z, Marquis M, et al. (2007) IPCC. Climate Change 2007:The Physical Science Basis. Contribution of Working Group I to the Fourth Assessment Report of the Intergovernmental Panel on Climate Change. Cambridge, UK, .
2. Petit JR, Jouzel J, Raynaud D, Barkov NI, Barnola J-M, et al. (1999) Climate and atmospheric history of the past 420,000 years from the Vostok ice core, Antarctica. Nature 399: 429–436.
3. Dynesius M, Jansson R (2000) Evolutionary consequences of changes in species' geographical distributions driven by Milankovitch climate oscillations. Proceedings of the National Academy of Sciences of the United States of America 97: 9115–9120.
4. Thomas CD, Cameron A, Green RE, Bakkenes M, Beaumont LJ, et al. (2004) Extinction risk from climate change. Nature 427: 145–148.
5. Moritz C, Patton JL, Conroy CJ, Parra JL, White GC, et al. (2008) Impact of a Century of Climate Change on Small-Mammal Communities in Yosemite National Park, USA. Science 322: 261–264.
6. Pease CM, Lande R, Bull JJ (1989) A model of population growth, dispersal, and evolution in a changing environment. Ecology 70: 1657–1664.
7. Parmesan C, Yohe G (2003) A globally coherent fingerprint of climate change impacts across natural systems. Nature 421: 37–42.
8. Fielding AH, Bell JF (1997) A review of methods for the assessment of prediction errors in conservation presence/absence models. . Environmental Conservation 24: 38–49.
9. Martínez-Meyer E, Peterson AT, Hargrove WW (2004) Ecological niches as stable distributional constraints on mammal species, with implications for pleistocene extinctions and climate change projections for biodiversity. . Global Ecology and Biogeography 13: 305–314.
10. Peterson AT (2003) Predicting the geography of species' invasions via ecological niche modeling. The Quarterly Review of Biology 78: 419–433.
11. Thuiller W, Richardson DM, Pysek P, Midgley GF, Hughes GO, et al. (2005) Niche-based modelling as a tool for predicting the risk of alien plant invasions at a global scale. Global Change Biology 11: 2234–2250.
12. Waltari E, Hijmans RJ, Peterson AT, Nyári AS, Perkins SL, et al. (2007) Locating Pleistocene Refugia: Comparing phylogeographic and ecological niche model predictions PLoS One 2: e563.
13. Davis AJ, Jenkinson LS, Lawton JL, Shorrocks B, Wood S (1998) Making mistakes when predicting shifts in species range in response to global warming. Nature 391: 783–786.
14. Davis AJ, Lawton JL, Shorrocks B, Jenkinson LS (1998) Individualistic species responses invalidate simple physiological models of community dynamics under global environmental change. Journal of Animal Ecology 67: 600–612.
15. Lawton JL, ed (2000) Concluding remarks: a review of some open questions. Cambridge: Cambridge University Press. pp 401–424.
16. Hutchinson GE (1957) Concluding Remarks. Cold Spring Harbor Symposium on Quantitative Biology 22: 415–457.
17. Pearson RG, Dawson TP (2003) Predicting the impacts of climate change on the distribution of species: are bioclimate envelope models useful? Global Ecology and Biogeography 12: 361–371.
18. Hijmans R, Graham CH (2006) The ability of climate envelope models to predict the effect of climate change on species distributions. Global Change Biology 12: 2272–2281.
19. Guisan A, Zimmermann NE (2000) Predictive habitat distribution models in ecology. Ecological Modelling 135: 147–186.
20. Prentice IC, Cramer W, Harrison SP, Leemans R, Monserud RA, et al. (1992) A global biome model based on plant physiology and dominance, soil properties and climate. Journal of Biogeography 19: 117–134.
21. Haxeltine A, Prentice IC (1996) BIOME3: An equilibrium terrestrial biosphere model based on ecophysical constraints, resourse availability, and competition among plant functional types. Global Biogeochemical Cycles 10: 693–709.
22. Kearney M, Porter W (2009) Mechanistic niche modelling: combining physiological and spatial data to predict species ranges. Ecology Letters 12: 1–17.
23. Vieites DR, Nieto-Roman S, Wake DB (2009) Reconstruction of the climate envelopes of salamanders and their evolution through time. Proceedings of the National Academy of Sciences of the United States of America 106: 19715–19722.
24. Graham CH, Ron SR, Santos JC, Schneider CJ, Moritz C (2004) Integrating phylogenetics and environmental niche models to explore speciation mechanisms in dendrobatid frogs. Evolution 58: 1781–1793.

25. Hardy CR, Linder HP (2005) Intraspecific variability and timing in ancestral ecology reconstruction: A test case from the Cape Flora. Systematic Biology 54: 299–316.
26. Nogués-Bravo D (2009) Predicting the past distribution of species climatic niches. Global Ecology and Biogeography 18: 521–531.
27. Fitzpatrick MC, Weltzin JC, Sanders NJ, Dunn RR (2007) The biogeography of prediction error: why does the introduced range of the fire ant over-predict its native range? Global Ecology and Biogeography 16: 24–33.
28. Broennimann O, Treier UA, Müller-Schärer H, Thuiller W, Peterson AT, et al. (2007) Evidence of climatic niche shift during biological invasion. Ecology Letters 8: 701–709.
29. Knouft JH, Losos JB, Glor RE, Kolbe JJ (2006) Phylogenetic analysis of the evolution of the niche in lizards of the *Anolis sagrei* group. Ecology 87: 29–38.
30. Pearman PB, Guisan A, Broennimann O, Randin CF (2007) Niche dynamics in space and time. Trends in Ecology & Evolution 23: 149–158.
31. Braconnot P, Otto-Bliesner B, Harrison S, Joussaume S, Peterschmitt J-Y, et al. (2007) Results of PMIP2 coupled simulations of the mid-Holocene and Last Glacial Maximum. Part 1: experiments and large-scale features. Climate of the Past 3: 261–277.
32. Otto-Bliesner BL, Marshall SJ, Overpeck JT, Miller GH, Hu A, et al. (2006) Simulating Arctic Climate Warmth and Icefield Retreat in the Last Interglaciation. Science 311: 1751–1753.
33. Ruddiman WF, Raymo ME, Martinson DG, Clement BM, Backman J (1989) Pleistocene evolution of Northern Hemisphere ice sheets and North Atlantic Ocean. Paleoceanography 4: 353–412.
34. Zachos J, Pagani M, Sloan L, Thomas E, Billups K (2001) Trends, rhythms, and aberrations in global climate 65 Ma to present. Science 292: 686–693.
35. Campbell JA, Lamar WW (2004) The Venomous Reptiles of the Western Hemisphere. IthacaNY: Cornell University Press.
36. Holman JA (2000) Fossil snakes of North America. Bloomington and Indianapolis: Indiana University Press.
37. Pook CE, Wüster W, Thorpe RS (2000) Historical biogeography of the western rattlesnake (Serpentes: Viperidae: *Crotalus viridis*), Inferred from mitochondrial DNA sequence information. Molecular Phylogenetics and Evolution 15: 269–282.
38. Ashton KG, de Queiroz A (2001) Molecular systematics of the western rattlesnake, *Crotalus viridis* (Viperidae), with comments on the utility of the D-Loop in phylogenetic studies of snakes. Molecular Phylogenetics and Evolution 21: 176–189.
39. Clark AM, Moler PE, Possardt EE, Savitzky AH, Brown WS, et al. (2003) Phylogeography of the timber rattlesnake (*Crotalus horridus*) based on mtDNA sequences. Journal of Herpetology 37: 145–154.
40. Douglas ME, Douglas MR, Schuett GW, Porras LW (2006) Evolution of rattlesnakes (Viperidae; *Crotalus*) in the warm deserts of western North America shaped by Neogene vicariance and Quaternary climate change. Molecular Ecology 15: 3353–3374.
41. Polly PD (2010) Tiptoeing through the trophics: geographic variation in carnivoran locomotor ecomorphology in relation to environment. In: Goswami A, Friscia A, eds. Carnivoran Evolution: New Views on Phylogeny, Form, and Function. Cambridge: Cambridge University Press.
42. Heikinheimo H, Fortelius M, Eronen J, Mannila H (2007) Biogeography of European land mammals shows environmentally distinct and spatially coherent clusters. Journal of Biogeography 34: 1053–1064.
43. Hijmans RJ, Cameron JL, Parra PG, Jones PG, Jarvis A (2005) Very high resolution interpolated climate surfaces for global land areas. International Journal of Climatology 25: 1965–1978.
44. Nix H (1986) A biogeographic analysis of Australian elapid snakes Atlas of elapid snakes of Australia 7: 4–15.
45. Austin MP, Cunningham RB, Fleming PM (1984) New approaches to direct gradient analysis using environmental scalars and statistical curve-fitting procedures. Vegetatio 55: 11–27.
46. Stockwell D, Peters D (1999) The GARP modelling system: problems and solutions to automated spatial prediction. International Journal of Geographical Information Science 13: 143–158.
47. Meik JM, Pires-DaSilva A (2009) Evolutionary morphology of the rattlesnake style. Bmc Evolutionary Biology 9: 35.
48. Castoe TA, Parkinson CL (2006) Bayesian mixed models and the phylogeny of pitvipers (Viperidae: Serpentes). Molecular Phylogenetics and Evolution 39: 91–110.
49. Murphy RW, Fu J, Lathrop A, Feltham JV, Kovac V (2002) Phylogeny of the rattlesnakes (*Crotalus* and *Sistrurus*) inferred from sequences of five mitochondrial DNA genes. In: Schuett GW, Hoggren M, Douglas ME, Greene HW, eds. Biology of the vipers. Eagle MountainUtah: Eagle Mountain Publishing. pp 69–92.
50. Castoe TA, Spencer CL, Parkinson CL (2007) Phylogeographic structure and historical demography of the western diamondback rattlesnake (*Crotalus atrox*): A perspective on North American desert biogeography. Molecular Phylogenetics and Evolution 42: 193–212.
51. Martins EP, Hansen TF (1997) Phylogenies and the comparative method: A general approach to incorporating phylogenetic information into the analysis of interspecific data. American Naturalist 149: 646–667.
52. Polly PD (2004) On the simulation of morphological shape: mutivariate shape under selection and drift. Palaeontologia Electronica: Coquina Press. pp 7A.
53. Polly PD (2008) Adaptive Zones and the Pinniped Ankle: A Three-Dimensional Quantitative Analysis of Carnivoran Tarsal Evolution. In: Sargis EJ, Dagosto M, eds. Mammalian Evolutionary Morphology. Dordrecht: Springer. pp 165–194.
54. Felsenstein J (1988) Phylogenies and quantitative characters. Annual Review of Ecology and Systematics 19: 445–471.
55. Losos JB (2008) Phylogenetic niche conservatism, phylogenetic signal and the relationship between phylogenetic relatedness and ecological similarity among species. Ecology Letters 11: 995–1007.
56. Bennett KD (1997) Biological Response: Evolution. Evolution and Ecology: The Pace of Life. Cambridge: Cambridge University Press. pp 154–177.
57. Davis MG, Shaw RG (2001) Range shifts and adaptive responses to Quaternary climate change. Science 292: 673–679.
58. Thomas CD, Bodsworth EJ, Wilson RJ, Simmons AD, Davies ZG, et al. (2001) Ecological and evolutionary processes at expanding range margins. Nature 411: 577–581.
59. Holt RD (2003) On the evolutionary ecology of species' ranges. Evolutionary Ecology Research 5: 159–178.
60. Loarie SR, Duffy PB, Hamilton H, Asner GP, Field CB, et al. (2009) The velocity of climate change. Nature 462: 1052–1055.
61. Means DB (1986) Special Concern: Eastern Diamondback Rattlesnake in Vertebrate Animals of Alabama in Need of Special Attention. Alabama Agricultural Experiment Station.
62. Bennett SH (1995) Ecology and Status of the Eastern Diamondback Rattlesnake (*Crotalus adamanteus*) in South Carolina. ColumbiaSouth Carolina: South Carolina Department of Natural Resources.
63. Guralnick RP (2007) Differential effects of past climate warming on mountain and flatland species distributions: a multispecies North American mammal assessment. Global Ecology and Biogeography 16: 14–23.
64. Hannah L (2008) Protected areas and climate change. Annals of the New York Academy of Sciences 1134: 201–212.
65. Hoegh-Guldberg O, Hughes L, McIntyre S, Lindenmayer DB, Parmesan C, et al. (2008) Assisted colonization and rapid climate change. Science 321: 345–346.

15

Shifts in Climate Foster Exceptional Opportunities for Species Radiation: The Case of South African Geraniums

Hugo I. Martínez-Cabrera[1]*, Pedro R. Peres-Neto[1,2]

1 Département des sciences biologiques, Université du Québec à Montréal, Montreal, Quebec, Canada, 2 Canada Research Chair in Spatial Modelling and Biodiversity, Université du Québec à Montréal Montreal, Quebec, Canada

Abstract

Climate change is often assumed to be a major driver of biodiversity loss. However, it can also set the stage for novel diversification in lineages with the evolutionary ability to colonize new environments. Here we tested if the extraordinary evolutionary success of the genus *Pelargonium* was related to the ability of its species to capitalize on the climate niche variation produced by the historical changes in southern Africa. We evaluated the relationship between rates of climate niche evolution and diversification rates in the main *Pelargonium* lineages and disentangled the roles of deep and recent historical events in the modification of species niches. *Pelargonium* clades exhibiting higher ecological differentiation along summer precipitation (SPP) gradients also experienced higher diversification rates. Faster rates of niche differentiation in spatially structured variables, along with lower levels of niche overlap among closely related species, suggest recent modification in species niches (e.g. dispersal or range shift) and niche lability. We suggest that highly structured SPP gradients established during the aridification process within southern Africa, in concert with niche lability and low niche overlap, contributed to species divergence. These factors are likely to be responsible for the extensive diversification of other lineages in this diversity hot spot.

Editor: David L. Roberts, University of Kent, United Kingdom

Funding: CONACYT and FQRNT provided postdoctoral support to HIM-C. PRP-N was supported by NSERC and CRC grants. Funding was partially provided by the "Plan Vert" of the Government of Québec via a grant by the Ouranos consortium (project 554010-107). Species field locations were determined during research that was funded by NSF-IRE (OISE-0623341) and NSF Dimensions of Biodiversity (1046328) to C. Schlichting, C. Jones, K. Holsinger and J. Silander (University of Connecticut). The funders had no role in the study design, data collection and analysis, decision to publish, or preparation of the manuscript.

Competing Interests: The authors have declared that no competing interests exist.

* E-mail: huismaca@yahoo.com.mx

Introduction

Patterns of species diversification associated with past climate shifts can offer insights into the processes generating biological diversity and may provide information on how lineages are likely to respond to future climate change. Under changing climate, lineages with rapidly evolving climatic preferences may have higher diversification success [1] because they are more likely to colonize and spread into new climate zones [2]. We now know that this capacity is mainly determined by the extent to which their climate niche is evolutionarily conserved [3] (but see [4,5] for exceptions), so that if phylogenetic niche conservatism is strong, species will probably fail to adapt and thus experience greater risks of extinction if the new environments are different from their ancestral state. When a new selective regime is established (e.g. new climate regime), niche lability may fuel species diversification [6] (but see [7]) by increasing the rate at which niche dimensions associated with the new regime evolve [3,6]. Niche conservatism, on the other hand, might hinder evolution of climate niche preferences even if change in climate has increased ecological opportunities for diversification [3].

Phylogenetic niche conservatism among closely related species originates due to phylogenetic [5,8] or spatial constraints [8,9] or a combination of both. This is because both, inherited environmental preferences due to common ancestry or dispersal constraints in relation to the ancestral range, can shape the opportunities for adaptation to new environments [8]. Spatial and phylogenetic components of niche variation may be equally important if they are intrinsically related (i.e. spatial distances among species have a strong relationship with their phylogenetic history). When the spatial component of niche variation is greater than the effect of phylogeny, it is suggested that species responses to recent evolutionary events were more important than deeper historical events for determining species niches [8]. For instance, Cooper et al. [10] found that the large spatial component of mammal niche variation could be mostly explained by a modification in species niches that evolved in response to recent environmental changes - possibly related to dispersal or shifts in geographical ranges.

Under rapid climate change, and particularly during the rapid establishment of seasonal gradients, ancestral ranges likely undergo climatic fragmentation (i.e. different populations occupy different climate zones in the new regime). This would cause reproductive isolation and species divergence and result in closely related species occupying distinct portions of the climate niche space. If true, then climate niche axes that have strong spatial signatures would also show higher rates of evolution because closely related species would show divergence in climate preferences. It follows that lineages exhibiting high niche lability would be expected to diversify more rapidly than lineages with highly conserved climate preferences, which are also more likely to go extinct. Due to such niche segmentation of the ancestral ranges, closely related species within lineages diversifying during the establishment of climate gradients should then also exhibit low degrees of niche overlap.

In the Cape Floristic Region (CFR) of South Africa, the rapid climate change which lead to the aridification and establishment of the Mediterranean type climate started during the Miocene and continued until the Pleistocene, and was driven by both changes in the position of high-pressure cells and the extent of the Antarctic ice sheet [11,12]. The subsequent widespread extinction of the western South Africa flora [13,14] is thought to have created a significant amount of vacant niches space and additional opportunities for diversification [15,16] among lineages adapted to summer aridity [14]. This climate change process increased niche availability in this system in two ways: 1) the establishment of a seasonal precipitation gradient with winter rainfall in the west *versus* summer rainfall in the east; and 2) within the winter rainfall regime, a sharp gradient in winter rainfall amounts [17,18]. The rapid establishment of climate gradients in highly segmented landscapes [19] would likely accelerate ecological differentiation among species within lineages showing climate niche lability. The wide range of habitats occupied by *Pelargonium*'s closely related species [20,21], as well as their diverse morphological adaptations [22], suggest that indeed this group has adaptively radiated [23]. This niche filling process in an adaptive radiation may lead to an acceleration in the accumulation of lineages, though as niche space becomes filled diversification rates may actually decline [6,24,25].

In this study we set out to determine the extent to which ecological differentiation has occurred in the genus and to know if this was associated with historical shifts in climate regimes. We also analyzed whether *Perlargonium* clades experiencing accelerated evolution of climate preferences and high species diversity underwent declines in diversification rates associated with the saturation of the climate niche space. Lastly, we explored the relationship between niche diversification rates, niche overlap and the relative importance of space *versus* phylogeny in determining climate niche structure. In particular, we tested whether those niche dimensions over which the genus had experienced greater ecological differentiation also exhibited greater spatial structure and low niche overlap. We expect that clades experiencing successful radiations would be associated with a greater ability to explore the newly established climatic regimes (niche lability). As a consequence, the rates of evolution of climate niche preferences should be positively related with clade diversification rates and richness.

Materials and Methods

Phylogeny, species distributions and environmental coverage

Rate diversification and comparative analyses were based on Bakker et al.'s 50% majority rule Bayesian consensus tree [26]. This phylogenetic hypothesis is based on four data partitions: cpDNA *trn*L-F, nrDNA ITS, mtDNA *nad*1b/c, and 30 recoded indels inferred from the plastid partition [22,26]. For the analysis of niche evolution, we pruned this phylogenetic tree (originally containing 150 species) to include only the 112 species with distributional information; distribution was obtained from digital databases [27,28], taxonomic monographs and our collection records. In total, we analyzed over 6500 occurrence points for 112 species (Table S1), representing roughly 40% of the species in the genus. Outlier occurrences for each species (i.e. those occurrences clearly beyond the distribution of the rest of the species occurrence data) were excluded from our analyses. The resultant pruned phylogeny was used to estimate a chronogram using penalized likelihood [29] in APE [30] assuming a root age of 30 MYR corresponding to the divergence time calculated based on the ancestral divergence of Geraniaceae [13].

Climate information at 1 arc minute resolution (1.55×1.85 km at South Africa latitude) was acquired from the South African Atlas of Climatology and Agrohydrology [31]. From an initial set of 33 climate variables, 10 were selected for use based on their loadings on the first three principal components and were relevant to the niche evolution and distribution of *Pelargonium* [21]: mean annual precipitation (mm), winter and summer precipitation (mm), annual heat units (°d), accumulated positive chill units (CU), winter and summer solar radiation ($MJ \cdot m^{-2} \cdot d^{-1}$), and winter and summer vapor pressure deficit (kPa).

Diversification analyses and rates of climate preference evolution

We estimated net diversification rates (speciation minus extinction) for each major clade (Fig. 1; A1, A2a, A2b, B, C1 and C2) using Magallón and Sanderson's method accounting for incomplete sampling [32] as implemented in the Geiger package for R [33]. We calculated diversification rates using three levels ($e=0$, 0.5 and 0.99) of extinction fractions (e = speciation/extinction), which dictate the probability of survival to the present. As clades in the large (A1, A2a, A2b and B) and small (C1, C2) chromosome clades differ in diversity and climate preferences, we tested for differences in diversification rates using the Mann-Whitney U-test. One of the consequences of rapid initial diversification is the potential for decreasing diversification rates due to species accumulation in clades [24,34]. Since *Pelargonium* and other cape lineages (e.g. *Muraltia*, some *Restionaceae* and *Satyrium* [14]) experienced explosive *in situ* radiation, we tested whether there was a decrease in diversification rates due to species accumulation or whether, despite rapid occupation of niches, *Pelargonium* experienced constant diversification rates. To test whether diversification rates decreased or increased over time the γ statistic [35] was used. A negative γ indicates a decrease in diversification rate relative to a constant diversification model through time. Given that incomplete phylogenetic sampling is known to bias γ [35], we conducted a Monte Carlo Constant Rate Test (MCMCR test) to account for incomplete sampling. We simulated 10 000 complete phylogenies (for each clade) under the null hypothesis of constant pure birth diversification; tips were randomly pruned to conform to actual clade sizes. The critical γ value (the 0.05 percentile of the null distribution) of the simulated replicates was contrasted with the observed γ value [35], with greater observed γ values indicating a rate decrease greater than expected by chance. This analysis was run using the R package LASER [36].

We also assessed which rate diversification model best fit the data for each clade using likelihood methods [36]. We explored two rate constant (RC) models, namely pure birth and birth-death models, for inferring diversification rates that do not vary over time [36,37], and four rate variable (RV) models: exponential (DX), linear (DL) density dependent models of diversification and two multi-rate models (Yule-2 rate and Yule-3) based on a variant of the Yule process [36]. To fit the diversification parameters, we used the ages estimated by Bakker et al. [13] for the genus (30 MY) and its main clades. To test whether the null hypothesis of the RC model could be rejected in support of an RV model, we estimated differences in fit using the Akaike Information Criterion as $\Delta AIC_{RC} = AIC_{RC} - AIC_{Rv}$ [36].

The rates of climate preference evolution for each clade were calculated using the species mean value for each climate variable. Climate evolution rates for the clades considered here were

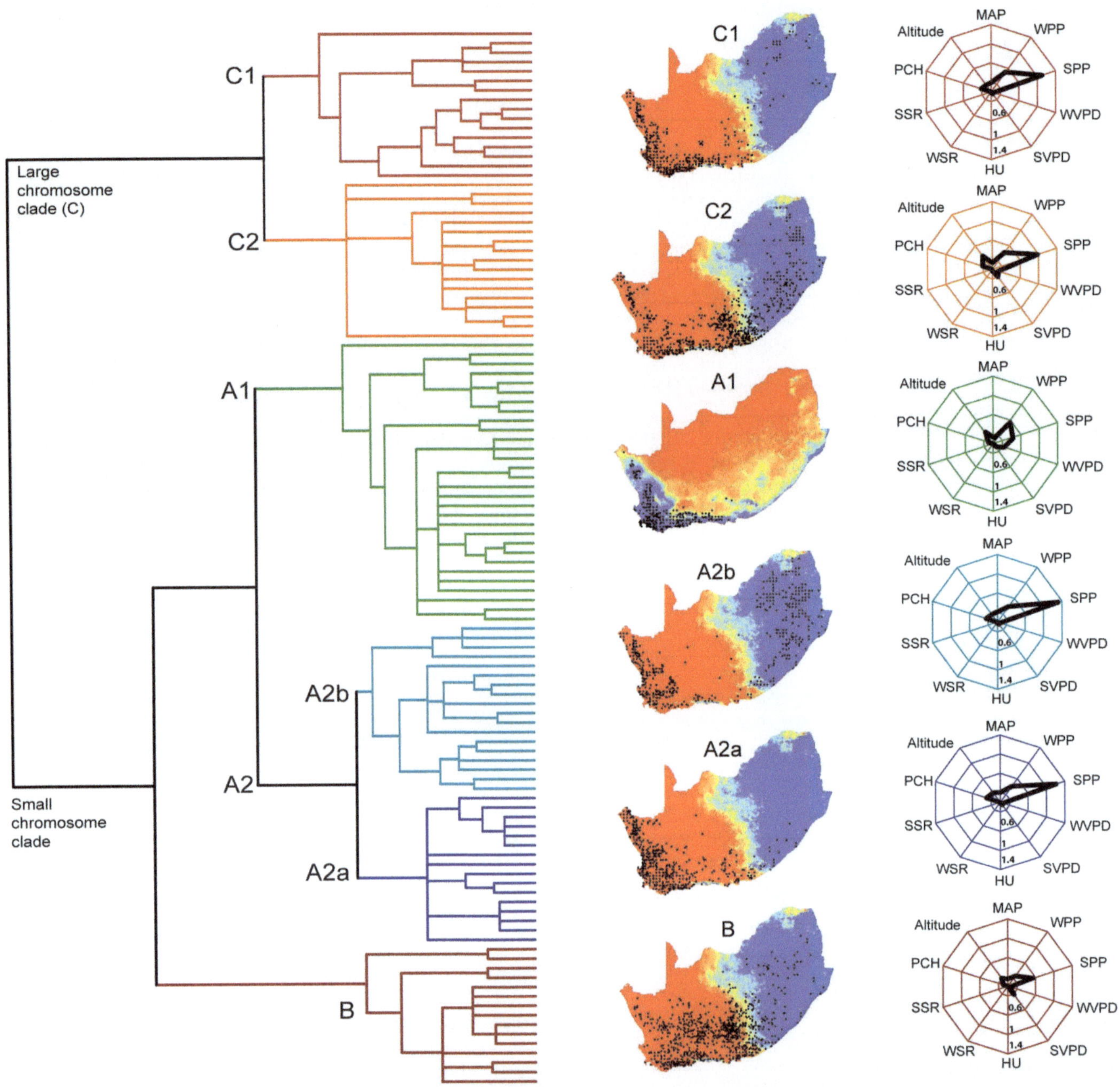

Figure 1. Chronogram, geographic distribution and climate rates of evolution for *Pelargonium* clades. Background colors of the maps represent summer precipitation, except for clade A1, which portrays winter precipitation. Blue and red regions represent high and low precipitation respectively. Radar plots show the rate of evolution in felsens for each climate variable in each clade; values closer to the center of the web indicate slower rates of evolution while larger β values indicate higher rates. For the actual values see Table S2. Major clades are colored in the phylogeny and radar plots. MAP = mean annual precipitation; WPP = mean winter precipitation; SPP = mean summer precipitation; WVPD = mean winter pressure deficit; SVPD = mean summer pressure deficit; HU = annual heat units; WSR = mean winter solar radiation; SSR = mean summer solar radiation; PCH = accumulated positive chill units.

calculated using felsens [38], which is a contrast-based metric that uses the time-calibrated phylogeny and ln-tranformed variables (here climatic variables). A felsen (β, the Brownian motion rate parameter) is an increase of one unit per million years in the trait (or niche in this case) variance among sister taxa [38]. Felsens represent the PIC variance describing the rate of divergence in the climate niche variables among species.

We used regression to test the hypothesis that highly diverse clades have higher rates of diversification and to determine if highly diverse, rapidly evolving clades experience declining diversification rates, possibly as a product of niche filling. Finally, using least-squares and polynomial regression, we explored whether mean clade evolution of individual climate niche preferences is related to clade diversification rates.

Space vs. phylogenetic signal in climate niche variables

For each clade, we evaluated the relative contribution of phylogenetic relationships versus spatial variation on evolution climate preferences across species using the Freckleton and Jetz's framework [8]. Their method modifies the expected variance of the phylogenetically independent contrast to simultaneously include a spatial and a phylogenetic signal. The relative contributions of the phylogenetic and spatial effects are estimated by $\emptyset$, which ranges in value between 0 (only phylogenetic structure is important) and 1 (only spatial structure is important). Based on Pagel's λ [39] and $\emptyset$, we calculated the spatially corrected version of Pagel's λ (λ') and the proportion of climate variation that is not explained by space or phylogeny (Γ). Pagel's λ [39] is a branch length transformation as it is simply a multiplier of the off-diagonal elements of the variance-covariance matrix that best fits the data [8,10]. λ measures phylogenetic signal where a value of 0 indicates no phylogenetic signal, 1 trait variation according to the Brownian motion model and a value of >1 indicates very strong phylogenetic conservatism. We can estimate the spatially corrected version of Pagel's λ as $\lambda' = (1-\emptyset)\ \lambda$), which measures the relative contribution of phylogeny independent of space, while $\Gamma = (1-\emptyset)(1-\lambda)$ measures the proportion of trait variation that is neither explained by space nor phylogeny [8]. We estimated $\emptyset$ and λ for each climate variable separately.

Niche overlap

Niche overlap among closely related species serves as a proxy for niche conservatism [40] and may explain constraints in species distributions. For each clade, we quantified species overlap for each climate variable using the metric D [41]. D varies between 0 and 1 indicating no overlap and identical niches, respectively. As there may be differences in sampling effort or design among regions, sampled occurrences may not characterize the entire range of climate suitable for species, which in turn may bias measures of niche overlap [42]. To avoid these biases, here we used the framework proposed by Broennimann et al. [42] which uses kernels to smooth density probability distributions of species occurrences in the environmental space. Since this method also uses smoothed densities of available environments, the niche overlap test within this framework compares the occurrence densities between species occurring in ranges where environments are not equally available [42]. In addition, the use of a kernel density estimator provides an unbiased estimate of D because it ensures independence between the environmental overlap and grid resolution [42]. The equivalency test [40] was used to test for significance in overlap for each environmental variable. The test consists of pooling the occurrence of any given species pair (pseudoreplicate datasets) and randomly partitioning it into two data sets (keeping the original occurrence size of the two species). This procedure was repeated for 1000 pseudoreplicate datasets and for each one, the niche overlap metric D was calculated to obtain the null distribution of similarity values. The observed value of D was then compared with its null distribution using a one-tailed test and the null hypothesis of niche equivalency was rejected if the observed D fell outside of the 95% of the simulated values, thus indicating niche differentiation.

Results

Diversification rates

Lineages within the small chromosome clade (A1, A2a, A2b and B) experienced higher net diversification rates (Table 1) compared to the lineages in the large chromosome clade (C1 and C2). These trends were robust across the three different extinction fractions tested ($e = 0$, 0.5 and 0.99). Net diversification rates were highest in clade A2b ($r_0 = 0.248$, $r_{0.5} = 0.229$, $r_{0.99} = 0.039$ speciation events per MY) and A2a ($r_0 = 0.186$, $r_{0.5} = 0.169$, $r_{0.99} = 0.018$), both in the xerophytic clade A2. In clades C1 ($r_0 = 0.093$, $r_{0.5} = 0.084$, $r_{0.99} = 0.007$) and C2 ($r_0 = 0.099$, $r_{0.5} = 0.089$, $r_{0.99} = 0.007$) diversification rates were the lowest. There was a marginally significant difference in diversification rates between small and large chromosome clades ($P = 0.06$, Mann-Whitney U-test). Declining diversification rates in all clades was suggested by negative values for the γ-statistic (except clade A1; Table 1). The steepest decline was detected in clade A2b ($\gamma = -2.99$, $P = 0.006$). Despite the negative values of γ for most clades, the null hypothesis of rate constancy, when accounting for an incomplete taxonomic sample, was only rejected for clade A2b (Table 1). In the rate variation analysis (Table S2), we found that RC models were rejected in favor of an RV model ($\Delta AIC_{RC} = 7.23$, DL) for clade A2b only. Positive ΔAIC_{RC} in clades A1 ($\Delta AIC_{RC} = 0.73$, Yule-2 rate), B ($\Delta AIC_{RC} = 0.27$, DL) and C2 ($\Delta AIC_{RC} = 1.51$, Yule-3 rate) indicate rate variation. However, for small phylogenies, the RC models can only be rejected with confidence when ΔAIC_{RC} is close to 4 [36] and as such, decreases in these last three clades cannot be asserted. In clades A2a ($\Delta AIC_{RC} = -0.52$) and C1 ($\Delta AIC_{RC} = -1.75$), RV models were rejected in favor of RC models (pure birth in both cases, Table S2). For the entire genus, however, there is clear evidence of rate variation ($\Delta AIC_{RC} = 5.83$, Yule-3 rate).

Rates of climate evolution

Mean rates of climate preference evolution varied from $\beta = 0.25$ felsens in clades A2a and A2b to $\beta = 0.15$ felsens in clade B based on all climate variables considered here (Table S3). As predicted, in most cases, rates related to precipitation variables were one order of magnitude greater than for the other climatic variables. Strikingly, even in clades mainly distributed in the winter rainfall region (A2a, A2b and C1, except for the small clade *Magnistipulacea* in clade A2a, that has radiated from the winter rainfall to the summer rainfall region, see Fig 1), the rate of evolution of summer precipitation preferences was faster than that for winter precipitation. Only in the evergreen shrubs of clade A1 (Fig. 1) winter precipitation preferences evolved faster than summer precipitation preferences. Rates for the evolutionary preference for summer precipitation observed the highest values in the winter rainfall clades A2a and A2b with $\beta = 1.21$ and 1.31 felsens, respectively. Preferences along seasonal solar radiation and vapor pressure deficit gradients evolved consistently slower across clades (Table S3, Fig. 1).

Spatial vs. phylogenetic effects

Across all clades and for all climate variables, the importance of the spatial component $\emptyset$ was greater than the phylogenetic component λ', indicating that spatially-related processes and not shared phylogenetic histories were most important in driving climate niche variation. Mean clade independent variation, Γ, was generally greater than λ' and lower than $\emptyset$. In most clades, variation in summer precipitation observed $\emptyset$ values closest to 1, followed by summer solar radiation and mean annual precipitation (Table S4). The spatial component of summer precipitation was consistently high throughout all clades ($\lambda' < \emptyset$). Phylogenetic effects were only greater than spatial effects ($\lambda' > \emptyset$) in niche variables related to environmental energy (heat units and seasonal solar radiation) in clades A1, A2a and A2b.

Table 1. Results of the diversification rate analyses for *Pelargonium* clades.

	Net diversification rate			Gamma					Clade diversity	Clade age (MY)
Clade	r_0	$r_{0.5}$	$r_{0.99}$	Observed γ	P-value	Simulated critical value γ	Observed- simulated γ	MCMCR corrected P-value		
A1	0.155	0.142	0.017	0.81	0.79	−2.06	2.88	0.91	34	20
A2a	0.186	0.169	0.018	−0.47	0.31	−1.32	0.85	0.23	20	15
A2b	0.248	0.229	0.039	−2.99	0.001	−2.96	−0.03	0.047	83	10
C1	0.093	0.084	0.007	−0.01	0.49	−1.76	1.74	0.57	23	26
C2	0.099	0.089	0.007	−0.44	0.32	−1.73	1.28	0.38	23	24
B	0.109	0.097	0.007	−0.8	0.2	−2.05	1.25	0.37	27	20

r_0, $r_{0.5}$, $r_{0.99}$ = net diversification rates with extinction fraction of 0, 0.5 and 0.99 respectively. MCMCR= Monte Carlo Constant Rate Test.

Niche Overlap

We used climate niche overlap among closely related species as a proxy for climate conservatism. Smaller overlaps were observed regarding summer and winter precipitation (Fig. 2A) than compared to any other climate variable. Interestingly, even for clades primarily distributed in the winter rainfall region (A1, A2a, A2b), closely related species overlapped least in their preferences for summer precipitation levels. This result strongly supports our finding that closely related species experienced large ecological differentiation (i.e. higher β) along precipitation gradients. Conversely, variables related to environmental energy such as solar radiation and vapor pressure deficit (Fig. 2A; Figure A in Figure S1) observed much greater levels of overlap and hence varied much less among closely related taxa. In all clades except A1, a larger number of pair-wise equivalence tests were rejected for precipitation variables (summer, winter or total precipitation) indicating that most of the ecological differentiation among closely related species has occurred along these niche axes (Fig. 2B; Figure B in Figure S1).

The drivers of the rate of climate preference evolution

Rapidly evolving climate niche preferences (i.e. precipitation) showed lower niche overlap (Fig 3A; Table S5) and were primarily structured by space (Fig 3A, C). The relationship between the rate of evolution and niche overlap was significant for all clades, except for a marginal effect for clade C2 ($P=0.06$). The relationship between Ø and climate evolution rate was only significant for clade B ($r^2=0.4$, $p=0.04$) and marginally significant for clades A2a and C2 ($P=0.07$ for both clades), indicating that (except for clade C1) fast evolving climate preferences are spatially structured (Fig. 3B). In general, the relationship between niche overlap and Ø was negative, indicating that niche overlap is smaller for those dimensions structured in space. Note, however, that this relationship was only significant for clade A1 ($r^2=0.45$, $p=0.032$, Fig. 3C).

Clade diversification rate and rates of climate evolution

Clade diversity was positively related to clade diversification rates ($r^2=0.91$ $p=0.011$). However these rich, rapidly diversifying clades, have experienced a tendency for decreases in diversification rates ($r^2=0.77$ $P=0.07$). Clades with higher net diversification rates are younger ($r^2=0.9$, $P=0.013$). Mean evolutionary rates for climate niche preferences (all climate variables) are independent of clade diversification rates ($r^2=0.009$, $P=0.85$), but the latter are significantly associated with rates of summer rainfall evolution (polynomial $r^2=0.915$, $P=0.024$; Fig. 4). The relatively high diversification rates of clades A2a and B are, however, not associated with high rates of summer rainfall evolution (Fig. 4).

Discussion

We showed that rapidly evolving climate preferences were mainly driven by spatial rather than phylogenetic variation. This result indicates that the influence of recent evolutionary events rather than earlier evolutionary events were the most important in shaping species niche preferences. Our results also indicate that closely related species in the studied system have low levels of niche overlap for those climate preferences that underwent rapid evolution. In the CFR, aridification and the rapid establishment of sharp seasonal rainfall gradients were likely the events that fueled this fast evolution in climate preferences. At one end, the rapidly evolving climate preferences in *Pelargonium* is represented by summer precipitation, while, on the opposite end, preferences for energy-related variables showed greater levels of niche overlap and a greater role of phylogenetic effects, indicating evolutionary conservatism along those climate axes. This could be due to the fact that energy-related variables changed less than rainfall regimes, thus creating fewer opportunities for ecological divergence and slower evolutionary rates. High divergence in summer precipitation preferences among closely related species across *Pelargonium* clades indicates that evolvability along this niche axis is not clade dependent, although greater rates of evolution in summer precipitation were associated with larger clade sizes and diversification rates. Interestingly, some variables (e.g. winter solar radiation and summer pressure deficit) showed the reverse pattern, with greater levels of spatial structure but comparatively slow evolutionary rates. However, despite high possibilities for evolution over these climate axes, suggested by their greater spatial structuring ($\lambda' < Ø$), they have not resulted in significant evolutionary differentiation among closely related species. In this case, low phylogenetic signal is not correlated with higher rates of evolution, which can be expected in some evolutionary circumstances (e.g. homogeneous rate genetic drift [43]).

Higher evolutionary rates in summer precipitation preferences can be understood in light of the climatic history of the region. During the early history of *Pelargonium* (late Oligocene-Early Miocene), a summer-wet climate was present in southern Africa. The subsequent aridification, which led to the extinction of the temperate – subtropical flora that dominated the CFR landscape [44], presumably generated a new adaptive zone over which ecological specialization occurred. Indeed, most CFR (winter rainfall) *Pelargonium* lineages (A2a and A2b) experienced even

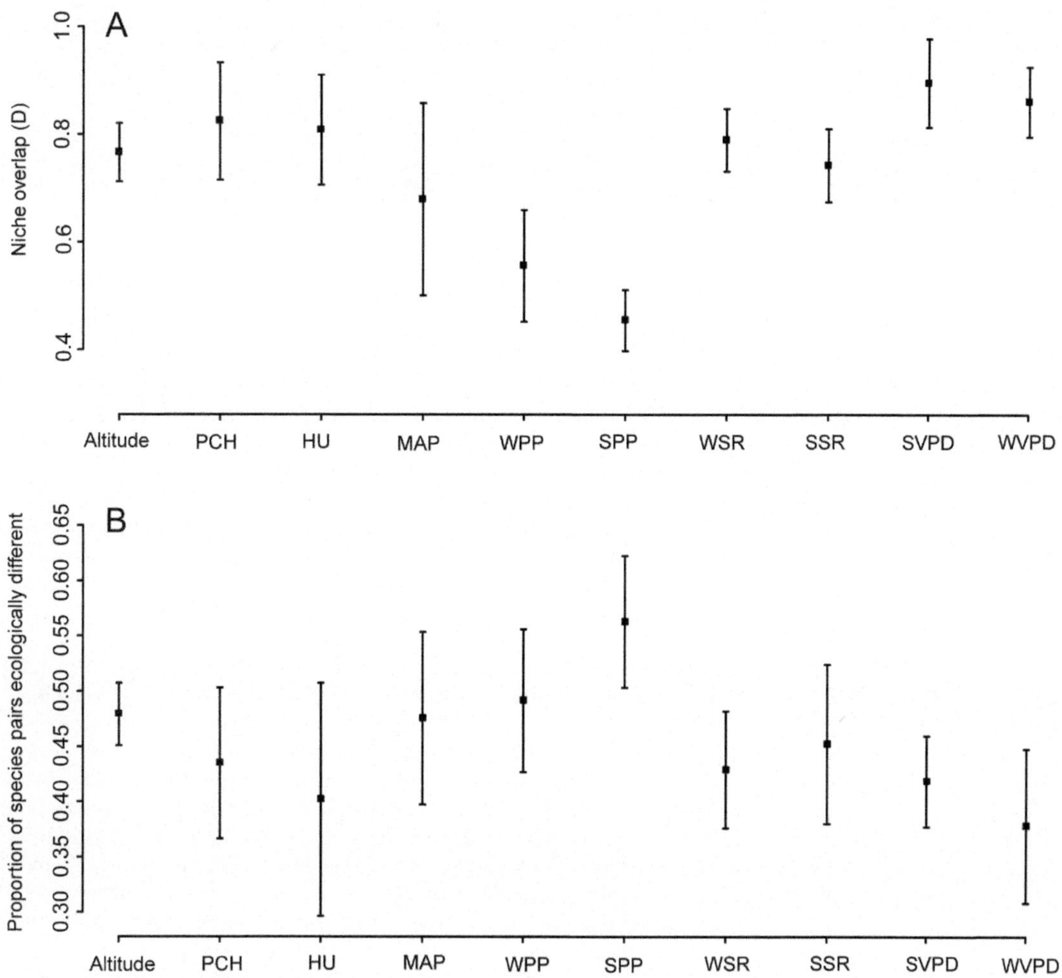

Figure 2. Results of the climate niche overlap analysis. The graphs show the mean and 95% confidence intervals across clades of the 2A) mean within clade pairwise species niche overlap and its 2B) proportion of rejected niche equivalency tests (proportion of species pairs ecologically distinct). D = Schoener niche overlap metric; PCH = accumulated positive chill units; HU = annual heat units; MAP = mean annual precipitation; WPP = mean winter precipitation; SPP = mean summer precipitation; WSR = mean winter solar radiation; SSR = mean summer solar radiation; WVPD = mean winter pressure deficit; SVPD = mean summer pressure deficit.

greater rates of evolution in summer rainfall preferences than clades distributed in the summer rainfall region or across rainfall regimes (clades C1, C2, B). This indicates that much of the evolution of the climate niche in the genus has occurred independent of the recently created winter precipitation gradients and that pre-adaptation to summer aridity might have played an important role in diversification [14]. Diversification in *Pelargonium* has therefore mainly occurred along the increased summer aridity gradients, agreeing with extended diversification in drought adapted growth forms, leaf attributes and phenologies in the genus [22,45]. The small chromosome clade A exhibits a much wider variation in morphological adaptations to cope with drought when compared to the large chromosome clade lineages [22]. On the other hand, the lower morphological diversity and lower diversification and climate evolutionary rates of clade C could be the result of remaining in the ancestral climate regime as evidenced by its higher diversity in the summer rainfall region. This differential capacity for diversification between major clades could also be linked to their divergent ecological strategies to cope with drought: morphologically mediated (probably throughout the development of drought avoidance, i.e. deciduousness) in clade A and through modification in their water use efficiency (development of drought resistance) in clade C [20,46].

During the course of an adaptive radiation, a burst of diversification should be facilitated by the ecological opportunity represented by empty niches or unexploited resources, which subsequently declines as niches become filled [6,24,25]. Such declines in diversification rates have been observed in some groups [24,25]. In some studies, clade age is positively related with diversity [47] but in others is instead independent (e.g. [48]). Here we found that young, highly diverse clades evolve at greater rates than old, species poor lineages. In the case of the highly diverse xerophytic clade A2b, the RC models were rejected in favor of a density dependent diversification model; this observation agrees with the significant decrease in diversification rate. Therefore for this clade, which originated about 10 MYA [13], the opening of empty niches that followed the aridification process, and its ability to evolve under increased summer drought, likely promoted an initial burst in diversification. As the radiation of this clade progressed, the niche space became saturated and the speciation rate started to decline. Niche saturation results in a decrease in opportunities for speciation [49] due to reduced chances for new populations to establish [50]. In members of clade C, we did not

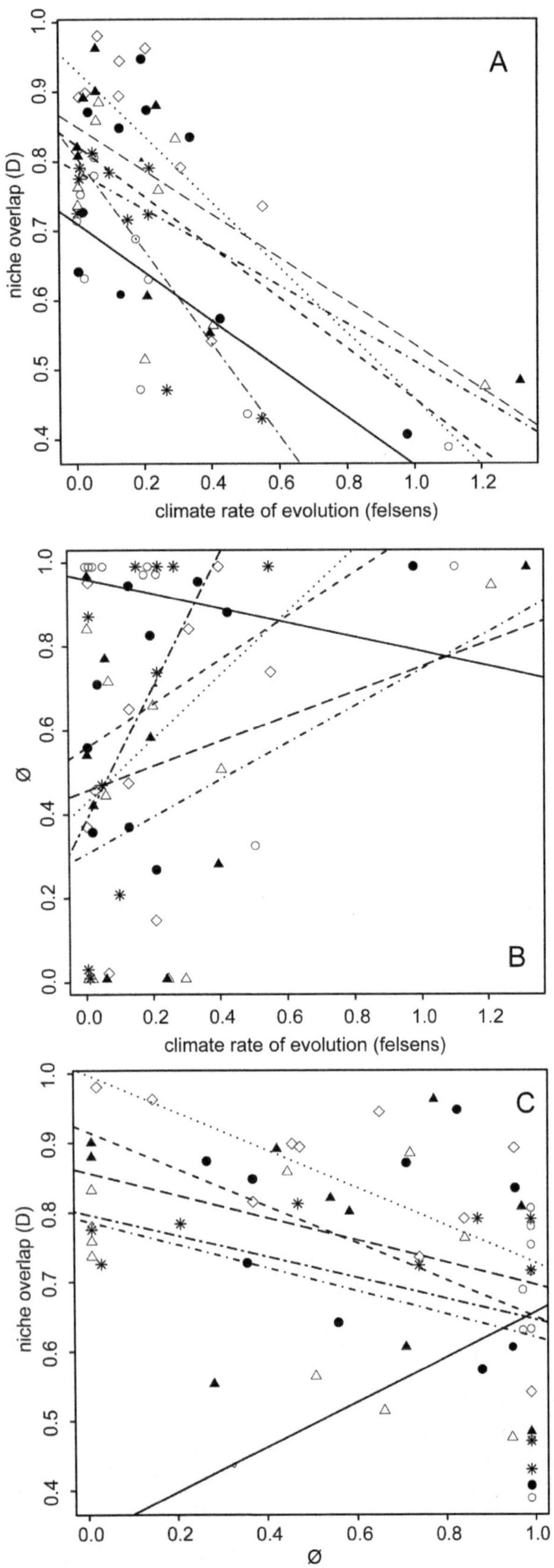

Figure 3. Plots showing a) niche overlap and b) *Ø* as a function of climate rate of evolution and c) between niche overlap and *Ø* for each of the *Pelargonium* clades. Each regression line represents a clade; data points are climate variables for each clade. Open circles and solid line = clade C1; filled circle and dashed line = C2; open triangles and dot-dashed line = A2a; closed triangles and long-dashed line = A2b; open diamonds and dotted line = A1 and; asterisks and two-dashed line = Clade B. For regression coefficients and associated *P-values* see Table S5.

detect any decreases in diversification rates and thus this old radiation has diversified at constant but comparatively slower rates. In this case, the shift in climate regime did not promote an increase in diversification likely because clade C mostly diversified in the summer rainfall region, its ecological zone of origin.

As the rapid radiations of many CFR lineages have occurred *in situ* [15], the scenario of quick niche filling in the region and increased competition is highly plausible for other highly diverse, young clades (e.g. *Satyrium*, *Schoeneae*). For some of these clades increased competition, a product of niche filling, likely decreased diversification rates, yet in others might have driven diversification as suggested by phylogenetic overdispersion in some groups [51]. Phylogenetic overdispersion at the geographic scale in *Pelargonium* [20] suggests that species divergences in the genus have occurred along the new climate gradients and, as a consequence, closely related species occupy different positions along these gradients. Our findings indicate that large species and phylogenetic turnover [20] is probably the consequence of species ecological differentiation over the highly spatially structured summer precipitation gradients. It is therefore likely that the evolution of *Pelargonium* clades, especially in the CFR, was a product of the genetic isolation of ancestral populations which then diverged into new species that differentiated ecologically in response to the establishment of new summer aridity gradients. Large spatial effects in precipitation variation coupled with greater rates of evolution and low niche overlap support such a scenario. In conclusion, we suggest that *Pelargonium* diversification is the result of rapid climate change in the region associated with a fast differentiation of lineages along aridity gradients. Accelerated ecological differentiation in precipitation preferences paired with low overlap, along with diverse morphologies, suggest that the radiation in some *Pelargonium* clades (A1 and A2) was adaptive. This climate shift, along with a segmented landscape, probably contributed to the diversification of other, drought pre-adapted, hyperdiverse CFR lineages.

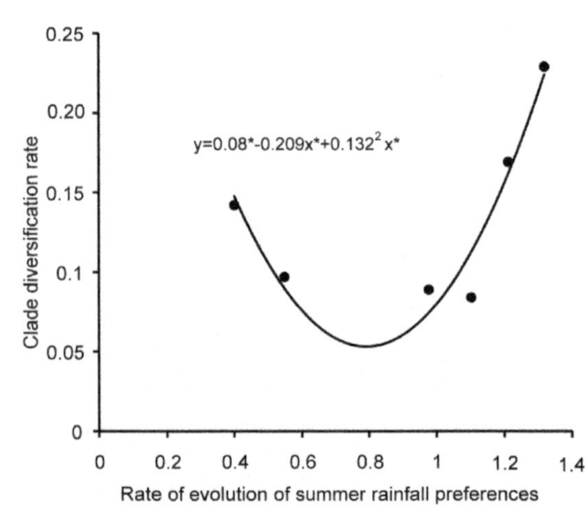

Figure 4. Plot showing the relationship between clade net diversification rate and rate of evolution of summer rainfall preferences (felsens). All the parameters in the polynomial regression equation are significant $P<0.05$. $r^2 = 0.91$, $P = 0.024$.

Supporting Information

Figure S1 Results of the climate niche overlap analysis for *Pelargonium* clades.

Table S1 Studied *Pelargonium* species and number of samples.

Table S2 Results of the diversification models analyses.

Table S3 Results of climate rate of evolution analysis.

Table S4 Results of the analysis describing the relative importance of spatial and phylogenetic effects in niche variation in *Pelargonium* clades.

Table S5 Results of the regression analysis among evolutionary rates, niche overlap and spatial effects in *Pelargonium* clades.

Acknowledgments

We thank Wagner Moreira for helping with the R code calculating niche overlap and the constructive criticism of Cynthia Jones, Bailey Jacobson and two anonymous reviewers. Rob Freckleton kindly provided their code for calculating rates of diversification and space vs. phylogenetic effects respectively. The authors thank Cape Nature, the Northern Cape Province, the Eastern Cape Parks and Tourism Agency, and the Province of the Eastern Cape for collection permits.

Author Contributions

Conceived and designed the experiments: HIMC PRPN. Performed the experiments: HIMC PRPN. Analyzed the data: HIMC PRPN. Contributed reagents/materials/analysis tools: HIMC PRPN. Wrote the paper: HIMC PRPN.

References

1. Holt RD (1990) Microevolutionary consequences of climate change. Trends in Ecology and Evolution 9: 311–315.
2. Kozak KH, Wiens JJ (2010) Accelerated rates of climatic-niche evolution underlie rapid species diversification. Ecology Letters 13: 1378–1389.
3. Losos JB (2008) Phylogenetic niche conservatism, phylogenetic signal and the relationship between phylogenetic relatedness and ecological similarity among species. Ecology Letters 11: 995–1007.
4. Kozak KH, Weisrock DW, Larson A (2006) Rapid lineage accumulation in a non-adaptive radiation: phylogenetic analysis of diversification rates in eastern North American woodland salamanders (Plethodontidae: Plethodon). Proceedings of the Royal Society of London B 273: 539–546.
5. Wiens JJ, Graham CH (2005) Niche conservatism:integrating evolution, ecology, and conservation biology. Annual Review of Ecology, Evolution and Systematics 36: 516–539.
6. Schluter D (2000) The ecology of adaptive radiation. Oxford: Oxford University Press.
7. Adams DC, Berns CM, Kozak KH, Wiens JJ (2009) Are rates of species diversification correlated with rates of morphological evolution? Proceedings of the Royal Society of London B 276: 2729–2738.
8. Freckleton RP, Jetz W (2009) Space versus phylogeny: disentangling phylogenetic and spatial signals in comparative data. Proceedings of the Royal Society of London B 276: 21–30.
9. Peres-Neto PR, Leibold MA, Dray S (2012) Assesing the effects of spatial contingency and environmental filtering on metacommunity phylogenetics. Ecology 93: 14–30.
10. Cooper N, Freckleton RP, Jetz W (2011) Phylogenetic conservatism of environmental niches in mammals. Proceedings of the Royal Society of London B 278: 2384–2391.
11. Siesser WG (1980) Late Miocene origin of Benguela upwelling system off northern Namibia. Science 208: 283–285.
12. Zachos J, Pagani M, Sloan L, Thomas E, Billups K (2001) Trends, rhythms, and aberrations in global climate 65 Ma to present. Science 292: 686–693.
13. Bakker FT, Culham A, Marais EM, Gibby M (2005) Nested radiation in Cape *Pelargonium*. In: Bakker FT, Chartrou LW, Gravendeel B, Pielser PB, Plant species-level systematics: New perspectives on pattern and process. Ruggell, Liechtstein: Ganter Verlag K. G.
14. Verboom GA, Archibald JK, Bakker FT, Bellstedt DU, Conrad F, et al. (2009) Origin and diversification of the Greater Cape flora: Ancient species repository, hot-bed of recent radiation, or both?. Molecular Phylogenetics and Evolution 51: 44–53.
15. Linder HP (2003) The radiation of the Cape flora, southern Africa. Biological Reviews of the Cambridge Philosophical Society 78: 597–638.
16. Verboom GA, Linder HP, Stock DW (2004) Testing the adaptive nature of radiation: growth form and life history divergence in the African grass genus *Ehrharta*. American Journal of Botany 91: 1364–1370.
17. Cowling RM, Richardson DM, Pierce SM (1997) Vegetation of southern Africa. Cape Town, South Africa: Cambridge University Press.
18. Linder HP (1991) Environmental correlates of patterns of species richness in the south-western Cape Province of South Africa. Journal of Biogeography 18: 509–518.
19. Thuiller W, Midgley GF, Rouget M, Cowling RM (2006) Predicting patterns of plant species richness in megadiverse South Africa. Ecography: 733–744.
20. Martínez-Cabrera HI (2010) Influence of climate in functional and species diversifi cation in South African *Pelargonium*. Storrs: University of Connecticut.
21. Martínez-Cabrera HI, Schlichting CD, Silander JA, Jones CS (2012) Low levels of climate niche conservatism may explain clade diversity patterns in the South African genus Pelargonium (Geraniaceae). American Journal of Botany 99: 954–960.
22. Jones CS, Bakker FT, Schlichting CD, Nicotra AB (2009) Leaf shape evolution in the South African genus *Pelargonium* L'Hér. (Geraniaceae). Evolution 63: 479–497.
23. Losos JB, Miles DB (2002) Testing the hypothesis that a clade has adaptively radiated: iguanid lizard clades as a case study. American Naturalist 160: 147–157.
24. Phillimore AB, Price TD (2008) Density-dependent cladogenesis in birds. PLoS Biology 3: e71.
25. Rabosky DL, Lovette IJ (2008) Density-dependent diversification in North American wood warblers. Proceedings of the Royal Society of London B 275: 2363–2371.
26. Bakker FT, Culham A, Touloumenidou T, Gibby M (2004) Phylogeny of *Pelargonium* (Geraniaceae) based on DNA sequences from three genomes. Taxon 53: 17–28.
27. Morris JW, Glen HF (1978) Précis, the National Herbarium of South Africa (PRE) Computerized Information System.
28. Rutherford MC, Powrie LW, Midgley GF (2003) ACKDAT: a digital spatial database of distributions of South African plant species and species assemblages. South African Journal of Botany 69: 99–104.
29. Sanderson MJ (2002) Estimating absolute rates of molecular evolution and divergence times: a penalized likelihood approach. Molecular Biology and Evolution 19: 101–109.
30. Paradis E, Claude J, Strimmer K (2004) APE: analyses of phylogenetics and evolution in R language. Bioinformatics 20: 289–290.
31. Schulze RE (2007) South African Atlas of Climatology and Agrohydrology. In: Commission WR, Pretoria.
32. Magallón S, Sanderson MJ (2001) Absolute diversification rates in angiosperm clades. Evolution 55: 1762–1780.
33. Harmon LJ, Weir J, Brock C, Glor RE, Challenger W (2008) GEIGER: Investigating evolutionary radiations. Bioinformatics 24: 129–131.
34. Wiens JJ, Pyron RA, Moen DS (2011) Phylogenetic origins of local-scale diversity patterns and the causes of Amazonian megadiversity. Ecology Letters 14: 643–652.
35. Pybus OG, Harvey PH (2000) Testing macro-evolutionary models using incomplete molecular phylogenies. Proceedings of the Royal Society of London B 267: 2267–2271.
36. Rabosky DL (2006) Likelihood Methods for Detecting Temporal Shifts in Diversification Rates Evolution 60: 1152–1164.
37. Nee S, Mooers AO, Harvey PH (1992) Tempo and mode of evolution revealed from molecular phylogenies. Proceedings of the National Academy of Sciences of the United States of America 89: 8322–8326.
38. Ackerly D (2009) Conservatism and diversification of plant functional traits: Evolutionary rates versus phylogenetic signal. Proceedings of the National Academy of Sciences of the United States of America 106: 19699–19706.
39. Pagel MD (1999) Inferring the historical patterns of biological evolution. Nature 401: 877–884.
40. Warren DL, Glor RE, Turelli M (2008) Environmental niche equivalency versus conservatism: Quantitative approaches to niche evolution. Evolution 62: 2868–2883.
41. Schoener TW (1968) Anolis lizards of Bimini: Resource partitioning in a complex fauna. Ecology 49: 704–726.
42. Broennimann O, Fitzpatrick MC, Pearman PB, Petitpierre B, Pellissier L, et al. (2012) Measuring ecological niche overlap from occurrence and spatial environmental data. Global Ecology and Biogeography 21: 481–497.

43. Revell LJ, Harmon LJ, Collar DC (2008) Phylogenetic signal, evolutionary process, and rate. Systematic Biology 57: 591–601.
44. Coetzee JA (1983) Intimations on the Tertiary vegetation of southern Africa. Bothalia 14: 345–354.
45. Jones CS, Pierce RA (1996) Diversity and evolution of seedling Bauplane in *Pelargonium* (Geraniaceae). Aliso 14: 281–295.
46. Jones CS, Martínez-Cabrera HI, Nicotra AB, Mocko K, Marais EM, et al. (2013) Phylogenetic infl uences on leaf trait integration in *Pelargonium* (Geraniaceae): Convergence, divergence, and historical adaptation to a rapidly changing climate. American Journal of Botany 100: 1306–1321.
47. Roncal J, Blach-Overgaard A, Borchsenius F, Balslev H, Svenning J (2011) A Dated Phylogeny Complements Macroecological Analysis to Explain the Diversity Patterns in Geonoma (Arecaceae). Biotropica 43: 324–334.
48. Ricklefs RE, Losos JB, Townsend TM (2007) Evolutionary diversification of clades of squamate reptiles. Journal of Evolutionary Biology 20: 1751–1762.
49. Walker TD, Valentine JW (1984) Equilibrium models of evolutionary species diversity and the number of empty niches. American Naturalist 124: 887–899.
50. Mayr E (1947) Ecological factors in speciation. Evolution 1: 263–288.
51. Slingsby JA, Verboom GA (2006) Phylogenetic relatedness limits co-occurrence at fine spatial scales: evidence from the schoenoid sedges (Cyperaceae: Schoeneae) of the Cape Floristic Region, South Africa. American Naturalist 168: 14–27.

The Contribution of Vegetation and Landscape Configuration for Predicting Environmental Change Impacts on Iberian Birds

Maria Triviño[1]*, Wilfried Thuiller[2], Mar Cabeza[1,3], Thomas Hickler[4,5], Miguel B. Araújo[1,6,7]

1 Department of Biodiversity and Evolutionary Biology, National Museum of Natural Sciences, CSIC, Madrid, Spain, **2** Laboratoire d'Ecologie Alpine, UMR CNRS 5553, Université Joseph Fourier, Grenoble, France, **3** Metapopulation Research Group, Department of Biological and Environmental Sciences, University of Helsinki, Helsinki, Finland, **4** Department of Physical Geography and Ecosystems Analysis, Geobiosphere Science Centre, Lund University, Lund, Sweden, **5** Biodiversity and Climate Research Centre (BiK-F) and Department of Physical Geography at Goethe-University and Senckenberg Gesellschaft für Naturforschung, Frankfurt/Main, Germany, **6** 'Rui Nabeiro' Biodiversity Chair, CIBIO, University of Évora, Évora, Portugal, **7** Center for Macroecology, Evolution and Climate, University of Copenhagen, Copenhagen, Denmark

Abstract

Although climate is known to be one of the key factors determining animal species distributions amongst others, projections of global change impacts on their distributions often rely on bioclimatic envelope models. Vegetation structure and landscape configuration are also key determinants of distributions, but they are rarely considered in such assessments. We explore the consequences of using simulated vegetation structure and composition as well as its associated landscape configuration in models projecting global change effects on Iberian bird species distributions. Both present-day and future distributions were modelled for 168 bird species using two ensemble forecasting methods: Random Forests (RF) and Boosted Regression Trees (BRT). For each species, several models were created, differing in the predictor variables used (climate, vegetation, and landscape configuration). Discrimination ability of each model in the present-day was then tested with four commonly used evaluation methods (AUC, TSS, specificity and sensitivity). The different sets of predictor variables yielded similar spatial patterns for well-modelled species, but the future projections diverged for poorly-modelled species. Models using all predictor variables were not significantly better than models fitted with climate variables alone for ca. 50% of the cases. Moreover, models fitted with climate data were always better than models fitted with landscape configuration variables, and vegetation variables were found to correlate with bird species distributions in 26–40% of the cases with BRT, and in 1–18% of the cases with RF. We conclude that improvements from including vegetation and its landscape configuration variables in comparison with climate only variables might not always be as great as expected for future projections of Iberian bird species.

Editor: Brian Gratwicke, Smithsonian's National Zoological Park, United States of America

Funding: MT was funded by predoctoral research grant FPI (Formación de Personal Investigador) from the Spanish Ministry of Science and Innovation (MICINN) (BES-2007-17331) URL: http://www.micinn.es/portal/site/MICINN/menuitem.dbc68b34d11ccbd5d52ffeb801432ea0/?vgnextoid=2d161e84ed136210VgnVCM1000001d04140aRCRD. MC and MBA were funded by the European Commission Seventh Framework Program project European RESPONSES to climate change (grant agreement number 244092). URL: http://www.responsesproject.eu/. WT, TH and MBA were funded by the European Commission Sixth Framework Program ECOCHANGE project (Challenges in Assessing and Forecasting Biodiversity and Ecosystem Changes in Europe, contract no. 036866-GOCE). URL: http://www.ecochange-project.eu/. TH was also supported by the Landesinitiative zur Entwicklung Wissenschaftlich-ökonomischer Exzellenz (LOEWE)-initiative for scientific and economic excellence of the German federal state of Hesse (http://www.hessen.de/irj/HMWK_Internet?cid=a2396939885f8c8c0a3ba39bda517cbd). MBA was further supported by the Spanish Ministry of Science and Innovation (complementary action no. CGL2008-01198-E/BOS). The funders had no role in study design, data collection and analysis, decision to publish, or preparation of the manuscript.

Competing Interests: The authors have declared that no competing interests exist.

* E-mail: m.trivino@mncn.csic.es

Introduction

Global environmental changes pose great challenges to biodiversity, with ongoing impacts on species distributions and abundances already being recorded (e.g. [1–3]). Attempts to estimate the future effects of global change on biodiversity have often relied on environmental envelope models [4]. These models relate known species distributions to environmental variables to project future altered potential distributions under global change scenarios (e.g. [5–7]). Most of the studies have used climatic factors alone to project species distributions into the future. Nevertheless, there are many factors other than climate that can affect the geographical distributions of species (e.g. [8,9]). This is particularly true for animal species for which climate is often used as a surrogate for resource availability or nesting suitability.

A large number of studies have included non-climatic factors for modelling contemporary species distributions. Such factors included, among others, land cover and land use [10–12], vegetation cover [13], topography [14], or a combination of all of them [15]. However, only a small number of assessments exploring the potential impacts of future global environmental changes have included predicted land use or vegetation changes to complement climatic information (but see [16–19]) because of the scarcity of relevant non-climatic data projected into the future. To our knowledge, none of these previous studies has incorporated vegetation dynamics modelled in a mechanistic way as we have done in this study. The question remains: how would changes in non-climatic environmental factors affect projections of future altered species distributions? We address this question using Iberian birds as a case study.

European bird species have already shown phenological (e.g. [20,21]) and distributional changes (e.g. [22,23]) and they are projected to shift their ranges substantially as a result of global change [24]. However, improvements of projections of future range shifts could be expected if information on vegetation dynamics was included because bird species distributions are known to be, at least partially, determined by vegetation and its spatial configuration (e.g. [25–27]). Variables characterizing aspects of vegetation have been used to model potential current distributions of birds (e.g. [13,28]), but they have rarely been incorporated in models projecting future range shifts under scenarios of global environmental change [29]. Furthermore, most attempts to incorporate vegetation dynamics into forecasts of species distributional changes have not considered vegetation dynamics, such as those simulated by Dynamic Vegetation Models (DVMs), but rather used statistical interpolation of vegetation patterns [18,30]. For example, Lawler et al [29] simulated changes in the vegetation distribution with the Mapped Atmospheric-Plant-Soil System (MAPSS), an equilibrium model that provides future static snapshots, but no year-to-year variability. The spatial configuration of vegetation cover is also thought to be important for explaining bird distributions (e.g. [31,32]), because it accounts for the amount of available habitat in the surrounding area, but again little attempts have been made to incorporate landscape dynamics in forecasts of biodiversity change.

In this study, we used distribution data for 168 breeding bird species in the Iberian Peninsula to fit models using combinations of climatic variables, vegetation characteristics, and their derived landscape configuration. Models were used to assess the importance of alternative aspects of the environment for projecting future potential bird ranges. Specifically, we address the following questions: (i) what sets of variables have greater predictive power: climate, vegetation or landscape configuration? (ii) Are projections using different environmental predictor variables coincident?

Table 1. Environmental variables used to build alternative models.

Variable name	Variable description
Climate data set	
1 mwintertmp	Mean winter temperature (°C)
2 annpre	Annual precipitation (mm)
3 acmgddaug	Accumulated degree days (January to August)
Vegetation data set	
Forest type	
1 Bet.pen	*Betula pendula*
2 Bet.pub	*Betula pubescens*
3 Car.bet	*Carpinus betulus*
4 Cor.ave	*Corylus avellana*
5 Fag.syl	*Fagus sylvatica*
6 Fra.exc	*Fraxinus excelsior*
7 Pic.abi	*Picea abies*
8 Pin.hal	*Pinus halepensis*
9 Que.ile	*Quercus ilex*
10 Que.pub	*Quercus pubescens*
11 Que.rob	*Quercus robur*
12 Til.cor	*Tilia cordata*
13 Total.Forest	Sum of all the forest types
Shrubland type	
14 MRS	Mediterranean Raingreen Shrub4
15 Jun.oxy	*Juniperus oxycedrus*
16 Que.coc	*Quercus coccifera*
Grassland type	
17 c3	Herbaceous
Landscape data set	
1 Forest.R30	Accumulated forest in a radius of 30 km
2 Shrub.R30	Accumulated shrubland in a radius of 30 km
3 Grass.R30	Accumulated grassland in a radius of 30 km

Materials and Methods

Species data

We used distributional records in the Iberian Peninsula for 168 native breeding bird species. Distribution data were extracted from the Spanish Atlas of Breeding Birds [33] and from the Portuguese Atlas of Nesting Birds [34] reporting the presence and absence of bird species in 5923 10×10 km resolution UTM cells. This is the highest-resolution animal distribution data available for the Iberian Peninsula. Our analyses of bird distributions excluded marine and aquatic species because modelling of their habitats would require information about variables that is not available to us. Species with less than 20 records were also excluded to avoid problems of modelling species with small sample sizes [35].

Environmental data for the baseline period

Variables were selected from a larger pool based on expert knowledge and data mining; the latter was done with the specific goal of reducing the number of variables and remove collinearity among them. Overall, four groups of continuous predictor variables were used to fit the models (Table 1): (i) *climatic* (3 variables), (ii) *vegetation* (17 variables), (iii) *landscape configuration* (3 variables) and (iv) *global* (including all previous variables).

For the (i) ***climatic group***, a set of aggregated climate parameters were derived from the Climate Research Unit at 10′ resolution. The CRU CL 2 and CRU CL 2.1 dataset at resolution of 10′ (~16 km at the latitude of the study) was chosen to represent current climate (average from 1971 to 1990). Average monthly temperature and precipitation in grid cells covering the mapped area of the Iberian Peninsula were used to calculate mean values of three different climate parameters: mean winter temperature, annual precipitation and accumulated degree days. These variables are considered ecologically important for explaining bird distribution patterns (e.g. [36–38]) and limit species distribution as a result of widely shared physiological constraints (e.g. [39,40]). Finally, variables were interpolated using kriging implemented within Geographical Information System (GIS) software ArcGIS 9.2 [41] to a resolution of 10 km to match the bird distribution datasets.

The (ii) ***vegetation group*** comprised potential natural vegetation composition and structure, simulated with the DVM LPJ-GUESS [42,43]. The model simulates the competition between main tree species and PFTs. Forest dynamics resemble successional patterns, adopting a forest "gap model" approach. The model has been parameterized to represent the main European tree species and a number of plant functional types

(PFTs) [44,45]. LPJ-GUESS reproduced the main general patterns in European potential vegetation at a coarse scale, but the model did not reproduce the fine-scale mosaic of different vegetation types existing in many areas. Discrepancies were, for example, caused by the fact that some real-world drivers, such as different soil nutrient levels, are not accounted for by the model. However, the model results we used present the first assessment of dynamic future vegetation changes at the level of important tree species and PFTs over continental Spain and Portugal. General vegetation features in the Iberian Peninsula, such as the distinction between forests, shrublands and grasslands, corresponded better with the potential natural vegetation in the Iberian Peninsula than in earlier studies with dynamic global vegetation models. [45]. The model also reproduced the main features of the coarse-scale distribution of major tree species covered by the Third Spanish Forest Inventory [46] (Figure S1). The PFTs were also grouped into three broad habitat types, reflecting the vegetation structure rather than individual tree species or PFTs: forest, shrubland, and grassland. The sum of the LAI of all species and PFTs belonging to each of the three broad habitat type group was then used in the analyses. Many bird species are rather dependent on such structural vegetation features than on individual tree species [25,26,47]. Furthermore, the model output for these structural ecosystem features is more robust than the simulated patterns for individual species or PFTs, and they are less likely to be fundamentally changed by forest management. A PCA was performed in order to investigate for collinearity among variables and potentially select a reduced set of variables. However, variables were not highly correlated so all were kept. The vegetation was represented by the continuous variable Leaf Area Index (LAI), which is the ratio of total upper projected leaf surface of vegetation divided by the surface area of the land on which the vegetation grows. LAI is a dimensionless value, typically ranging from 0 to 8 for a dense forest. The variables were originally at 10′ (~16 km at the latitude of the study) resolution and were interpolated at 10 km resolution to match the bird distribution datasets.

Because potential vegetation cover variables modelled with LPJ-GUESS do not account for current and future land use, we combined them with land use information derived from CORINE Land Cover (CLC) as follows [48]. Categories from CLC were aggregated and represented by 6 land cover classes: *Urban*, *Cropland*, *Permanent Crops*, *Grasslands*, *Forest* and *Others* (for a complete description of the methodology see [49], despite in this reference they use the PELCOM dataset, the analyses were re-done using CORINE dataset and are the ones used for this study). The percentage of each land use type within the UTM grid cells was calculated using the Zonal Statistics tool implemented in ArcGIS 9.2. Grid cells were classified as forested when 10% or more of their surface were covered by *Forest*. If, for example, the vegetation model predicted forest but less than 10% of the grid cell was forested according to the land cover data, non-forest vegetation cover was assumed in the analysis. From the grid cells classified as shrublands we excluded the ones in which the sum of non-compatible land use types (*Permanent Croplands*, *Croplands* and *Urban*) represented 90% or more of the grid area. Finally, cells were classified as grasslands when their area was covered by at least 10% of *Grasslands*. Thus, we assume that a certain fraction of available habitat within a grid cell is sufficient for populations to persist. Different classes were not exclusive between each other and grid cells could hold more than one vegetation type at the same time. If, for example, a grid cell was covered by 17% of forest and 16% of grassland according to land cover data and was occupied by *Quercus ilex* (PFT of forest type) and c3 (PFT of grassland type) according to the vegetation model, that grid cell was considered both as "forest" and as "grassland".

The (iii) ***landscape configuration group*** was calculated based on the accumulated sum of the different PFTs values included in each habitat type: forest, shrubland, and grassland. Using ArcGIS 9.2., three concentric bands, each 10 km wide, were delimited around each grid cell for the three habitat types. Within each band and for each habitat type, the accumulated vegetation abundance was calculated. These data provided information of the spatial arrangement and composition of the landscape around each grid cell. From the nine variables created only the three variables of radius equal to 30 km were retained due to the high correlation between the three different radiuses (Spearman's correlations, $r = 0.8$–0.9) and also because they capture a broader range of landscape and were the variables least correlated with the original habitat types.

Finally, the (iv) ***global group*** included the three previous data sets.

Environmental data for the future

We used a European climate scenario from the EU framework program Assessing Large-scale environmental Risks for biodiversity with tested Methods (ALARM) at a resolution of 10′ for the period 2051–2080 [50]. The climate scenario was derived from a simulation with the global climate model HadCM3, using the BAMBU (Business As Might Be Usual) scenario (which corresponds to A2 SRES) of the ALARM project. Scenarios for future potential natural vegetation were developed by a previous study [45] as well as the scenarios for future land use change [51]. Land use projections used to constrain potential vegetation cover from LPJ-GUESS were based on the BAMBU scenario [52] (for details see [51,53]).

Data analysis

The models were built using the BIOMOD library [54] in R [55] (version 1.15), using the default settings and parameters. Two ensemble modelling techniques were selected: Random Forests (RF) [56,57] and Boosted Regression Trees (BRT) [58,59]. Both techniques are effective in dealing with non-linearities and interactions among variables. Random forest uses a bootstrap aggregation algorithm by fitting multiple un-pruned classification trees on sub-samples of the original data. The prediction is then the average of the predictions of all trees weighted by their internal predictive accuracy (out-of-bag estimator). We fitted random forest using a maximum of 700 trees and using a random half of the predictor variables for each tree. BRT is a boosting algorithm in which very short classification trees (seven nodes) are repeatedly built on the residuals from the previous tree to improve the fit using cross-validation to stop the process. In BRT models the maximum number of trees was set to 3000, the learning-rate was 0.001 and the interaction-depth was 4 as suggested by Elith et al. [58]. The full dataset for the 168 breeding bird species was randomly partitioned into two subsets (calibration and evaluation), with 70% and 30% respectively, and this overall procedure was repeated five times to make sure that the evaluation procedure was independent of the random splitting procedure. Future projections were made assuming unlimited dispersal, which is a more likely scenario among birds at the geographical extent of the study area than the alternative no dispersal scenario.

Models were assessed using four evaluation methods: the area under curve (AUC) of the receiver operating characteristic (ROC) [60], the true skill statistics (TSS) [61], sensitivity that measures the percentage of presences correctly predicted and specificity that measure the percentage of absences correctly predicted. The specificity and sensitivity were determined separately after using an AUC and TSS protocol to convert probabilities of occurrence into presences and absences (Figure 1).

There is a large number of statistical techniques available to fit environmental envelope models and they are known to produce markedly different future projections of species range shifts when projections are made into the future [62–64]. Commonly used evaluation metrics measuring agreement between predicted potential and observed distributions are useful to verify the models' discrimination ability [63]. However, discrimination between predicted potential and observed distributions is known to be a relatively poor surrogate of the models' ability to predict future distributions well [65]. Therefore, there are little guidelines for selection of the models to use under future scenarios [66]. A possible approach to handle inter-model variability and reduce uncertainty is to use ensemble forecasting by generating multiple copies of the models and combining them using consensus techniques (see for review [66]). In this study, a consensus approach based on the mean of the probabilities from the sets of projections made by RF and BRT was selected (see also [67–69]) and TSS method was chosen to convert probabilities values into presence-absence data.

The relative importance of environmental variables was also calculated for RF and BRT. In Random Forests, variable importance is determined by comparing the misclassification error rate of a tree with the error rate that occurs if the values of a predictor variable are randomly permuted [57]. In Boosted Regression Trees variable importance is based on the number of times a variable is selected for binary splitting, weighted by the squared improvement to the model as a result of each split, and averaged over all the individual trees [70]. Because measures of variable importance are calculated differently in RF (Mean Decrease Accuracy and Mean Decrease Gini) and BRT, a ranking system was created to compare environmental selection among the different model types. Environmental variables were ranked from 1 (most important) to 23, although only the three first ones were analysed to compare across all groups of variables (only three variables for the climatic group).

Bird species were classified into eight categories based on their main habitat use: Forest, Shrubland, Grassland, Grassland/Forest, Shrubland/Forest, Grassland/Shrubland, Grassland/Shrubland/Forest and Others (including bird's species which do not depend on any vegetation type such as those specialized on urban areas or cliffs). In order to define the degree of habitat specialization of species we counted the number of habitat types used for breeding or feeding and considered that the more habitats used the less specialized are the species. The information was gathered from the Spanish Atlas of Breeding Birds [33] and complemented by consultation with experts (Table S1).

Results

Average discrimination ability of models based on cross validated AUC and TSS values differed statistically among the

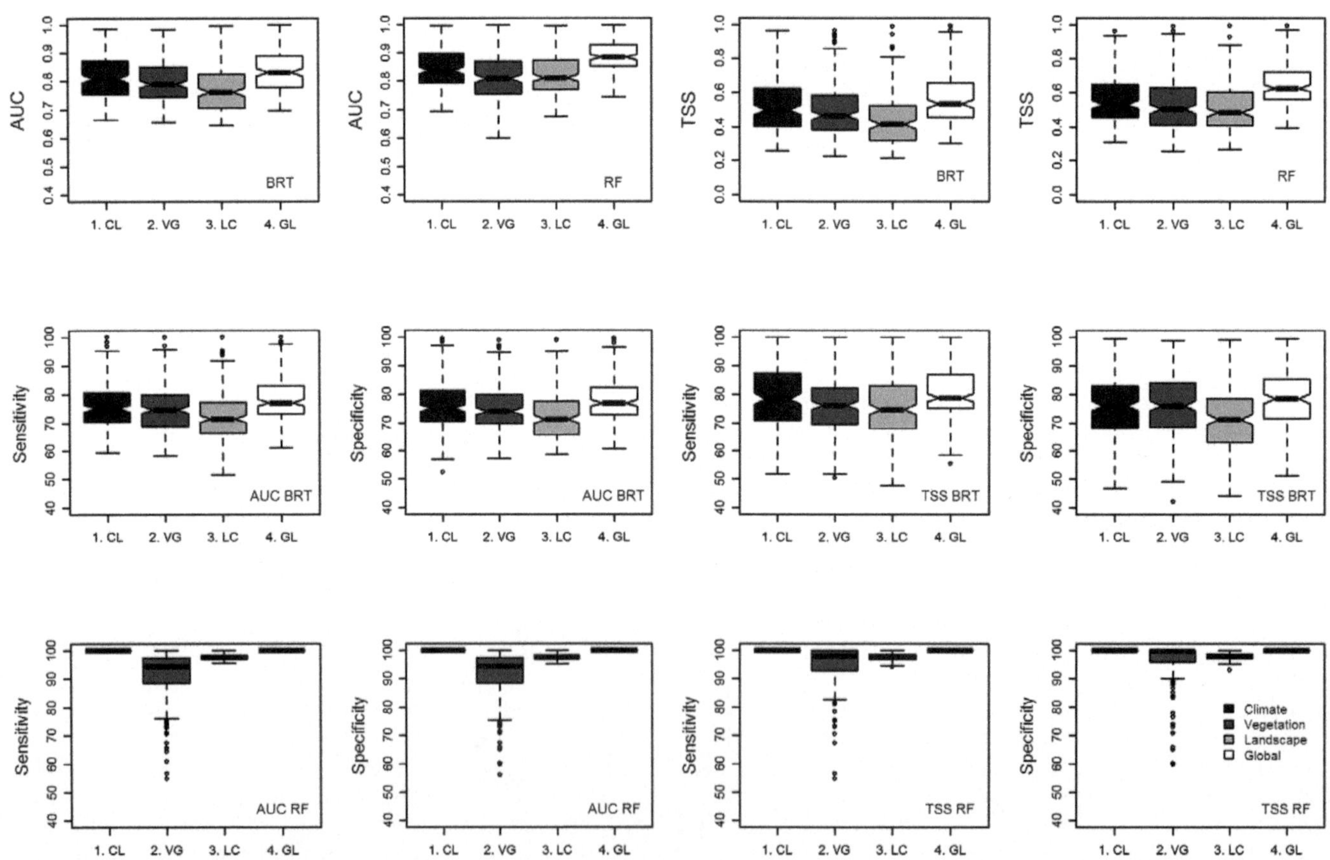

Figure 1. Four evaluation methods to compare model performance using different predictor variables. Boxplot summarizing results of measures of performance (AUC and TSS) of each dataset used (Climate, Vegetation, Landscape and Global) for the cross validation results for BRT and RF models. Percentage of presence and absence correctly predicted (sensitivity and specificity) were also provided. Median values (line across box), range excluding outliers (error bars), interquartile range containing 50% of values (box) and outliers (circles) from results. Untransformed values have been used.

Table 2. Results of pairwise Wilcoxon test of the effect of predictor variables (climate, vegetation and landscape configuration) on model performance estimated by AUC and TSS.

Predictor variables	**Climate-Vegetation**				**Climate-Landscape**				**Vegetation-Landscape**			
Evaluation method	**AUC**		**TSS**		**AUC**		**TSS**		**AUC**		**TSS**	
Model technique	**BRT**	**RF ****	**BRT**	**RF**	**BRT *****	**RF ****	**BRT *****	**RF ****	**BRT ****	**RF**	**BRT ****	**RF**
Wilcoxon test	$W=13512$	$W=14459$	$W=13326$	$W=13410$	$W=15853$	$W=14168$	$W=15727$	$W=14155$	$W=14528$	$W=11319$	$W=14595$	$W=12470$
	$p=0.0575$	$\mathbf{p=0.0019}$	$p=0.096$	$p=0.077$	$\mathbf{p=1.1e\text{-}06}$	$\mathbf{p=0.0063}$	$\mathbf{p=2.5e\text{-}06}$	$\mathbf{p=0.0066}$	$\mathbf{p=0.0014}$	$p=0.3801$	$\mathbf{p=0.001}$	$p=0.562$

Table 3. Results of pairwise Wilcoxon test comparison between each individual model (climate, vegetation and landscape configuration) and the global model based on performance estimated by AUC and TSS.

Predictor variables	**Climate-Global**				**Vegetation-Global**				**Landscape-Global**			
Evaluation method	**AUC**		**TSS**		**AUC**		**TSS**		**AUC**		**TSS**	
Model technique	**BRT**	**RF *****	**BRT**	**RF*****	**BRT *****	**RF *****	**BRT *****	**RF *****	**BRT *****	**RF *****	**BRT ****	**RF *****
Wilcoxon test	$W=9839$	$W=7566$	$W=10030$	$W=7666$	$W=8258$	$W=5940$	$W=860$	$W=6878$	$W=6396$	$W=5737$	$W=6568$	$W=5889$
	$p=0.0059$	$\mathbf{p=1.8e\text{-}08}$	$p=0.01203$	$\mathbf{p=3.6e\text{-}08}$	$\mathbf{p=2.0e\text{-}06}$	$\mathbf{p=1.4e\text{-}14}$	$\mathbf{p=1.6e\text{-}05}$	$\mathbf{p=7.7e\text{-}11}$	$\mathbf{p=1.1e\text{-}12}$	$\mathbf{p=1.8e\text{-}15}$	$\mathbf{p=5.2e\text{-}12}$	$\mathbf{p=8.5e\text{-}15}$

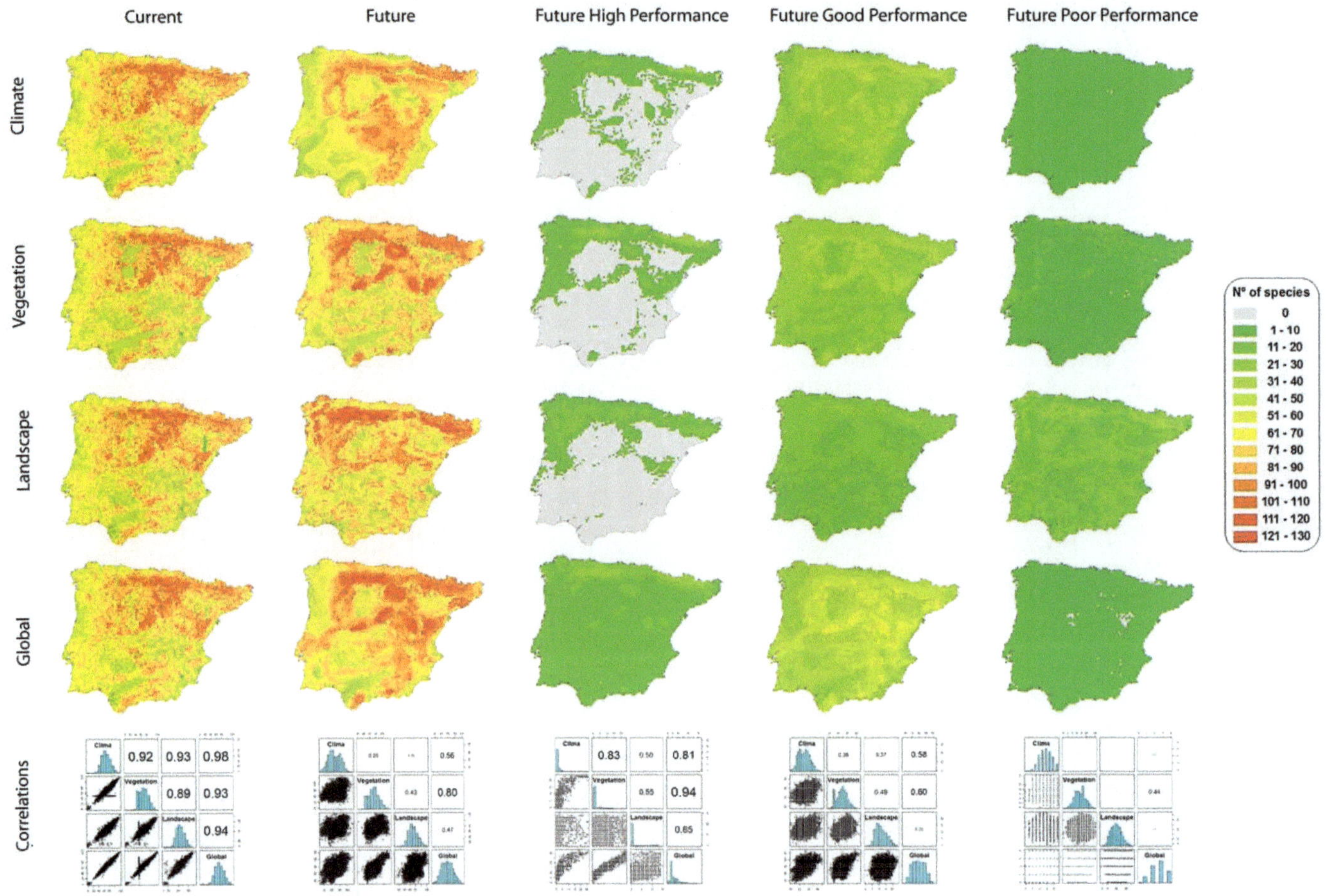

Figure 2. Spatial pattern comparison of bird distributions. The maps represent the total number of species per each 10 km cell for the four model types (Climate, Vegetation, Landscape and Global) and for two time periods (current and future projections). The correlation graphs indicate the level of agreement between the four model types for each column. The calculations for the first two columns (current and future) were done using the total number of bird species (N = 168) whereas the last three columns illustrate subsets of the future projection based on model performance categories (AUC method): high (N = 32), good (N = 63) and poor (N = 37).

different groups of predictor variables (Friedman test, $p<0.001$), being lower for landscape models and higher for models including all predictor variables together (Figure 1). Models including climatic variables alone were generally better than models fitted solely with vegetation or landscape variables, although not always significantly better than models including vegetation (Wilcoxon test, $p<0.05$) (Table 2). The comparison between the models including all variables and the models including climate, vegetation or landscape showed that the all-variables models were significantly better than any other model, except for the models

Table 4. Number of species from the 168 species classified in different accuracy classes of AUC and TSS based on two modelling techniques.

Predictor variables	Climate				Vegetation				Landscape			
Evaluation method	**AUC**		**TSS**		**AUC**		**TSS**		**AUC**		**TSS**	
Model technique	**BRT**	**RF**	**BRT**	**RF**	**BRT**	**RF**	**BRT**	**RF**	**BRT**	**RF**	**BRT**	**RF**
High-performance	28	35	17	19	23	27	15	21	17	28	5	10
Good-performance	61	78	28	32	53	60	21	25	35	71	19	30
Fair-performance	70	53	79	90	78	46	80	74	75	59	66	85
Poor-performance	9	2	44	27	14	28	51	48	41	5	78	43
Fail	0	0	0	0	0	7	1	0	0	5	0	0

AUC: High = AUC>0.9, Good = 0.9<AUC<0.8; Fair = 0.7<AUC<0.8; Poor = 0.6<AUC<0.7. Fail AUC<0.6.
TSS: High = TSS>0.8, Good = 0.8<TSS<0.6; Fair = 0.6<TSS<0.4 and Poor = 0.2<TSS<0.4. Fail TSS<0.4.

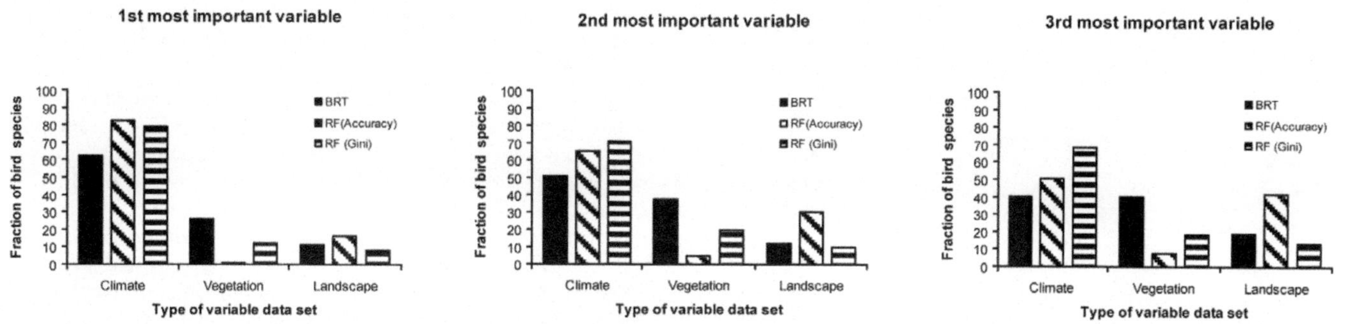

Figure 3. Ranking of variable importance for BRT and RF models. Fraction of the 168 bird species for which the model selected climatic, vegetation or landscape variables as the first, the second or the third most important variable.

fitted with climatic variables alone for which the all-variables-model was significantly better only in 50% of the cases (Table 3). Regarding the differences in discrimination ability between modelling techniques, we found that Random Forests adjusted projections to the data more closely than Boosted Regression Trees in almost all of the cases and regardless of the four evaluation techniques used (Figure 1).

Spatial correspondence among projections of species richness for the four sets of models was very high for the baseline period, but substantially variable for future scenarios. Inter-model variability was constrained by model performance (Figure 2). That is, species for which models performed notably well (*high-performance species*) had lower inter-model variability than species for which models performed well (*good-performance species*) and poorly (*poor-performance species*) (Table 4). Overall, the pairwise correlation among future projections for the 168 species varies considerably (Spearman's correlations, $r = 0.26$–0.8). However, pairwise comparisons for groups of species with models of similar accuracy (grouped according to AUC values) showed that higher correlation between model predictions was obtained for the models with higher accuracy: *high-performance species* (Spearman's correlations, $r = 0.5$–0.94; maximum number of species = 32); *good-performance species* ($r = 0.37$– 0.6; maximum number of species = 63); and *poor-performance species* ($r = 0.17$–0.44; maximum number of species = 37).

After ranking the relative importance of all the environmental variables, we calculated the fraction of species for which the models selected climatic, vegetation or landscape variables among the three most important ones. Results were different depending on the method used (Figure 3). Using the procedure for assessment of variable importance in BRT, we found that vegetation was selected as important for a larger fraction of bird species (26.2–40.5%) than that estimated with RF models (Accuracy 1.2–7.7%, and Gini index 12.5–18.4%). For the three measures of variable importance used (BRT, Accuracy and Gini index), the fraction of species for which the models selected non-climatic variables increased from the first most important variable (1.2–26.2%) to the second (4.8–37.5%) and third variable selected (7.7–40.5%).

The main type of habitat used by the bird species was not associated with the choice of variables entering into the models (Figure 4) neither did the degree of habitat specialization (Table 5). As it can be seen in figure 4, vegetation variables were selected as the first, second, or third most important variable for a constant fraction of bird species. For example, vegetation was associated with ~35% of forest bird specialists in all cases. Unlike the expectation, no clear variable discrimination emerged in models using vegetation variables among forest, shrubland or grassland birds.

Discussion

In this study we asked whether adding vegetation and landscape configuration variables in environmental envelope models would significantly increase discrimination ability of models and whether different sets of variables would affect the spatial representation of climate change impacts on bird species. We showed that models using climatic variables generally fit the data better than models using vegetation or landscape configuration variables. However, improvements of discrimination with the climate models, as compared with the two alternative models, were significant in all

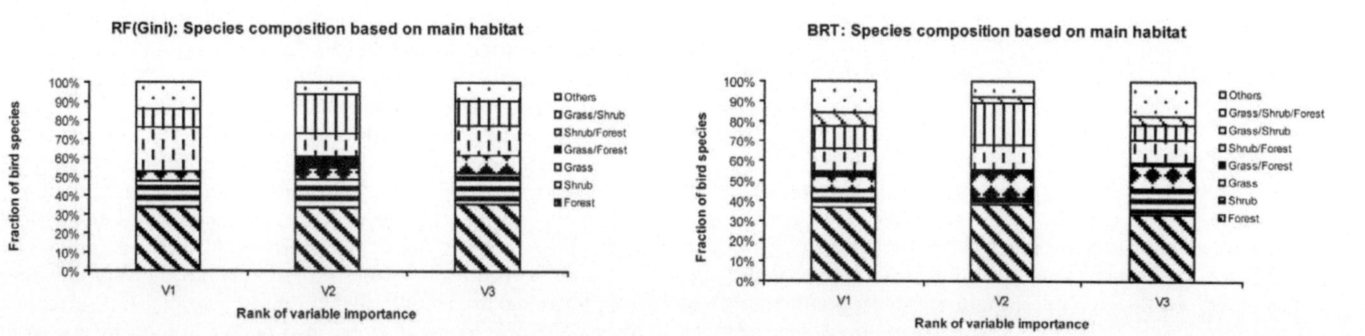

Figure 4. Importance of vegetation variables among bird species with different habitat preferences. Species composition based on the main habitat used by the bird species selecting vegetation variables as the first, second or third most important for explaining their distribution. For BRT model species number for V1 = 44, V2 = 63 and V3 = 69 whereas for RF model (Mean Decrease Gini measure) the species number for V1 = 21, V2 = 33 and V3 = 31.

Table 5. Fraction of bird species for which the model included vegetation variables as the first (V1), second (V2) or third (V3) most important variables.

Model technique	**BRT**			**RF (Gini index)**		
Specialization level	**V1**	**V2**	**V3**	**V1**	**V2**	**V3**
High specialization	52.3%	54%	58%	52.4%	54.5%	61.3%
Mid specialization	25%	34.9%	20.3%	33.3%	39.4%	29%
Low specialization	6.8%	3.2%	4.3%	0%	0%	0%

Species are grouped by their degree of habitat specialization based on the number of habitat types they use for breeding and feeding. **High specialization** means the species use one habitat type (N = 90), **mid specialization** means the species use two habitat types (N = 51) and **low specialization** means the species use three habitat types (N = 9).

cases only for the climatic-landscape model comparison. Disagreement existed between future projections using different predictors, but the discrepancy decreased when species with high levels of discrimination ability in ensembles of forecasts were retained. Finally, the importance of variables appeared to be species specific and, despite the importance of climatic variables, vegetation and landscape configuration were also important for explaining the distribution patterns of a number of bird species.

Climatic variables perform better than non-climatic variables when predicting potential distributions of birds

Authors have repeatedly suggested that greater care should be given to the choice of environmental predictors when modelling the potential distributions of species (e.g. [71]). Previous studies have suggested that non-climatic variables should be incorporated in bioclimatic models for projecting future range shifts (e.g. [13,72]), but the impossibility of validating future projections [65,73] makes it complicated to measure the relative importance of non-climatic variables. It is well-established that the configuration and composition of vegetation are good predictors of bird species distributions because they are associated with many of their breeding, feeding or nesting requirements (e.g. [74] and references therein). For example, Seoane et al. [13] found that vegetation models were significantly more accurate than topo-climatic models. However, our results showed that vegetation or landscape models did not outperform climatic models. Indeed, for half of the modelled species consideration of all variables did not result in better discrimination than that obtained with models only accounting for climate variation. Possible explanations for this result are that: (i) the relative importance of climatic versus non-climatic predictors is scale dependent (e.g. [75]). For example, in a previous study, land cover data did not improve model accuracy at coarse resolution (50 km) in Europe [11]. In another study, using a finer resolution (10 and 1 km), the inclusion of land use improved model discrimination ability [12]. In effect, the resolution and extent of our study might be too coarse to capture the dependence of birds on vegetation; (ii) vegetation in Mediterranean countries has been modified by humans for millennia. The human impact is not represented by the simulated potential vegetation. We sought to address this issue by tailing vegetation to land use, but the land cover data used herein is still a rather coarse approximation of real land cover and its associated habitat characteristics. However, the correspondence between species potential distributions and simulated potential vegetation might be higher in regions where the actual vegetation has been little influenced by human activities; (iii) the vegetation model used here was parameterized to represent the main dominant tree species and vegetation types across Europe, but it did not include all important trees in the Iberian Peninsula. Furthermore, as with any process-based vegetation model, simulated vegetation patterns do not always correspond well with real patterns; (iv) the coarse vegetation and land use variables used in this study do not account for all important habitat characteristics, such as forest age and size structure in plantations and the amount of deadwood.

Discrepancies between future projections could be partly explained by the expected decrease in the correlation between climate and simulated vegetation across time. This is because, firstly, the vegetation model accounts for potential effects of increasing atmospheric CO_2 on productivity and water cycling [44,76]. "CO_2 fertilization" and reductions in stomatal conductance and water losses might alleviate some of the negative effects of increasing drought on vegetation [44,77]. Secondly, the vegetation model simulates transient vegetation shifts, not the equilibrium response to the climatic forcing. Over a few decades, only a small fraction of the long-term equilibrium response of the vegetation can be expected [45]. This non-equilibrium is much more important for the discrepancies in the projections than the CO_2 effects [44,45].

Species characteristics influence model accuracy

Species characteristics have been shown to influence model accuracy and many biological traits such as body size or dispersion rate and also population trends have been measured for evaluating their influence on modelling results [78]. Species with narrower or spatially more aggregated ranges (e.g. [79,80]) and higher habitat specialization (e.g. [81,82]) can generally be predicted with higher accuracy. Our results support the conclusions from these studies, as the species with the highest accuracy values across all model types (climate, vegetation, landscape configuration and global) included high-mountain species with very narrow ranges and low prevalence, such as Tengmalm's owl *Aegolius funereus*, bearded vulture *Gypaetus barbatus*, rock ptarmigan *Lagopus mutus*, capercaillie *Tetrao urogallus* and ring ouzel *Turdus torquatus*. In our study, the ranking of species by accuracy values was similar across models as it was shown when future projections for the subgroup of species with good model performance were compared (Figure 2). Therefore, other relevant environmental or biological predictors might be required for those species that were difficult to model.

The importance of predictors is species specific

It is difficult to determine what are the most important environmental variables constraining species distributions, especially when a large number of species is considered. Nevertheless, we note that most of the divergence in future projections was caused by species that were difficult to model with our predictors, i.e., that performed poorly with the measures of discrimination ability used to verify model performance. Models discriminating data well yielded less variable projections into the future. More work is needed to identify whether animal species can be grouped based on their response to global environmental changes as well as identify which functional traits made them more resistant to these changes.

We conclude that the discrimination ability of envelope models is not always improved by inclusion of vegetation and landscape configuration variables. In the particular case of bird species in the

Iberian Peninsula, climate was sufficient to describe current distributions for ca. 50% of the species and in some of the remaining cases vegetation could help improving the fit of the models but not landscape configuration. With our data and analysis, no general patterns emerged with regards to the selection of vegetation variables by models of different guilds of species. So, the decision as to whether to include specific non-climatic factors in the models requires case specific considerations based on the auto-ecology of the species.

Supporting Information

Figure S1 (A) Comparison between the simulated LAI of the first five main tree species (*Betula pendula*, *Corylus avellana*, *Fagus sylvatica*, *Fraxinus excelsior* and *Quercus robur*) and presence data from the Third Spanish Forestry Inventory (IFN = Inventario Forestal Nacional). Inventory data was not available for all simulated tree species. The first column of maps represents the model outputs, the second column the result from the combination of LPJ-GUESS results with a land use dataset (see Materials and Methods for further details), and the third column represents the presence data of the IFN. The model reproduced the broad distinction between northern and southern trees, but the simulated distribution of more northerly distributed species generally expanded further to the south than according to the inventory data. This was too some extent expected as the model represented potential natural vegetation. The Mediterranean region has a long history of large-scale anthropogenic impacts. Most areas once occupied by forest were transformed into croplands and pastures hundreds and in many cases even thousands of years ago (e.g. [83]), while the rest of the remaining forest has been intensively managed [84]. Also the imposition of real land use patterns could only partly remove this mismatch because the land use data only distinguished forest and non-forest areas, without tree species-specific information. As a result, the simulated distribution was maintained in the simulated data as long as the land use data indicated that the forest cover was, at least, 10% (see Materials and Methods). Another explanation for the wider simulated ranges might be that the inventory might not cover all small outlier populations. (**B**) Comparison between the simulated LAI of the last five main tree species (*Picea abies*, *Pinus halepensis*, *Quercus ilex*, *Quercus pubescens* and *Tilia cordata*) and presence data from the Third Spanish Forestry Inventory (IFN = Inventario Forestal Nacional).

Table S1 Main habitats (G = grassland, S = shrubland, F = forest, O = others) for the 168 bird species included in the study. The information was gathered from the Spanish Atlas of Breeding Birds [33] and complemented by consultation with the following experts: Carlos Ponce, Sergio Pérez Gil and Alejandro Aparicio Valenciano.

Acknowledgments

We thank D. Alagador, R. García-Valdés, S. Varela, S. Calvo and H. Nenzen for earlier comments on the manuscript. C. Ponce and S. Perez for helping with the bird classification. We greatly appreciate the input of K. Böhning-Gaese and two anonymous reviewers which improved this manuscript.

Author Contributions

Conceived and designed the experiments: MT WT MBA. Analyzed the data: MT WT. Contributed reagents/materials/analysis tools: TH. Wrote the paper: MT MBA MC TH WT.

References

1. Lenoir J, Gegout JC, Marquet PA, de Ruffray P, Brisse H (2008) A significant upward shift in plant species optimum elevation during the 20th century. Science 320: 1768–1771.
2. Parmesan C (2006) Ecological and evolutionary responses to recent climate change. Annual Review of Ecology Evolution and Systematics 37: 637–669.
3. Walther GR, Roques A, Hulme PE, Sykes MT, Pysek P, et al. (2009) Alien species in a warmer world: risks and opportunities. Trends in Ecology & Evolution 24: 686–693.
4. Heikkinen RK, Luoto M, Araújo MB, Virkkala R, Thuiller W, et al. (2006) Methods and uncertainties in bioclimatic envelope modelling under climate change. Progress in Physical Geography 30: 751–777.
5. Thuiller W, Lavergne S, Roquet C, Boulangeat I, Lafourcade B, et al. (2011) Consequences of climate change on the Tree of Life in Europe. Nature.
6. Araújo MB, Thuiller W, Pearson RG (2006) Climate warming and the decline of amphibians and reptiles in Europe. Journal of Biogeography 33: 1712–1728.
7. Pompe S, Hanspach J, Badeck F, Klotz S, Thuiller W, et al. (2008) Climate and land use change impacts on plant distributions in Germany. Biology Letters 4: 564–567.
8. Hampe A (2004) Bioclimate envelope models: what they detect and what they hide. Global Ecology and Biogeography 13: 469–471.
9. Melles SJ, Fortin MJ, Lindsay K, Badzinski D (2011) Expanding northward: influence of climate change, forest connectivity, and population processes on a threatened species' range shift. Global Change Biology 17: 17–31.
10. Luoto M, Virkkala R, Heikkinen RK (2007) The role of land cover in bioclimatic models depends on spatial resolution. Global Ecology and Biogeography 16: 34–42.
11. Thuiller W, Araújo MB, Lavorel S (2004) Do we need land-cover data to model species distributions in Europe? Journal of Biogeography 31: 353–361.
12. Pearson RG, Dawson TP, Liu C (2004) Modelling species distributions in Britain: a hierarchical integration of climate and land-cover data. Ecography 27: 285–298.
13. Seoane J, Bustamante J, Diaz-Delgado R (2004) Competing roles for landscape, vegetation, topography and climate in predictive models of bird distribution. Ecological Modelling 171: 209–222.
14. Luoto M, Heikkinen RK (2008) Disregarding topographical heterogeneity biases species turnover assessments based on bioclimatic models. Global Change Biology 14: 483–494.
15. Brotons L, Herrando S, Pla M (2007) Updating bird species distribution at large spatial scales: applications of habitat modelling to data from long-term monitoring programs. Diversity and Distributions 13: 276–288.
16. Jetz W, Wilcove DS, Dobson AP (2007) Projected impacts of climate and land-use change on the global diversity of birds. Plos Biology 5: 1211–1219.
17. Araújo MB, Nogués-Bravo D, Reginster I, Rounsevell M, Whittaker RJ (2008) Exposure of European biodiversity to changes in human-induced pressures. Environmental Science & Policy 11: 38–45.
18. Preston K, Rotenberry JT, Redak RA, Allen MF (2008) Habitat shifts of endangered species under altered climate conditions: importance of biotic interactions. Global Change Biology 14: 2501–2515.
19. Kissling WD, Field R, Korntheuer H, Heyder U, Bohning-Gaese K (2010) Woody plants and the prediction of climate-change impacts on bird diversity. Philosophical Transactions of the Royal Society B-Biological Sciences 365: 2035–2045.
20. Møller AP, Fiedler W, Berthold P (2004) Birds and climate change. London: Advances in Ecological Research Elsevier Academic Press.
21. Lehikoinen E, Sparks TH, Zalakevicius M (2004) Arrival and departure dates. In: Moller AP, Fielder W, Berthold P, eds. Birds and Climate Change. pp 1–31.
22. Brommer JE (2004) The range margins of northern birds shift polewards. Annales Zoologici Fennici 41: 391–397.
23. Thomas CD, Lennon JJ (1999) Birds extend their ranges northwards. Nature 399: 213–213.
24. Huntley B, Green RE, Collingham YC, Willis SG (2007) A Climatic atlas of European Breeding Birds. Durham University, The RSPB and Lynx Edicions, Barcelona.
25. Rotenberry JT, Wiens JA (1980) Habitat Structure, Patchiness, and Avian Communities in North American Steppe Vegetation: A Multivariate Analysis. Ecology 61: 1228–1250.
26. Root T (1988) Environmental-Factors Associated with Avian Distributional Boundaries. Journal of Biogeography 15: 489–505.
27. Julliard R, Clavel J, Devictor V, Jiguet F, Couvet D (2006) Spatial segregation of specialists and generalists in bird communities. Ecology Letters 9: 1237–1244.
28. Peterson AT, Ball LG, Cohoon KP (2002) Predicting distributions of Mexican birds using ecological niche modelling methods. Ibis 144: E27–E32.
29. Lawler JJ, White D, Neilson RP, Blaustein AR (2006) Predicting climate-induced range shifts: model differences and model reliability. Global Change Biology 12: 1568–1584.

30. Hughes GO, Thuiller W, Midgley GF, Collins K (2008) Environmental change hastens the demise of the critically endangered riverine rabbit (Bunolagus monticulairis). Biological Conservation 141: 23–34.
31. Saab V (1999) Importance of spatial scale to habitat use by breeding birds in riparian forests: A hierarchical analysis. Ecological Applications 9: 135–151.
32. Pearson SM (1993) The spatial extent and relative influence of landscape-level factors on wintering bird populations. Landscape Ecology 8: 3–18.
33. Martí R, del Moral JC (2003) Atlas de las aves reproductoras de España. Madrid: Dirección General de Conservación de la Naturaleza & Sociedad Española de Ornitología.
34. Equipa Atlas (2008) Atlas das aves nidificantes em Portugal; Alvim A, ed. Lisboa.
35. Stockwell DRB, Peterson AT (2002) Effects of sample size on accuracy of species distribution models. Ecological Modelling 148: 1–13.
36. Araújo MB, Thuiller W, Yoccoz NG (2009) Reopening the climate envelope reveals macroscale associations with climate in European birds. Proceedings of the National Academy of Sciences 106: E45–E46.
37. Gregory RD, Willis SG, Jiguet F, Vorisek P, Klvanova A, et al. (2009) An Indicator of the Impact of Climatic Change on European Bird Populations. PLoS ONE 4: e4678.
38. Huntley B, Collingham YC, Willis SG, Green RE (2008) Potential Impacts of Climatic Change on European Breeding Birds. PLoS ONE 3: e1439.
39. Crick HQP (2004) The impact of climate change on birds. Ibis 146: 48–56.
40. Whittaker RJ, Nogues-Bravo D, Araújo MB (2007) Geographical gradients of species richness: a test of the water-energy conjecture of Hawkins et al. (2003) using European data for five taxa. Global Ecology and Biogeography 16: 76–89.
41. ESRI (2006) Redlands, CA.
42. Hickler T, Smith B, Sykes MT, Davis MB, Sugita S, et al. (2004) Using a generalized vegetation model to simulate vegetation dynamics in northeastern USA. Ecology 85: 519–530.
43. Smith B, Prentice IC, Sykes MT (2001) Representation of vegetation dynamics in the modelling of terrestrial ecosystems: comparing two contrasting approaches within European climate space. Global Ecology and Biogeography 10: 621–637.
44. Hickler T, Fronzek S, Araújo MB, Schweiger O, Thuiller W, et al. (2009) An ecosystem model-based estimate of changes in water availability differs from water proxies that are commonly used in species distribution models. Global Ecology and Biogeography 18: 304–313.
45. Hickler T, Vohland K, Feehan J, Miller PA, Smith B, et al. (2012) Projecting the future distribution of European potential natural vegetation zones with a generalized, tree species-based dynamic vegetation model. Global Ecology & Biogeography 21: 50–63.
46. Villanueva JA (2004) Tercer Inventario Forestal Nacional (1997–2007). Madrid: Ministerio de Medio Ambiente y Medio Rural y Marino.
47. Karr JR, Roland RR (1971) Vegetation Structure and Avian Diversity in Several New World Areas. The American Naturalist 105: 423–435.
48. European Commission (1993) Corine land cover map and technical guide. Technical report, European Union Directorate General Environment (Nuclear Safety and Civil Protection).
49. Dendoncker N, Rounsevell M, Bogaert P (2007) Spatial analysis and modelling of land use distributions in Belgium. Computers Environment and Urban Systems 31: 188–205.
50. Fronzek S, Carter TR, Jylhä K (in press) Representing two centuries of past and future climate for assessing risks to biodiversity in Europe. Global Ecology & Biogeography doi:10.1111/j.1466-8238.2011.00695.x.
51. Rounsevell MDA, Reginster I, Araújo MB, Carter TR, Dendoncker N, et al. (2006) A coherent set of future land use change scenarios for Europe. Agriculture, Ecosystems & Environment 114: 57–68.
52. Spangenberg JH (2007) Integrated scenarios for assessing biodiversity risks. Sustainable Development 15: 343–356.
53. Dendoncker N, Bogaert P, Rounsevell M (2006) A statistical method to downscale aggregated land use data and scenarios. Journal of Land Use Science 1: 63–82.
54. Thuiller W, Lafourcade B, Engler R, Araújo MB (2009) BIOMOD - a platform for ensemble forecasting of species distributions. Ecography 32: 369–373.
55. R (Development Core Team 2009) R: a language and environment for statistical computing. R- Foundation for Statistical Computing.
56. Breiman L (2001) Random forests. Machine Learning 45: 5–32.
57. Cutler DR, Edwards TC, Beard KH, Cutler A, Hess KT (2007) Random forests for classification in ecology. Ecology 88: 2783–2792.
58. Elith J, Leathwick JR, Hastie T (2008) A working guide to boosted regression trees. Journal of Animal Ecology 77: 802–813.
59. Friedman JH (2001) Greedy function approximation: A gradient boosting machine. Annals of Statistics 29: 1189–1232.
60. Swets JA (1988) Measuring the accuracy of diagnostic systems. Science 240: 1285–1293.
61. Allouche O, Tsoar A, Kadmon R (2006) Assessing the accuracy of species distribution models: prevalence, kappa and the true skill statistic (TSS). Journal of Applied Ecology 43: 1223–1232.
62. Pearson RG, Thuiller W, Araújo MB, Martinez-Meyer E, Brotons L, et al. (2006) Model-based uncertainty in species range prediction. Journal of Biogeography 33: 1704–1711.
63. Araújo MB, Guisan A (2006) Five (or so) challenges for species distribution modelling. Journal of Biogeography 33: 1677–1688.
64. Thuiller W, Araújo MB, Pearson RG, Whittaker RJ, Brotons L, et al. (2004) Biodiversity conservation - Uncertainty in predictions of extinction risk. Nature 430.
65. Araújo MB, Pearson RG, Thuiller W, Erhard M (2005) Validation of species-climate impact models under climate change. Global Change Biology 11: 1504–1513.
66. Araújo MB, New M (2007) Ensemble forecasting of species distributions. Trends in Ecology & Evolution 22: 42–47.
67. Araújo MB, Whittaker RJ, Ladle RJ, Erhard M (2005) Reducing uncertainty in projections of extinction risk from climate change. Global Ecology and Biogeography 14: 529–538.
68. Marmion M, Parviainen M, Luoto M, Heikkinen RK, Thuiller W (2009) Evaluation of consensus methods in predictive species distribution modelling. Diversity and Distributions 15: 59–69.
69. Araújo MB, Alagador D, Cabeza M, Nogués-Bravo D, Thuiller W (2011) Climate change threatens European conservation areas. Ecology letters 14: 484–492.
70. Friedman JH, Meulman JJ (2003) Multiple additive regression trees with application in epidemiology. Statistics in Medicine 22: 1365–1381.
71. Austin MP, Van Niel KP (2011) Improving species distribution models for climate change studies: variable selection and scale. Journal of Biogeography 38: 1–8.
72. Araújo MB, Luoto M (2007) The importance of biotic interactions for modelling species distributions under climate change. Global Ecology and Biogeography 16: 743–753.
73. Araújo MB, Rahbek C (2006) How does climate change affect biodiversity? Science 313: 1396–1397.
74. Lee PY, Rotenberry JT (2005) Relationships between bird species and tree species assemblages in forested habitats of eastern North America. Journal of Biogeography 32: 1139–1150.
75. Whittaker RJ, Willis KJ, Field R (2001) Scale and species richness: towards a general, hierarchical theory of species diversity. Journal of Biogeography 28: 453–470.
76. Hickler T, Smith B, Prentice IC, Mjofors K, Miller P, et al. (2008) CO2 fertilization in temperate FACE experiments not representative of boreal and tropical forests. Global Change Biology 14: 1531–1542.
77. Gerten D, Lucht W, Schaphoff S, Cramer W, Hickler T, et al. (2005) Hydrologic resilience of the terrestrial biosphere. Geophysical Research Letters 32.
78. McPherson JM, Jetz W (2007) Effects of species' ecology on the accuracy of distribution models. Ecography 30: 135–151.
79. Segurado P, Araújo MB (2004) An evaluation of methods for modelling species distributions. Journal of Biogeography 31: 1555–1568.
80. Seoane J, Carrascal LM, Alonso CL, Palomino D (2005) Species-specific traits associated to prediction errors in bird habitat suitability modelling. Ecological Modelling 185: 299–308.
81. Seoane J, Carrascal LM (2008) Interspecific differences in population trends of Spanish birds are related to habitat and climatic preferences. Global Ecology and Biogeography 17: 111–121.
82. Poyry J, Luoto M, Heikkinen RK, Saarinen K (2008) Species traits are associated with the quality of bioclimatic models. Global Ecology and Biogeography 17: 403–414.
83. Kaplan JO, Krumhardt KM, Zimmermann N (2009) The prehistoric and preindustrial deforestation of Europe. Quaternary Science Reviews 28: 3016–3034.
84. Bohn U, Neuhäusle R, Gollub G, Hettwer C, Neuhäuslová Z, et al. (2003) Map of the natural vegetation of Europe. Explanatory text with CD-ROM. German Federal Agency for Nature Conservation. Bonn, Germany.

Community Functional Responses to Soil and Climate at Multiple Spatial Scales: When Does Intraspecific Variation Matter?

Andrew Siefert*, Jason D. Fridley, Mark E. Ritchie

Department of Biology, Syracuse University, Syracuse, New York, United States of America

Abstract

Despite increasing evidence of the importance of intraspecific trait variation in plant communities, its role in community trait responses to environmental variation, particularly along broad-scale climatic gradients, is poorly understood. We analyzed functional trait variation among early-successional herbaceous plant communities (old fields) across a 1200-km latitudinal extent in eastern North America, focusing on four traits: vegetative height, leaf area, specific leaf area (SLA), and leaf dry matter content (LDMC). We determined the contributions of species turnover and intraspecific variation to between-site functional dissimilarity at multiple spatial scales and community trait responses to edaphic and climatic factors. Among-site variation in community mean trait values and community trait responses to the environment were generated by a combination of species turnover and intraspecific variation, with species turnover making a greater contribution for all traits. The relative importance of intraspecific variation decreased with increasing geographic and environmental distance between sites for SLA and leaf area. Intraspecific variation was most important for responses of vegetative height and responses to edaphic compared to climatic factors. Individual species displayed strong trait responses to environmental factors in many cases, but these responses were highly variable among species and did not usually scale up to the community level. These findings provide new insights into the role of intraspecific trait variation in plant communities and the factors controlling its relative importance. The contribution of intraspecific variation to community trait responses was greatest at fine spatial scales and along edaphic gradients, while species turnover dominated at broad spatial scales and along climatic gradients.

Editor: Suzannah Rutherford, Fred Hutchinson Cancer Research Center, United States of America

Funding: AS was supported by National Science Foundation grant DEB-1210609. JDF was supported by National Science Foundation grant DEB-1119743. The funders had no role in study design, data collection and analysis, decision to publish, or preparation of the manuscript.

Competing Interests: The authors have declared that no competing interests exist.

* Email: asiefert@syr.edu

Introduction

Understanding and predicting how communities respond to environmental variation is a central goal of ecology, and ecologists are increasingly adopting trait-based approaches to study these responses [1–3]. Because an organism's functional traits directly influence its responses to and effects on the environment [4], information about the traits of individuals in a community (i.e. community trait distributions) offers insights into community assembly mechanisms and can be used to predict community composition and ecosystem functioning [3,5,6]. Understanding how community trait distributions, particularly community-weighted mean trait values (CWMs), respond to environmental variation has therefore become a major focus in community ecology [5,7,8].

Community trait distributions may change along environmental gradients through a combination of species turnover (changes in species presence and relative abundance) and intraspecific trait responses, including genetic adaptation and phenotypic plasticity [8,9]. Most studies examining trait-environment relationships in plant communities have assigned a single, fixed trait value to each species, thereby accounting only for trait variation due to species turnover and ignoring intraspecific variation [10]. However, recent studies have shown that intraspecific variation may also play an important role in community trait responses to environmental variation. For example, Jung et al. [11] found that intraspecific variation accounted for up to 44% of the change in CWMs of several key functional traits across an elevation gradient in flood meadow communities. Similarly, Lepš et al. [10] found that community-level responses of multiple traits to fertilization and mowing in grassland communities were primarily driven by intraspecific responses. Results of these and other recent studies [12–14] demonstrate that accounting for intraspecific trait variation may be crucial for quantifying community trait responses to the environment, but the relative magnitude of intraspecific variation has varied strongly among and within studies. The next step beyond simply quantifying intraspecific trait variation is to understand the factors controlling its relative role and importance in communities. Determining when and where intraspecific variation matters at the community level is a major concern for plant ecology, with important implications for predicting community and ecosystem responses to global change [15].

One factor that may influence the relative importance of intraspecific trait variation is the spatial scale of the studied communities. Previous studies have shown that interspecific trait variation increases with increasing spatial scale due to species turnover driven by dispersal limitation and environmental filtering along gradients of increasing breadth [16,17]. Intraspecific variation is also expected to increase with increasing spatial scale, as more genetic and plastic variability within species is included, but it is expected to saturate at large scales once the entire range and thus potential trait variation of individual species is reached [15,18]. The relative magnitude of interspecific vs. intraspecific variation is therefore expected to increase at broad spatial scales encompassing strong environmental gradients [15]. This hypothesis has not been tested to our knowledge, in part due to the lack of studies measuring intraspecific variation at broad scales.

Another factor that may influence the relative importance of intraspecific trait variation is the type of environmental gradient considered. Previous studies have shown that intraspecific variation may be important for community responses to local-scale edaphic variation [8,11], but its role in responses to broad-scale climatic variation has not been examined. Determining whether the relative importance of intraspecific variation differs between edaphic and climatic gradients is a useful step towards a more general understanding of when intraspecific variation matters at the community level. Knowledge of the role of intraspecific variation is also relevant for predicting responses of communities to climate change [19]. If community trait responses to climate are driven by species turnover, climate change will result in large changes in community composition and species distributions. On the other hand, if species are able to cope with climatic variation through genetic adaptation or phenotypic plasticity, community composition may remain stable [20,21]. Assessing the degree to which intraspecific variation contributes to community trait responses to strong spatial climatic gradients will provide insights into which of these scenarios is most likely in the face of future climate change.

In this study, we examined community functional responses to environmental variation in old-field plant communities across eastern North America. The study was conducted on a broad spatial extent (1200 km in latitude), allowing us to test the relative importance of species turnover vs. intraspecific variation to community trait patterns along strong edaphic and climatic gradients at multiple spatial scales. Specifically, we addressed the following questions: 1) What is the relative importance of species turnover vs. intraspecific variation to among-site trait variation, and how is this influenced by spatial scale? We hypothesized that the relative importance of intraspecific trait variation would decrease with increasing spatial scale and breadth of environmental gradients. 2) How do community mean trait values respond to edaphic and climatic variation, and what are the relative contributions of species turnover and intraspecific variation to these responses? We hypothesized that intraspecific variation would be more important for community responses to edaphic compared to climatic factors.

Materials and Methods

Study site

We surveyed vegetation and functional traits in 22 old fields across the eastern United States in June-August, 2012. The study area extended from central South Carolina (30°40″N) to central New York (43°10″N), spanning approximately 1200 km of latitude (see Fig. S1 for map of study sites). The study sites were located on private and public land. A list of authorities who granted permission to conduct field work and should be contacted for future permissions is found in the Supporting Information (Table S5). The study area encompassed strong variation in both climatic and edaphic factors (Table 1), making it a useful system for comparing the influence of these factors on community functional composition. Moving from south to north, there was a strong decrease in mean annual temperature (17.9 to 6.9°C) and growing season length (263 to 156 annual frost-free days) and a weaker decrease in mean annual precipitation (1330 to 976 mm). In addition, with increasing latitude there was a strong increase in soil fertility and shift from coarse to fine-textured soils driven primarily by recent glaciation history [22].

The fields sampled had different histories of agricultural land use, but all had been abandoned for at least 5 years prior to sampling. Fields were maintained in early stages of succession by mowing in late summer or fall once every 1–2 years. Time since mowing was not significantly related to any environmental variable or response variable measured and was not included in the analyses. No burning, livestock grazing, or herbicide application had occurred in any of the fields within the past 5 years. Vegetation in the fields was almost entirely herbaceous, including a mix of grasses and forbs. Dominant species included goldenrods (e.g., *Solidago altissima*, *S. rugosa*) and grasses (e.g., *Andropogon virginicus*, *Schedonorus pratensis*, *Poa pratensis*). While there was considerable turnover in species composition across the study area, many of the dominant species were widely distributed, creating the potential for intraspecific variation to play an important role in community trait patterns.

Vegetation and environmental data

In each field, we recorded the percent cover of vascular plant species in 20 1-m^2 quadrats arrayed along transects. The number and arrangement of transects and spacing between quadrats varied depending on the size and shape of the field, with the goal of obtaining a representative sample of the overall vegetation. Cover values in the 20 quadrats were pooled to obtain the relative cover of each species in each field. We recorded a total of 227 species in the 22 sampled sites, with a mean richness of 21.4 species per site. Most species occurred in three or fewer fields, and there were large differences in species composition among sites (mean Bray-Curtis dissimilarity based on relative cover = 0.76; based on presence/absence = 0.75).

We collected soil samples in four randomly selected quadrats in each field and pooled samples for physical and chemical analysis. Percent sand, silt, and clay were determined using the hydrometer method [23]. Percent organic matter was measured as loss on ignition at 360°C. Soil samples were analyzed for cation exchange capacity (CEC), pH, available nitrogen (nitrate plus ammonium; KCl extraction/cadmium reduction method), available (Bray II) phosphorus [24], and available (Mehlich 3 extractant) sulfur, calcium, magnesium, potassium, iron, manganese, and aluminum [25]. Soil analyses were performed by Brookside Laboratories, Inc., New Bremen, OH, USA. We accessed daily precipitation and temperature data (1980–2010) for each site from Daymet (http://www.daymet.org). Using these data, we derived mean annual temperature, mean temperature of the coldest month, mean temperature of the warmest month, temperature seasonality (standard deviation of monthly mean temperature), annual frost-free days, annual growing-degree days (base of 5°C and cap of 30°C), annual precipitation, precipitation in the driest month, precipitation in the wettest month, and precipitation seasonality

Table 1. Mean, standard deviation, and range of environmental variables included in regression analyses.

Variable	Abbreviation	Unit	Mean	Standard deviation	Range (max-min)
Mean annual temperature	MAT	°C	11.95	3.0	11.0
Mean annual precipitation	MAP	mm	1132	99.2	354
Cation exchange capacity	CEC	mEq kg^{-1}	79.4	30.0	11.23
Soil pH	pH		5.6	0.37	1.5
Soil available phosphorus	P	mg kg^{-1}	49.2	53.8	200.0
Soil available nitrogen	N	ppm	6.31	2.87	10.4
Soil organic matter	OM	%	4.98	1.73	7.38
Sand	Sa	%	41.75	18.82	71.91

(CV of monthly precipitation). Environmental data were log transformed as necessary to improve normality.

Trait data

We focused on four traits that relate to different aspects of plant functional strategy: vegetative height, leaf area, specific leaf area (SLA), and leaf dry matter content (LDMC). Vegetative height is related to light acquisition and competitive ability [26,27]. Leaf area relates to energy and water balance and tolerance to environmental stress [28]. SLA is a central component of the leaf economics spectrum, which captures the tradeoff between rapid growth and resource conservation [29]. LDMC is also associated with the leaf economics spectrum as well as leaf water balance and resistance to physical stress [30,31].

In each field, we measured functional traits of species that collectively accounted for 80–100% of the total vegetation cover. This sampling threshold has been shown to provide robust estimates of community mean trait values [32] but excludes some rare species. In each field, we selected five mature- and healthy-looking individuals of each species from different areas of the field for trait measurements. Vegetative height was measured as the distance (cm) from the base to the highest part of the general canopy of the plant. We selected one young, fully-expanded, upper canopy leaf per individual for leaf trait measurements. We measured the one-sided surface area and fresh mass of each leaf following full rehydration [33] and dry mass after oven drying at 80°C for 48 hours [31]. SLA was calculated as leaf area divided by dry mass (mm^2/mg), and LDMC was calculated as leaf dry mass divided by fresh mass. We measured traits on 61 species total (mean 7.5 per field), representing on average 87% of the total cover of each field.

Data analysis

Our first analysis partitioned the contributions of species turnover and intraspecific variation to among-site variation in functional traits following the approach of Lepš et al. [10]. For each field, we calculated three types of community-weighted mean trait values. 1) "Total CWM", calculated as the abundance-weighted average of site-specific species mean trait values:

$$\text{Total CWM}_j = \sum_{i=1}^{S} p_{ij}x_{ij}$$

where p_{ij} is the relative cover of species i in site j, x_{ij} is the mean trait value of species i measured in site j, and S is the number of species sampled in the site. Among-site variation in total CWMs may be generated by a combination of changes in species composition (i.e. species turnover) and changes in species mean trait values (i.e. intraspecific variation) among sites. 2) "Interspecific CWM", calculated as the abundance-weighted average of overall species mean trait values:

$$\text{Interspecific CWM}_j = \sum_{i=1}^{S} p_{ij}x_i$$

where x_i is the mean trait value of species i across all sites where it is measured. Variation in interspecific CWMs is generated by species turnover only. 3) "Intraspecific CWM", calculated as the difference between the total CWM and interspecific CWM for a given site:

$$\textit{Intraspecific } CWM_j = \textit{Total } CWM_j - \textit{Interspecific } CWM_j$$

Variation in intraspecific CWMs is generated by intraspecific trait variation only. Using the total, interspecific, and intraspecific CWMs, we partitioned trait variation among sites into species turnover, intraspecific variation, and covariation components using the sum of squares decomposition approach of Lepš et al. [10]. Positive covariation indicates that species turnover and intraspecific variation reinforce each other (i.e., sites dominated by species with high trait values also have individuals with high trait values for their species), while negative covariation indicates that species turnover and intraspecific variation oppose each other (i.e., sites dominated by species with high trait values have individuals with low trait values for their species).

Our second analysis assessed community trait responses to edaphic and climatic variation and partitioned the contributions of species turnover and intraspecific variation to these responses. Due to strong correlations among variables, we attempted to reduce the dimensionality of the environmental data using principal components analysis, but the primary axes identified left much unexplained variation and were poorly related to community trait values. We therefore selected subsets of climatic (mean annual temperature and annual precipitation) and edaphic variables (sand, pH, CEC, organic matter, available nitrogen, and available phosphorus) that were expected to be important drivers of community functional structure and were not strongly correlated with each other ($r<0.5$). We modeled relationships between

Table 2. Percentages of total variation in community-weighted mean trait values due to species turnover, intraspecific variation, and their covariation.

Trait	Species turnover	Intraspecific variation	Covariation
Height	52%	22%	26%
Leaf area	89%	7%	4%
SLA	70%	28%	2%
LDMC	52%	27%	21%

Positive covariation indicates that sites dominated by species with high trait values also have individuals with higher than average trait values for their species.

community-weighted mean trait values and edaphic and climatic factors using multiple linear regressions with the full subset of edaphic or climatic variables as predictors. We then performed stepwise model selection by AIC_c to select the best edaphic and climatic model for each trait. To quantify the contributions of species turnover, intraspecific variation, and their covariation to overall community trait responses, we partitioned the variance explained by the edaphic and climatic models using the sum of squares decomposition described above [10]. To examine community trait responses to specific environmental factors, we conducted a similar variance partitioning analysis using single environmental variables as predictors. We also used regression analyses to quantify and compare trait responses of the five most abundant and widely distributed species in the study area: forbs *Solidago altissima* and *Solidago rugosa*, and grasses *Schedonorus pratensis*, *Poa pratensis*, and *Andropogon virginicus*.

Our third analysis tested whether the between-site trait variation and the relative contribution of species turnover vs. intraspecific variation increased with increasing spatial and environmental distance. First, we calculated the geographic distance and environmental distance between each pair of sites in the study area (22 sites, resulting in 231 pairs). Geographic distance between sites was calculated as great circle distance and ranged from 6.5 to 1,151 km (see Fig. S2 for distribution of between-plot distances). Environmental distance was calculated as Euclidean distance using scaled environmental variables (Fig. S2). Next, for each pair of plots we calculated the total dissimilarity in CWMs, dissimilarity due to species turnover, and dissimilarity due to intraspecific variation by applying the sum of squares decomposition described above [10] to each plot pair. We quantified the relative importance of species turnover vs. intraspecific variation by taking the log of the ratio of the species turnover and intraspecific variation components. This created a symmetric measure of the relative contribution of species turnover vs. intraspecific variation to between-site trait dissimilarity, with positive values indicating a greater contribution of species turnover and negative values a greater contribution of intraspecific variation. We tested whether total between-site trait dissimilarity, dissimilarity due to species turnover, dissimilarity due to intraspecific variation, and the relative importance of turnover vs. intraspecific variation varied as a function of geographic and environmental distance using Mantel tests. Because geographic and environmental distance were strongly correlated (Mantel $r = 0.70$), we also used partial Mantel tests to test for the effect of either geographic or environmental distance while controlling for the other.

All analyses were conducted in R [34] using the fields [35], vegan [36], and ecodist [37] packages.

Results

Partitioning among-site trait variation

Among-site variation in community mean trait values was generated by a combination of species turnover and intraspecific variation, with species turnover making the greater contribution for each trait (Table 2). For height, SLA, and LDMC the contribution of species turnover was 2–2.5 times greater than that of intraspecific variation. For leaf area, almost all variation was due to species turnover (Table 2). There was positive covariation between species turnover and intraspecific variation for height and LDMC, indicating that the effects of species turnover and intraspecific variation reinforced each other (i.e., sites dominated by species with high values of those traits also tended to have individuals with high trait values for their species). There was little covariation between species turnover and intraspecific variation for leaf area or SLA (Table 2).

Community trait responses to soil and climate and relative contributions of species turnover vs. intraspecific variation

The best linear models relating environmental variables to community mean trait values varied among traits, but for all traits edaphic factors explained more variation (16–50%) than climatic factors (<20%; Table 3). For example, CWM height was best explained by a combination of soil pH, CEC, organic matter, and available phosphorus, and the best model for CWM leaf area included soil organic matter and available phosphorus (Table 3). CWM SLA was the trait most strongly influenced by climate, showing a significant decrease with increasing mean annual temperature (Table 3).

Community trait responses to the environment were generated primarily by species turnover (Fig. 1), but the importance of intraspecific variation depended on the trait and environmental factor. Intraspecific variation contributed most strongly to community responses of height. For all traits, intraspecific variation was more important, both in terms of total magnitude and relative to species turnover, for community responses to edaphic compared to climatic factors (Fig. 1). Similar results were obtained when using single environmental variables as predictors (Table S2). For example, the only cases in which intraspecific variation contributed more than species turnover were the responses of CWM height to soil CEC and available P (Table S2). Positive covariation between species turnover and intraspecific variation contributed strongly to community responses of height and to a lesser extent SLA, indicating that changes due to species turnover and intraspecific variation reinforced each other (Fig. 1). In contrast, there was weak or even negative covariation between

Table 3. Relationships between community-weighted mean trait values and environmental variables measured in the study.

Trait	Model	Predictors	R^2
Height	Edaphic	CEC (+), pH (−), OM (−), P (+)	0.50
	Climatic	MAP (-)	0.17
Leaf area	Edaphic	OM (−), P (−)	0.25
	Climatic	MAT (+)	0.16
SLA	Edaphic	pH (−), OM (+)	0.26
	Climatic	MAT (−)	0.20
LDMC	Edaphic	N (−), sand (+)	0.16
	Climatic	ns	-

Results are shown for the best edaphic and climatic models for each trait as determined by stepwise selection, including the predictor variables retained in each model and the direction of their effects (negative or positive) on the community trait value. Abbreviations: SLA, specific leaf area; LDMC, leaf dry matter content; predictor variables abbreviated as in Table 1.

species turnover and intraspecific variation effects for leaf area and LDMC.

Individual species had strong trait responses to environmental variables in many cases, but these responses tended to be highly idiosyncratic, differing in strength and direction among species (Fig. 2; Table S3). There were no obvious patterns in the strength or consistency of intraspecific trait responses across species, traits, or environmental variables.

Effects of geographic and environmental distance on between-site trait dissimilarity and species turnover vs. intraspecific variation effects

The influence of geographic and environmental distance on community trait dissimilarity varied among traits. Between-site dissimilarity in CWM SLA increased significantly with increasing geographic ($r = 0.23$; $P = 0.02$) and environmental distance ($r = 0.20$; $P = 0.04$; Fig. 3), indicating that nearby and environmentally similar sites had similar mean SLA. There was also a marginally significant increase in between-site dissimilarity in CWM leaf area with increasing environmental distance ($r = 0.17$; $P = 0.07$; Fig. 3). For both traits, increases in functional dissimilarity were driven by increases in species turnover effects, whereas intraspecific variation was insensitive to both geographic and environmental distance (Table S4). As a result, the relative importance of species turnover vs. intraspecific variation increased with increasing geographic ($r = 0.33$; $P<0.01$) and environmental distance ($r = 0.24$; $P<0.01$) for SLA and environmental distance for leaf area ($r = 0.12$; $P = 0.08$; Fig. 3). In contrast, between-site dissimilarity in CWMs and the relative importance of turnover vs. intraspecific variation were not related to geographic or environmental distance for height or LDMC (Fig. 3).

Due to the strong correlation between geographic and environmental distance, it was generally not possible to separate the effects of space vs. environment on community trait dissimilarity (partial Mantel tests not statistically significant; Table S4). The only exception was a positive relationship between geographic distance and the relative importance of species turnover for SLA even after removing the effect of environmental distance (partial $r = 0.24$; $P = 0.01$; Table S4).

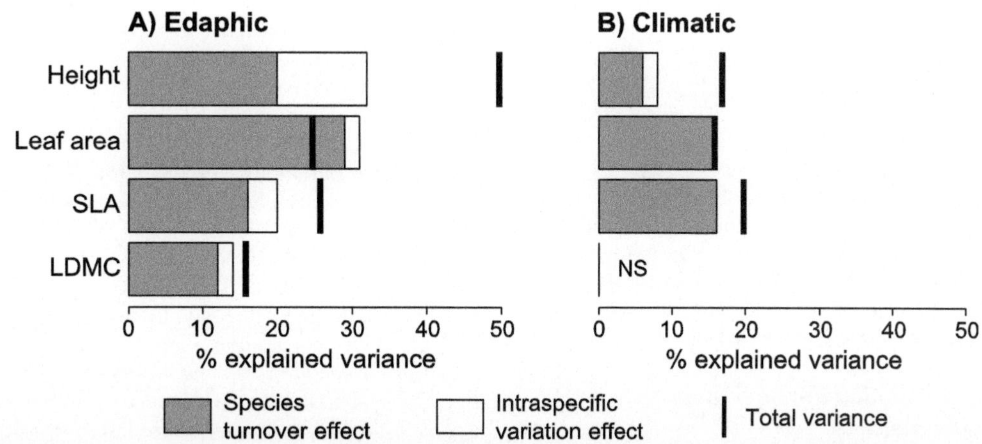

Figure 1. Decomposition of community trait responses to environment. Among-site variance in community-weighted mean trait values explained by (A) edaphic and (B) climatic variables is partitioned into species turnover, intraspecific variation, and covariation effects. Covariation is represented by the difference between the total variance and the sum of the species turnover and intraspecific variation effects. Total variance greater than the sum of species turnover and intraspecific variation effects indicates positive covariance. Total variance less than the sum of species turnover and intraspecific variation effects indicates negative covariance. Abbreviations: SLA, specific leaf area; LDMC, leaf dry matter content.

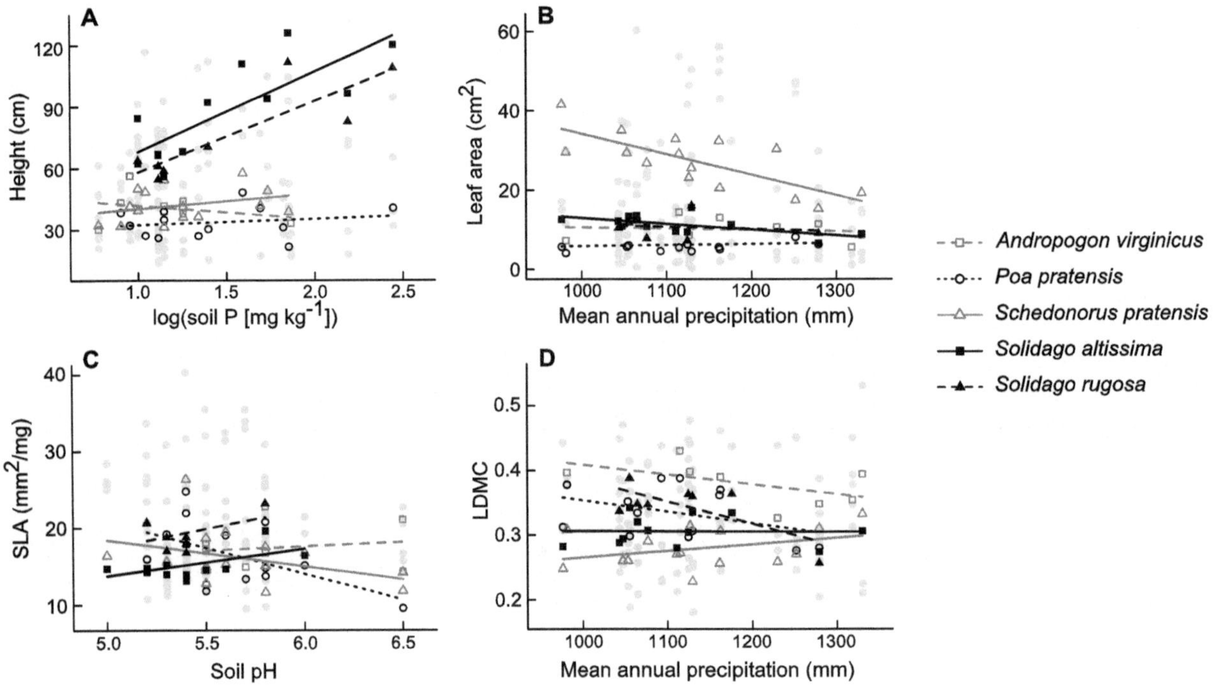

Figure 2. Intraspecific trait responses to environment. Examples of relationships between trait values of individual species and important environmental predictors are shown for each trait measured in the study: (A) height vs. soil P; (B) leaf area vs. mean annual precipitation; (C) specific leaf area vs. soil pH; (D) leaf dry matter content vs. mean annual precipitation. Points represent site-specific mean trait values of all species sampled in each site. Mean trait values and best fit lines from linear regressions are indicated for the five most abundant and widespread species in the study area: *Andropogon virginicus*, *Poa pratensis*, *Schedonorus pratensis*, *Solidago altissima*, and *Solidago rugosa*.

Discussion

Recent studies have shown that intraspecific trait variation may play a fundamental role in responses of plant communities to environmental variation [10–12,19], but many questions remain about when intraspecific variation matters at the community level and the factors controlling its relative importance. By examining intraspecific trait variation in plant communities at an unprecedented spatial extent (1200 km), we were able to address unresolved questions about the relative role of intraspecific variation in community responses to the environment. We found that community-level trait variation was driven primarily by species turnover, but the relative importance of intraspecific variation depended strongly on the trait, environmental factor, and spatial scale considered. In particular, intraspecific variation was more important for responses to edaphic compared to climatic factors, and its relative importance decreased with increasing spatial scale and the related increase in breadth of environmental gradients for two of the four traits examined. These findings contribute to a more general understanding of the role intraspecific trait variation in plant communities, with implications for understanding community assembly and predicting community responses to global change.

Role of intraspecific variation in community trait responses to environment

Community trait responses to environmental variation were primarily driven by species turnover, but intraspecific variation tended to play a larger role in community trait responses to edaphic compared to climatic variation. This finding may reflect differences in the spatial scale on which edaphic and climatic factors vary. Edaphic factors tend to vary on finer spatial scales than climatic factors [38], so the geographic ranges of individual species are likely to include a relatively larger proportion of the total edaphic variation than the total climatic variation found in a region. As a result of species being distributed widely along edaphic gradients but narrowly along climatic ones, there is greater potential for strong intraspecific trait responses to drive community-level trait shifts along edaphic gradients. In addition, fine-scale environmental heterogeneity is known to promote the evolution of adaptive phenotypic plasticity [39,40], raising the possibility that plants have stronger plastic trait responses to fine-scale edaphic compared to broad-scale climatic factors. Assessing the strength of intraspecific trait responses to different environmental factors and how this relates to the spatial scale at which those factors vary is an important aim for future research.

The relative importance of intraspecific variation also varied among traits. The contribution of intraspecific variation to total among-site variation and responses to environmental factors was greatest for vegetative height. Intraspecific variation contributed to community-level variation in height directly and through strong positive covariation with species turnover effects, consistent with the expectation that interspecific and intraspecific trait responses should be in the same direction to increase individual plant fitness [8]. Intraspecific variation played a particularly large role in responses of height to edaphic factors such as available phosphorus. This finding is in line with previous studies showing that plant height is highly responsive to soil resource availability within species and that intraspecific variation contributes strongly to community-level variation in height [10,41,42]. In contrast, the relative contribution of intraspecific variation was smallest for leaf area and LDMC, traits that are known to be less plastic and for which intraspecific variation tends to be much smaller than interspecific variation [30,43].

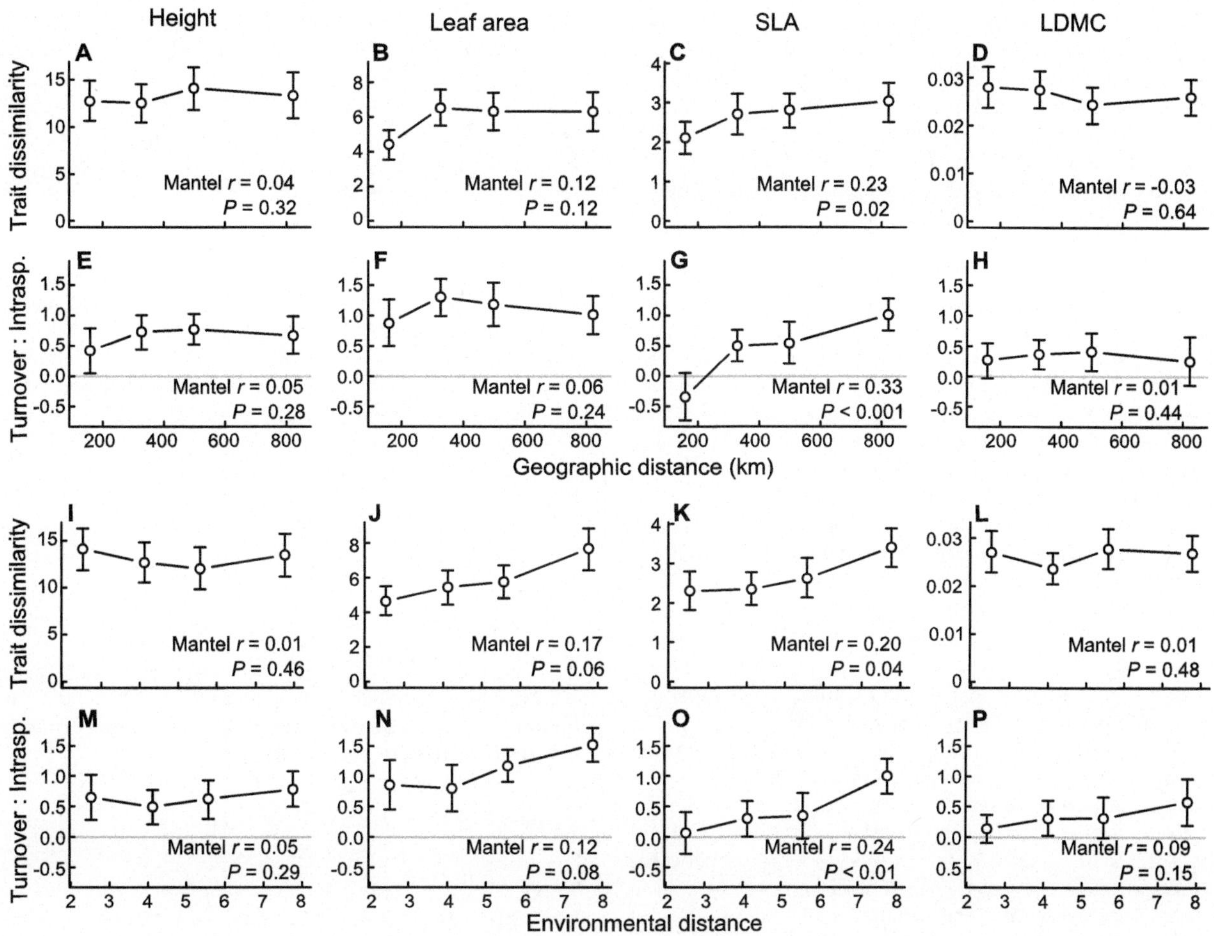

Figure 3. Effects of geographic and environmental distance on between-site variation in plant functional traits. Panels A-C and I-L show the trait dissimilarity (difference in community-weighted mean trait values) between sites as a function of geographic or environmental distance. Panels E-H and M-P show the relative contribution of species turnover vs. intraspecific variation (measured as log of species turnover effect divided by intraspecific variation effect) to the total trait dissimilarity between sites. Values were calculated for each pair of sites in the study area and binned by geographic or environmental distance (n = 57–58 pairs per bin) for ease of interpretation. Points represent mean values for each distance bin; error bars represent 95% bootstrap confidence intervals. Abbreviations: SLA, specific leaf area; LDMC, leaf dry matter content.

What explains the relatively weak contribution of intraspecific variation to most community-level trait-environment relationships observed in this study? The effect of intraspecific trait responses seen at the community level is the aggregate of responses of many individual species. Although traits of individual species responded strongly to the environment in many cases, these responses were highly idiosyncratic, differing in both strength and direction. As a result, they often cancelled out and generally failed to scale up to the community level. Previous studies examining trait variation in multiple species along an environmental gradient have also found intraspecific trait responses to be highly variable [12,44,45]. This variability may have several causes. First, species may respond to environmental variation by altering their multivariate functional strategies rather than single traits in isolation [19]. Functional tradeoffs can result in multiple trait combinations that are equally adaptive in a given environment [46], so responses of single traits may be variable. Second, traits respond not only to the abiotic environment, but also to biotic interactions [47]. The trait response of a particular species to an environmental change will therefore depend not only on the direct influence of the environment, but also on changes in biotic interactions mediated by traits of the target species and its neighbors [12]. Third, some traits may have unimodal responses to environmental gradients, such that trait values are maximized at a species' environmental optimum. The observed direction of the intraspecific trait responses may therefore vary depending on the part of the gradient examined in relation to the environmental ranges and optima of the sampled species [44].

It should be noted that community trait-environment relationships observed in this study were generally weak, with much unexplained variation. This is probably due in part to a strong influence of site history on community assembly [48]. Sites with similar edaphic and climatic conditions were often dominated by functionally dissimilar species, suggesting that unknown past management regimes, along with dispersal history and stochastic effects may have played a large role in determining the functional composition of these early-successional communities. In addition, it is likely that key environmental variables, such as nitrogen supply rate and midsummer water potential, were not measured.

Influence of spatial scale on the relative importance of species turnover vs. intraspecific variation

The broad spatial extent of our study allowed us to examine the influence of spatial scale on the relative importance of species turnover vs. intraspecific variation. Albert et al. [15] hypothesized that the relative importance of intraspecific variation should decrease with increasing spatial scale and the accompanying increase in environmental heterogeneity. We found limited support for this hypothesis for two of the four traits examined in the study, leaf area and SLA. Trait variation due to species turnover increased with increasing geographic and environmental distance for both traits, reflecting the effects of environmental filtering, dispersal limitation, or both [17]. In contrast, intraspecific trait variation was insensitive to both geographic and environmental distance, indicating that intraspecific variation in the studied communities was for the most part spatially random or driven by unmeasured fine-scale environmental variation. This agrees with the findings of Albert et al. [45] that most intraspecific variance in plant traits was found at fine spatial scales, i.e. within rather than between populations along an environmental gradient.

This study is the first to our knowledge to assess the contribution of intraspecific variation to community-level trait variation across a broad geographic extent encompassing strong climatic variation. The relative importance of intraspecific variation observed in this study was generally less than that observed for the same traits in previous local-scale studies of grassland communities. For example, Li et al. (unpublished data) showed that intraspecific variation was the primary source of community-level changes in height, SLA, and LDMC along a soil moisture gradient in subalpine meadow communities. Similarly, Kichenin et al. [12] found that intraspecific variation drove community shifts in SLA along an elevation gradient. At relatively small distances (<200 km), we also found that intraspecific variation contributed more than species turnover to among-site variation in SLA. For height, leaf area, and LDMC, species turnover was the primary source of between-community variation even at the smallest distances resolvable in this study (6.5 km), but it is possible that a similar transition from intraspecific variation to species turnover as the main source of among-community trait variation occurred at finer spatial scales.

Implications

Our results have several potentially interesting implications for plant community ecology. First, with the growing recognition that intraspecific variation may play an important role in plant communities [49], and the great effort required to measure it [50], there is a need for information to help researchers decide when it should be considered in plant ecology studies. To this end, Albert et al. [15] proposed that intraspecific variation might be negligible and therefore safely ignored at very broad spatial scales encompassing strong environmental gradients. Our results provide limited support for this recommendation by showing for the first time that the relative importance of intraspecific variation in some traits decreases with increasing geographic and environmental distance between sites, so that species turnover dominates at very broad scales. In such cases, it may be acceptable to use species mean trait values to characterize community-level trait patterns, whereas measuring each species in each site where it occurs may be necessary for studies conducted on finer scales and shorter gradients [50]. Second, community trait-environment relationships are commonly used to infer community assembly mechanisms, particularly trait-mediated environmental filtering [8], but most studies have only considered interspecific trait variation. Our findings suggest that different environmental filters or selection pressures operate at different levels of organization within communities. Specifically, climatic factors may filter species based on their mean trait values, resulting in species turnover along broad-scale climatic gradients, whereas intraspecific variation may be more important for responses to edaphic filters. Third, the dominance of species turnover in driving community trait responses to spatial climatic gradients suggests that species turnover will also play a large role in responses of the studied communities to future predicted climate change [51].

To conclude, intraspecific trait variation may play an important role in community trait responses to the environment in some situations, and there is a need for empirical data to generalize when and to what extent it matters at the community level. We found that functional responses of old-field plant communities to environmental variation at broad spatial scales were primarily driven by species turnover, but several factors influenced the relative importance of intraspecific variation. Specifically, intraspecific variation was more important for responses of vegetative height compared to leaf traits, responses to edaphic compared to climatic gradients, and at fine compared to broad spatial scales. Future research should examine whether our findings extend to other plant communities and types of environmental gradients.

Supporting Information

Figure S1 Map of study sites and location of study area within North America.

Figure S2 Distributions of geographic and environmental distances between study sites.

Table S1 Mean trait values of sampled species.

Table S2 Decomposition of variation in community-weighted mean trait values explained by single environmental variables.

Table S3 Slopes of intraspecific trait-environment relationships for the five most frequently occurring study species.

Table S4 Results of Mantel and partial Mantel tests for effects of geographic and environmental distance on between-site trait dissimilarity.

Table S5 List of authorities issuing permits for field work at study sites.

Acknowledgments

We are grateful to Josh Lynn and Luka Negoita for assistance with field work and data collection. We thank the following agencies and organizations for providing access to field sites: National Park Service, South Carolina State Park Service, North Carolina Division of Parks and Recreation, Duke Forest, VCU Rice Center, Virginia Department of Conservation and Recreation, Pennsylvania Department of Conservation and Natural Resources, Hawk Mountain Sanctuary, SUNY Cortland, Cary Institute of Ecosystem Studies, Mohonk Preserve and Cornell Biological Field Station. We thank Francesco de Bello for providing helpful comments on the manuscript.

Author Contributions

Conceived and designed the experiments: AS JDF MER. Performed the experiments: AS. Analyzed the data: AS. Contributed to the writing of the manuscript: AS JDF MER.

References

1. Lavorel S, Garnier E (2002) Predicting changes in community composition and ecosystem functioning from plant traits: revisiting the Holy Grail. Funct Ecol 16: 545–556.
2. McGill BJ, Enquist BJ, Weiher E, Westoby M (2006) Rebuilding community ecology from functional traits. Trends Ecol Evol 21: 178–185.
3. Suding KN, Lavorel S, Chapin FS III, Cornelissen JHC, et al. (2008) Scaling environmental change through the community-level: a trait-based response-and-effect framework for plants. Glob Chang Biol 14: 1125–1140.
4. Diaz S, Cabido M (2001) Vive la différence: plant functional diversity matters to ecosystem processes. Trends Ecol Evol 16: 646–655.
5. Shipley B, Vile D, Garnier E (2006) From plant traits to plant communities: a statistical mechanistic approach to biodiversity. Science 314: 812–814.
6. Chapin FS III, Zavaleta ES, Eviner VT, Naylor RL, Vitousek PM, et al. (2000) Consequences of changing biodiversity. Nature 405: 234–242.
7. Ricotta C, Moretti M (2011) CWM and Rao's quadratic diversity: a unified framework for functional ecology. Oecologia 167: 181–188.
8. Cornwell WK, Ackerly DD (2009) Community assembly and shifts in plant trait distributions across an environmental gradient in coastal California. Ecol Monogr 79: 109–126.
9. Ackerly DD (2003) Community assembly, niche conservatism, and adaptive evolution in changing environments. Int J Plant Sci 164: S165–S184.
10. Lepš J, de Bello F, Šmilauer P, Doležal J (2011) Community trait response to environment: disentangling species turnover vs intraspecific trait variability effects. Ecography 34: 856–863.
11. Jung V, Violle C, Mondy C, Hoffmann L, Muller S (2010) Intraspecific variability and trait-based community assembly. J Ecol 98: 1134–1140.
12. Kichenin E, Wardle DA, Peltzer DA, Morse CW, Freschet GT (2013) Contrasting effects of plant inter- and intraspecific variation on community-level trait measures along an environmental gradient. Funct Ecol 27: 1254–1261.
13. Auger S, Shipley B (2012) Inter-specific and intra-specific trait variation along short environmental gradients in an old-growth temperate forest. J Veg Sci 24: 419–428.
14. Pérez-Ramos I, Roumet C, Cruz P, Blanchard A, Autran P, et al. (2012) Evidence for a "plant community economics spectrum" driven by nutrient and water limitations in a Mediterranean rangeland of southern France. J Ecol 100: 1315–1327.
15. Albert CH, Grassein F, Schurr FM, Vieilledent G, Violle C (2011) When and how should intraspecific variability be considered in trait-based plant ecology? Perspect Plant Ecol Evol Syst 13: 217–225.
16. Swenson NG, Anglada-Cordero P, Barone JA (2011) Deterministic tropical tree community turnover: evidence from patterns of functional beta diversity along an elevational gradient. Proc R Soc B 278: 877–884.
17. Siefert A, Ravenscroft C, Weiser MD, Swenson NG (2013) Functional beta-diversity patterns reveal deterministic community assembly processes in eastern North American trees. Glob Ecol Biogeogr 22: 682–691.
18. Violle C, Jiang L (2009) Towards a trait-based quantification of species niche. J Plant Ecol 2: 87–93.
19. Jung V, Albert CH, Violle C, Kunstler G, Loucougaray G, et al. (2014) Intraspecific trait variability mediates the response of subalpine grassland communities to extreme drought events. J Ecol 102: 45–53.
20. Lloret F, Escudero A, Iriondo JM, Martínez-Vilalta J, Valladares F (2012) Extreme climatic events and vegetation: the role of stabilizing processes. Glob Chang Biol 18: 797–805.
21. Grime JP, Fridley JD, Askew AP, Thompson K, Hodgson JG, et al. (2008) Long-term resistance to simulated climate change in an infertile grassland. Proc Natl Acad Sci USA 105: 10028–10032.
22. Wright JP, Fridley JD (2010) Biogeographic synthesis of secondary succession rates in eastern North America. J Biogeogr 37: 1585–1596.
23. Bouyoucos GJ (1962) Hydrometer method improved for making particle size analyses of soils. Agron J 54: 464–465.
24. Bray RH, Kurtz LT (1945) Determination of total, organic, and available forms of phosphorus in soils. Soil Sci 59: 39–46.
25. Mehlich A (1984) Mehlich 3 soil test extractant: A modification of Mehlich 2 extractant. Commun Soil Sci Plant Anal 15: 1409–1416.
26. Gaudet CL, Keddy PA (1988) A comparative approach to predicting competitive ability from plant traits. Nature 334: 242–243.
27. Wright IJ, Reich PB, Cornelissen JHC, Falster DS, Groom PK, et al. (2005) Modulation of leaf economic traits and trait relationships by climate. Glob Ecol Biogeogr 14: 411–421.
28. Ackerly D, Reich P (1999) Convergence and correlations among leaf size and function in seed plants: a comparative test using independent contrasts. Am J Bot 86: 1272–1281.
29. Wright IJ, Reich PB, Westoby M, Ackerly DD, Baruch Z, et al. (2004) The worldwide leaf economics spectrum. Nature 428: 821–827.
30. Wilson P, Thompson K, Hodgson J (1999) Specific leaf area and leaf dry matter content as alternative predictors of plant strategies. New Phytol 143: 155–162.
31. Perez-Harguindeguy N, Diaz S, Garnier E, Lavorel S, Poorter H, et al. (2013) New handbook for standardised measurement of plant functional traits worldwide. Aust J Bot 61: 167–234.
32. Pakeman RJ, Lepš J, Kleyer M, Lavorel S, Garnier E (2009) Relative climatic, edaphic and management controls of plant functional trait signatures. J Veg Sci 20: 148–159.
33. Garnier E, Shipley B, Roumet C, Laurent G (2001) A standardized protocol for the determination of specific leaf area and leaf dry matter content. Funct Ecol 15: 688–695.
34. R Core Development Team (2012) R: A Language and Environment for Statistical Computing. Available: http://www.r-project.org.
35. Furrer R, Nychka D, Sain S (2012) fields: Tools for spatial data. Available: http://cran.r-project.org/package=fields.
36. Oksanen J, Blanchet FG, Kindt R, Legendre P, Minchin P, et al. (2012) vegan: Community ecology package. Available: http://cran.r-project.org/package=vegan.
37. Goslee S, Urban D (2007) The ecodist package for dissimilarity-based analysis of ecological data. J Stat Softw 22: 1–19.
38. Lechowicz MJ, Bell G (1991) The ecology and genetics of fitness in forest plants II. Microspatial heterogeneity of the edaphic environment. J Ecol 79: 687–696.
39. Via S, Lande R (1985) Genotype-environment interaction and the evolution of phenotypic plasticity. Evolution 39: 505–522.
40. Baythavong BS (2011) Linking the spatial scale of environmental variation and the evolution of phenotypic plasticity: selection favors adaptive plasticity in fine-grained environments. Am Nat 178: 75–87.
41. Dantas VDL, Pausas JG, Batalha MA, Loiola PDP, Cianciaruso MV (2013) The role of fire in structuring trait variability in Neotropical savannas. Oecologia 171: 487–494.
42. Gross N, Börger L (2013) Uncovering multiscale effects of aridity and biotic interactions on the functional structure of Mediterranean shrublands. J Ecol: 637–649.
43. Roche P, Díaz-Burlinson N, Gachet S (2004) Congruency analysis of species ranking based on leaf traits: which traits are the more reliable? Plant Ecol 174: 37–48.
44. Albert CH, Thuiller W, Yoccoz NG, Soudant A, Boucher F, et al. (2010) Intraspecific functional variability: extent, structure and sources of variation. J Ecol 98: 604–613.
45. De Frenne P, Graae B (2013) Latitudinal gradients as natural laboratories to infer species' responses to temperature. J Ecol 101: 784–795.
46. Marks CO, Lechowicz MJ (2006) Alternative designs and the evolution of functional diversity. Am Nat 167: 55–66.
47. Callaway RRM, Pennings SSC, Richards CLC (2003) Phenotypic plasticity and interactions among plants. Ecology 84: 1115–1128.
48. Chase JM (2003) Community assembly: when should history matter? Oecologia 136: 489–498.
49. Violle C, Enquist BJ, McGill BJ, Jiang L, Albert CH, et al. (2012) The return of the variance: intraspecific variability in community ecology. Trends Ecol Evol 27: 244–252.
50. Baraloto C, Paine CET, Patiño S, Bonal D, Hérault B, et al. (2010) Functional trait variation and sampling strategies in species-rich plant communities. Funct Ecol 24: 208–216.
51. Northeast Climate Impacts Assessment (2006) Climate change in the US Northeast. Union of Concerned Scientists Publications, Cambridge, MA.

18

Vegetation Controls on Weathering Intensity during the Last Deglacial Transition in Southeast Africa

Sarah J. Ivory[1]*, Michael M. McGlue[2], Geoffrey S. Ellis[3], Anne-Marie Lézine[4], Andrew S. Cohen[5], Annie Vincens[6]

1 Brown University, Providence, Rhode Island, United States of America, **2** University of Kentucky, Lexington, Kentucky, United States of America, **3** U.S. Geological Survey, Denver, Colorado, United States of America, **4** LOCEAN, CNRS, Paris, France, **5** University of Arizona, Tucson, Arizona, United States of America, **6** CEREGE, CNRS, Aix-en-Provence, France

Abstract

Tropical climate is rapidly changing, but the effects of these changes on the geosphere are unknown, despite a likelihood of climatically-induced changes on weathering and erosion. The lack of long, continuous paleo-records prevents an examination of terrestrial responses to climate change with sufficient detail to answer questions about how systems behaved in the past and may alter in the future. We use high-resolution records of pollen, clay mineralogy, and particle size from a drill core from Lake Malawi, southeast Africa, to examine atmosphere-biosphere-geosphere interactions during the last deglaciation (~18–9 ka), a period of dramatic temperature and hydrologic changes. The results demonstrate that climatic controls on Lake Malawi vegetation are critically important to weathering processes and erosion patterns during the deglaciation. At 18 ka, afromontane forests dominated but were progressively replaced by tropical seasonal forest, as summer rainfall increased. Despite indication of decreased rainfall, drought-intolerant forest persisted through the Younger Dryas (YD) resulting from a shorter dry season. Following the YD, an intensified summer monsoon and increased rainfall seasonality were coeval with forest decline and expansion of drought-tolerant miombo woodland. Clay minerals closely track the vegetation record, with high ratios of kaolinite to smectite (K/S) indicating heavy leaching when forest predominates, despite variable rainfall. In the early Holocene, when rainfall and temperature increased (effective moisture remained low), open woodlands expansion resulted in decreased K/S, suggesting a reduction in chemical weathering intensity. Terrigenous sediment mass accumulation rates also increased, suggesting critical linkages among open vegetation and erosion during intervals of enhanced summer rainfall. This study shows a strong, direct influence of vegetation composition on weathering intensity in the tropics. As climate change will likely impact this interplay between the biosphere and geosphere, tropical landscape change could lead to deleterious effects on soil and water quality in regions with little infrastructure for mitigation.

Editor: Olivier Boucher, Université Pierre et Marie Curie, France

Funding: Lake Malawi Drilling Project-Earth System History Program (NSF-EAR-0602404) funded field operations, logistics, and some laboratory analysis. NSF Graduate Research Fellowship (2009078688) provided student salary and tuition and some travel support for laboratory analysis. The funders had no role in study design, data collection and analysis, decision to publish, or preparation of the manuscript.

Competing Interests: The authors have declared that no competing interests exist.

* Email: sarah_ivory@brown.edu

Introduction

Tropical climate is rapidly changing resulting in altered atmospheric and oceanic circulation, as well as changing variability of important climatic modes like the El Niño–South Oscillation [1]. Much work is being done within the framework of the IPCC to better understand climate sensitivity and climate-induced changes to the biosphere; however, integrating the effects on and feedbacks from the geosphere has been largely unexplored [2]. Alterations to the geosphere, an important component of the Earth's critical zone, could have costly, hazardous implications for water and soil quality, as well as a host of ecosystems services provided by inland waters (e.g., serving as fisheries, rookeries, and zones of stormwater retention).

Weathering and climate have hypothetically been linked via feedback loops that regulate the Earth system [3]. However, the low resolution of most marine sedimentary records and the strong influence of long time-scale (10^6–10^7 yrs) processes such as orogeny do not provide a scalable framework for evaluating future changes in the critical zone. Additionally, although climate has long been thought to play a strong, direct role, a number of datasets indicate that the role of climate for weathering and erosion may in fact be indirect, mediated by vegetation and soil storage of organic acids [4–7].

Some studies have generated geochemical indicator records of weathering or produced denudation rates for tropical watersheds based on mass balance models; however, such studies are unable to determine how these rates vary over long time-scales [8–11]. Due to the strong interrelation of vegetation and climate, it has been particularly difficult to separate the individual influences of each mechanism. However, it is expected that vegetation plays an important role. Other studies that have examined the relationship between land use and sediment yield in tropical catchments have demonstrated that heavily forested areas generate less sediment than grasslands or areas modified for agriculture or ranching (e.g., [12]). Furthermore, human land-use change could have strong but unclear effects on weathering and the carbon cycle via alteration of vegetation with no climatic change [13]. To answer the question of how climate change influences weathering and erosion in the tropics, well-dated, high-resolution paleo-records from the critical zone are needed. However, records of tropical lowland vegetation and climate are exceedingly rare, despite the great need to better understand these regions and their growing populations, as articulated by the IPCC [1]. Existing Quaternary paleoenvironmental records from tropical Africa attest to the fact that both vegetation and climate have varied greatly over the last 20 ka [14]. However, it is as yet uncertain what the relative importance of their effects on the geosphere may have been.

Our study seeks to address this important knowledge gap using a scientific drill core from Lake Malawi, southeast Africa (Figure 1). The lake is situated at the current southern extent of the Intertropical Convergence Zone (ITCZ) and has been found to be climatically and ecologically sensitive, making it a remarkable natural laboratory to investigate the relationships and feedbacks among vegetation, climate, weathering, and erosion during the last deglaciation (18–9 ka). This period is of interest because of high-amplitude changes in both climate and vegetation that are both progressive and abrupt. This record thus has the potential to assess the influence of climate change on weathering in the tropics.

Modern Setting

Lake Malawi is the southernmost rift lake in the western branch of the East African Rift System (Figure 1). The physical geography of Lake Malawi and its watershed is controlled by Cenozoic extensional tectonics and volcanism [15]. The lake occupies a series of half-graben basins that are linked *en echelon*, such that steeply dipping border faults with opposing polarity are connected by accommodation zones [16]. Subsidence and deformation patterns within individual half-graben basins control lake bathymetry, with maximum depths (~700 m) achieved adjacent to border faults [17].

The drill core used in this study was collected from the northern basin of the lake (Figure 1). Onshore, distinct topography and drainage patterns follow major tectonic features. A border fault margin to the east of the lake is characterized by high-elevation escarpments which form the Livingstone Mountains. The topography of this faulted margin precludes all but short, steep river systems from forming [18]. In contrast, flexural margins and accommodation zones exhibit comparatively lower altitudes and relief, and large river systems are common in these settings [19]. North of the lake, a well-developed axial delta system building along low-gradient plains is sourced by the Songwe-Kiwira River, which drains the Rungwe highlands. Because of the sloping topography and presence of the delta, Johnson and McCave [20] considered the Songwe-Kiriwa River to be the only significant sediment source to the northern basin. The Songwe-Kiriwa delta has a relatively gentle subaqueous slope (1:70) with a broad shelf, reaching 2–3 km offshore [19]. The drill core site is located ~40 km from the mouth of the river. Although the nearest shoreline is 20 km from the drill site, a sub-lacustrine trough to the east probably prevents significant clastic input from the eastern part of the watershed [20]. Offshore, the lake deepens adjacent to the border fault to ~500 m, and sedimentary processes are dominated by debris and turbidity flows [21].

The smaller watersheds that drain into the lake consist primarily of Neogene alkaline volcanic bedrock, with abundant olivine and alkali basalts, phonolites, trachytes, and nephelinites [22]. Permo-Triassic and Cretaceous sedimentary rocks also crop out north and west of the lake (Figure 1). Soils in the woodlands and forests of the lowlands dominantly consist of pellic vertisols and mollic andosols [23]. The higher-gradient watersheds along the northeastern margin of the basin draining the Livingstone Mountains consist primarily of the Neoproterozoic Mozambique belt, with abundant biotite-horneblende-pyroxene gneiss, charnockites, and minor schists and quartzites [22]. Soils in this area are typically thin and weakly developed, consisting of lithosols, chromic cambisols and dystric regosols [23].

Climate within the watershed is primarily controlled by the yearly passage of the ITCZ, imparting a highly seasonal rainfall regime in which rainfall occurs from November to March. A single, long dry season lasts from April to October when little to no rainfall occurs and prevailing winds are strong southeasterlies [24]. Mean annual precipitation varies from ~800 mm/yr in the lowlands to over 2400 mm/yr in the Rungwe highlands [25]. In addition to a marked N-S gradient of precipitation, the rift escarpments bordering the lake create a pronounced local orographic effect. This results in substantial variability in local rainfall over relatively short spatial scales throughout the watershed.

Vegetation within the watershed is controlled by rainfall and dry season length, although temperature plays a role in the subalpine and alpine zones of the highlands [26–28]. Four principal biome types are observed within the Malawi watershed. In the lowlands (<1500 masl), Zambezian miombo woodlands, a low-diversity deciduous tropical woodland, dominate the landscape. These woodlands primarily comprise *Uapaca*, *Brachystegia*, *Isoberlinia*, *Julbernardia*, and species of Combretaceae and are relatively open, with 30–70% canopy cover. Tropical seasonal forests, closed-canopy semi-deciduous forests— with trees such as *Myrica*, *Macaranga*, Ulmaceae, and Moraceae—are not common in the lowlands; however, they are typically found in areas with locally moist conditions and as riparian corridors along streams and rivers. In the highlands (>1500 masl), afromontane vegetation is found. Today, this region consists of discontinuous patches of afromontane forest separated by high-elevation grasslands. Within the forests, composition is controlled by elevation and rainfall with lower montane forests having moister forest taxa such as *Olea capensis* (1500–2000 mm/yr; 0–3 dry months) from 1500 to 2500 m, whereas *Podocarpus*, *Juniperus*, Ericaceae are common above 2000 m in drier sites (800–1700 mm/yr; ~4 dry months) [27]. The most open biome occurs in the southernmost part of the watershed, where rainfall is the lowest (<800 mm/yr) and contains wooded grasslands with Zambezian affinities.

Methods

Core MAL05-2A was collected from the northern basin of Lake Malawi (10°1.1′S, 34°11.2′E; 359 m water depth) during the Lake Malawi Drilling Project (LMDP) in 2005 (permits issued by Malawi Geological Survey) [29]. Coring site, stratigraphy, and age model details can be found in Figures 1 and 2 as well as in Brown et al. [11] and Scholz et al. [29]. We studied a total of 40 samples, which were taken every ~5 cm from a 3-meter core section (~6–9 meters below lake floor [mblf]). The only gap occurred at ~8.1 mblf, where not enough sediment was available in the core archive to maintain the routine sample interval. The age model of the upper 22 m of the core is based on 24 calibrated accelerator mass spectrometry ^{14}C dates fit with a second-order polynomial

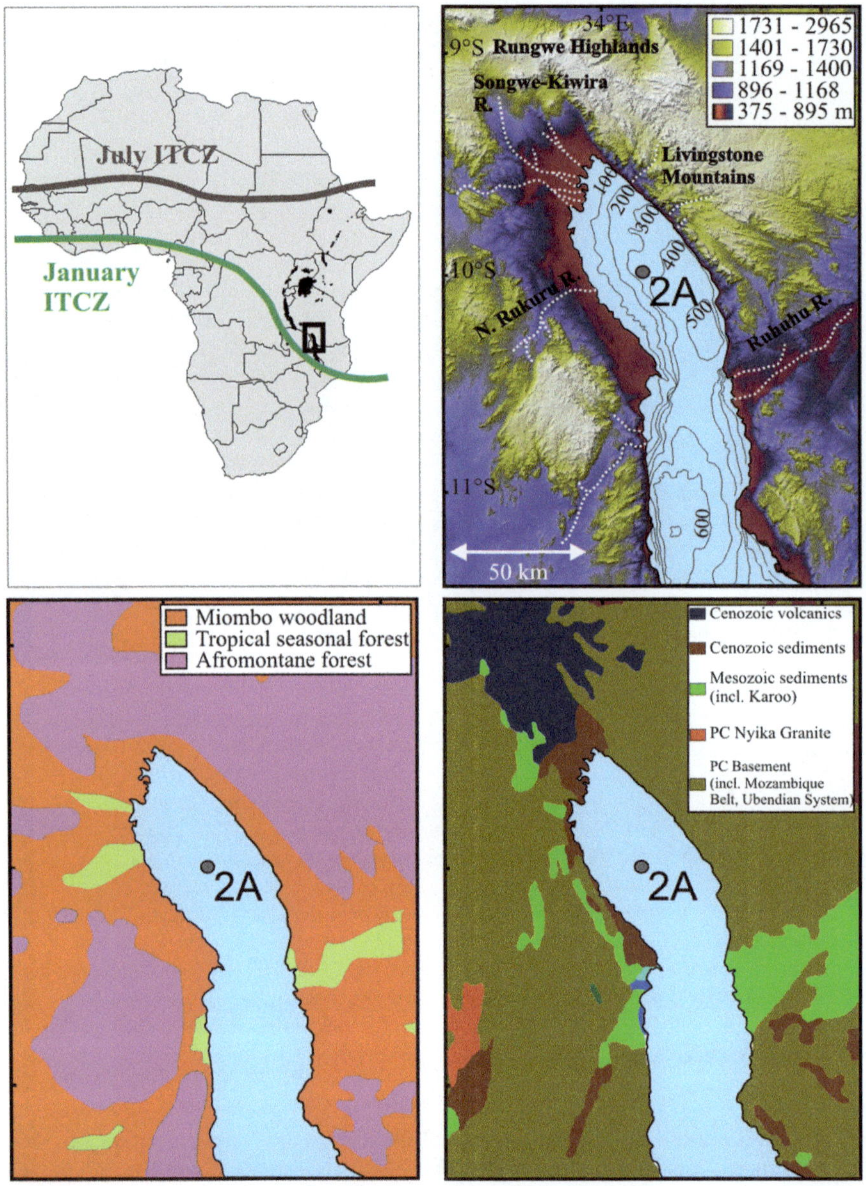

Figure 1. Geography, vegetation, and geology of the study site. (A) showing the July and January positions of the Intertropical Convergence Zone (ITCZ); rectangle represents inset area for watershed maps (modified from Nicholson, 1996). (B) map of the northern basin of Lake Malawi showing topography and bathymetry (isobaths [in meters] modified from Scholz et al., 1989), (C) modern potential vegetation distribution (modified from White, 1983), and (D) bedrock geology (modified from Schlüter, 2006). Scale for panels C and D is the same as for panel B. 2A identifies the location of drill core MAL05-2A.

(calibrated with the "Fairbanks 0107" calibration curve [30] so as to make ages comparable to those of previous studies). The average temporal resolution within the studied section is 208 years, but is ~100 years from 14–11 ka.

Terrigenous mass accumulation rates (TMAR) are a metric of the amount of sediment entering the lake from the watershed and provide insights on the erosion of Lake Malawi's northern watershed. Calculation of TMAR relied on several datasets. Linear sedimentation rates (cm/yr) were calculated using the radiocarbon age model presented in earlier LMDP publications [11]. Dry bulk density was calculated for MAL05-2A sediment horizons using the formulas presented in Dadey et al. [31]. We calculated water content for this purpose and used the gamma-ray density curves from GEOTEK multi-sensor logging of the core [32]. Sediment mass accumulation rates ($g/cm^2/yr$) were calculated by multiplying the linear sedimentation rate and dry bulk density. Weight percent terrigenous sediment was determined from X-ray patterns analyzed using the RockJock quantitative mineralogy computer program [33].

Terrigenous particle size data (sand: >62.5 μm; silt: 3.9–62.5 μm; clay: <3.9 μm), interpreted in conjunction with core lithostratigraphy, provide an indicator of landscape sediment flushing, sub-lacustrine hydrodynamic energy, and depositional processes [e.g., [34]]. Particle size analysis was conducted on the terrigenous fraction of the core MAL05-2A samples using a Malvern laser-diffraction particle size analyzer coupled to a Hydro 2000S dispersion bench (see Methods S1).

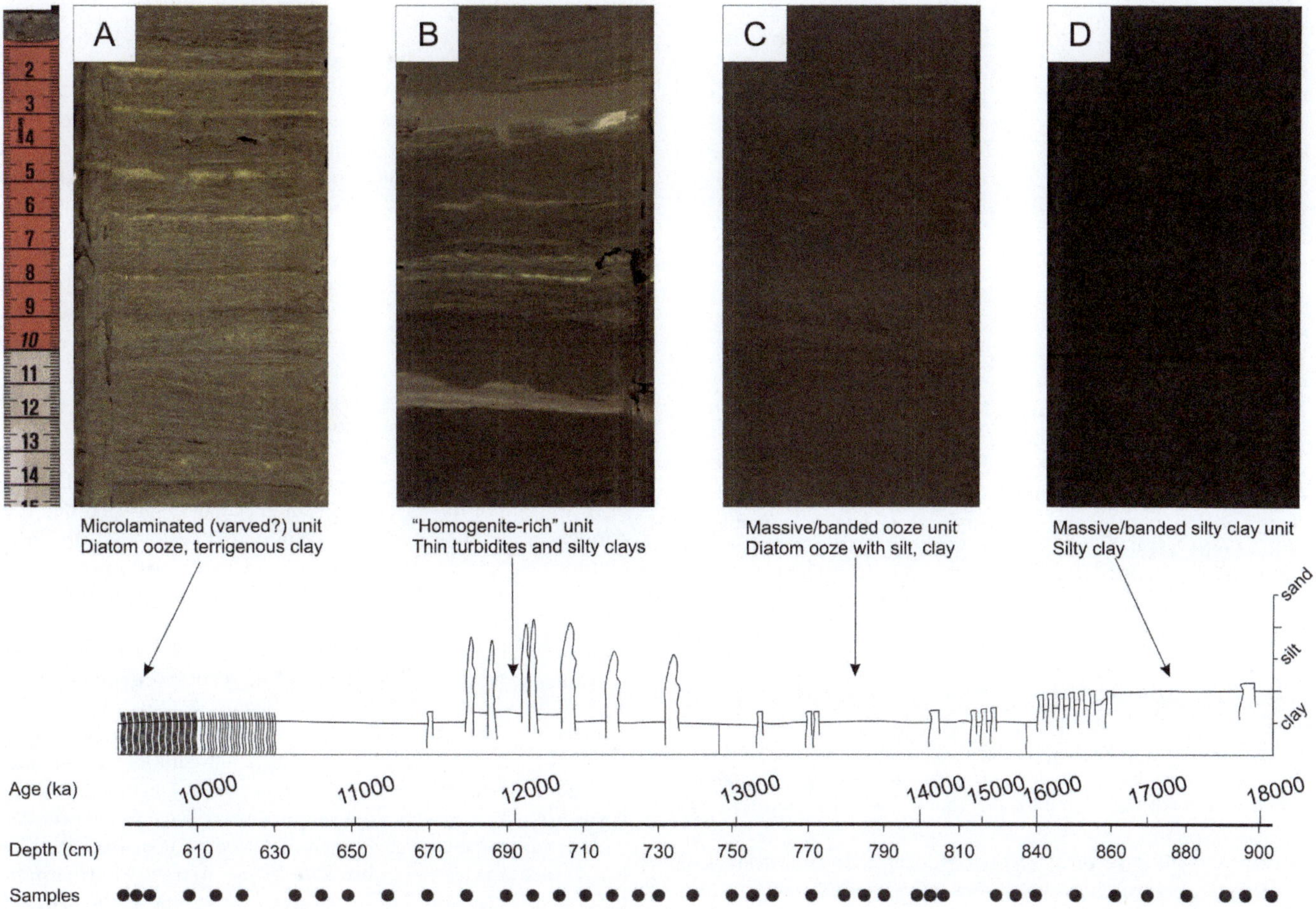

Figure 2. Core MAL05-2A lithostratigraphy within the study interval. Sediment composition and structures vary significantly over the deglacial interval.

Detrital clay minerals in lake sediments provide evidence for the alteration of watershed parent lithologies by physical (disintegration) and chemical (compositional alteration) weathering processes. Quantitative mineralogy was determined using powder X-ray diffraction (XRD) and the computer program RockJock v.11, which has been successfully used for the analysis of fine-grained Cenozoic sediments at a number of locales globally (e.g., [35–37]). For more information on the preparation and processing of these samples, see Methods S1.

In our interpretation of clay mineralogy, we focus on the ratio of kaolinite to smectite (K/S) as a proxy for chemical weathering intensity, which has been used in other similar studies in Africa [38–39]. Kaolinite is produced by the process of leaching in tropical regions marked by high rainfall [40]. In contrast, smectite is typical of tropical semiarid regions and points to increased rainfall seasonality, as it frequently forms during the dry season from the concentration of chemical elements transported to downstream areas by runoff [41–43]. We interpret elevated percentages of micas (illite plus chlorite) to reflect soil formation influenced by physical weathering processes [44].

Pollen samples were taken at the same depths as those for particle size and clay mineralogy and were processed following the standard methods of Faegri and Iverson [45]. More details about sample preparation and pollen identification can be found in Ivory et al. [46]. Vegetation groupings presented in this study are based on biomes of White [27], as well as prior pollen studies within the watershed by Debusk [25] and Vincens et al. [47].

Results

Vegetation

Detailed descriptions of the ecological dynamics surrounding vegetation change in the Malawi watershed during the last deglaciation appear in Ivory et al. [46]. In this study, we present pollen taxa grouped into biomes (afromontane forest, tropical seasonal forest, Zambezian miombo woodland) in order to present the composition and physiognomy of vegetation on the landscape over time. Following the Last Glacial Maximum (LGM) at 18.1–16.4 ka, afromontane forest taxa percentages were relatively stable at around 20% (Figure 3). Their subsequent decline began in a stepwise manner until 14 ka, at which time, the high-elevation arboreal taxa stabilized at percentages of 7 to 10%. Coeval with the decline of montane taxa, percentages of both lowland arboreal vegetation types increased. Beginning at 14 ka, both tropical seasonal forest and miombo woodland increased progressively in abundance until ~11.8 ka from percentages of <5% up to ~13%. Similarly, a progressive increase in grasses is observed in the record from around 30% to over 45% following the decline of afromontane forest; however, grasses once again returned to lower values around 30% relatively abruptly at 13 ka before the increase of the lowland arboreal vegetation types. The lowland arboreal vegetation continued to increase in a similar manner to a maximum value of 13.5% until 11.8 ka when trends in the tropical seasonal forest and miombo woodland diverged. Forest taxa began a slow decline until a minimum of 3.5 to 7% at 10 ka. At this same time, grasses began to increase at the expense of lowland trees with abundances of nearly 60% by 11 ka. Miombo woodland continued to increase until the end of the record at a maximum value of around 15%.

Mineralogy and Particle Size

Qualitative analysis of X-ray diffraction patterns generated from oriented clay mounts indicated the presence of smectite, kaolinite, illite, and minor chlorite in MAL05-2A (for more information, see the Methods S1). These results are in accord with the findings of Kalindekafe et al. [48] and Branchu et al. [49] for the modern clay mineralogy of northern Lake Malawi. Using RockJock, we determined that smectite was the most abundant clay in the deglacial sequence at all times, ranging from 9 to 29%. Kaolinite ranges from 6 to 13%, and illite ranges from 4 to15%. At the end of the LGM, kaolinite:smectite (K/S) within the lake sediments is relatively high, with values varying from 0.55 to 0.70 from 18.1 to 15 ka. The maximum value (0.79) of K/S within the lake is reached at 14.5 ka. After this time, sediment K/S begins a two-step decline with values of 0.55 to 0.65 from 14.5 to 13 ka, followed by highly variable values from 0.45 to 0.70 from 13 to 11.8 ka. It is only at the onset of the early Holocene at 11.8 ka that stable minimum values are reached (0.40 to 0.50). In addition, illite plus chlorite (maximum = 18%) increased sharply for the first time in the record at 11.8 ka and remained at stable values for the rest of the record.

Over the entire record (~18.1 to 9.5 ka), there is a decreasing trend in the ratio of silt to clay (silt/clay) which is marked by a two-step transition towards finer material. In the interval from ~18.1 to 14.6 ka, we observe high relative silt/clay values, with a mean of 1.69. Here, particles in the silt-size class (~4.0 to 62.5 μm) dominated the terrigenous fraction. The form of the silt/clay curve is blocky during this early interval, with prolonged periods of invariant conditions on the order of 1000–2000 years in duration. Values of silt/clay from ~18.1 to 16.9 ka range from ~1.84 to 1.95, whereas values from ~16.9 to 14.6 ka range from 1.48 to 1.59. A transition in the record after ~14.6 ka is marked by the saw-tooth form of the silt/clay curve until ~11.9 ka. During this more variable interval, values of silt/clay range from 0.87 to 1.72 with a mean of 1.21. The highest silt/clay values in this interval correlate with high relative values of sand (up to ~3.0%); this was the only instance very fine sand was detected by our analysis. The second important transition in the record took place after ~11.9 ka, where silt/clay ranges from 0.68 to 1.0, with a mean of ~0.82. Over this interval, particles in the clay-size fraction (<4.0 μm) generally exceeded 50% of the terrigenous fraction.

Overall, the record of TMAR is marked by a major transition from low to relatively high rates after 14.3 ka. From ~18.1 to 14.3 ka, the average was ~0.004 gm/cm^2/yr, and TMAR never exceeded ~0.006 gm/cm^2/yr. The transition after ~14.3 ka was initially marked by an abrupt two-fold increase in TMAR. At a finer scale, the TMAR record shows a stepwise change from ~14.3 to 9.5 ka. The initial step, from ~14.3 to 12.8 ka, was characterized by TMAR values ranging from ~0.007 to 0.012 gm/cm^2/yr. This period is followed by brief decline to lower values from ~12.8 to 11.9 ka, when the average TMAR was ~0.006 gm/cm^2/yr. The second TMAR increase was after ~11.9 ka, with a mean value of 0.012 gm/cm^2/yr until ~10.1 ka.

Interpretation

Dense forest and chemical weathering

At the end of the LGM, paleoclimate records from throughout Africa and at Lake Malawi suggest cooler temperatures and aridity [14,50–51]. Although it might be expected that these environmental conditions would slow reaction rates and mineral transformation during the early deglacial period, this does not seem to be the case in northern Malawi. Throughout this phase, K/S is relatively high in our core record (Figure 3). Higher proportions of kaolinite, a clay mineral leached of mobile cations typically associated with soils of the humid tropics, suggests that chemical weathering was relatively intense. The increased delivery of heavily weathered clays may appear counterintuitive given the comparatively cool temperatures and aridity during this time as recorded in the same core based on TEX-86 and leaf wax $\delta^{13}C$, respectively [50–51]; however, the clay mineralogy record is similar to changes in vegetation during the early deglacial period.

At the end of the LGM, many studies have suggested that East African afromontane forest communities expanded to lower altitudes than at present because of cooler temperatures and low atmospheric CO_2 [52–54]. This was also the case within the Lake Malawi watershed, where afromontane taxa, dominated by *Juniperus*, Ericaceae, and *Podocarpus*, were found in high percentages at the earliest part of our record beginning around 18.1 ka and lasting until ~15.5 ka (Figure 3) [46]. Although it is unlikely that the afromontane forest reached the lakeshore, Ivory et al. [46] suggest that these normally high-altitude communities expanded down to ~900 masl within several kilometers of the lakeshore. Dense forest communities must have been prevalent within the watershed, given the presence of taxa typical of the lower montane moist forest (*Olea* spp., *Myrica*). Furthermore, grass percentages during this interval are very low, the lowest of the record (Figure 3). In modern lake-floor sediment samples, grass percentages within Lake Malawi contribute ~55% of the total pollen [25], supporting the idea of a much denser canopy cover and larger contribution of the arboreal taxa throughout the watershed and into the lowlands. This may suggest that the presence of dense vegetation on the landscape as a result of lowering of the afromontane belt resulted in higher concentrations of organic acids and increased soil moisture, counteracting any slowdown in weathering due to climate.

The first marked change in both the vegetation and sedimentary records began at 15.5 ka, a period which is coeval with the end of Heinrich Stadial 1 (H1) [55]. Within Africa and at Lake Malawi, this period was characterized by progressive warming [50–51]. During this phase from 15.5 to 14.3 ka, K/S reached the highest values of the record (Figure 3). This suggests that, as in the previous phases of afromontane forest dominance, chemical weathering within the northern Malawi watershed remained relatively intense. Although some wetting occurs in East Africa during this post-H1 period, organic geochemical analysis from leaf wax $\delta^{13}C$ implies that Malawi remained relatively arid with respect to modern at this time, despite indications of warming [50–51]. Once again, although chemical weathering reactions would be expected to slow during an arid period, intense weathering continued in the northern Lake Malawi watershed.

Following H1, all of the principal afromontane arboreal taxa began to decline [46]. This decline has been attributed to increasing temperatures which forced afromontane forest to retreat to higher elevations [25]. However, the transition was stepwise. Although the principal afromontane forest taxa that were dominant just following the LGM (Ericaceae, *Juniperus*, and *Podocarpus*) began to decline, from 15.5 to 14.3 ka, the dense

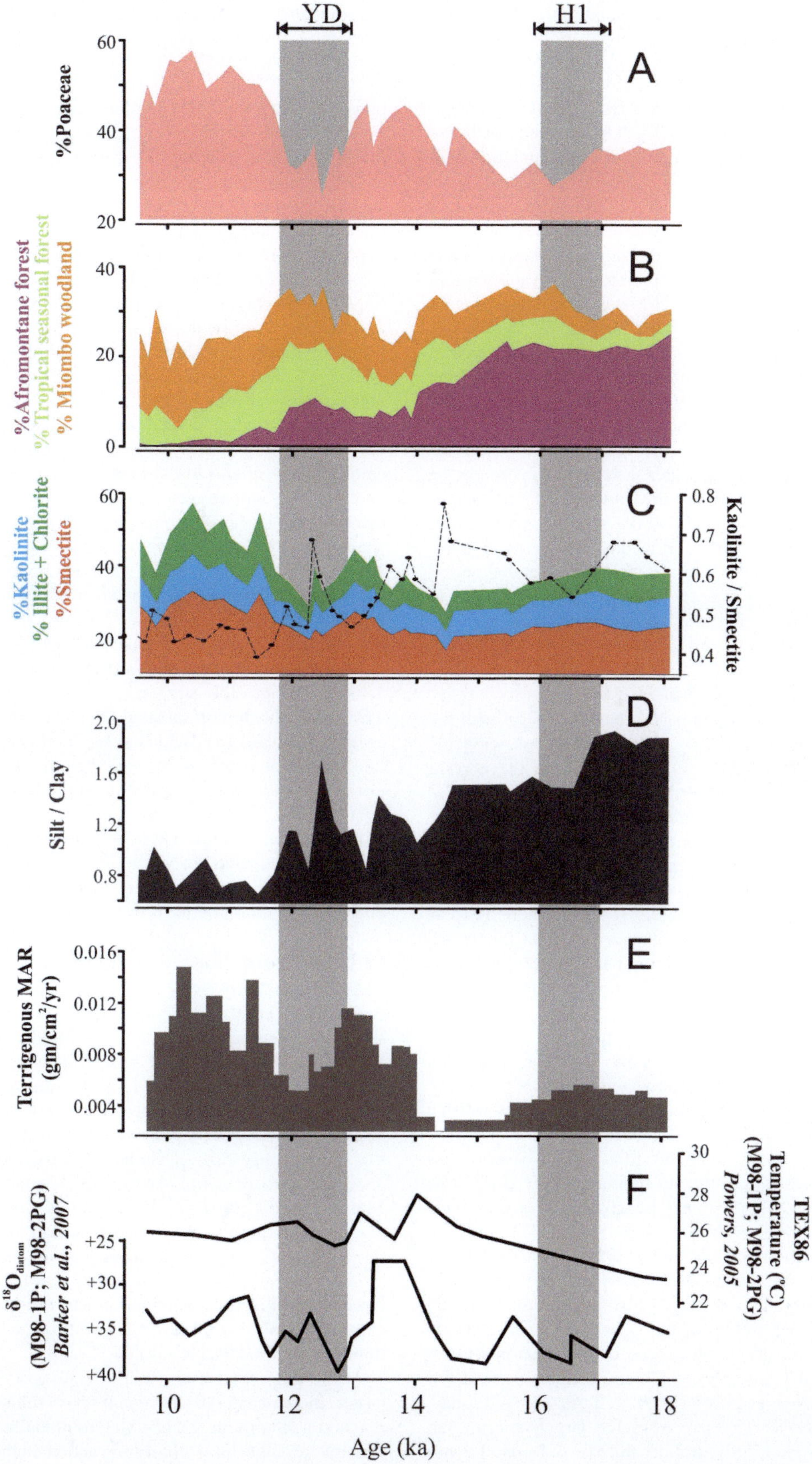

Figure 3. Vegetation and weathering indicators from drill core MAL05-2A.YD, Younger Dryas. H1, Heinrich event 1. (A) pollen percentages of Poaceae and (B) arboreal vegetation types (afromontane forest, tropical seasonal forest, miombo woodland), (C) XRD clay mineral percentages of smectite, kaolinite, and illite+chlorite, ratio of kaolinite to smectite (K/S), (D) ratio of silt to clay (silt/clay), and (E) terrigenous mass

accumulation rate, and (F) $\delta^{18}O$ from diatoms and TEX-86 temperature from piston cores M98-1P and M98-2PG from the northern basin of Lake Malawi from Barker et al. (2007) and Powers (2005), respectively.

lower montane moist forest taxa, such as *Myrica* and *Olea*, known pioneer taxa, continued to expand [46]. The presence of dense forest in the lowlands is the best explanation for the continued high values of K/S at this time.

Throughout the early deglacial period, minimum values of TMAR suggest that erosion was low (Figure 3). Johnson and McCave [20] suggest that TMAR in the northern basin of Lake Malawi is indicative of moisture, and therefore low values would suggest a reduced riverine transport. However, our dataset makes clear that a more likely mechanism for abating the flushing of weathered parent material from the landscape is the presence of dense lowland forests. During the LGM, a slight lake-level regression at Malawi of ~40 to 100 m would have decreased the distance from the mouth of the Songwe-Kiriwa River and other important drainages to the core site [56]. A basinward progradation of the delta during a lake-level lowstand might be expected to show an increase in the amount of terrigenous material deposited; however, we observe the opposite. Thus, despite noted cooler temperature and aridity relative to modern, intense weathering and low erosion are observed over this period by K/S and TMAR. This regime strongly suggests that dense moist forest in the watershed had a strong control on clay transformation and delivery of terrigenous siliciclastic material to the basin.

Transition to the Holocene (14.3 to 11.8 ka)

At 14.3 ka, a transitional period began that was coeval with the onset of the Bolling-Allerod (BA) [57]. In contrast with the predominantly silty detritus in our record lower in the core, a broad range of grain sizes produces highly variable values of silt/clay (1.8 to 0.6) during this time (Figure 3). Over this interval, lake levels were highly variable. Evidence from diatom assemblages implies high-frequency changes on the order of 50 to 100 m [56]. Sand is present in the samples from this interval, but only in significant percentages in one sample. However, there is clear evidence of turbidity flows here as early as ~13.2 ka (Figure 2). The presence of time-equivalent turbidites has also been described in other cores in northern Lake Malawi ("homogenites" of Barry et al. [58]). Thus, the variable silt content is expected to be caused by a change in sub-lacustrine depositional processes and energy. Thin, dominantly fine-grained turbidites may have been produced by wave remobilization of distal deltaic sediments, which move downslope at low velocity under the force of gravity as benthic nepheloid plumes [20].

At this time, the moist montane forest taxa became less abundant in the lowlands, and the afromontane forest as a biome reached very low percentages values, suggesting trees may have retreated to higher elevations by this time (Figure 3). Within the lowlands, however, arboreal taxa of the tropical seasonal forest began to increase in abundance. Miombo woodland abundances began increasing as well; however, the response time is much slower than the other arboreal taxa and values remained relatively low. Although the lowland closed-canopy tropical seasonal forest began expanding at this time, the percentages remained at <10%, values which are not significantly larger than those observed today, suggesting that dense forests were restricted largely to riparian corridors [25]. Ivory et al. [46] attribute this rise of lowland dense forest to the higher temperatures and reinforced monsoon recorded at the time in the basin. The lack of other arboreal vegetation on the landscape away from river banks in the lowlands indicates a marked change of physiognomy from the previous phase following the LGM to more open vegetation. This openness in the lowlands is reflected in the high percentages of grass pollen during this phase (Figure 3).

Within the basin, the increase in grass is tracked by rising TMAR values in core MAL05-2A (Figure 3). The increase in terrigenous material as the landscape opens supports the argument that TMAR represents erosion from the hinterland into the basin when vegetation is open and forests are scarce. This suggests that the relatively open character of vegetation caused by the retreat of montane forests prevented water storage on the landscape and promoted flushing of clastic material into the basin.

The presence of turbidites makes the clay record in this interval more complex. From 14.3 to 12.7 ka, an overall declining trend in K/S is observed, followed by an abrupt increase at 12.7 to 12.3 ka. Dense tropical seasonal forest and miombo woodland continued to expand in the lowlands at Lake Malawi between 13 and 11.7 ka, and tropical seasonal forest reached its maximum extent at the end of this interval (Figure 3). The beginning of this zone is coeval with the inception of the Younger Dryas (YD) in the Northern Hemisphere [59]. Although many other sites in East Africa record a retreat or slowdown of deglacial forest expansion during this time due to increased aridity, the dense lowland forest became more prevalent at Lake Malawi [60-64].

This indication of dense forest in conjunction with higher K/S values could possibly suggest a return to more intense chemical weathering beginning around 12.7 ka. However, independent particle size evidence, which shows elevated silt/clay and minor sand in the core during the period from 12.7 to 12.3 ka, suggests instead that the elevated K/S may be related to minor reworking by turbidites. Therefore, we interpret the multiple sedimentary indicators from this interval to be influenced by mass wasting. This may be the result of flushing from enhanced precipitation promoting sediment transport (and hyperpycnal sub-lacustrine flows) rather than chemical weathering.

Early Holocene open woodlands

Although records from farther north in tropical and subtropical Africa show an abrupt resumption of the African monsoon and significant wetting during the early Holocene, including at Lake Rukwa (~200 km north of Lake Malawi), the situation at Lake Malawi is slightly different [14,65]. Organic geochemical studies of $\delta^{13}C$ agree that the watershed was wetter than during the late Pleistocene; however, a ~100 m lake-level regression occurred at this time, suggesting a change in hydrology [66–67]. Pollen analysis on a core from Lake Masoko, a small maar lake within the Malawi watershed, suggests that the expansion of open miombo woodlands indicates high summer rainfall, but heightened rainfall seasonality [47]. This conclusion is in agreement with Zr:Ti and biomarker studies from Lake Malawi, which demonstrate wind direction and air mass changes at this time consistent with major reorganization of tropical circulation and a more northerly ITCZ than during the late Pleistocene [11,68].

At this time, total abundance of clays was very high, resulting in a net increase in all dominant clay minerals; however, K/S reached a minimum (Figure 3). Additionally, for the first time in this record, illite plus chlorite reached higher abundances than kaolinite (Figure 3). The dominance of detrital clays retaining mobile cations suggests a very strong decrease in the intensity of chemical weathering beginning at 11.8 ka. Furthermore, the importance of illite, a clay mineral that is only present in low abundances in the modern lake today, is suggestive of increasing physical weathering and reworking during the early Holocene despite increased wetness [48]. In addition, TMAR during this interval reaches maximum values. The agreement of these two independent datasets seems to indicate peak erosion from 11.8 to 10 ka.

This transition in clay mineralogy at 11.8 ka is coeval with the opening of the lowland vegetation (Figure 3). At 11.8 ka, near the onset of the Holocene, an abrupt transition occurred from denser lowland forests to more open miombo woodland in the watershed (Figure 3). Additionally, for the first time in the record, afromontane forest tree pollen is nearly absent. Given the high abundance of woodland and grasses, it is likely that this early Holocene transition represented that largest change in vegetation physiognomy in Malawi during the studied interval.

This increase in erosion was likely caused by the more intense summer rains of the African monsoon and the open character vegetation, conditioning the landscape for the flushing of physically weathered material into the basin. Cecil and Edgar [69] noted a similar relationship between greater coarse siliciclastic sediment transport and strongly seasonal rainfall.

Discussion

Traditional models for the tropics suggest strong climatic controls such as precipitation and temperature on chemical and physical weathering and erosion [44,70]. However, our record from Lake Malawi suggests a more indirect role of climate. During the early part of the last deglaciation, climate within the watershed varied dramatically in mean state and variability, including cooler temperatures and relative aridity from 18.1 to 14.5 ka. This was followed by successive wetting and warming from 14.5 to 12.5 ka, a return to slightly drier conditions with much reduced rainfall seasonality during the YD, and finally an increase in rainfall seasonality at the time of the resumption of the African monsoon at the early Holocene. However, despite the changes in climate, both highland and lowland forests were common in the watershed throughout this time until 11.8 ka. Although changes in rainfall and temperature lead to compositional vegetation changes during this period as high-altitude forest was replaced successively by lowland forest, the physiognomy of the landscape, with dominance of arboreal taxa, remained until the early Holocene. Similarly, although chemical weathering should be more intense during the moister periods, instead we see relative stability in the clay minerals with higher values of kaolinite throughout this period regardless of climatic regime. During the early Holocene, the physiognomic change to more open woodland is coeval with higher mean annual precipitation because of heightened seasonality following the resumption of the African monsoon. It might be expected that this high temperature and high rainfall should have resulted in intense chemical weathering, but the available data do not support this idea. At this time, kaolinite decreases, and high abundance of illite suggests not only less-intense chemical weathering, but higher physical weathering and transport of less-weathered micas into the basin.

The strong, direct control of vegetation on mineral transformation and erosion is not surprising for several reasons as many processes mediated by plants influence chemical and physical breakdown of bedrock. First, although heat and moisture are necessary for driving chemical breakdown of aluminosilicate rocks, organic acids, particularly low-weight organic acids such as those produced by plants and mycorrhizae associated with plant roots, are essential for biological mediation of weathering in soils [71–72]. These acids not only provide H+ for the transformation of primary materials, but also act as ligands that form strong complexes with trivalent cations such as Al^{3+} and Fe^{3+} [6]. This function of organic acids has the potential to dramatically intensify weathering rates by increasing the solubility of minerals containing these elements. Second, root networks, particularly in dense, old-growth forests, are critical for the mechanical breakdown of bedrock [73]. Furthermore, the removal of trees results in an increased response of erosion rates and closer control of abiotic erosion mechanisms as seen in our record.

Perhaps the most important biotic control on weathering is the role of forests and large forest trees in regulating the hydrologic cycle locally within a watershed and at the scale of a soil [74]. In dense forests, canopies intercept rainfall. During intense monsoon rains, this function acts to reduce runoff and increase storage on the landscape, increasing the residence time of moisture available for chemical reactions in soils. Additionally, evapotranspiration within a forest often creates a microclimate that results in higher local precipitation over dark, dense trees as well as more stored water on the landscape and cooler temperatures from evaporation [75]. More importantly however, evapotranspiration controlled by forest trees limits the drainage depth within the weathering profile of a soil, reducing physical weathering processes and allowing time for chemical alteration [76]. Vegetation transitions, as we see in the Malawi record, however, are key as long-lived dense forests, such as those found in the Congo Basin or in the Amazon exhibit very little chemical weathering [77–78]. Early stages of succession, such as the period of intense chemical weathering observed during the deglaciation following the retreat of the afromontane forest as lowland forest colonized the lake shore, are in fact typically the period of most intense chemical weathering as large trees colonize previously unweathered substrate [79–80].

TMAR values show a positive correlation with grass abundances throughout our record. Although TMAR has been interpreted as an indication of moisture availability, we instead suggest that delivery of terrigenous material to the basin is strongly regulated by vegetation on the landscape. This means that more open, grassy woodlands are less effective at trapping material on the landscape leading to greater erosion and deposition of terrigenous material than when forests are dominant in the lowlands [12,81]. This first order control of erosion on the landscape is in agreement with cosmogenic nuclide studies in the Alps by Vanacker et al. [82] who show exponential increases in denudation rates due to modern land use change. Although this suggests rapid landscape conversion once forest is gone, this study, as well as that of Vanacker et al. [82], suggests that once forest returns, stabilization is possible as denudation rates return to background levels. Reducing the residence time of water on the landscape likewise influences chemical weathering intensity, and we note that the inverse correlation of K/S with TMAR and grass percentages, which appears to support this hypothesis.

In our analysis of core records from Lake Malawi, we have observed coarser sediments (high silt/clay or % sand) around the inception of lake-level highstands, which also show high ratios of trees to herb pollen. Several factors may influence this particle size trend. For example, storage release (i.e., flushing) of coarser sediment from low-to-moderate gradient flexural and axial margins concomitant with transgression likely helps to explain some of the variability we have observed. Additionally, increased fluvial discharge implied by the presence of afromontane forest is consistent with what is known about the efficacy of coarse siliciclastic transport [34]. Another potential influence relates to the shortening of fluvial transport networks associated with the higher base level (i.e., less reduction of siliciclastic particle size by transport abrasion and downstream fining), but because fining upward patterns are characteristic of the highstand strata we have examined, the absolute influence of this mechanism is not well known. By contrast, during lowstand intervals such as the early Holocene, when dry season length was long, fluvial networks lengthen and appear to generate finer siliciclastic detritus. Physical weathering of parent rocks and soils, captured in the clay mineral record, may serve as a feedback on reducing particle size during these intervals. Indeed, the lack of forest cover to intercept rainfall during these intervals may also serve to help flush fine sediment from hillslopes and floodplains into river channels. Variability in depositional processes can complicate the interpretation of silt/clay, which is why evidence of mass wasting is critical to identify in the lithostratigraphy. For example, the transition from lowstand to highstand in rift lakes occurs rapidly (10^2 to 10^5 yrs) [83], and as such flexural margins can become prone to gravity flows as pore pressures change and slopes destabilize. Therefore, our study employed particle size and stratigraphic data in support of pollen, clay minerals, and TMAR to provide the most rigorous assessment of weathering and erosion patterns for the deglacial period yet developed for this region of East Africa.

Conclusions

Vegetation composition and structure at Lake Malawi and elsewhere does not unequivocally track simple precipitation amount [28,46–47]. Our record attests to a strong negative relationship between increased rainfall seasonality and vegetation density that results in a specific depositional signature within the lake. When dense forests occupy the watershed, chemical weathering is intense and erosion is low. Leaching within soils leads to generation of highly altered clay minerals like kaolinite. However, during times of open vegetation, chemical weathering is less intense, but erosion is increased. Smectite dominates during these periods, but flashy precipitation on the open landscape leads to flushing of siliciclastics into the lake and thus high relative TMAR.

Finally, these results have important implications for better understanding how weathering and erosion may change in the future. Although little work has been done to quantify potential future alterations in weathering, most models do not take ecosystem change related to climatic change into account, leading to significant potential biases. Only in the last few years has vegetation been added to projections of soil weathering, thus leading to the suggestion of high sensitivity of weathering to future climate states [84–86]. Although paleostudies hint at the role of biological mediation in weathering over long time scales, no other studies have used paleo-records to quantify the effects of vegetation on weathering processes in southeast Africa [7]. This study points to the importance of vegetation for mediating weathering in the past; however, more and longer records are needed in order to better quantify this effect over a larger range of climatic and vegetation variability in watersheds throughout the tropics.

Supporting Information

Table S1 Pollen counts from core MAL05-2A.

Table S2 Particle size and clay mineralogy from core MAL05-2A.

Methods S1 Materials and methods details of particle size and x-ray diffraction analysis.

Acknowledgments

Disclaimer: Any use of trade, product, or firm names is for descriptive purposes only and does not imply endorsement by the U.S. Government.

Thanks to Jean-Pierre Cazet, Guillaume Buchet, Owen Davis, and John Logan for help with sample processing and pollen identification, William Benzel and Adam Boehlke for assistance with x-ray diffraction, and the National Lacustrine Core Repository (LacCore) at the University of Minnesota for sub-sampling and core curation. We would also like to thank three reviewers for helpful comments that improved the manuscript.

Author Contributions

Conceived and designed the experiments: SJI MMM. Performed the experiments: SJI MMM. Analyzed the data: SJI MMM. Contributed reagents/materials/analysis tools: SJI MMM GE AC AML AV. Wrote the paper: SJI MMM.

References

1. IPCC (2013) Climate Change 2013: The Physical Science Basis. Contribution of Working Group I to the Fifth Assessment Report of the Intergovernmental Panel on Climate Change [Stocker TF, Qin D, Plattner GK, Tignor M, Allen SK, et al. (eds.)]. Cambridge University Press, Cambridge, United Kingdom and New York, NY, USA, 1535 pp.
2. Brantley SL, Megonigal JP, Scatena FN, Balogh-Brunstad Z, Barnes RT, et al. (2011) Twelve testable hypotheses on the geobiology of weathering. Geobiology 9: 140–165.
3. Walker JC, Hays PB, Kasting JF (1981) A negative feedback mechanism for the long-term stabilization of Earth's surface temperature. J Geophys Res: Oceans 86: 9776–9782.
4. Langbein WB, Schumm SA (1958) Yield of sediment in relation to mean annual precipitation. Eos 39: 1076–1084.
5. Egli M, Mirabella A, Sartori G (2008) The role of climate and vegetation in weathering and clay mineral formation in late Quaternary soils of the Swiss and Italian Alps. Geomorphology 102: 307–324.
6. Goddéris Y, Roelandt C, Schott J, Pierret M.C, François LM (2009) Towards an integrated model of weathering, climate, and biospheric processes. Rev Mineral Geochem 70: 411–434.
7. Dosseto A, Hesse PP, Maher K, Fryirs K, Turner S (2010) Climatic and vegetation control on sediment dynamics during the last glacial cycle. Geology 38: 395–398.
8. Einsele G, Hinderer M (1998) Quantifying denudation and sediment-accumulation systems (open and closed lakes): basic concepts and first results. Palaeogeogr Palaeoclimatol Palaeoecol 140: 7–21.
9. Sémah AM, Sémah F, Moudrikah R, Fröhlich F, Djubiantono T (2004) A late Pleistocene and Holocene sedimentary record in Central Java and its palaeoclimatic significance. Modern Quaternary Research in SE Asia, Balkema 19: 63–88.
10. Felton AA, Russell JM, Cohen AS, Baker ME, Chesley JT, et al. (2007) Paleolimnological evidence for the onset and termination of glacial aridity from Lake Tanganyika, Tropical East Africa. Palaeogeogr Palaeoclimatol Palaeoecol 252: 405–423.
11. Brown ET, Johnson TC, Scholz CA, Cohen AS, King JW (2007) Abrupt change in tropical African climate linked to the bipolar seesaw over the past 55,000 years. Geophys Res Lett 34.
12. Dunne T (1979) Sediment yield and land use in tropical catchments. J Hydrol 42: 281–300.
13. Dixon RK, Solomon AM, Brown S, Houghton RA, Trexier MC, et al. (1994) Carbon Pools and Flux of Global Forest Ecosystems. Science 263: 185–190.
14. Gasse F (2000) Hydrological changes in the African tropics since the Last Glacial Maximum. Quat Sci Rev 19: 189–211.
15. Ebinger CJ (1989) Tectonic development of the western branch of the East African rift system. Geol Soc Am Bull 101: 885–903.
16. Rosendahl BR (1987) Architecture of continental rifts with special reference to East Africa. Annu Rev Earth Planet Sci 15: 445–503.
17. Scholz CA, Rosendahl BR, Versfelt JW, Kaczmarick J, Woods LD (1989) Seismic Atlas of Lake Malawi (Nyasa), East Africa. Project PROBE Geophysical Atlas Series, Vol. 2. Duke University, Durham, NC.
18. Soreghan MJ, Scholz CA, Wells JT (1999) Coarse-grained, deep-water sedimentation along a border fault margin of Lake Malawi, Africa: seismic stratigraphic analyses. J Sediment Res 69: 832–846.
19. Scholz CA (1995) Deltas of the Lake Malawi rift, East Africa: seismic expression and exploration implications. AAPG Bull 79: 1679–1697.

20. Johnson T, McCave I (2008) Transport mechanism and paleoclimatic significance of terrigenous silt deposited in varved sediments of an African rift lake, Limnol Oceanogr 5: 1622–1632.
21. Scott DL, Ng'ang'a P, Johnson TC, Rosendahl BR (1991) High-resolution acoustic character of Lake Malawi (Nyasa), East Africa, and its relationship to sedimentary processes. Special Publication of the International Association of Sedimentologists 13: 129–145.
22. Schlüter T (2006) Geological Atlas of Africa. Springer, Berlin. 272 pp.
23. FAO (1988) Soil Map of the World, Revised legend, World Resources Report, 60, FAO, Rome, Italy.
24. Malawi Government (1983) The National Atlas of Malawi. National Atlas Committee and Department of Surveys of Malawi. Natl. Atlas Comm. Dep. Surv. Malawi.
25. DeBusk GH (1994) Transport and stratigraphy of pollen in Lake Malawi, Africa. PhD Thesis, Duke University.
26. Polhill RM (1966) Flora of Tropical East Africa. In: Hubbard, C.E., Milne-Redhead, E. (Eds.), Crown Agents for Oversea Governments and Administrations, London.
27. White F (1983) Vegetation of Africa—a descriptive memoir to accompany the Unesco/ AETFAT/UNSO vegetation map of Africa. Natural Resources Research Report XX. UNESCO, Paris, France.
28. Hély C, Bremond L, Alleaume S, Smith B, Sykes M, Guiot J (2006) Sensitivity of African biomes to changes in the precipitation regime. Global Ecol Biogeogr 15: 258–270.
29. Scholz CA, Johnson TC, Cohen AS, King J, Peck J, et al. (2007) East African megadroughts between 135 and 75 thousand years ago and bearing on early-modern human origins. PNAS 104: 16416–16421.
30. Fairbanks R, Mortlock R, Chiu T, Cao L, Kaplan A, et al. (2005) Radiocarbon calibration curve spanning 0 to 50,000 years BP based on paired 230Th/234U/238U and 14C dates on pristine corals. Quat Sci Rev 25: 1781–1796.
31. Dadey KA, Janecek T, Klaus A (1992) Dry-bulk density: its use and determination. In Proceedings of the Ocean Drilling Program, Scientific Results (Vol. 126, pp. 551–554) College Station, TX: Ocean Drilling Program.
32. Lyons WB, Carey AE, Hicks DM, Nezat CA (2005) Chemical weathering in high-sediment yielding watersheds, New Zealand. J Geophys Res: Earth Surface (2003–2012) 110.
33. Eberl DD (2003) User's guide to RockJock – A program for determining quantitative mineralogy from powder X-ray diffraction data: U.S. Geological Survey Open-File Report 2003–78, 47 p.
34. Heins Keiro (2007), in Sedimentary Provenance and Petrogenesis: Perspectives from petrography and geochemistry, in Arribas, J, Critelli, S. and Johnson, M., eds., GSA Special paper 420, p. 345–379
35. Andrews JT (2008) The role of the Iceland Ice Sheet in the North Atlantic during the late Quaternary: a review and evidence from Denmark Strait. J Quat Sci 23: 3–20.
36. Refsnider KA (2010) Dramatic increase in late Cenozoic alpine erosion rates recorded by cave sediment in the southern Rocky Mountains. Earth Planet Sci Lett 297: 505–511.
37. Dühnforth M, Anderson RS, Ward D J, Blum A (2012) Unsteady late Pleistocene incision of streams bounding the Colorado Front Range from measurements of meteoric and in situ 10Be. J Geophys Res: Earth Surface (2003–2012) 117.
38. Pastouret L, Chamley H, Delibrias G, Duplessy JC, Thiede J (1978) Late Quaternary climatic changes in western tropical Africa deduced from deep-sea sedimentation off Niger Delta. Oceanol Acta 1: 217–232.
39. Lézine AM, Duplessy JC, Cazet JP (2005) West African monsoon variability during the last deglaciation and the Holocene: Evidence from fresh water algae, pollen and isotope data from core KW31, Gulf of Guinea. Palaeogeogr Palaeoclimatol Palaeoecol 219: 225–237.
40. Birkeland PW, (1984) Soils and Geomorphology. Oxford University Press, Oxford. 310 pp.
41. Chamley H (1989) Clay sedimentology (Vol. 623) New York: Springer-Verlag. 623 pp.
42. Weaver CE (1989) Clays, muds, and shales. Elsevier. 818 pp.
43. Alizai A, Hillier S, Clift PD, Giosan L, Hurst A, et al. (2012) Clay mineral variations in Holocene terrestrial sediments from the Indus Basin. Quat Res 77: 368–381.
44. Thiry M (2000) Palaeoclimatic interpretation of clay minerals in marine deposits; an outlook from the continental origin. Earth Sci Rev 49: 201–221
45. Faegri K, Iversen J (1989) Textbook of Pollen Analysis, Wiley, Chichester, UK. 328 pp.
46. Ivory S, Lézine AM, Vincens A, Cohen A (2012) Effect of aridity and rainfall seasonality on vegetation in the southern tropics of East Africa during the Pleistocene/Holocene transition, Quat Res 77: 77–76.
47. Vincens A, Garcin Y, Buchet G (2007) Influence of rainfall seasonality on African lowland vegetation during the late Quaternary: pollen evidence from Lake Masoko, Tanzania. J Biogeog 34: 1274–1288.
48. Kalindekafe LN, Dolozi MB, Yuretich R (1996) Distribution and origin of clay minerals in the sediments of Lake Malawi, in Johnson, T.C., and Odada, E., eds., The Limnology, Climatology and Paleoclimatology of the East African Lakes: Amsterdam, Gordon and Breach, p. 443–460.
49. Branchu P, Bergonzini L, Delvaux D, De Batist M, Golubev V, et al. (2005) Tectonic, climatic and hydrothermal control on sedimentation and water chemistry of northern Lake Malawi (Nyasa), Tanzania. J African Earth Sci 43: 433–446.
50. Powers LA, Johnson TC, Werne JP, Castañeda I, Hopmans E, et al. (2005) Large temperature variability in the southern African tropics since the Last Glacial Maximum. Geophys Res Lett 32: L08706.
51. Castañeda IS, Werne JP, Johnson TC (2007) Wet and arid phases in the southeast African tropics since the Last Glacial Maximum. Geology 35: 823–826.
52. Street-Perrott FA, Huang Y, Perrott RA, Eglinton G, Barker P, et al. (1997) Impact of lower atmospheric carbon dioxide on tropical mountain ecosystems. Science 278: 1422–1426.
53. Wu H, Guiot J, Brewer S, Guo Z (2007) Climatic changes in Eurasia and Africa at the last glacial maximum and mid-Holocene: reconstruction from pollen data using inverse vegetation modelling. Clim Dyn 29: 211–229.
54. Woltering M, Johnson TC, Werne JP, Schouten S, Sinninghe Damsté JS (2011) Late Pleistocene temperature history of southeast Africa: A TEX86 temperature record from Lake Malawi. Palaeogeogr Palaeoclimatol Palaeoecol 303: 93–102.
55. Hemming SR (2004) Heinrich events: Massive late Pleistocene detritus layers of the North Atlantic and their global climate imprint. Rev Geophys 42.
56. Stone JR, Westover KS, Cohen AS (2011) Late Pleistocene paleohydrography and diatom paleoecology of the central basin of Lake Malawi, Africa. Palaeogeogr Palaeoclimatol Palaeoecol 303: 51–70.
57. Wohlfarth B (1996) The chronology of the last termination: A review of radiocarbon date, high resolution terrestrial stratographies. Quat Sci Rev 15: 267–284.
58. Barry SL, Filippi ML, Talbot MR, Johnson TC (2002) Sedimentology and geochronology of late Pleistocene and Holocene sediments from northern Lake Malawi. In The East African Great Lakes: Limnology, Palaeolimnology and Biodiversity, Springer Netherlands, 369–391.
59. Alley RB, Meese DA, Shuman CA, Gow AJ, Taylor KC, et al. (1993) Abrupt increase in Greenland snow accumulation at the end of the Younger Dryas event. Nature 362: 527–527.
60. Vincens A (1991) Late quaternary vegetation history of the South-Tanganyika basin. Climatic implications in south central Africa. Palaeogeogr Palaeoclimatol Palaeoecol 86: 207–226.
61. Vincens A (1993) Nouvelle sequence pollinique du Lac Tanganyika: 30,000 ans d'histoire botanique et climatique du Bassin Nord. Rev Palaeobot Palyno 78: 381–394.
62. Bonnefille R, Riollet G, Buchet G, Icole M, Lafont R, et al. (1995) Glacial-interglacial record from intertropical Africa, high resolution pollen and carbon data at Rusaka, Burundi. Quat Sci Rev 14: 917–936.
63. Beuning KR, Talbot MR, Kelts K (1997) A revised 30,000-year paleoclimatic and paleohydrologic history of Lake Albert, East Africa. Palaeogeography, Palaeoclimatology, Palaeoecology 136: 259–279.
64. Ryner M, Gasse F, Rumes B, Verschuren D (2007) Climatic and hydrological instability in semi-arid equatorial East Africa during the late glacial to Holocene transition: a multi-proxy reconstruction of aquatic ecosystem response in northern Tanzania. Palaeogeogr Palaeoclimatol Palaeoecol 248: 440–458.
65. Cohen AS, Van Bocxlaer B, Todd JA, McGlue M, Michel E, et al. (2013) Quaternary ostracodes and molluscs from the Rukwa Basin (Tanzania) and their evolutionary and paleobiogeographic implications. Palaeogeogr Palaeoclimatol Palaeoecol 392: 79–97.
66. Johnson TC, Brown ET, McManus J, Barry S, Barker P, et al. (2002) A high-resolution paleoclimate record spanning the past 25,000 years in southern East Africa. Science 296: 113–132.
67. Barker PA, Leng MJ, Gasse F, Huang Y (2007) Century-to-millennial scale climatic variability in Lake Malawi revealed by isotope records. Earth Planet Sci Lett 261: 93–103.
68. Konecky BL, Russell JM, Johnson TC, Brown ET, Berke MA, et al. (2011) Atmospheric circulation patterns during late Pleistocene climate changes at Lake Malawi, Africa. Earth Planet Sci Lett 312: 318–326.
69. Cecil CB, Edgar NT (2003) Climate Controls on Stratigraphy: Society for Sedimentary Geology Special Publication 77, 275 p.
70. Burnett AP, Soreghan MJ, Scholz CA, Brown ET (2011) Tropical East African climate change and its relation to global climate: a record from Lake Tanganyika, Tropical East Africa, over the past 90+ kyr. Palaeogeogr Palaeoclimatol Palaeoecol 303: 155–167.
71. Landeweert R, Hoffland E, Finlay RD, Kuyper TW, van Breemen N (2001) Linking plants to rocks: ectomycorrhizal fungi mobilize nutrients from minerals. Trends Ecol Evol 16: 248–254.
72. Bonneville S, Smits MM, Brown A, Harrington J, Leake JR, et al. (2009) Plant-driven fungal weathering: early stages of mineral alteration at the nanometer scale. Geology 37: 615–618.
73. Roering JJ, Marshall J, Booth AM, Mort M, Jin Q (2010) Evidence for biotic controls on topography and soil production. Earth Planet Sci Lett 298: 183–190.
74. Rodríguez-Iturbe I, Porporato A (2004) Ecohydrology of water-controlled ecosystems. In Soil Moisture and Plant Dynamics.
75. Spracklen DV, Arnold SR, Taylor CM (2012) Observations of increased tropical rainfall preceded by air passage over forests. Nature 489: 282–285.
76. Roelandt C, Goddéris Y, Bonnet MP, Sondag F (2010) Coupled modeling of biospheric and chemical weathering processes at the continental scale. Global Biogeochem Cy 24: GB2004.

77. Gaillardet J, Dupré B, Louvat P, Allegre CJ (1999) Global silicate weathering and CO2 consumption rates deduced from the chemistry of large rivers. Chem Geo159: 3–30.
78. Moquet JS, Crave A, Viers J, Seyler P (2011) Chemical weathering and atmospheric/soil CO2 uptake in the Andean and foreland Amazon basins. Chem Geo 287: 1–26.
79. Moulton KL, West J, Berner RA (2000) Solute flux and mineral mass balance approaches to the quantification of plant effects on silicate weathering. Am J Sci 300: 539–570.
80. Goddéris Y, Donnadieu Y, Tombozafy M, Dessert C (2008) Shield effect on continental weathering: implication for climatic evolution of the Earth at the geological timescale. Geoderma 145: 439–448.
81. Hay WW (1998) Detrital sediment fluxes from continents to oceans. Chem Geo 145: 287–323.
82. Vanacker V, von Blanckenburg F, Govers G, Molina A, Poesen J, et al. (2007) Restoring dense vegetation can slow mountain erosion to near natural benchmark levels. Geology 35: 303–306.
83. McGlue MM, Lezzar KL, Cohen AS, Russell JM, Tiercelin JJ, et al. (2008) Seismic records of late Pleistocene aridity in Lake Tanganyika, tropical East Africa. J Paleolimn 40: 635–653.
84. Beaulieu E, Goddéris Y, Labat D, Roelandt C, Oliva P, et al. (2010) Impact of atmospheric CO2 levels on continental silicate weathering. Geochem Geophys Geosyst 11.
85. Beaulieu E, Goddéris Y, Donnadieu Y, Labat D, Roelandt C (2012) High sensitivity of the continental-weathering carbon dioxide sink to future climate change. Nature Climate Change 2: 346–349.
86. Goddéris Y, Brantley S, François LM, Schott J, Pollard D, et al. (2013) Rates of consumption of atmospheric CO 2 through the weathering of loess during the next 100 yr of climate change. Biogeosciences 10: 135–148.

PERMISSIONS

All chapters in this book were first published in PLOS ONE, by The Public Library of Science; hereby published with permission under the Creative Commons Attribution License or equivalent. Every chapter published in this book has been scrutinized by our experts. Their significance has been extensively debated. The topics covered herein carry significant findings which will fuel the growth of the discipline. They may even be implemented as practical applications or may be referred to as a beginning point for another development.

The contributors of this book come from diverse backgrounds, making this book a truly international effort. This book will bring forth new frontiers with its revolutionizing research information and detailed analysis of the nascent developments around the world.

We would like to thank all the contributing authors for lending their expertise to make the book truly unique. They have played a crucial role in the development of this book. Without their invaluable contributions this book wouldn't have been possible. They have made vital efforts to compile up to date information on the varied aspects of this subject to make this book a valuable addition to the collection of many professionals and students.

This book was conceptualized with the vision of imparting up-to-date information and advanced data in this field. To ensure the same, a matchless editorial board was set up. Every individual on the board went through rigorous rounds of assessment to prove their worth. After which they invested a large part of their time researching and compiling the most relevant data for our readers.

The editorial board has been involved in producing this book since its inception. They have spent rigorous hours researching and exploring the diverse topics which have resulted in the successful publishing of this book. They have passed on their knowledge of decades through this book. To expedite this challenging task, the publisher supported the team at every step. A small team of assistant editors was also appointed to further simplify the editing procedure and attain best results for the readers.

Apart from the editorial board, the designing team has also invested a significant amount of their time in understanding the subject and creating the most relevant covers. They scrutinized every image to scout for the most suitable representation of the subject and create an appropriate cover for the book.

The publishing team has been an ardent support to the editorial, designing and production team. Their endless efforts to recruit the best for this project, has resulted in the accomplishment of this book. They are a veteran in the field of academics and their pool of knowledge is as vast as their experience in printing. Their expertise and guidance has proved useful at every step. Their uncompromising quality standards have made this book an exceptional effort. Their encouragement from time to time has been an inspiration for everyone.

The publisher and the editorial board hope that this book will prove to be a valuable piece of knowledge for researchers, students, practitioners and scholars across the globe.

LIST OF CONTRIBUTORS

Eike Luedeling
World Agroforestry Centre (ICRAF), Nairobi, Kenya

Evan H. Girvetz
The Nature Conservancy, Seattle, Washington, United States of America

Mikhail A. Semenov
Rothamsted Research, Harpenden, United Kingdom

Patrick H. Brown
Department of Plant Sciences, University of California Davis, Davis, California, United States of America

Martin de Luis, Klemen Novak and Miguel Angel Saz
Departamento de Geografía y Ordenación del Territorio, Universidad de Zaragoza, Zaragoza, Spain
Instituto de Investigación en Ciencias Ambientales (IUCA), Universidad de Zaragoza, Zaragoza, Spain

Katarina Čufar
Department of Wood Science and Technology, University of Ljubljana, Ljubljana, Slovenia

Alfredo Di Filippo and Gianluca Piovesan
DendrologyLab (DAFNE), Università Degli Studi della Tuscia, Viterbo, Italy

Andreas Papadopoulos
Department of Forestry and Natural Environment Management, Technological Education Institute of Lamia, Karpenissi, Greece

Cyrille B. K. Rathgeber
Laboratoire d'Etude des Ressources Forèt-Bois (LERFoB), Centre INRA de Nancy, Champenoux, France

José Raventós
Departamento de Ecología, Universidad de Alicante, San Vicente del Raspeig, Spain

Kevin T. Smith
Northern Research Station, USDA Forest Service, Durham, North Carolina, United States of America

Subhashni Taylor, Lalit Kumar and Nick Reid
Ecosystem Management, School of Environmental and Rural Science, University of New England, Armidale, New South Wales, Australia

Darren J. Kriticos
CSIRO Ecosystem Sciences, Canberra, Australian Capital Territory, Australia
E. H. Graham Centre, Charles Sturt University, Wagga Wagga, New South Wales, Australia

Chun Qin and Bao Yang
Key Laboratory of Desert and Desertification, Cold and Arid Regions Environmental and Engineering Research Institute, Chinese Academy of Sciences, Lanzhou, China

Thomas M. Melvin and Keith R. Briffa
Climatic Research Unit, School of Environmental Sciences, University of East Anglia, Norwich, United Kingdom

Zexin Fan
Key Laboratory of Tropical Forest Ecology, Xishuangbanna Tropical Botanical Garden, Chinese Academy of Sciences, Kunming, China

Yan Zhao
Institute of Geographic Sciences and Natural Resources Research, Chinese Academy of Sciences, Beijing, China

Stijn De Schepper
Department of Earth Science, University of Bergen, Bergen, Norway
Geosciences Department, University of Bremen, Bremen, Germany

Jeroen Groeneveld
MARUM – Center for Marine Environmental Sciences, University of Bremen, Bremen, Germany

B. David A Naafs
Alfred Wegener Institute for Polar and Marine Research, Bremerhaven, Germany

Cédéric Van Renterghem and Stephen Louwye
Research Unit Palaeontology, Ghent University, Ghent, Belgium

Jan Hennissen
Department of Earth Sciences, University of Toronto, Toronto, Ontario, Canada

Martin J. Head
Department of Earth Sciences, University of Toronto, Toronto, Ontario, Canada
Department of Earth Sciences, Brock University, St. Catharines, Ontario, Canada

Karl Fabian
Norwegian Geological Survey, Trondheim, Norway

Ulf Büntgen and wDavid Frank
Swiss Federal Research Institute WSL, Birmensdorf, Switzerland
Oeschger Centre for Climate Change Research, Bern, Switzerland

Christoph C. Raible and Dominik Hofer
Oeschger Centre for Climate Change Research, Bern, Switzerland
Climate and Environmental Physics, Physics Institute University of Bern, Bern, Switzerland

Samuli Helama
Arctic Centre, University of Lapland, Rovaniemi, Finland

Laura Cunningham
School of Geography and Geosciences, University of St Andrews, St Andrews, Scotland

Daniel Nievergelt and Anne Verstege
Swiss Federal Research Institute WSL, Birmensdorf, Switzerland

Mauri Timonen
The Finnish Forest Research Institute, Rovaniemi, Finland

Nils Chr. Stenseth
Centre for Ecological and Evolutionary Synthesis CEES, University of Oslo, Blindern, Norway

Jan Esper
Department of Geography, Johannes Gutenberg University of Mainz, Mainz, Germany

Alicia Carrillo-Bastos, Gerald A. Islebe and Nuria Torrescano-Valle
El Colegio de la Frontera Sur, Unidad Chetumal, Herbario, Chetumal, Quintana Roo, México

Xiaoyun Wu
Laboratory for Conservation and Utilization of Bio-resource, Key Laboratory for Animal Genetic Diversity and Evolution of High Education in Yunnan Province, School of Life Sciences, Yunnan University, Kunming, Yunnan, China
Key Laboratory of Animal Models and Human Disease Mechanisms, Chinese Academy of Sciences and Yunnan Province, Kunming Institute of Zoology, Chinese Academy of Sciences, Kunming, Yunnan, China

Jing Luo, Ziming Chen and Heng Xiao
Laboratory for Conservation and Utilization of Bio-resource, Key Laboratory for Animal Genetic Diversity and Evolution of High Education in Yunnan Province, School of Life Sciences, Yunnan University, Kunming, Yunnan, China

Song Huang
Kunming Zoo, Kunming, Yunnan, China

Yaping Zhang
Laboratory for Conservation and Utilization of Bio-resource, Key Laboratory for Animal Genetic Diversity and Evolution of High Education in Yunnan Province, School of Life Sciences, Yunnan University, Kunming, Yunnan, China
State Key Laboratory of Genetic Resources and Evolution, Kunming Institute of Zoology, Chinese Academy of Sciences, Kunming, Yunnan, China

Raul R. Cordero, Alessandro Damiani, Jorge Ferrer and Mario Tobar
Departamento de Física, Universidad de Santiago de Chile, Santiago, Chile

Jose Jorquera
Departamento de Física, Universidad de Santiago de Chile, Santiago, Chile
Escuela Superior Politécnica del Litoral, Guayaquil, Ecuador

Fernando Labbe
Departamento de Ingeniería Mecánica, Universidad Técnica Federico Santa María, Valparaíso, Chile

Jorge Carrasco
Dirección Meteorológica de Chile, Santiago, Chile

David Laroze
Instituto de Alta Investigación, Universidad de Tarapacá , Arica, Chile

Danijela Markovic, Jörg Freyhof and Christian Wolter
Leibniz-Institute of Freshwater Ecology and Inland Fisheries, Berlin, Germany

Daniel E. Impoinvil
LUCINDA Group, Department of Epidemiology and Population Health, Institute of Infection and Global Health, University of Liverpool, Leahurst Campus, Neston, Cheshire, United Kingdom
Brain Infections Group, Department of Clinical Infection, Microbiology and Immunology, Institute of Infection and Global Health, University of Liverpool, Liverpool, Merseyside, United Kingdom

Tom Solomon
Brain Infections Group, Department of Clinical Infection, Microbiology and Immunology, Institute of Infection and Global Health, University of Liverpool, Liverpool, Merseyside, United Kingdom

W. William Schluter
Programme for Immunization Preventable Diseases, World Health Organization Country Office for Nepal, Kathmandu, Nepal

Ajit Rayamajhi
Brain Infections Group, Department of Clinical Infection, Microbiology and Immunology, Institute of Infection and Global Health, University of Liverpool, Liverpool, Merseyside, United Kingdom
Department of Pediatrics, National Academy of Medical Sciences, Kanti Children's Hospital, Maharajgunj, Kathmandu, Nepal

Ram Padarath Bichha
Child Health Division, Department of Health Services, Ministry of Health and Population, Government of Nepal, Kathmandu, Nepal

Geeta Shakya
National Public Health Laboratory, Department of Health Services, Ministry of Health and Population, Government of Nepal, Kathmandu, Nepal

Cyril Caminade
Department of Earth's Changing Environment, School of Environmental Science, University of Liverpool, Liverpool, Merseyside, United Kingdom

Matthew Baylis
LUCINDA Group, Department of Epidemiology and Population Health, Institute of Infection and Global Health, University of Liverpool, Leahurst Campus, Neston, Cheshire, United Kingdom

Xiaojun Kou
State Key Laboratory of Earth Surface Processes and Resource Ecology, Beijing Normal University, Beijing, PR China
College of Life Sciences, Beijing Normal University, Beijing, PR China

Qin Li
Department of Botany, University of British Columbia, Vancouver, B.C., Canada

Carl Beierkuhnlein
Department of Biogeography, University of Bayreuth, Bayreuth, Germany

Yiheng Zhao
College of Life Sciences, Beijing Normal University, Beijing, PR China

Shirong Liu
Institute of Forest Environment and Ecology, Chinese Academy of Forestry, Beijing, PR China

Marie-Hélène Moncel, Jackie Despriée, Pierre Voinchet, Davinia Moreno, Jean- Jacques Bahain, Gilles Courcimault and Christophe Falguères
Department of Prehistory-UMR 7194 CNRS, National Museum of Natural History, Institut de Paléontologie Humaine, Paris, France

Hélène Tissoux
Department of Prehistory-UMR 7194 CNRS, National Museum of Natural History, Institut de Paléontologie Humaine, Paris, France
Bureau de Recherches Géologiques et Minières, GEO/G2R, BP 36009, Orléans, France

A. Michelle Lawing
Department of Geological Sciences, Indiana University, Bloomington, Indiana, United States of America
Department of Biology, Indiana University, Bloomington, Indiana, United States of America

P. David Polly
Department of Geological Sciences, Indiana University, Bloomington, Indiana, United States of America

Hugo I. Martínez-Cabrera,
Département des sciences biologiques, Université du Québec à Montré al, Montreal, Quebec, Canada

Pedro R. Peres-Neto
Département des sciences biologiques, Université du Québec à Montré al, Montreal, Quebec, Canada
Canada Research Chair in Spatial Modelling and Biodiversity, Université du Québec à Montré al Montreal, Quebec, Canada

Maria Triviño
Department of Biodiversity and Evolutionary Biology, National Museum of Natural Sciences, CSIC, Madrid, Spain

Wilfried Thuiller
Laboratoire d'Ecologie Alpine, UMR CNRS 5553, Université Joseph Fourier, Grenoble, France

Mar Cabeza
Department of Biodiversity and Evolutionary Biology, National Museum of Natural Sciences, CSIC, Madrid, Spain

Metapopulation Research Group, Department of Biological and Environmental Sciences, University of Helsinki, Helsinki, Finland

Thomas Hickler
Department of Physical Geography and Ecosystems Analysis, Geobiosphere Science Centre, Lund University, Lund, Sweden
Biodiversity and Climate Research Centre (BiK-F) and Department of Physical Geography at Goethe-University and Senckenberg Gesellschaft fu¨ r Naturforschung, Frankfurt/Main, Germany

Miguel B. Araújo
Department of Biodiversity and Evolutionary Biology, National Museum of Natural Sciences, CSIC, Madrid, Spain
'Rui Nabeiro' Biodiversity Chair, CIBIO, University of Évora, Évora, Portugal
Center for Macroecology, Evolution and Climate, University of Copenhagen, Copenhagen, Denmark

Andrew Siefert, Jason D. Fridley and Mark E. Ritchie
Department of Biology, Syracuse University, Syracuse, New York, United States of America

Sarah J. Ivory
Brown University, Providence, Rhode Island, United States of America

Michael M. McGlue
University of Kentucky, Lexington, Kentucky, United States of America

Geoffrey S. Ellis
U.S. Geological Survey, Denver, Colorado, United States of America

Anne-Marie Lézine
LOCEAN, CNRS, Paris, France

Andrew S. Cohen
University of Arizona, Tucson, Arizona, United States of America

Annie Vincens
CEREGE, CNRS, Aix-en- Provence, France

Index